Oilseeds: Health Attributes and Food Applications

Beenu Tanwar • Ankit Goyal
Editors

Oilseeds: Health Attributes and Food Applications

Springer

Editors
Beenu Tanwar
Department of Dairy Technology
Mansinhbhai Institute of Dairy
and Food Technology
Mehsana, Gujarat, India

Ankit Goyal
Department of Dairy Chemistry
Mansinhbhai Institute of Dairy
and Food Technology
Mehsana, Gujarat, India

ISBN 978-981-15-4193-3 ISBN 978-981-15-4194-0 (eBook)
https://doi.org/10.1007/978-981-15-4194-0

This Springer imprint is published by the registered company Springer Nature Singapore Pte Ltd.
The registered company address is: 152 Beach Road, #21-01/04 Gateway East, Singapore 189721, Singapore

Preface

Oilseeds are grown all over the world and are considered important economic crops. Recently, oilseeds have garnered the attention of researchers and food scientists owing to their phytonutrients, which confer various health benefits, particularly in prevention and treatment of several non-communicable diseases. Oilseed processing yields valuable edible oil and protein-rich meal (cake) utilized as feed. Advancement in science and technology research suggests that each oilseed has one or the other phytochemical and/or antinutrient component/s which depending on several factors are positively as well as negatively associated with human health and well-being. The beneficial health effects of omega-3 fatty acids, lignans (phytoesterogens) and phytosterols present in various oilseeds have been evidenced in cardiovascular diseases, rheumatoid arthritis, obesity, diabetes, cancer, hypertension, etc. possibly due to their antioxidative and anti-inflammatory mechanisms. Consequently, the ever-increasing demand for functional foods has steered the development and production of novel foods supplemented with functional and bioactive ingredients from oilseeds. In this book, 20 oilseeds (including non-conventional sources) are discussed with special emphasis on their bioactive components/phytonutrients and their health effects. The possible mechanisms behind the physiological/metabolic effects as well as the physicochemical and sensory changes in fortified foods have also been discussed. The contributing authors are national and international experts in their respective fields, and we are grateful for their valuable contribution and suggestions. All the chapters consist of a brief history, proximate composition (including antinutrients and phytonutrients, if present), associated health effects with major findings of the clinical and epidemiological studies and food applications. This book will be of interest to a wide spectrum of professionals from academia of food and nutrition, pharmaceuticals, food science and technology, agriculture, etc. It will also serve as a unique reference book for the research scholars conducting their research in the field of oilseeds and olive fruit (from food point of view), value-added foods and new product development. Furthermore, the book will be useful for the undergraduate and postgraduate students to understand the basic composition of oilseeds and their health effects.

Mehsana, Gujarat, India Beenu Tanwar
Mehsana, Gujarat, India Ankit Goyal

Contents

About the Editors

Beenu Tanwar is currently working as an Assistant Professor in the Department of Dairy Technology, Mansinhbhai Institute of Dairy and Food Technology, Mehsana, Gujarat. She holds a Master of Science in Food Technology from Sri Sathya Sai Institute of Higher Learning, Anantapur, India; and a Doctorate in Food Science and Nutrition from Himachal Pradesh Agricultural University, Palampur, India. She is also a recipient of the UGC-JRF (University Grants Commission-Junior Research Fellowship) Award in 2011. She is actively involved in teaching and research and has published numerous peer-reviewed articles, participated in several seminars, conferences, and workshops, and has received awards for best paper/poster presentations. She is an active member of the Indian Association for Parenteral and Enteral Nutrition (IAPEN), Indian Science Congress Association (ISCA), and Association of Food Scientists and Technologists India (AFSTI).

Ankit Goyal is an Assistant Professor in the Department of Dairy Chemistry at Mansinhbhai Institute of Dairy and Food Technology, Mehsana, Gujarat, India. He has published more than 30 research articles in leading journals, conference proceedings, and books. He is a member of various professional bodies, including the Indian Science Congress Association (ISCA), American Oil Chemist's Society (AOCS), and Dairy Technology Society of India (DTSI). Dr. Goyal holds a Bachelor of Science in Biochemistry from Kurukshetra University; a Master of Science in Food Science and Technology from Chaudhary Charan Singh Haryana Agricultural University, Hisar, India; and a Doctorate in Dairy Chemistry from the National Dairy Research Institute (ICAR-NDRI), Karnal, India. Dr. Goyal's research interests include the development and formulation of functional food and dairy products, and he is dynamically involved in teaching (graduate and postgraduate students) and research.

Chapter 1
Soybean (*Glycine max*)

Rajni Modgil, Beenu Tanwar, Ankit Goyal, and Vikas Kumar

Abstract Soybean (*Glycine max*), also called as soja bean or soya bean, holds tremendous economic importance owing to its high amount of oil content (18%), high-quality proteins (~40%), contribution toward soil fertility, high productivity, and profitability; and, thus, is rightly referred to as the miracle crop. Soybeans are also a significant source of polysaccharides, soluble fibers, phytosterols, lecithins, saponins, and phytochemicals mainly isoflavones which either individually or collaboratively help in promoting health by reducing the incidence of debilitating diseases like hyperglycemia, hypertension, dyslipidemia, obesity, inflammation, cancer, etc. Century-old literature shows that soybean seeds have been primarily used in Asia to prepare a variety of fresh, fermented, and dried foods, viz., soy milk, tofu, soy paste, soy sauce, miso, natto, etc. which have now become popular all over the world. Furthermore, soybean and its products find various non-food applications such as in the production of papers, plastics, pharmaceuticals, inks, paints, varnishes, pesticides, cosmetics, and, more recently, biodiesel.

Keywords Soybean (*Glycine max*) · Isoflavones · Soy products · Antiadipogenic · Anticancer · Hypoglycemic · Soy protein isolates

---

R. Modgil
Department of Food Science, Nutrition and Technology, College of Home Science, Himachal Pradesh Agriculture University, Palampur, Himachal Pradesh, India

B. Tanwar
Department of Dairy Technology, Mansinhbhai Institute of Dairy and Food Technology, Mehsana, Gujarat, India

A. Goyal (✉)
Department of Dairy Chemistry, Mansinhbhai Institute of Dairy and Food Technology, Mehsana, Gujarat, India

V. Kumar
Department of Food Science and Technology, College of Agriculture, Punjab Agricultural University, Ludhiana, Punjab, India
e-mail: vikaschopra@pau.edu

B. Tanwar, A. Goyal (eds.), *Oilseeds: Health Attributes and Food Applications*,
https://doi.org/10.1007/978-981-15-4194-0_1

1.1 Origin and History

The cultivation of soybeans (*Glycine max*) was started some 4000–5000 years ago, with its origin being traced back to China (Qiu and Chang 2010), Japan, and Korea. Kwon et al. (2005) reported occurrence of carbonized soybeans at some historic sites in Korean Peninsula, thus providing a strong evidence for its cultivation from the Bronze Age. Engelbert Kaempfer, a German botanist, has been credited with the introduction of soybean to Europe in the eighteenth century, but poor climatic and soil conditions limited its production in the continent. Soybean cultivation began in the United States only in the nineteenth century and has been a significant commercial crop since then. Apart from the US Food and Drug Administration (FDA) approving the health claims which linked soybean to reduction in the coronary heart diseases in 1999, the availability of transgenic soybean, the existence of regulatory biosafety frameworks, and the incorporation of new land and innovative agricultural technologies by farmers provided the impetus to the popularization of soybean in the United States and world at large (Qiu and Chang 2010).

1.2 Production

Three countries, viz., the United States, Brazil, and Argentina, dominate global production and account for approximately 80% of the world's soybean supply. United States accounts for 34% and 18% of the world's production of soybean and soya oil, respectively (Table 1.1). Owing to the fact that soybean holds 25% share of the global vegetable oil production, two-thirds of the world's protein concentrate, and an essential ingredient in the fish and poultry feeds, it has now acquired the prominent place as the world's most important seed legume. This fact is evidently seen in the data presented in Fig. 1.1, wherein, the increasing production

Table 1.1 World production of soya grain and oil (2018–2019)

	Production (million metric tons)	
Country	Soybean grain	Soya oil
USA	123.664	10.093
Brazil	117.0	8.195
Argentina	55.00	8.415
China	15.90	15.949
India	11.00	1.620
Paraguay	9.50	0.740
Canada	7.30	0.366
Mexico	0.34	0.946
European Union	2.70	3.154
Others	18.589	6.591
Total	360.993	56.069

Source: USDA (2019)

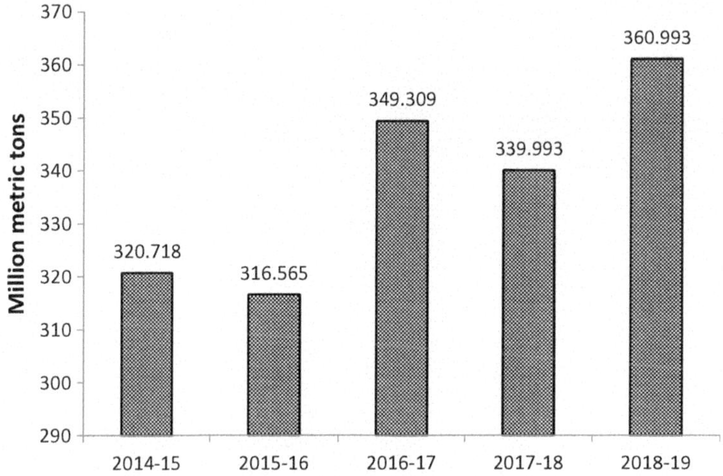

Fig. 1.1 Global production of soybean (USDA 2019)

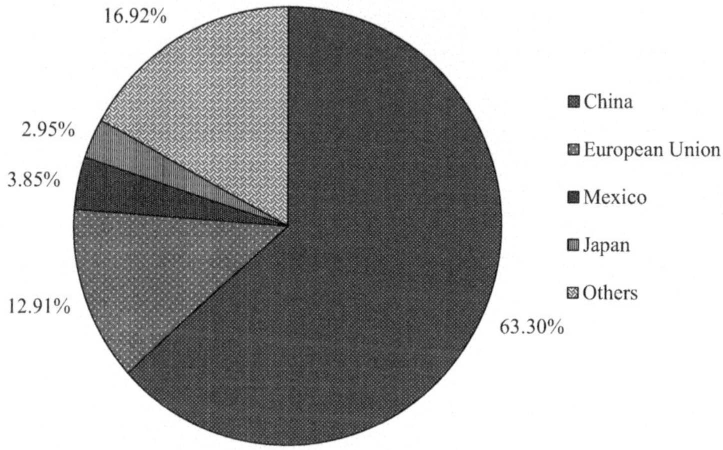

Fig. 1.2 Soybean-importing countries and their share (USDA 2019)

trend of soybean at global level for the last 5 years is illustrated. According to the Product Complexity Index (PCI), soybeans are the 44th most traded product and the 978th most complex product. The data pertaining to the export and import of soybean is depicted in Figs. 1.2 and 1.3, respectively. Globally, every year approximately 85% of the soybeans are processed into soybean meal and oil, of which 95% oil is consumed as edible oil, and 5% is utilized in the industrial production of biodiesel, soaps, and fatty acids. On the other hand, approximately 98% of the soybean meal is utilized in animal feed and the rest in soy proteins and flour (USDA 2019).

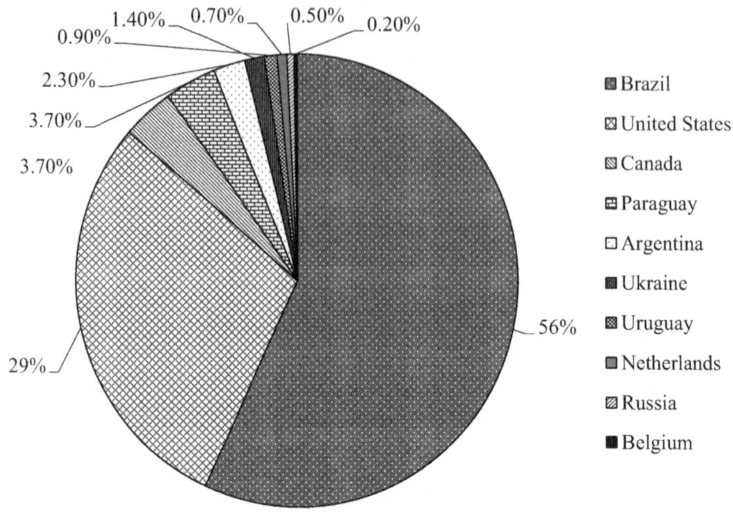

Fig. 1.3 Soybean-exporting countries and their share (USDA 2019)

1.3 Nutritional Composition

Among legume seeds, soybeans contain not only the maximum crude protein but also the best amino acid composition and, hence, are considered the best vegetable protein source for humans as well as animals. However, it is important to note that the composition of whole soybean and its structural parts is highly dependent on geographic location, growing season, environmental stress, and variety of the seed.

1.3.1 Carbohydrates

Carbohydrates are the second largest component (22.11–33.18%) in soybeans (Warle et al. 2015; Alghamdi et al. 2018). Previously, soy carbohydrates held much less economic importance than soy protein and oil due to the thought that it just provides calories. Also, its digestion was better in monogastric animals and, hence, was primarily used in ruminant feeds. However, the picture is changing now due to the recent emphasis on the role of dietary oligosaccharides and dietary fiber in preventing colon cancer and other diseases (Benito-González et al. 2019).

The hull comprises 8% of the seed and contains approximately 86% carbohydrates, whereas hypocotyl axis and cotyledons comprise 2% and 90% of the whole seeds and contain 43% and 29% of carbohydrates, respectively (Liu 1997). Soybeans contain sucrose (2.5–8.2%), raffinose (0.1–0.9%), stachyose (1.4–4.1%), and trace amounts of glucose and arabinose (Hymowitz et al. 1972). Raffinose and stachyose are the limiting factors for its utilization as food due to their poor digestion, flatulence, and abdominal discomfort in humans (Steggerda et al. 1966). Also, it has been observed that these oligosaccharides are poorly metabolized by ruminants and, thus, decrease its nutritive value (Coon et al. 1988). Nevertheless, efforts have been made to reduce these negative effects by fermentation and germination (Neus et al. 2005).

The cell walls of the soybeans contain insoluble carbohydrates, viz., cellulose (20%), hemicellulose (50%), pectin (30%), and some amount of starch (12%) (Warle et al. 2015). Polysaccharides, oligosaccharides, lignin, and associated plant substances are termed dietary fiber and range between 9% and 16% in whole soybeans (Esteves et al. 2010). Insoluble fraction (74–78%) is mainly composed of arabinose, galactose, glucose, xylose, and uronic acids, and soluble fiber (22–26%) comprises of arabinose, galactose, and uronic acids (Redondo-Cuenca et al. 2007). Although they do not contribute nutritionally, they are now considered as an integral part of the diet as they confer many health-protective effects, viz., reducing the risk of coronary heart disease (Liu et al. 1999), stroke (Steffen et al. 2003), hypertension (Whelton et al. 2005), diabetes (Montonen et al. 2003), obesity (Lairon et al. 2005), and certain gastrointestinal disorders (Petruzziello et al. 2006).

1.3.2 Proteins

Protein content of soybean ranges from 33% to 45% (Machado et al. 2008). Soy protein has garnered a lot of attention lately owing to it being at par with animal proteins both in quantitative and qualitative terms and being the only vegetable food containing all eight essential amino acids (Dudek 2013; Hark and Morrison 2000). Methionine and cysteine are the only limiting factors in soybean; nevertheless, the Protein Digestibility Corrected Amino Acid Score (PDCAAS) for whole soybeans is 0.92 which indicates that soy protein is excellent for human nutrition in terms of both the amino acid pattern and protein digestibility. Thus, suggesting that soy food consumption is the best way to increase protein consumption in vegetarian population (Burssens et al. 2011). In terms of protein characterization, just like other legumes, the bulk of soybean proteins are the globulins. The four major fractions are β-conglycinin (principal component), glycinin (principal protein), trypsin inhibitors, and enzymes and hemagglutinins (Burssens et al. 2011). Among these, glycinin and α-, β-, and γ-conglycinin constitute the major legume storage proteins (Wolf 1970).

1.3.3 Lipids

The total fat content of soybean ranges from 16% to 22%. Among all the other legumes, it contains highest amount of fat and contributes nearly 47% to its energy

Table 1.2 Nutritional profile of soybeans (per 100 g dry matter)

Composition	Soybean seeds	Composition	Soybean seeds
Carbohydrate	30.16 g	**Ash**	4.87 g
Total dietary fiber	9.3 g	**Minerals**	
Total sugars	7.33 g	Calcium, Ca	277 mg
Protein	36.49 g	Iron, Fe	15.7 mg
Amino acids		Magnesium, Mg	280 mg
Tryptophan	0.591 g	Phosphorus, P	704 mg
Threonine	1.766 g	Potassium, K	1797 mg
Isoleucine	1.971 g	Sodium, Na	2 mg
Leucine	3.309 g	Zinc, Zn	4.89 mg
Lysine	2.706 g	Copper, Cu	1.658 mg
Methionine	0.547 g	Manganese, Mn	2.517 mg
Cystine	0.655 g	Selenium, Se	17.8 μg
Phenylalanine	2.122 g	**Vitamins**	
Tyrosine	1.539 g	Vitamin C	6 mg
Valine	2.029 g	Thiamin	0.874 mg
Arginine	3.153 g	Riboflavin	0.87 mg
Histidine	1.097 g	Niacin	1.623 mg
Alanine	1.915 g	Pantothenic acid	0.793 mg
Aspartic acid	5.112 g	Vitamin B_6	0.377 mg
Glutamic acid	7.874 g	Folate, total	375 μg
Glycine	1.88 g	Choline, total	115.9 mg
Proline	2.379 g	Betaine	2.1 mg
Serine	2.357 g	Vitamin B_{12}	0 μg
Total lipid	19.94 g	Vitamin A, RAE*	1 μg
Fatty acids (FA)		Retinol	0 μg
Total saturated FA	2.884 g	β-Carotene	13 μg
Myristic acid (C14:0)	0.055 g	Vitamin A, IU	22 IU
Palmitic acid (C16:0)	2.116 g	Vitamin E (α-tocopherol)	0.85 mg
Stearic acid (C18:0)	0.712 g	Vitamin K (phylloquinone)	47 μg
Total monounsaturated FA	4.404 g		
Palmitoleic acid (C16:1)	0.055 g		
Oleic acid (C18:1)	4.348 g		
Total polyunsaturated FA	11.255 g		
Linoleic acid (C18:2)	9.925 g		
Linolenic acid (C18:3)	1.33 g		
Total trans FA	0 g		
Cholesterol	0 mg		
Phytosterols	161 mg		

Source: USDA (2019)
*RAE: Retinol Activity Equivalents

value (Liu 1999; Messina 1999; Burssens et al. 2011). Triglycerides have the highest percentage (96%), followed by phospholipids or lecithin (2%), unsaponifiable lipids (1.6%), and free fatty acids (0.5%). As shown in Table 1.2, it is a good source of mono- and polyunsaturated fatty acids and essential fatty acids and poor source of saturated fatty acids and, thus, has been approved 'Heart-healthy' by the American Food and Drug Administration, the British Joint Health Claims Initiative, and the American Heart Association (Sacks et al. 2006).

1.3.4 Vitamins

The vitamin content of soybeans is presented in Table 1.2. In soybean, vitamin E is available in substantial amount with α-tocopherol (0.4–8 mg/100 g dried weight), γ-tocopherol (4–80 mg/100 g), δ-tocopherol (1–50 mg/100 g), and trace amount of β-tocopherol (Kasim et al. 2010; Li et al. 2010).

1.3.5 Minerals

Ash content in soybean ranges from 4.5% to 6.0% (Monteiro et al. 2003; Burssens et al. 2011). Potassium is found in the highest concentration, followed by phosphorus, magnesium, and calcium; whereas, iron, sodium, zinc, copper, and manganese are present in minor and selenium in trace amounts (Please refer to the mineral composition in Table 1.2).

1.3.6 Antinutrients and Phytonutrients

Soybean is a good source not only of protein, oil, and minerals but also of numerous bioactive/phytonutrient compounds, viz., isoflavones, saponins, protease inhibitors, and phytic acid. These phytonutrients are proven to be beneficial in the prevention and/or treatment of numerous diseases or physiological disorders by exerting antioxidant, hypolipidemic, antiallergic, anti-spasmodic, anti-microbial, hypotensive, and anti-inflammatory effects (Burssens et al. 2011; Gupta and Prakash 2014).

1.3.6.1 Phytate

Phytate [inositol hexaphosphate (IP6)] is the storage form of phosphorus and a natural plant antioxidant in leguminous seeds like soybean. However, it reduces not only the bioavailability of phosphorus, zinc, magnesium, calcium, potassium, and iron (Davidsson et al. 1994; Hurrell 2003) but also the bioavailability of protein and carbohydrates by decreasing the enzymatic activity of pepsin, trypsin, and amylase (Sebastian et al. 1998; Selle et al. 2000).

Nevertheless, through numerous scientific studies, phytate has demonstrated its role as a bioactive agent and holds a great promise in cancer treatments. Phytic acid intake helps in reducing the blood glucose and cholesterol levels (Lee et al. 2006; Lee et al. 2007); increases bone mineral density (López-González et al. 2008); inhibits the crystallization of calcium salts, thus avoiding kidney stone formation (Grases et al. 2000); exerts protective effect in Parkinson's by avoiding excess iron accumulation (Xu et al. 2008); inhibits iron-mediated lipid peroxidation and Fenton oxidative reaction (Graf and Eaton 1990; Rimbach and Pallauf 1998); reduces cell proliferation; and induces differentiation of malignant cells, thus controlling tumor growth, progression, and metastasis (Shamsuddin 2002; Vucenik and Shamsuddin 2004).

Soybean and its products contain almost 1–1.5 g phytic acid/100 g of dry matter (Mikić et al. 2009; Agarwal 2014; Yasothai 2016). Liener (2000) reported that almost two-thirds of the phosphorus in soybean is bound as phytate and unavailable to animals. Several studies have thus been carried out to observe the effects of processing methods on the phytate content of soybean, and it was reported that sprouting, roasting, and pressure cooking lead to a 32%, 88%, and 55% decrease, respectively (Pele et al. 2016; Agarwal 2014). Also, research work is being carried out to develop low phytic acid content in soybean genotypes with better yield and seed viability (Spear and Fehr 2007).

1.3.6.2 Protease Inhibitors

Antiproteolytic substances were first noted by Read and Haas (1938), and since then numerous studies have been undertaken either to estimate the soybean protease inhibitors (SBPI) or to evaluate their antinutritional and health-promoting effects. However, two major forms of SBPI present in soybeans, namely, Kunitz trypsin inhibitor (KTI) and Bowman-Birk inhibitor (BBI), represent 6% of the protein present in soybean seed. It is reported that KTI inhibits trypsin, whereas, BBI affects the enzymatic activity of both trypsin and chymotrypsin, and thus, decreases the biological quality of the soy proteins (Lajolo and Genovese 2002). Also, KTI reportedly leads to hypersecretion of pancreatic enzymes leading to hypertrophy and hyperplasia, and thus, raw soybean is recommended not to be fed to monogastric animals (Kassell 1970; Rackis and Gumbmann 1981; Birk 1985; DiPietro and Liener 1989; Werner and Wemmer 1992). On the contrary, recent research suggests that BBI concentrate (BBIC) can be used as a potential cancer chemopreventive agent and can prevent the development of coronary diseases in humans without toxicity (Kennedy 1998; Dia et al. 2008).

Just like the nutritional attributes, antinutrient and phytonutrient composition is also affected by various genetic and environmental factors. Consequently different authors have reported different values for trypsin inhibitor in soybean seeds. Esteves et al. (2010) pointed out that per gram of soy protein comprises between 30 and

125 mg trypsin inhibitors. De Toledo et al. (2007) found it ranging between 42.6 and 71.6 UTI/mg; Carvalho et al. (2002) reported much higher concentration, i.e., 122–206 UTI/mg; and Gu et al. (2010) observed that trypsin inhibitors range from 3000 to 6000 mg/100 g of raw soybean grain samples.

1.3.6.3 Lectins

Lectins (hemagglutinin or agglutinin) are the carbohydrate-binding proteins which are resistant to digestion and are often considered one of the most toxic constituents of the pulses/beans. Lectins possess high affinity to cellular and intracellular membrane-associated carbohydrates (glycoprotein and glycolipids) [especially N-acetyl-D-galactosamine] and thus, by binding, not only reduce their absorption but can even lead to agglutination of red blood cells, leading to hemolytic anemia (Liener 1994), impairment of the key enzymes of metabolism, and growth retardation, and if administered orally or intraperitoneally can even lead to death (Grant et al. 1988). Numerous studies suggest that lectin consumption can also cause the atrophy of microvilli, intestinal epithelial injury, diarrhea, increased intestinal permeability, and increased proliferation of pathogenic bacteria in the gut (Grant et al. 1995; Pusztai et al. 1991; Pusztai et al. 1995; Machado et al. 2008).

However, recent research suggests that lectins have potent in vivo biological activities and exhibit anticarcinogenic and antitumor activity by either reduction of cell proliferation, induction of tumor-specific cytotoxicity of macrophages, or by having a strong effect on the immune system by production of various interleukins (de Mejia et al. 2003; de lumen 2005). Soybean seeds contain 300–600 mg/100 g lectin (Gu et al. 2010), and approximately 6.5 g lectin/kg is reported in defatted soy meal (Vasconcelos et al. 2001) and approx. 0.2–2% of the soybean protein mass (de Mejia et al. 2003; Rizzi et al. 2003; Anta et al. 2010). Treatments like soaking (Hernandez-Infante et al. 1998), heat treatment, and fractionation during food processing have been reported to be efficient in eliminating lectins and abolishing its hemagglutinating activity, thus improving the nutritional quality of soybeans (Vasconcelos et al. 1997).

1.3.6.4 Oxalate

Oxalate, the simplest dicarboxylic acid is synthesized by the body or absorbed from the gut. It cannot be further metabolized by human beings and must be excreted in the urine (Massey et al. 1993). Oxalate inhibits calcium absorption in the kidneys and increases the risk of developing kidney stones, renal edema, and calcification along with mineral deficiency (Al-Wahsh et al. 2005; Horner et al. 2005). American Dietetic Association categorizes foods containing more than 10 mg oxalate per serving as high-oxalate foods (Al-Wahsh et al. 2005). Soybeans contain moderate amount ranging between 0.67 and 3.5 g oxalate per 100 g (dry weight) (Massey et al. 2001; Horner et al. 2005), which is much lower than the classic high-oxalate foods like spinach (1145 mg/100 g) and chocolate (155–485 mg/100 g dry matter) (Massey et al. 1993) but raises concern over its consumption. Al-Wahsh and authors reported 2–58 mg oxalate per serving in 40 soy foods. Soy flour, textured vegetable soy protein, roasted soybeans, soy nuts, tempeh, and soya butter reportedly had higher than 10 mg oxalate per serving and, thus, come under high-oxalate foods (Al-Wahsh et al. 2005; Massey et al. 2001). However, domestic processing methods like salt treatment, roasting, etc. can help decreasing the oxalate content up to 40% and 20%, respectively (Maidala et al. 2013), and can help in efficient nutrient utilization.

1.3.6.5 Phenolics

Decreased incidence of cardiovascular diseases and different types of cancer in Asian population vis-à-vis Europeans and Americans drove scientists to search for valid reasons behind and reported the intake of soybeans as the significant factor. More comprehensive studies made an observation that the polyphenolic compounds present in soy which also possess estrogenic activity could be the contributing factors (Lampe 2003). De Toledo et al. (2007) observed phenolic compounds and tannin concentration in the range between 6.60 and 8.07 mg/g and 0.28 and 0.39 mg/g, respectively, in different soybean cultivars.

1.3.6.6 Isoflavones

Among all the other flavonoids present in legumes, isoflavones are the major ones in soybeans and occur as β-glucosides (30–35%; genistin, daidzin, and glycitin), aglycones (4–12%; daidzein, genistein, and glycitein), 6″-*O*-acetyl-β-glucosides (0–5%), 6″-*O*-malonyl-β-glucosides (50–65%), and 4′-methyl ethers of daidzein and genistein, formononetin, and biochanin A (Franke et al. 1999; Genovese et al. 2006; Villares et al. 2011). Whole soy contains approx. 50–450 mg/100 g dried weight of total isoflavones (Kim et al. 2006; Ciabotti et al. 2016) with majority (80–90%) of total seed isoflavones concentrated in the cotyledons (Tsukamoto et al. 1995). Soy isoflavones have been reported to reduce cholesterol levels and, thus, risk of cardiovascular diseases (Rivas et al. 2002; Wiseman et al. 2000; Harland and Haffner 2008); inhibit cell proliferation; and have, thus, anticancer property (Shu et al. 2001; Lamartiniere et al. 2000; Jayachandran and Xu 2019) in addition to antioxidant (Lee et al. 2008), anti-aging (Oyama et al. 2012; Kim et al. 2015), anti-inflammatory (García-Lafuente et al. 2009), and antiallergic properties (Masilamani et al. 2011).

1.3.6.7 Saponins

Saponins, the triterpenes, or steroid aglycones (sapogenin) with one or more sugar chains occur in a wide variety of plants. Soybeans are the prime dietary source of saponins and are often referred to soyasaponins. Soyasaponins are usually concentrated in the hypocotyl of the seed, and the total value ranges between 140 and 975 mg/100 g dried weight of the seeds (Fenwick and Oakenfull 1981; Lin and Wang 2004). On the basis of chemical structure, they are classified into three groups, viz., soyasapogenol A (hydroxyl group at the C-21 position), soyasapogenol B (hydrogen atom at the C-21 position), and soyasapogenol E (carbonyl group at C-22 and are oxidation products from group B) (Yoshiki et al. 1998). They impart bitter taste, astringency, foaming properties, and hemolytic activity to the plant material (Ridout et al. 1988 and, thus, were once considered as antinutrients.

However, research has established that soyasaponins are not only safe to be used as food and feed (Ishaaya et al. 1969) but also possess plethora of health benefits, viz., hypocholesterolemic (Chávez-Santoscoy et al. 2013), immune-modulatory (Sun et al. 2014; Qiao et al. 2014), anti-inflammatory (Francis et al. 2002; Mudryj et al. 2014), antiobesity (Kim et al. 2014), hepatoprotective (Kuzuhara et al. 2000), anticarcinogenic (Gurfinkel and Rao 2003; Du et al. 2014), and antimutagenic (Berhow et al. 2002) activities.

1.4 Health Attributes

Numerous epidemiological studies have established that consumption of soybean and its various phytonutrients aids in the prevention and treatment of cancer, cardiovascular disease, and metabolic, musculoskeletal, gynecological, endocrine, and renal outcomes especially in perimenopausal women (Li et al. 2019; Watanabe and Uehara 2019; Abo-Elsoud et al. 2019; Latorraca et al. 2019). Detailed information on dosage, route of administration, the model used, and the results based on the experimental research study both in vitro and in vivo are depicted in Table 1.3.

1.4.1 In Hyperglycemia

Numerous animal and human studies have reported the antidiabetic potential of soybean and its bioactive components (soy isoflavones, protein, fiber, saponins, etc.) by regulating blood glucose levels and improving insulin resistance and kidney filtration (Holt et al. 1996; Chandalia et al. 2000; Jenkins et al. 2003a, b; Trujillo et al. 2005; Azadbakht et al. 2007; Pipe et al. 2009; Kwon et al. 2010; Chalvon-Demersay et al. 2017). In a recent study, Tatsumi et al. (2013) reported lower type 2 diabetes incidence in Japanese men with BMI >23.6 kg/m^2 on consumption of ≥ 4 servings/week of soybean products.

Table 1.3 Summary of numerous in vivo and in vitro studies indicating health effects of soybean and its bioactive components

Experimental model	The form of soya bean	Dose and route of administration	Investigation	Major finding/s	Reference
Human (DM patients)	Roasted soya bean powder	69 g/d for 4 weeks	Assay of glycemic control and lipid metabolism parameters	Hypolipidemic, hypoglycemic, and antioxidant activity	Chang et al. (2008)
Sprague-Dawley rats	Isolated soy protein (ISP) and genistein	STZ-genistein, 600 mg/kg diet, and STZ-ISP, 200 g/kg diet for 3 weeks	Assay of blood glucose, lipid metabolism parameters, and antioxidant enzymes	Hypolipidemic, hypoglycemic, and antioxidant activity	Lee (2006)
Human (DM patients)	Soy-based dietary supplement	Abalon (50 g soy protein, 165 mg isoflavones, and 20 g cotyledon fiber)/day for 6 weeks	Assay of blood glucose, lipid metabolism parameters	Hypolipidemic	Hermansen et al. (2001)
Human (hypercholes-terolemic patients)	Soy-based dietary supplement	Abalon (30 g soy protein, 9 g cotyledon fiber, and 100 mg isoflavones)/day for 24 weeks	Assay of LDL cholesterol and other cardiovascular risk fac-tors (including endothelial function)	Hypolipidemic	Hermansen et al. (2005)
Human (postmeno-pausal diabetic women)	Phytoestrogens	Phytoestrogens (soy protein 30 g/d, isoflavones 132 mg/d)/day for 12 weeks	Assay of glycemic control and cardiovascular risk markers	Hypolipidemic, hypoglycemic, and cardioprotective effect	Jayagopal et al. (2002)
Human (diabetic women)	Soy isoflavones	435 mg/d for 2 months	Assay of blood glucose, lipid metabolism parameters	Hypolipidemic and reduced risk of diabetes	Chi et al. (2016)
Human (obese diabetic patients)	Soy-based meal replace-ment (MR) plan	12 months	Assay of weight loss and met-abolic profile	Weight reduction and hypoglycemic effect	Li et al. (2005)
Human (postmeno-pausal women)	Soy isoflavones	40–80 mg/d for 1 year	Assay of blood glucose, lipid metabolism parameters	Hypoglycemic	Ho et al. (2007a, b)

Human (DM patients)	Soy protein isolate (SPI)	80 mg/d for 57 d	Assay of lipid metabolism parameters	Hypolipidemic and cardioprotective effect	Pipe et al. (2009)
Human	Soy protein isolate (SPI)	Low-isoflavone SPI (1.64 ± 0.19 mg/d) and high-isoflavone SPI (61.7 ± 7.4 mg/d) for 57 d	Assay of lipid metabolism parameters	Hypolipidemic and cardioprotective effect	McVeigh et al. (2006)
Human (hyperlipidemic and diabetic patients)	Soya bean dietary supplement	D-LeciVita (12% lecithin, 35% soy protein) 15 g/d for 12 weeks	Assay of blood glucose, lipid metabolism parameters	Hypolipidemic and cardioprotective effect	Medić et al. (2006)
Human (men and Postmenopausal women)	Soy protein isolate (SPI)	SPI (40 g soy protein, 118 mg isoflavones)/d for 3 months	Assay of blood pressure and lipid metabolism parameters	Improvement in blood pressure and hypolipidemic effect	Liang et al. (2006)
Human (hyperlipidemic)	Soybean product diet	2%/d for 4 weeks	Assay of blood pressure and lipid metabolism parameters	Hypolipidemic and cardioprotective effect	Kurowska et al. (1997)
Human (hypercholesterolemic renal transplant recipients)	Soy protein diet	25 g soy protein/day for 5 week	Assay of blood lipid metabolism parameters	Hypolipidemic	Cupisti et al. (2004)
Human (obese)	Soy-based low-calorie diet	Soy protein (only protein source) for 8 weeks	Assay of weight control, body composition, and blood lipid profile	Weight and body fat reduction along with hypolipidemic effect	Liao et al. (2007)
Human (hypercholesterolemic)	Isolated soy protein (ISP)	30–50 g ISP and 10–16.6 g cotyledon fiber/d for 16 weeks	Assay of lipid, lipoprotein, and homocysteine concentrations	Hypolipidemic and antiatherosclerotic effect	Tonstad et al. (2002)
Human (postmenopausal)	Soy protein isolate (SPI)	SPI (40 g/d) for 6 weeks	Assay of weight control, body composition, and blood lipid profile	Hypolipidemic and cardioprotective effect	Hanson et al. (2006)
Human (diabetics)	Soy nut diet	60 g/d for 8 weeks	Assay of blood glucose, lipid parameters, and antioxidant enzymes	Hypolipidemic, hypoglycemic, and cardioprotective effect	Sedaghat et al. (2019)

(continued)

Table 1.3 (continued)

Experimental model	The form of soya bean	Dose and route of administration	Investigation	Major finding/s	Reference
Sprague-Dawley rats	Phenolic-rich soy husk powder extract (SHPE)	250 mg SHPE/kg BW or 500 mg SHPE/kg BW	Assay of blood glucose, lipid metabolism parameters	Hypoglycemic and antiadipogenic	Tan et al. (2019)
Human (diabetic)	Soy protein with or without isoflavones (SPI, SP)	7.5 g (15 g daily) of 70% isolated soy protein powder with or without added isoflavones (Solgen 16 mg per bar, 32 mg in total daily) for 8 weeks	Assay of blood glucose, lipid metabolism parameters	Weight and body fat reduction along with hypolipidemic effect	Konya et al. (2019)
Goto-Kakizaki rats	Soy isoflavones (SIF)	SIF (150 mg/kg BW) for 16 weeks	Assay of blood glucose, lipid metabolism parameters	Hypoglycemic	Jin et al. (2018)
Wistar rats	Soy isoflavones	80 mg/kg BW/d for 4 weeks	Assay of plasma insulin, blood glucose, and hepatic glycogen	Antidiabetic and hypolipidemic	Hamden et al. (2011)
In vitro		–	Analysis of insulino-secretory effect of isoflavones and α-amylase inhibitory activity		
Sprague-Dawley rats	Soy hull soluble dietary fiber (SHSDF)	4% of SHSDF for 4 weeks	Hypocholesterolemic Activity assay	Hypocholesterolemic	Liu et al. (2016)
In vitro		–	Assay of in vitro cholesterol-binding capacity, bile acid-binding capacity, and glucose dialysis retardation index		
Sprague-Dawley rats	Soy isoflavones	0.2% soy isoflavones rich powder for 5 weeks	Assay of blood lipid parameters and antioxidant enzymes	Antioxidant and hypocholesterolemic	Kawakami et al. (2004)
Wistar adult rats	Soybean β-conglycinin (>90% protein, 0.4% isoflavone, and 0.2% saponin)	20% soybean β-conglycinin for 4 weeks	Assay of carbohydrate and lipid metabolism parameters	Hypolipidemic	Inoue et al. (2015)

Human (hyperlipidemic)	Soy protein-based cookies	30 g/d for 5 months	Assay of blood glucose, lipid metabolism parameters	Hypolipidemic	Borodin et al. (2009)
Hamsters	Soy pinitol	Pinitol supplementation (0.05% P-I and 0.1% pinitol, P-II) with an HFHC diet (10% coconut oil plus 0.2% cholesterol) for 10 weeks	Assay of blood glucose, lipid metabolism parameters	Lipid-lowering, antioxidant, and hepatoprotective effects	Choi et al. (2009)
Sprague-Dawley rats	Non-dialyzed soybean protein hydrolysate (NSPH)	14.7% casein +5% NSPH for 12 weeks	Assay of plasma and liver lipid profiles	Hypolipidemic	Yang et al. (2007)
New Zealand male rabbits	Soy isoflavones	0.73 or 7.3 mg of isoflavones/kg/day for 180 d	Assay of blood lipid metabolism parameters	Atheroprotective	Damasceno et al. (2007)
New Zealand white rabbits	Soy isoflavones	2.5 or 5 mg/kg B.W. doses of isoflavones for 13 weeks	Assay of blood glucose, lipid parameters, and antioxidant enzymes	Hypolipidemic and antioxidant activity	Yousef et al. (2004)
Sprague-Dawley rats	Soy protein isolate (SPI)	195 g/kg SPI protein/d for 6 weeks	Bone analyses, serum bone turnover markers, and NEFAs estimation	Bone-protective effect	Chen et al. (2013)
Rattus norvegicus albinus	Soy isoflavones (ISO)	ISO (150 mg/kg by gavage) for 30 d	Assay of collagen I (ColI) and sulfated glycosaminoglycans (GAGs) in the bone matrix	Decreased bone loss	Carbonel et al. (2019)
Human (postmenopausal women)	Soy product	–	Estimation of intake of soy products, folate, methionine, and vitamins B-6 and B-12 by a semiquantitative food frequency questionnaire	Favorable effect on homocysteine metabolism	Nagata et al. (2003)
Human (prostate cancer)	Soy bread	2 slices/d (60 mg aglycone equivalents of isoflavones/day) for 8 weeks	Evaluation of plasma isoflavonoids and isoflavonoids in urine	Cancer-protective effect	Ahn-Jarvis et al. (2015)

(continued)

Table 1.3 (continued)

Experimental model	The form of soya bean	Dose and route of administration	Investigation	Major finding/s	Reference
Prostate cancer cells (PC3 and DU145)	D-Pinitol	(0, 1, 3, 10, and 30 μM) for 24 h,	Assay of cell viability, TUNEL, caspase 3 activity, migration and invasion, and wound-healing migration	Reduced metastatic activity of human prostate cancer cells	Lin et al. (2013)
C57BL/6 × FVB F_1 TRAMP male pups	Soy isoflavone	–	Assay of hepatic aromatase and 5α-reductase; expression of AR, AR-regulated genes, FOXA1, UGT weight, and tumor progression; and upregulated protective FOXO3	Chemopreventive	Christensen et al. (2013)
Transgenic adenocarcinoma of the mouse prostate (TRAMP) model	Soy germ powder and tomato powder	2% soy germ (SG) powder or 10% tomato powder with 2% soy germ powder (TP + SG) for 14 weeks	Assay of isoflavone and carotenoid analysis in serum, prostate, and tissues	Prostate cancer-preventive effect	Zuniga et al. (2013)
LNCaP and C4-2B cells	Cooked and in vitro digested soy extracts	500,1000, and 2000 μg/mL	Apoptosis and cytotoxicity assays	Anticancer effect	Dong et al. (2012)
BALB/c mice (4T1 breast tumor model)	Soy isoflavone	100 mg/kg diet	Analysis of NF-κBp65′ vascular endothelial growth factor receptor 2 (VEGFR2) and Pgp gene and protein expressions	Anticarcinogenic effects	Hejazi et al. (2017)

A diet high in fiber, especially soluble fiber, can reduce the carbohydrate absorption rate and, hence, decrease the plasma glucose concentration in diabetic patients (Messina 1999; Chandalia et al. 2000). Similarly, another studies showed that intake of soybean dietary fiber increased fecal bile excretion, thus decreased fat absorption (Jenkins et al. 2003a, b), and, hence, a protective effect on hyperglycemia. Liu et al. (2016) reported that the physicochemical properties and in vitro binding capacity of soluble fibers extracted from soy hulls are similar to oat β-glucan which possesses proven glucose- and cholesterol-lowering properties.

Lee (2006) investigated the effect of soy protein and genistein on the blood glucose, lipid profile, and antioxidant enzyme activities in streptozotocin-induced diabetic Sprague-Dawley rats. The results implicated the beneficial role of soy protein and genistein in diabetes as their supplementation not only increased the glucokinase level, hepatic superoxide dismutase, catalase, and glutathione peroxidase activities but also decreased the HbA1c level of the STZ-induced diabetic rats. Ascencio et al. (2004) reported that the soy protein intake not only decreases the accumulation of triglycerides in the liver but also reduces the damaging effects of lipotoxicity in the liver, which had been recognized as the primary cause of obesity and related disorders, viz., insulin resistance, heart failure, and type 2 diabetes (Unger 2003; Sharma et al. 2004). In addition, soy protein intake in diabetic and non-diabetic patients has been reported to reduce the kidney damage and inflammation by reducing glomerular-filtration rate and improving creatinine clearance and, thus, holds the potential to be used as a therapeutic agent in the chronic kidney diseases (Azadbakht et al. 2003; Teixeira et al. 2004; Stephenson et al. 2005).

Recently, soy isoflavones especially daidzein, commonly found in fermented soybeans, have been reported to be beneficial in the therapeutic management of type 2 diabetes (Usui et al. 2013). Several in vitro studies have examined the antidiabetic and hypoglycemic effects and observed a dose-dependent effect of daidzein on intracellular glucose uptake in absence of insulin (Cheong et al. 2014) and inhibitory effect on α-glucosidase and α-amylase activities (Choi et al. 2010) and on the mRNA expression of CCL2 and IL6 (pro-inflammatory cytokines) in the adipocytes (Sakamoto et al. 2016). Also, its supplementation in the lean mice diet led not only to an increased glucose uptake but also glycogen synthesis in the liver, heart, and red blood cells (Meezan et al. 2005). Several clinical studies have also established the role of daidzein and its metabolite equol in the treatment of type 2 diabetes (Ho et al. 2007a, b; Villegas et al. 2008; Nguyen et al. 2017).

Lu et al. (2012) have reported the antidiabetic potential of aglycin, a bioactive peptide isolated from soybeans, in diabetic BALB/c mice. The authors reported that by increasing insulin receptor signalling pathway in the skeletal muscle, aglycin controlled hyperglycemia and improved oral glucose tolerance in the diabetic mice. Sivakumar and Subramanian (2009) investigated the effect of D-pinitol, a bioactive component isolated from soybeans, in diabetic rats, and observed that it alters the activities of key hepatic enzymes involved in carbohydrate metabolism and, thus, attenuates the hyperglycemic effect in diabetic rats. Soyasaponins have also been reported to possess hypoglycemic effect by exerting inhibitory effect on α-glucosidase enzyme (Quan et al. 2003). A more recent in vitro and in vivo study by Wang et al. (2017) suggests that stigmasterol (phytosterol derived from soybean oil) has potential therapeutic effect in type 2 diabetes. In the in vitro study, stigmasterol exhibited a mild GLUT4 translocation activity and enhanced glucose uptake in L6 cells. Furthermore, when stigmasterol was orally administered to KK-Ay mice, it led not only to a significant reduction in the fasting blood glucose level, triglyceride, and cholesterol but also an improvement in insulin resistance and oral glucose tolerance.

1.4.2 In Cardiovascular Diseases

Several epidemiological studies have investigated the role of soybean in the incidence of cardiovascular disease and have reported an inverse relationship owing to the presence of soy proteins (Torres et al. 2006), bioactive peptides (Friedman and Brandon 2001; Choi et al. 2002), soy isoflavones (Nagata et al. 2016; Liu et al. 2014), polyphenols (Huang et al. 2016a, b), phospholipids (Sahebkar 2013), stanols and lecithins (Spilburg et al. 2003), and soy phytosterols (Anderson et al. 1995; Ostlund Jr 2004; Escurriol et al. 2010; Genser et al. 2012).

Shimazu et al. (2007) reported an inverse association between soybean intake and CVD mortality. A meta-analysis of randomized controlled trials by Tokede et al. (2015) also found that soy product intake led to a significant decrease in total cholesterol, LDL-C, HDL-C, and triglycerides. However, interestingly Nagata et al. (2016) observed that different soy foods may present different biological efficacy and protective effects.

Cholesterol-lowering effect of soy protein was first studied in 1967 (Hodges et al. 1967), and since then, numerous epidemiological surveys and nutritional interventions have suggested the possible cardioprotective role of soy proteins (Radcliffe and Czajka-Narins 1998; Jenkins et al. 2003a, b; Merritt 2004; Anderson and Bush 2011; Zhan and Ho 2005). Homocysteine (Hcy) is one of the risk factors for cardiovascular diseases, and since methionine is a precursor of Hcy, hence, the intake of soy protein which is low in methionine helps in reducing the coronary heart disease risk (Tovar et al. 2002). Schmitt et al. (1998) reported that high ratio of insulin/glucagon is positively associated with hyperlipidemic and atherogenic effects and long-term soy protein intake reduces the insulin/glucagon ratio and, hence, exhibits hypolipidemic effect. β-Conglycinin, a bioactive peptide present in soybean, has been reported to possess greater cholesterol- and triglyceride-lowering effects when compared with soy protein isolate (Bringe 2001) owing to the decreased intestinal cholesterol absorption, bile acid uptake (Nagaoka et al. 1999), reduced aortic accumulation of cholesteryl esters (Adams et al. 2004), and increased cholecystokinin levels which suppress food intake and gastric emptying (Nishi et al. 2003).

It is believed that soy isoflavones by activating the estrogen receptors and intracellular kinase signalling cascades exert anti-inflammatory responses and modulate the vascular reactivity (Li et al. 2006) and could lower the bile acid synthesis, hepatic lipid synthesis, and cholesterol reabsorption (Ricketts et al. 2005). However, recent studies like that of Engelbert et al. (2016) and Taku et al. (2008), wherein no significant changes in the total cholesterol were reported in postmenopausal women taking isoflavone supplements/extracts, have led to a wide disagreement regarding the hypolipidemic role of soy isoflavones. Moreover, recent research suggests that a synergistic interaction of soy isoflavones and soy protein augments the blood lipid profile and, hence, exerts hypolipidemic effect (Xiao et al. 2014; Kobayashi et al. 2014).

1.4.3 In Hypertension

Soybean and its bioactive components (soy protein, isoflavones, etc.) have been reported to mitigate hypertension by mechanisms involving vasodilation and inhibition of key enzyme involved in the blood pressure regulation (Jackson et al. 2011). In the postmenopausal pre-diabetic hypertensive women, soy protein and isoflavone intake attenuated the blood pressure (Welty et al. 2007; Liu et al. 2013). Colacurci et al. (2005) studied the effect of isoflavone supplementation in postmenopausal women for 6 months and reported improvement in endothelial vasodilation along with reduction in cellular adhesion molecules. However, other studies have found that soy isoflavone supplementation can exert hypotensive effect in hypertensive but not in normotensive adults (Taku et al. 2010; Patten et al. 2016). Since soybean is a rich source of arginine which in turn is a precursor to nitric oxide in the L-arginine pathway, thus, it improves endothelial function and demonstrates hypotensive effect (Bai et al. 2009; Dong et al. 2011). Also, in vitro studies have showed that soy pulp containing oligopeptides and fiber in high amounts exhibited anti-angiotensin-converting enzyme activity and, hence, hypotensive effect (Nishibori et al. 2017).

1.4.4 In Obesity

Overweight and obesity have a profound impact on global health, and it is a risk factor for several chronic diseases like diabetes and hypertension. Epidemiological evidence suggests a positive association between soy consumption and weight management by enhancing insulin resistance and subsiding lipoprotein lipase activity (Velasquez and Bhathena 2007; Ørgaard and Jensen 2008; Muscogiuri et al. 2016).

Kurrat et al. (2015) reported that lifelong intake of soy isoflavones reduced the body weight, serum leptin, and visceral fat mass and resulted in smaller adipocytes in female Wistar rats. In another study, diet-induced obese male rats when fed with soy isoflavones showed enhanced lipolysis and β-oxidation along with suppressed lipogenesis, adipogenesis, and decreased body weight (Huang et al. 2016a, b). However, some studies have also reported that soy isoflavone supplementation resulted in increased adipose tissue (Zanella et al. 2015), increased total cholesterol, and leptin concentrations in mice (Giordano et al. 2015).

Since soy protein is a major constituent and possesses high biological value in addition to the presence of bioactive compounds, its role in obesity cannot be overlooked. Soy protein isolate and its hydrolysate were found to be effective in reducing the body fat and perirenal fat pads when compared with whey protein isolate in the treatment of obese male Sprague-Dawley rats (Aoyama et al. 2000). Nagasawa et al. (2002) observed decreased body fat content and plasma glucose levels in obese mice fed with soy protein than the mice fed with casein protein diet. Anderson and Hoie (2005) investigated the effects of soy- versus milk-based meal replacement in obese women (BMI: 27–40 kg/m^2) for 12 weeks and reported modest weight loss coupled with significant reduction in blood lipids of the subjects. Neacsu et al. (2014) in a randomized crossover trial reported appetite control and weight loss among obese men fed with soy-based high-protein weight-loss diets.

1.4.5 In Inflammation

Han et al. (2015) suggested the anti-inflammatory effect of genistein (a soy isoflavone) in homocysteine (Hcy)-induced endothelial cell inflammatory injury. It is reported that soy isoflavones exhibited the anti-inflammatory effects by showing a reduction in the release of reactive oxygen species (ROS), inhibited NF-kB activation; down-regulating the expression of cytokine IL-6 and adhesion molecules ICAM-1, avoiding inflammatory cells and platelet adhesion, and thus, balanced the endothelial cell proliferation and apoptosis. Sakamoto et al. (2016) concluded from their in vitro study that daidzein or soy consumption can be helpful in suppressing chronic inflammation which in turn can alleviate obesity-related insulin resistance. Wang and Wu (2017) reported that dietary soy isoflavones hold the potential to alleviate dextran sulfate sodium (DSS)-induced inflammation in mice by enhancing antioxidant function and inhibiting the TLR4/MyD88 signal.

1.4.6 Effects on Menopausal Symptoms

Since 1991, several studies have been undertaken to investigate the role of soybeans and its bioactive components on the menopausal symptoms and have reported their efficacy in the same (Lockley 1991; Adlercreutz et al. 1992; Murkies et al. 1995; Lethaby et al. 2007; Howes et al. 2006). However, Newton and Grady (2011) have commented that the results may have been misinterpreted, and Messina (2014) observed that the studies did not sub-analyze the data according to the isoflavone profile of the intervention product. Recently, Furlong et al. (2019) investigated the effects of a commercially available soy drink containing 10–60 mg/d dose of isoflavones for 12 weeks on the cognitive function and menopausal symptoms in postmenopausal women. The authors reported no change in the cognitive function but significant reduction in the vasomotor symptoms in subjects with severe symptoms at baseline.

1.4.7 In Bone Health

Zhang et al. (2005) conducted a prospective cohort study in Shanghai to investigate the efficacy of soy food consumption among 24,403 postmenopausal women, and the results revealed that consumption of soy protein (>10 g/d) was associated with almost one-third reduction in the fracture risk in the subjects. Similar results were reported in the postmenopausal women participating in the Singaporean study wherein 63,257 Chinese adults (45–72 years) were studied. However, Levis et al. (2011) and Tai et al. (2012) did not find any positive effect.

1.4.8 Anticarcinogenic Activities

Numerous experimental models have revealed the anticancer activity of soy-based diets and its bioactive compounds (Zhou et al. 1999; Mentor-Marcel et al. 2001; Trottier et al. 2010; Zuniga et al. 2013). The consumption of foods rich in soy isoflavones is associated with reduction in the occurrence and mortality of breast cancer (Valachovicova et al. 2004; He and Chen 2013; Applegate et al. 2018). Applegate et al. (2018) conducted a meta-analysis and reported that soy foods and their isoflavones (genistein and daidzein) resulted in the reduction of prostate cancer (PCa) risk. Soy isoflavones are assumed to effect PCa aggression through various pathways such as inhibition of tumor growth factor signalling (Wang et al. 2003), cell cycle inhibition (Zhou et al. 1999), metastasis (Pavese et al. 2014), and anti-angiogenesis (Fotsis et al. 1993; Guo et al. 2007). Yu et al. (2016) performed a meta-analysis of 17 epidemiological studies consisting of 13 case controls and 4 prospective cohort studies to investigate the association between colorectal cancer (CRC) risk and soy isoflavone consumption in humans. The study revealed that intake of soy foods containing soy isoflavones by Asian population in the case-control studies resulted in a decreased CRC risk. It is believed that soy isoflavones can exert the antitumor effects via their roles in antioxidation, DNA repair, antagonism of estrogen- and androgen-mediated signalling pathways, inhibition of angiogenesis and metastasis, and potentiation of radio- and chemotherapeutic agents (Bilir et al. 2017; Mahmoud et al. 2014).

As discussed above the consumption of soy products rich in isoflavones has been associated with decreased cancer risks, but in the recent times, some studies have raised concerns on the deleterious health effects of isoflavones, predominantly on the carcinogenic activity (Poschner et al. 2017; Wei et al. 2015; Andrade et al. 2015; Shike et al. 2014) and reproductive toxicity (Patel et al. 2016; Chinigarzadeh et al. 2017), adverse effects on growth and development (Harlid et al. 2016; Yin et al. 2014; D'Aloisio et al. 2013), and impacts on immune functioning (Wynn et al. 2013; Gaffer et al. 2018).

A recent study suggests that soyasapogenol B (Soy B) through inducing apoptotic and autophagic cell death, thus, attenuates laryngeal carcinoma progression in human laryngeal carcinoma cell lines HeP-2 and TU212 (Zhi et al. 2019). Wang et al. (2019) investigated the effect of Soy B in the prevention and treatment of CRC; and reported that it promoted apoptosis and autophagy in both in vitro and in vivo assays, by triggering endoplasmic reticulum stress, and, hence, can be utilized as a chemotherapeutic agent in CRC.

Since Liener (1991) reported that soybean agglutinin holds the potential to inhibit the growth of transplanted tumor in rats, numerous studies have established the possible antitumor and anticarcinogenic potential of plant lectins (Suzuki et al. 1999; Jakab et al. 2000; Pryme and Bardocz 2001; Evans et al. 2002). Soybean lectins have been found to suppress the tumor growth, Dalton's lymphoma, macrophages, peripheral blood lymphocytes (Ganguly and Das 1994) and cytoagglutination/aggregation in SW 1222 human colon cancer, HT29 human colon cancer (Jordinson et al. 1999), SP2 myeloma; and Lox-2 Ab-producing hybridoma (Takamatsu et al. 1999). Additionally, recent research has pointed out toward a newly discovered soybean peptide, lunasin, as a new and novel cancer chemopreventive agent (de Mejia et al. 2003).

1.5 Food Applications

1.5.1 Soybean Oil

Soybean oil, one of the most consumed edible oils (26.7% of the total), is the largest commercial source of essential fatty acids and is utilized in many food products, viz., as salad and cooking oil, shortening, margarine, mayonnaise, and salad dressing. Its easy availability and functionality coupled with cheap price are the reason behind its massive acceptance as edible oil in the world (Medina-Juarez et al. 1998; List 2016). From the nutritional point of view, it contains linoleic acid (50–60%), oleic acid (22–30%), palmitic acid (7–10%), linolenic acid (5–9%), stearic acid (2–5%), and arachidic acid (1–3%) and polyunsaturated/saturated ratio of 4.1. Apart from this it also contains lecithin (phosphatidylcholine), phospholipids, tocopherols, and phytosterols (Fan and Eskin 2015).

However, high amounts of polyunsaturated acids and omega-3 fatty acids in soybean oil limit its commercial functionality as they present oxidative stability problems and limit the fry life, respectively (List 2016).

1.5.2 Soy Products

1.5.2.1 Soy Protein Products

Cereals form an important source of energy and nutrients in the majority of the world population. However, the low protein quality due to imbalanced amino acid composition is a major concern. Hence, soybean protein in various forms, viz., flour and grits (full fat, medium fat, low fat, defatted, and lecithinated), soy protein concentrate, soy protein isolate, and textured soy protein, is utilized by the food industry for its good-quality protein, low cost, and numerous functional and nutraceutical properties (Kulkarni et al. 1992).The chemical composition of various soy products is depicted in Table 1.4.

1.5.2.2 Soy Flour and Grits

After oil is removed from the soybean, the proteinaceous material left is referred to as soybean flakes which are then ground into flour (100 mesh or finer). Grits are coarser than the flour (>100 mesh). Soy flour and grits are the least refined form and possess varying amount of fat, particle size, texture, saponins and isoflavones which result in the typical beany flavor (Singh et al. 2008). Full-fat soy flour is used in the production of soy milk and tofu and as an economic extender in developing countries for non-fat dry milk in beverages (Ohr 1997) and baked goods (Rakosky 1974). The lipoxidase enzyme-active full-fat soy flour helps in improving the whiteness of bread dough (French 1977), acts as a good emulsifier and stabilizer, and, hence, helps in

Table 1.4 Chemical requirements of soy products

Soy product	Protein (Nx6.25)%	Carbohydrates (%)	Fat (%)	Moisture (%)	Crude fiber (%)	Ash (%)	PDCAAS*
Full-fat soy flour[a]	35–40	20.0	18–20.5	10	4.0	6.5	–
High fat[a]	46	–	14.5	6.0	–	–	–
Low fat[a]	52.5	–	4.0	6.0	–	–	–
Lecithinated[a]	51	–	6.5	7.0	–	–	–
Defatted soy flour[b]	50	20	1.5	9.0	3.5	7.0	0.90
Soy protein concentrate[b]	65	20	1.0	6.0	4.0	6.0	0.95
Soy protein isolates[b]	90	4.0	1.0	6.0	0.2	4.5	0.90
Texturized soy protein[b]	50	20	3.0	10	4.0	6.5	0.90

*PDCAAS: Protein digestibility-corrected amino acid score
[a]Campbell et al. (1985)
[b]ASA (2000)

homogenizing milk in cakes and improves mixing tolerance when mixed with ready flour mixes (Onayemi and Lorenz 1978; Lutzow 1996; Stauffer 2006).

On the other hand, defatted soy flour and grits can be utilized in the fortification of cereals and processed foods like baked goods. Its fortification not only helps in the enhancement of the protein content in the processed foods but also improves the crumb body in the baked foods, enhances shelf life of cookies by reducing the moisture content (Marques et al. 2000), and improves water holding capacity and sheeting properties of the dough, thus resulting in tender finished product (Golbitz 2000). The toasted soy flour and grits are preferred in cookies/crackers, cereal applications (like breads), ground meats, and fermentation media because of their better texture than the untoasted counterparts (Endres 2001; Jideani 2011).

1.5.2.3 Soy Protein Isolates (SPI)

Soy protein isolates are generally made from defatted soy meal and are the most refined form. They should possess good emulsifying capacity, gelling capacity, viscosity, and water and fat absorption to be utilized efficiently in various food systems (Singh et al. 2008). Since, they have a very high protein content (>90%) and are prepared by removing the water-insoluble polysaccharides and oligosaccharides, they do not result in flatulence and have bland flavor. Thus, SPI are ideally suited ingredients in meat systems (due to the mouthfeel and texture); dairy foods as milk replacer, nutritional supplements, and infant formulas; beverages; soups; sauces; snacks; etc. aimed for people having high protein needs owing to various circumstances like growth in children, famine, and chronic diseases, viz., AIDS, tuberculosis, etc. (Singh et al. 2008).

1.5.2.4 Soy Protein Concentrates (SPC)

Soy protein concentrates contain approximately 70% protein and are prepared from the defatted soy flour by removing oligosaccharides, some amount of ash, and other minor components (ASA 2000). The concentration process improves the dispersibility and water and fat holding capacity and improves its flavor profile which modifies the viscosity and textural characteristics of the food system. Hence, SPC are utilized in the production of baby foods, dry food mixes, nutritional powder drinks, emulsion-type meat products, bakery products, milk replacers, snacks, and pet foods (Singh et al. 2008).

1.5.2.5 Textured Soy Protein (TSP)

Textured soy proteins are commercially prepared by the thermoplastic extrusion of soy flour/grits/concentrates under heat and pressure to yield chunks/chips/flakes or any other shape. The high amount of heat and pressure along with extrusion not only increases the water absorption index and water hydration capacity but also hardens and expands the product to impart a fibrous texture (Horan 1974; Grasso et al. 2019; Toldrà et al. 2019) and, thus, be used majorly as meat alternatives (Macedo-Silva et al. 2001; Wong et al. 2019). In the dry form, they are incorporated in the applications wherein during processing the juices liberated by the meats need to be absorbed for a firm final product, for example, beef patties, sausages, pizza toppings, frozen dinners, packaged dinners and soups, taco fillings, vegetarian foods, pet foods, etc. On the other hand, the hydrated forms are handled just like any other meat/perishable food (Lin et al. 2000; Hennenger 2002).

1.5.3 Fermented Soy Foods

The classification of traditional soy products is represented in Fig. 1.4.

1.5.3.1 Soy Sauce

It is a fermented soybean condiment which originated approximately 2500 years ago in China. *Aspergillus oryzae* and/or *Aspergillus sojae* molds are usually used for fermenting soybean paste (Chen et al. 2012). Soy sauce was first used in only

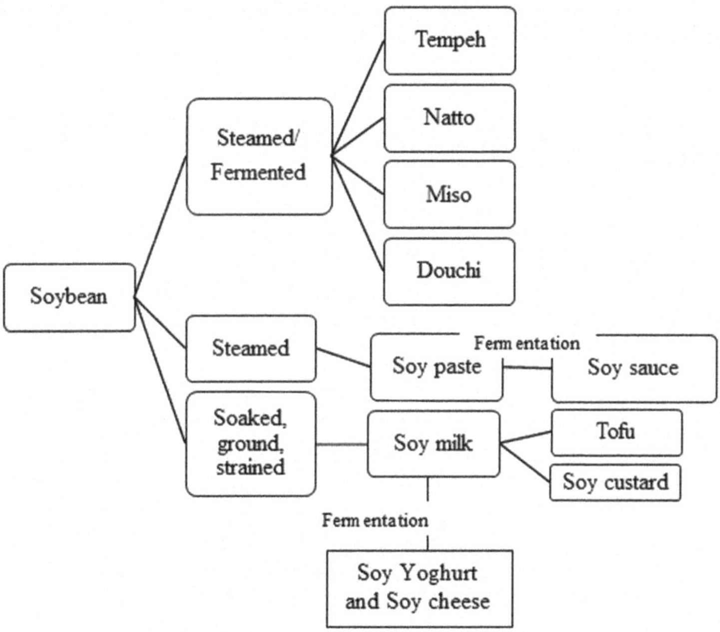

Fig. 1.4 Classification of traditional soybean products

Oriental cuisine but now has gained popularity and is now a major ingredient in American cuisine.

1.5.3.2 Soy Sprouts

Since ancient times, sprouted soybeans have been included in the Korean cuisine as the development of food products from germinated soybeans further increases the versatility and utility owing to it having zero cholesterol, low saturated fatty acid, low calorie, and high fiber content (Hwang 1997; Kwon et al. 2005).

1.5.3.3 Tempeh

Fermentation of whole soybeans along with rice/millets yields a smoky- or nutty-flavored, chunky, and tender soybean cake. This product is called tempeh and is a traditional Indonesian food. It protects against diarrhea and chronic degenerative diseases and, thus, is slowly gaining importance as an important functional food ingredient (Vital et al. 2018).

1.5.3.4 Natto

Natto, a traditional Japanese health food, is prepared by *Bacillus subtilis* var. *natto* (*Bacillus natto*)-fermented whole soybeans and is reported to be having numerous health benefits (Fujiwara et al. 2008).

1.5.3.5 Miso

Miso, a fermented soybean paste, is a healthy Japanese seasoning which has numerous health benefits. In the recent times, owing to its health attributes and superb taste, it is utilized in numerous food applications like soups, sauces, dressings, marinade, and pastes (Mani and Ming 2017).

1.5.4 Non-fermented Soy Foods

1.5.4.1 Soy Milk/Beverages

Soybeans when soaked, ground, and strained yield an aqueous, white, and creamy extract which is similar in appearance and consistency to cow's milk. It is a traditional drink of the Eastern world and can be consumed by lactose-intolerant people (Rivas et al. 2002). It contains nearly 3–4% protein (same as cow milk, although amino acid composition differs), 1.5–2.0% fat, and 8–10% carbohydrates (Kohli et al. 2017). The soy milk serves as a base material for tofu, soy yogurt, custard, cheese, etc. (Favaro Trindade et al. 2001; Liu et al. 2006).

1.5.4.2 Tofu (Soy Paneer)

Tofu is also known as soy curd which is usually served as a dessert and side dish. It is a soft cheese made by curdling the fresh hot soy milk by addition of calcium or magnesium salts. Tofu is a good source of proteins and isoflavones and can be stored up to 1 year under ambient conditions (Chen et al. 2012).

1.5.4.3 Soy Cheese

Soy cheese is made from soy milk. Its creamy texture makes it an easy substitute for animal protein and can be utilized as a low-cost protein source (Ibironke and Alakija 2018).

1.5.4.4 Non-dairy Soy Frozen Dessert

The production of non-dairy frozen desserts is a novel trend in the functional food industry. Soy frozen dessert is made from either soy milk or soy yogurt and is one of the most popular healthy desserts made from soybeans (Atallah and Hassan 2017; Norouzi et al. 2019; Rezaei et al. 2019).

1.5.4.5 Soy Nut Butter

It is made from roasted and crushed whole soybeans which are then blended with soy oil and other ingredients for a creamy and crunchy texture. It is a tasty and healthy alternative to peanut butter and provides 7 g of soy protein per serving. Recently, owing to the risk of the peanut allergies, many schools in the United States are introducing soy nut butter in the school lunch programs (Shurtleff and Aoyagi 2012).

1.5.4.6 Soy Fiber (Okara, Soy Bran)

The solid residue left after the extraction in the production of soy milk or tofu is called okara. It is rich in proteins (24.5–37.5%), fiber (14.5–55.4%), and fats (9.3–22.3%) and has a neutral taste unlike the other soy products. The essential amino acid composition and functional properties like emulsification, foaming, fat binding, and fat absorption capacity are similar to the commercial soy protein isolates, and thus, it holds a great potential to be utilized as a functional food ingredient (Ma et al. 1996). The outer covering of soybean removed during initial processing is referred to as soy bran. Psodorov et al. (2015) reported that soy bran particles (<50 μm) possess ideal textural and sensorial characteristics to be used as a fat replacer in the development of gluten-free cookies and cakes.

1.5.4.7 Green Vegetable Soybean (Edamame)

Edamame, a specialty soybean, is harvested when the beans are still immature, green, and sweet-tasting and have expanded 80–90% of the pod and is served as a snack or a main vegetable dish in East Asia (Konovsky et al. 1994).

1.5.5 Soy-Based Infant Formulas

Soy-based infant formulas constitute a soy protein isolate powder as the milk substitute and are aimed for the infants suffering with galactosemia and lactase deficiency. The soy-based infant formulas are quite popular and constitute nearly 25% of the formula market in the United States (Bhatia and Greer 2008).

1.5.6 Hydrolyzed Vegetable Protein (HVP)

HVP, a flavor protein produced by the hydrolysis of untoasted defatted soybean flour, is commonly used as a flavor enhancer in numerous food applications such as in soups, broths, sauces, gravies, etc. The typical taste (umami or the fifth taste) of HVP is contributed mostly by the presence of free amino acids (glutamic acid), smaller peptides, salts, and various other volatile compounds (Aaslyng et al. 1998).

1.5.7 Lecithin

Soybean lecithin, a by-product of soybean processing, is an important emulsifier used in the food, pharmaceutical, feed, and technical industries. It contains 37% neutral oils, 18% phosphatidylcholine (PC), 14% phosphatidylethanolamine (PE), 11% glycolipids, 9% phosphatidylinositol (PI), 5% phosphatidic acid (PA), 5% complex sugars, and 2% phospholipids (PL) (Wu and Wang 2003).

1.6 Alternative Applications

1.6.1 Animal Feed

Soybean meal is a major protein, mineral, and vitamin source in animal feed, and around 90–95% of the total soybean meal produced is utilized for livestock feed. Soy hulls, a by-product resulting from the processing of soybean, are also used widely as animal feed (Horan 1974; Peisker 2001).

1.6.2 Soybean Protein Fiber (SPF)

Soybean protein fiber (SPF) is the protein fiber produced from soybean cake and is quite similar to synthetic fiber. It is used as a blend with cashmere to give smoothness, with wool to reduce the shrinkage, and with silk to prevent stickiness when wet. Apart from these it also provides strength, comfort, easy to care properties, absorbency, and luster (Rijavec and Zupin 2011).

1.6.3 Soy Oil

Soybean oil finds numerous applications in the production of drying oil, inks for newspaper and offset printing, plasticizer, surfactant, dimmer acids, hydraulic fluids, insecticides and fungicides, solvents and cleaners, water-dispersible poly resins, and biodiesel (Honary 1996; Sonntag 1985; Kinney and Clemente 2005).

1.6.4 Soy Protein

Soy protein concentrate is used as a nutrient base for fermentation in the production of pharmaceuticals, gums, and gels. It is also utilized in the production of plastics, cosmetics, and wood replacers (Kato 2002; Huang and Sun 2000; Wang et al. 2007).

1.6.5 Soy Lecithin

Soy lecithin is used in the automotive industry for cleaning as well as a chelating agent. In the cosmetic and pharmaceutical industry, it finds its use in controlling and modification of fat crystal structure. It is used as an emollient in the shampoo products, as a lubricator in the shock absorbers and hydraulics, as a wetting agent, and softening and curing agent in leather tanning agent (Szuhaj 1983; Ghosh and Bhattacharyya 1997; Xu et al. 2011; Cerminati et al. 2019).

1.7 Conclusion

Soybean is a rich source of lipids and proteins and contains numerous phytonutrients such as saponins, bioactive peptides, phytosterols, and phenolic compounds; thus conferring a plethora of health benefits like hypoglycemic, hypolipidemic, antiobesity, hepatoprotective, anticancer, etc. The processing conditions like soaking, roasting, germination, autoclaving, etc. alleviate antinutrients like oxalates, phytic acids, saponins, etc. and, thus, enhance functionality of soybean in food industry. Numerous products like soybean oil, soy protein isolates, soy protein concentrates, soy textured protein, soy flour/grits, and traditional products like soy milk, tofu, natto, miso, etc. are prepared from soybeans which are not only nutritionally superior but also possess therapeutic properties. However, long-term human studies are required to further claim the health benefits of soybean, soy bioactive agents, and soy products and to elucidate their mechanism as therapeutic agents.

References

Aaslyng MD, Martens M, Poll L, Nielsen PM, Flyge H, Larsen LM (1998) Chemical and sensory characterization of hydrolyzed vegetable protein, a savory flavoring. J Agric Food Chem 46 (2):481–489

Abo-Elsoud MA, Hashem NM, El-Din AN, Kamel KI, Hassan GA (2019) Soybean isoflavone affects in rabbits: effects on metabolism, antioxidant capacity, hormonal balance and reproductive performance. Anim Reprod Sci 203:52–60

Adams MR, Golden DL, Franke AA, Potter SM, Smith HS, Anthony MS (2004) Dietary soy β-conglycinin (7S globulin) inhibits atherosclerosis in mice. J Nutr 134(3):511–516

Adlercreutz H, Hämäläinen E, Gorbach S, Goldin B (1992) Dietary phyto-oestrogens and the menopause in Japan. Lancet 339(8803):1233

Agarwal K (2014) Comparative biochemical analysis of raw and processed soybean. J Sci Technol 3(1):61–66

Ahn-Jarvis JH, Clinton SK, Grainger EM, Riedl KM, Schwartz SJ, Lee ML, Cruz-Cano R, Young GS, Lesinski GB, Vodovotz Y (2015) Isoflavone pharmacokinetics and metabolism after consumption of a standardized soy and soy–almond bread in men with asymptomatic prostate cancer. Cancer Prev Res 8(11):1045–1054

Alghamdi S, Migdadi H, Khan M, El-Harty EH, Ammar M, Farooq M, Afzal M (2018) Phytochemical profiling of soybean (Glycine max (L.) Merr.) genotypes using GC-MS analysis. In: Asao T, Asaduzzaman M (eds) Phytochemicals-source of antioxidants and role in disease prevention. IntechOpen, London, UK

Al-Wahsh IA, Horner HT, Palmer RG, Reddy MB, Massey LK (2005) Oxalate and phytate of soy foods. J Agric Food Chem 53(14):5670–5674

Anderson JW, Bush HM (2011) Soy protein effects on serum lipoproteins: a quality assessment and meta-analysis of randomized, controlled studies. J Am Coll Nutr 30(2):79–91

Anderson JW, Hoie LH (2005) Weight loss and lipid changes with low-energy diets: comparator study of milk-based versus soy-based liquid meal replacement interventions. J Am Coll Nutr 24 (3):210–216

Anderson JW, Johnstone BM, Cook-Newell ME (1995) Meta-analysis of the effects of soy protein intake on serum lipids. N Engl J Med 333(5):276–282

Andrade JE, Ju YH, Baker C, Doerge DR, Helferich WG (2015) Long-term exposure to dietary sources of genistein induces estrogen-independence in the human breast cancer (MCF-7) xenograft model. Mol Nutr Food Res 59(3):413–423

Anta L, Marina ML, García MC (2010) Simultaneous and rapid determination of the anticarcinogenic proteins Bowman-Birk inhibitor and lectin in soybean crops by perfusion RP-HPLC. J Chromatogr A 1217(45):7138–7143

Aoyama T, Fukui K, Nakamori T, Hashimoto Y, Yamamoto T, Takamatsu K, Sugano M (2000) Effect of soy and milk whey protein isolates and their hydrolysates on weight reduction in genetically obese mice. Biosci Biotechnol Biochem 64(12):2594–2600

Applegate CC, Rowles JL, Ranard KM, Jeon S, Erdman JW (2018) Soy consumption and the risk of prostate cancer: an updated systematic review and meta-analysis. Nutrients 10(1):40

ASA (2000) American Soybean Association. Soy stats guide. https://soygrowers.com/education-resources/publications/soy-stats/. Accessed 19 Oct 2019

Ascencio C, Torres N, Isoard-Acosta F, Gómez-Pérez FJ, Hernández-Pando R, Tovar AR (2004) Soy protein affects serum insulin and hepatic SREBP-1 mRNA and reduces fatty liver in rats. J Nutr 134(3):522–529

Atallah AA, Hassan B (2017) Preparation of non-dairy soft ice milk with soy milk. J Adv Dairy Res 5(172):2

Azadbakht L, Shakerhosseini R, Atabak S, Jamshidian M, Mehrabi Y, Esmaill-Zadeh A (2003) Beneficiary effect of dietary soy protein on lowering plasma levels of lipid and improving kidney function in type II diabetes with nephropathy. Eur J Clin Nutr 57(10):1292

Azadbakht L, Kimiagar M, Mehrabi Y, Esmaillzadeh A, Padyab M, Hu FB, Willett WC (2007) Soy inclusion in the diet improves features of the metabolic syndrome: a randomized crossover study in postmenopausal women. Am J Clin Nutr 85(3):735–741

Bai Y, Sun L, Yang T, Sun K, Chen J, Hui R (2009) Increase in fasting vascular endothelial function after short-term oral L-arginine is effective when baseline flow-mediated dilation is low: a meta-analysis of randomized controlled trials. Am J Clin Nutr 89(1):77–84

Benito-González I, Martínez-Sanz M, Fabra MJ, López-Rubio A (2019) Health effect of dietary fibers. In: Galanakis CM (ed) Dietary fiber: properties, recovery, and applications. Academic Press, London, UK, pp 125–163

Berhow MA, Cantrell CL, Duval SM, Dobbins TA, Maynes J, Vaughn SF (2002) Analysis and quantitative determination of group B saponins in processed soybean products. Phytochem Anal 13(6):343–348

Bhatia J, Greer F (2008) Use of soy protein-based formulas in infant feeding. Pediatrics 121 (5):1062–1068

Bilir B, Sharma NV, Lee J, Hammarstrom B, Svindland A, Kucuk O, Moreno CS (2017) Effects of genistein supplementation on genome wide DNA methylation and gene expression in patients with localized prostate cancer. Int J Oncol 51(1):223–234

Birk Y (1985) The Bowman-Birk inhibitor. Trypsin-and chymotrypsin-inhibitor from soybeans. Int J Pept Protein Res 25(2):113–131

Borodin EA, Menshikova IG, Dorovskikh VA, Feoktistova NA, Shtarberg MA, Yamamoto T, Takamatsu K, Mori H, Yamamoto S (2009) Effects of two-month consumption of 30 ga day of soy protein isolate or skimmed curd protein on blood lipid concentration in Russian adults with hyperlipidemia. J Nutr Sci Vitaminol 55(6):492–497

Bringe NA (2001) Inventor; Monsanto Co, assignee. High beta-conglycinin products and their use. United States patent US 6,171,640

Burssens S, Pertry I, Ngudi DD, Kuo YH, Van Montagu M, Lambein F (2011) Soya, human nutrition and health. In: Yin Y, Fatufe AA, Blachier F (eds) Soya bean meal and its extensive use in livestock feeding and nutrition. INTECH Open Access Publisher, pp 157–180

Campbell MF, Kraut CW, Yackel WC, Yang HS (1985) Soy protein concentrate. In: Altschul AM, Wilcke HL (eds) New protein foods: seed storage proteins. Academic Press, Orlando, FL, pp 301–337. https://doi.org/10.1016/B978-0-12-054805-7.50016-6

Carbonel AA, Vieira MC, Simões RS, Lima PD, Fuchs LF, Girão ER, Cicivizzo GP, Sasso GR, de Moraes LC, Soares Junior JM, Baracat EC (2019) Isoflavones improve collagen I and glycosaminoglycans and prevent bone loss in type 1 diabetic rats. Climacteric 28:1–9

Carvalho MR, Kirschnik PG, Paiva KC, Aiura FS (2002) Avaliação da atividade dos inibidores de tripsina após digestão enzimática em grãos de soja tratados termicamente. Rev Nutr 1:267–272

Cerminati S, Paoletti L, Aguirre A, Peirú S, Menzella HG, Castelli ME (2019) Industrial uses of phospholipases: current state and future applications. Appl Microbiol Biotechnol 103 (6):2571–2582

Chalvon-Demersay T, Azzout-Marniche D, Arfsten J, Egli L, Gaudichon C, Karagounis LG, Tome D (2017) A systematic review of the effects of plant compared with animal protein sources on features of metabolic syndrome. J Nutr 147(3):281–292

Chandalia M, Garg A, Lutjohann D, Von Bergmann K, Grundy SM, Brinkley LJ (2000) Beneficial effects of high dietary fiber intake in patients with type 2 diabetes mellitus. N Engl J Med 342 (19):1392–1398

Chang JH, Kim MS, Kim TW, Lee SS (2008) Effects of soybean supplementation on blood glucose, plasma lipid levels, and erythrocyte antioxidant enzyme activity in type 2 diabetes mellitus patients. Nutr Res Pract 2(3):152–157

Chávez-Santoscoy RA, Gutiérrez-Uribe JA, Serna-Saldívar SO (2013) Effect of flavonoids and saponins extracted from black bean (Phaseolus vulgaris L.) seed coats as cholesterol micelle disruptors. Plant Foods Hum Nutr 68(4):416–423

Chen KI, Erh MH, Su NW, Liu WH, Chou CC, Cheng KC (2012) Soyfoods and soybean products: from traditional use to modern applications. Appl Microbiol Biotechnol 96(1):9–22

Chen JR, Zhang J, Lazarenko OP, Cao JJ, Blackburn ML, Badger TM, Ronis MJ (2013) Soy protein isolates prevent loss of bone quantity associated with obesity in rats through regulation of insulin signaling in osteoblasts. FASEB J 27(9):3514–3523

Cheong SH, Furuhashi K, Ito K, Nagaoka M, Yonezawa T, Miura Y, Yagasaki K (2014) Antihyperglycemic effect of equol, a daidzein derivative, in cultured L6 myocytes and ob/ob mice. Mol Nutr Food Res 58(2):267–277

Chi XX, Zhang T, Zhang DJ, Yu W, Wang QY, Zhen JL (2016) Effects of isoflavones on lipid and apolipoprotein levels in patients with type 2 diabetes in Heilongjiang Province in China. J Clin Biochem Nutr 59(2):134–138

Chinigarzadeh A, Muniandy S, Salleh N (2017) Combinatorial effect of genistein and female sex-steroids on uterine fluid volume and secretion rate and aquaporin (AQP)−1, 2, 5, and 7 expression in the uterus in rats. Environ Toxicol 32(3):832–844

Choi SK, Adachi M, Utsumi S (2002) Identification of the bile acid-binding region in the soy glycinin A1aB1b subunit. Biosci Biotechnol Biochem 66(11):2395–2401

Choi MS, Lee MK, Jung UJ, Kim HJ, Do GM, Park YB, Jeon SM (2009) Metabolic response of soy pinitol on lipid-lowering, antioxidant and hepatoprotective action in hamsters fed-high fat and high cholesterol diet. Mol Nutr Food Res 53(6):751–759

Choi CW, Choi YH, Cha MR, Yoo DS, Kim YS, Yon GH, Hong KS, Kim YH, Ryu SY (2010) Yeast α-glucosidase inhibition by isoflavones from plants of Leguminosae as an in vitro alternative to acarbose. J Agric Food Chem 58(18):9988–9993

Christensen MJ, Quiner TE, Nakken HL, Lephart ED, Eggett DL, Urie PM (2013) Combination effects of dietary soy and methylselenocysteine in a mouse model of prostate cancer. Prostate 73 (9):986–995

Ciabotti S, Silva AC, Juhasz AC, Mendonça CD, Tavano OL, Mandarino JM, Conçalves C (2016) Chemical composition, protein profile, and isoflavones content in soybean genotypes with different seed coat colors. Embrapa Soja-Artigo em periódico indexado (ALICE). Int Food Res J 23(2):621–629

Colacurci N, Chiàntera A, Fornaro F, de Novellis V, Manzella D, Arciello A, Chiantera V, Improta L, Paolisso G (2005) Effects of soy isoflavones on endothelial function in healthy postmenopausal women. Menopause 12(3):299–307

Coon CN, Obi I, Hamre ML (1988) Use of barley in laying hen diets. Poult Sci 67(9):1306–1313

Cupisti A, D'Alessandro C, Ghiadoni L, Morelli E, Panichi V, Barsotti G (2004) Effect of a soy protein diet on serum lipids of renal transplant patients. J Ren Nutr 14(1):31–35

D'Aloisio AA, DeRoo LA, Baird DD, Weinberg CR, Sandler DP (2013) Prenatal and infant exposures and age at menarche. Epidemiology (Cambridge, MA) 24(2):277

Damasceno NRT, Apolinário E, Flauzino FD, Fernandes I, Abdalla DSP (2007) Soy isoflavones reduce electronegative low-density lipoprotein (LDL−) and anti-LDL− autoantibodies in experimental atherosclerosis. Eur J Nutr 46(3):125–132

Davidsson L, Kastenmayer P, Hurrell RF (1994) Sodium iron EDTA [NaFe (III) EDTA] as a food fortificant: the effect on the absorption and retention of zinc and calcium in women. Am J Clin Nutr 60(2):231–237

De Lumen BO (2005) Lunasin: a cancer-preventive soy peptide. Nutr Rev 63(1):16–21

De Mejia EG, Bradford T, Hasler C (2003) The anticarcinogenic potential of soybean lectin and lunasin. Nutr Rev 61(7):239–246

De Toledo TC, Canniatti-Brazaca SG, Arthur V, Piedade SM (2007) Effects of gamma radiation on total phenolics, trypsin and tannin inhibitors in soybean grains. Radiat Phys Chem 76 (10):1653–1656

Dia VP, Berhow MA, Gonzalez De Mejia E (2008) Bowman−birk inhibitor and genistein among soy compounds that synergistically inhibit nitric oxide and prostaglandin E2 pathways in lipopolysaccharide-induced macrophages. J Agric Food Chem 56(24):11707–11717

DiPietro CM, Liener IE (1989) Heat inactivation of the Kunitz and Bowman-Birk soybean protease inhibitors. J Agric Food Chem 37(1):39–44

Dong JY, Qin LQ, Zhang Z, Zhao Y, Wang J, Arigoni F, Zhang W (2011) Effect of oral L-arginine supplementation on blood pressure: a meta-analysis of randomized, double-blind, placebo-controlled trials. Am Heart J 162(6):959–965

Dong X, Xu W, Sikes RA, Wu C (2012) Apoptotic effects of cooked and in vitro digested soy on human prostate cancer cells. Food Chem 135(3):1643–1652

Du JR, Long FY, Chen C (2014) Research progress on natural triterpenoid saponins in the chemoprevention and chemotherapy of cancer. In: Tamanoi F, Bathaie S (eds) The enzymes, vol 36. Academic Press, London, UK, pp 95–130

Dudek SG (2013) Nutrition essentials for nursing practice. Lippincott Williams & Wilkins, Philadelphia, PA

Endres JG (2001) Soy protein products: characteristics, nutritional aspects, and utilization. AOCS Publishing, Champaign, IL

Engelbert AK, Soukup ST, Roth A, Hoffmann N, Graf D, Watzl B, Kulling SE, Bub A (2016) Isoflavone supplementation in postmenopausal women does not affect leukocyte LDL receptor and scavenger receptor CD36 expression: a double-blind, randomized, placebo-controlled trial. Mol Nutr Food Res 60(9):2008–2019

Escurriol V, Cofán M, Moreno-Iribas C, Larrañaga N, Martínez C, Navarro C, Rodríguez L, González CA, Corella D, Ros E (2010) Phytosterol plasma concentrations and coronary heart disease in the prospective Spanish EPIC cohort. J Lipid Res 51(3):618–624

Esteves EA, Martino HS, Oliveira FC, Bressan J, Costa NM (2010) Chemical composition of a soybean cultivar lacking lipoxygenases (LOX2 and LOX3). Food Chem 122(1):238–242

Evans RC, Fear S, Ashby D, Hackett A, Williams E, van der Vliet M, Dunstan FD, Rhodes JM (2002) Diet and colorectal cancer: an investigation of the lectin/galactose hypothesis. Gastroenterology 122(7):1784–1792

Fan L, Eskin NM (2015) The use of antioxidants in the preservation of edible oils. In: Handbook of antioxidants for food preservation. Woodhead Publishing, Cambridge, UK, pp 373–388

Favaro Trindade CS, Terzi SC, Trugo LC, Della Modesta RC, Couri S (2001) Development and sensory evaluation of soy milk based yoghurt. Arch Latinoam Nutr 51(1):100–104

Fenwick DE, Oakenfull D (1981) Saponin content of soya beans and some commercial soya bean products. J Sci Food Agric 32(3):273–278

Fotsis T, Pepper M, Adlercreutz H, Fleischmann G, Hase T, Montesano R, Schweigerer L (1993) Genistein, a dietary-derived inhibitor of in vitro angiogenesis. Proc Natl Acad Sci 90 (7):2690–2694

Francis G, Kerem Z, Makkar HP, Becker K (2002) The biological action of saponins in animal systems: a review. Br J Nutr 88(6):587–605

Franke AA, Hankin JH, Yu MC, Maskarinec G, Low SH, Custer LJ (1999) Isoflavone levels in soy foods consumed by multiethnic populations in Singapore and Hawaii. J Agric Food Chem 47 (3):977–986

French F (1977) Bakery uses of soy products. Bakers Dig 51(5):98–103

Friedman M, Brandon DL (2001) Nutritional and health benefits of soy proteins. J Agric Food Chem 49(3):1069–1086

Fujiwara K, Miyaguchi Y, Toyoda A, Nakamura Y, Yamazaki M, Nakashima K, Abe H (2008) Effect of fermented soybean "Natto" supplement on egg production and qualities. Asian Australas J Anim Sci 21(11):1610–1615

Furlong ON, Parr HJ, Hodge SJ, Slevin MM, Simpson EE, McSorley EM, McCormack JM, Magee PJ (2019) Consumption of a soy drink has no effect on cognitive function but may alleviate vasomotor symptoms in post-menopausal women; a randomised trial. Eur J Nutr 12:1–2

Gaffer GG, Elgawish RA, Abdelrazek HM, Ebaid HM, Tag HM (2018) Dietary soy isoflavones during pregnancy suppressed the immune function in male offspring albino rats. Toxicol Rep 5:296–301

Ganguly C, Das S (1994) Plant lectins as inhibitors of tumour growth and modulators of host immune response. Chemotherapy 40(4):272–278

García-Lafuente A, Guillamón E, Villares A, Rostagno MA, Martínez JA (2009) Flavonoids as anti-inflammatory agents: implications in cancer and cardiovascular disease. Inflamm Res 58 (9):537–552

Genovese MI, Davila J, Lajolo FM (2006) Isoflavones in processed soybean products from Ecuador. Braz Arch Biol Technol 49(5):853–859

Genser B, Silbernagel G, De Backer G, Bruckert E, Carmena R, Chapman MJ, Deanfield J, Descamps OS, Rietzschel ER, Dias KC, März W (2012) Plant sterols and cardiovascular disease: a systematic review and meta-analysis. Eur Heart J 33(4):444–451

Ghosh M, Bhattacharyya DK (1997) Soy lecithin-monoester interchange reaction by microbial lipase. J Am Oil Chem Soc 74(6):761–763

Giordano E, Dávalos A, Crespo M, Tomé-Carneiro J, Gómez-Coronado D, Visioli F (2015) Soy isoflavones in nutritionally relevant amounts have varied nutrigenomic effects on adipose tissue. Molecules 20(2):2310–2322

Golbitz P. (2000) Soyfoods: state of the industry and market. Soy foods, Orlando, FL, Feb 16–18

Graf E, Eaton JW (1990) Antioxidant functions of phytic acid. Free Radic Biol Med 8(1):61–69

Grant G, Edwards JE, Pusztai A (1995) α-Amylase inhibitor levels in seeds generally available in Europe. J Sci Food Agric 67(2):235–238

Grases F, March JG, Prieto RM, Simonet BM, Costa-Bauzá A, García-Raja A, Conte A (2000) Urinary phytate in calcium oxalate stone formers and healthy people: dietary effects on phytate excretion. Scand J Urol Nephrol 34(3):162–164

Grasso S, Smith G, Bowers S, Ajayi OM, Swainson M (2019) Effect of texturised soy protein and yeast on the instrumental and sensory quality of hybrid beef meatballs. J Food Sci Technol 56 (6):3126–3135

Gu C, Pan H, Sun Z, Qin G (2010) Effect of soybean variety on anti-nutritional factors content, and growth performance and nutrients metabolism in rat. Int J Mol Sci 11(3):1048–1056

Guo Y, Wang S, Hoot DR, Clinton SK (2007) Suppression of VEGF-mediated autocrine and paracrine interactions between prostate cancer cells and vascular endothelial cells by soy isoflavones. J Nutr Biochem 18(6):408–417

Gupta C, Prakash D (2014) Phytonutrients as therapeutic agents. J Complement Integr Med 11 (3):151–169

Gurfinkel DM, Rao AV (2003) Soyasaponins: the relationship between chemical structure and colon anticarcinogenic activity. Nutr Cancer 47(1):24–33

Hamden K, Jaouadi B, Carreau S, Aouidet A, Elfeki A (2011) Therapeutic effects of soy isoflavones on α-amylase activity, insulin deficiency, liver–kidney function and metabolic disorders in diabetic rats. Nat Prod Res 25(3):244–255

Han S, Wu H, Li W, Gao P (2015) Protective effects of genistein in homocysteine-induced endothelial cell inflammatory injury. Mol Cell Biochem 403(1-2):43–49

Hanson LN, Engelman HM, Alekel DL, Schalinske KL, Kohut ML, Reddy MB (2006) Effects of soy isoflavones and phytate on homocysteine, C-reactive protein, and iron status in postmenopausal women. Am J Clin Nutr 84(4):774–780

Hark LA, Morrison G (2000) Development of a case-based integrated nutrition curriculum for medical students. Am J Clin Nutr 72(3):890S–897S

Harland JI, Haffner TA (2008) Systematic review, meta-analysis and regression of randomised controlled trials reporting an association between an intake of circa 25 g soya protein per day and blood cholesterol. Atherosclerosis 200(1):13–27

Harlid S, Adgent M, Jefferson WN, Panduri V, Umbach DM, Xu Z, Stallings VA, Williams CJ, Rogan WJ, Taylor JA (2016) Soy formula and epigenetic modifications: analysis of vaginal epithelial cells from infant girls in the IFED study. Environ Health Perspect 125(3):447–452

He FJ, Chen JQ (2013) Consumption of soybean, soy foods, soy isoflavones and breast cancer incidence: differences between Chinese women and women in Western countries and possible mechanisms. Food Sci Human Wellness 2(3-4):146–161

Hejazi E, Tavakoli M, Jeddi-Tehrani M, Kimiagar M, Hejazi J, Houshyari M, Amiri Z, Edalatkhah H, Nasrollahzadeh J, Idali F (2017) Investigating the antiangiogenic, anti-drug resistance and apoptotic effects of soy isoflavone extract alone or in combination with docetaxel on murine 4t1 breast tumor model. Nutr Cancer 69(7):1036–1042

Hennenger C (2002) Scanner data. Pet Food Ind 44(12):11–14

Hermansen K, Søndergaard M, Høie L, Carstensen M, Brock B (2001) Beneficial effects of a soy-based dietary supplement on lipid levels and cardiovascular risk markers in type 2 diabetic subjects. Diabetes Care 24(2):228–233

Hermansen K, Hansen B, Jacobsen R, Clausen P, Dalgaard M, Dinesen B, Holst JJ, Pedersen E, Astrup A (2005) Effects of soy supplementation on blood lipids and arterial function in hypercholesterolaemic subjects. Eur J Clin Nutr 59(7):843–850

Hernandez-Infante M, Sousa V, Montalvo I, Tena E (1998) Impact of microwave heating on hemagglutinins, trypsin inhibitors and protein quality of selected legume seeds. Plant Foods Hum Nutr 52(3):199–208

Ho SC, Chan AS, Ho YP, So EK, Sham A, Zee B, Woo JL (2007a) Effects of soy isoflavone supplementation on cognitive function in Chinese postmenopausal women: a double-blind, randomized, controlled trial. Menopause 14(3):489–499

Ho SC, Chen YM, Ho SS, Woo JL (2007b) Soy isoflavone supplementation and fasting serum glucose and lipid profile among postmenopausal Chinese women: a double-blind, randomized, placebo-controlled trial. Menopause 14(5):905–912

Hodges RE, Krehl WA, Stone DB, Lopez A (1967) Dietary carbohydrates and low cholesterol diets: effects on serum lipids of man. Am J Clin Nutr 20(2):198–208

Holt S, Muntyan I, Likver L (1996) Soya-based diets for diabetes mellitus. Altern Complement Ther 2(2):79–82

Honary LA (1996) An investigation of the use of soybean oil in hydraulic systems. Bioresour Technol 56(1):41–47

Horan FE (1974) Soy protein products and their production. J Am Oil Chem Soc 51(1Part1):67A–73A

Horner HT, Cervantes-Martinez T, Healy R, Reddy MB, Deardorff BL, Bailey TB, Al-Wahsh I, Massey LK, Palmer RG (2005) Oxalate and phytate concentrations in seeds of soybean cultivars [Glycine max (L.) Merr.]. J Agric Food Chem 53(20):7870–7877

Howes LG, Howes JB, Knight DC (2006) Isoflavone therapy for menopausal flushes: a systematic review and meta-analysis. Maturitas 55(3):203–211

Huang W, Sun X (2000) Adhesive properties of soy proteins modified by urea and guanidine hydrochloride. J Am Oil Chem Soc 77(1):101–104

Huang C, Pang D, Luo Q, Chen X, Gao Q, Shi L, Liu W, Zou Y, Li L, Chen Z (2016a) Soy isoflavones regulate lipid metabolism through an AKT/mTORC1 pathway in diet-induced obesity (DIO) male rats. Molecules 21(5):586

Huang H, Krishnan HB, Pham Q, Yu LL, Wang TT (2016b) Soy and gut microbiota: interaction and implication for human health. J Agric Food Chem 64(46):8695–8709

Hurrell RF (2003) Influence of vegetable protein sources on trace element and mineral bioavailability. J Nutr 133(9):2973S–2977S

Hwang JH (1997) Angiotensin converting enzyme inhibitory effect of doenjang fermented by B. subtilis SCB-3 isolated from meju, Korean traditional food. Kor J Food Sci Nutr 26:775–783

Hymowitz T, Collins FI, Panczner J, Walker WM (1972) Relationship between the content of oil, protein, and sugar in soybean seed 1. Agron J 64(5):613–616

Ibironke SI, Alakija O (2018) Production of soy cheese from vegetable protein using different coagulants. J Nutr Health Sci 5(1):107

Inoue N, Fujiwara Y, Kato M, Funayama A, Ogawa N, Tachibana N, Kohno M, Ikeda I (2015) Soybean β-conglycinin improves carbohydrate and lipid metabolism in Wistar rats. Biosci Biotechnol Biochem 79(9):1528–1534

Ishaaya I, Birk Y, Bondi A, Tencer Y (1969) Soyabean saponins IX. Studies of their effect on birds, mammals and cold-blooded organisms. J Sci Food Agric 20(7):433–436

Jackson RL, Greiwe JS, Schwen RJ (2011) Emerging evidence of the health benefits of S-equol, an estrogen receptor β agonist. Nutr Rev 69(8):432–448

Jakab F, Mayer A, Hoffmann A, Hidvegi M (2000) First clinical data of a natural immunomodulator in colorectal cancer. Hepatogastroenterology 47(32):393–395

Jayachandran M, Xu B (2019) An insight into the health benefits of fermented soy products. Food Chem 271:362–371

Jayagopal V, Albertazzi P, Kilpatrick ES, Howarth EM, Jennings PE, Hepburn DA, Atkin SL (2002) Beneficial effects of soy phytoestrogen intake in postmenopausal women with type 2 diabetes. Diabetes Care 25(10):1709–1714

Jenkins DJ, Kendall CW, Marchie A, Faulkner DA, Wong JM, de Souza R, Emam A, Parker TL, Vidgen E, Lapsley KG, Trautwein EA (2003a) Effects of a dietary portfolio of cholesterol-lowering foods vs lovastatin on serum lipids and C-reactive protein. JAMA 290(4):502–510

Jenkins DJ, Kendall CW, Marchie A, Jenkins AL, Augustin LS, Ludwig DS, Barnard ND, Anderson JW (2003b) Type 2 diabetes and the vegetarian diet. Am J Clin Nutr 78(3):610S–616S

Jideani VA (2011) Functional properties of soybean food ingredients in food systems. Soybean-Biochem Chem Physiol 26:345–366

Jin M, Shen MH, Jin MH, Jin AH, Yin XZ, Quan JS (2018) Hypoglycemic property of soy isoflavones from hypocotyl in Goto-Kakizaki diabetic rats. J Clin Biochem Nutr 62(2):148–154

Jordinson M, El-Hariry I, Calnan DA, Calam J, Pignatelli M (1999) Vicia faba agglutinin, the lectin present in broad beans, stimulates differentiation of undifferentiated colon cancer cells. Gut 44 (5):709–714

Kasim NS, Gunawan S, Yuliana M, Ju YH (2010) A simple two-step method for simultaneous isolation of tocopherols and free phytosterols from soybean oil deodorizer distillate with high purity and recovery. Sep Sci Technol 45(16):2437–2446

Kassell B (1970) [66b] Bovine trypsin-kallikrein inhibitor (kunits inhibitor, basic pancreatic trypsin inhibitor, polyvalent inhibitor from bovine organs). In: Perman GE, Lorand L (eds) Methods in enzymology, vol 19. Academic Press, New York, pp 844–852

Kato AK (2002) Industrial applications of Maillard-type protein-polysaccharide conjugates. Food Sci Technol Res 8(3):193–199

Kawakami Y, Tsurugasaki W, Yoshida Y, Igarashi Y, Nakamura S, Osada K (2004) Regulative actions of dietary soy isoflavone on biological antioxidative system and lipid metabolism in rats. J Agric Food Chem 52(6):1764–1768

Kennedy AR (1998) Cancer prevention by Bowman–Birk inhibitor concentrate (BBIC). In: Prasad KN, Cole WC (eds) Cancer nutrition. IOS Press, Amsterdam, pp 93–97

Kim EH, Kim SH, Chung JI, Chi HY, Kim JA, Chung IM (2006) Analysis of phenolic compounds and isoflavones in soybean seeds (Glycine max (L.) Merill) and sprouts grown under different conditions. Eur Food Res Technol 222(1–2):201

Kim HJ, Hwang JT, Kim MJ, Yang HJ, Sung MJ, Kim SH, Park S, Gu EJ, Park Y, Kwon DY (2014) The inhibitory effect of saponin derived from Cheonggukjang on adipocyte differentiation In vitro. Food Sci Biotechnol 23(4):1273–1278

Kim YM, Huh JS, Lim Y, Cho M (2015) Soy isoflavone glycitin (4′-hydroxy-6-methoxyisoflavone-7-D-glucoside) promotes human dermal fibroblast cell proliferation and migration via TGF-β signaling. Phytother Res 29(5):757–769

Kinney AJ, Clemente TE (2005) Modifying soybean oil for enhanced performance in biodiesel blends. Fuel Process Technol 86(10):1137–1147

Kobayashi M, Egusa S, Fukuda M (2014) Isoflavone and protein constituents of lactic acid-fermented soy milk combine to prevent dyslipidemia in rats fed a high cholesterol diet. Nutrients 6(12):5704–5723

Kohli D, Kumar S, Upadhyay S, Mishra R (2017) Preservation and processing of soymilk: a review. Int J Food Sci Nutr 2(6):66–70

Konovsky J, Lumpkin TA, McClary D (1994) Edamame: the vegetable soybean. In: O’Rourke AD (ed) Understanding the Japanese food and agrimarket: a multifaceted opportunity. The Haworth Press, Binghamton, NY, pp 173–181

Konya J, Sathyapalan T, Kilpatrick ES, Atkin SL (2019) The effects of soy protein and cocoa with or without isoflavones on glycemic control in type 2 diabetes. A double-blind, randomized, placebo-controlled study. Front Endocrinol 10:1–6. https://doi.org/10.3389/fendo.2019.00296

Kulkarni SD, Wijeratne WB, Wei TM (1992) Production of medium fat soyflour by dry-extrusion-expelling of raw soybean and its use in bread fortification. J Food Sci Technol 29(4):220–223

Kurowska EM, Jordon J, Spence JD, Wetmore S (1997) Effects of substituting dietary soybean protein and oil for milk protein and fat in subjects with hypercholesterolemia. Clin Invest Med 20(3):162–170

Kurrat A, Blei T, Kluxen FM, Mueller DR, Piechotta M, Soukup ST, Kulling SE, Diel P (2015) Lifelong exposure to dietary isoflavones reduces risk of obesity in ovariectomized Wistar rats. Mol Nutr Food Res 59(12):2407–2418

Kuzuhara H, Nishiyama S, Minowa N, Sasaki K, Omoto S (2000) Protective effects of soyasapogenol A on liver injury mediated by immune response in a concanavalin A-induced hepatitis model. Eur J Pharmacol 391(1-2):175–181

Kwon T, Kim S, Kim W, Park K, Son H, Seung J, Shin D, Shin S, Lee K, Lee Y, Lee C, Lee H, Cho S, Cho J, Cho H, Ji K, Hong E (2005) Soybean. In: Utilization history of soybean. Korea University Press, Seoul, Korea. ISBN: 89-764-156-20

Kwon DY, Daily JW III, Kim HJ, Park S (2010) Antidiabetic effects of fermented soybean products on type 2 diabetes. Nutr Res 30(1):1–3

Lairon D, Arnault N, Bertrais S, Planells R, Clero E, Hercberg S, Boutron-Ruault MC (2005) Dietary fiber intake and risk factors for cardiovascular disease in French adults. Am J Clin Nutr 82(6):1185–1194

Lajolo FM, Genovese MI (2002) Nutritional significance of lectins and enzyme inhibitors from legumes. J Agric Food Chem 50(22):6592–6598

Lamartiniere CA, Zhao YX, Fritz WA (2000) Genistein: mammary cancer chemoprevention, in vivo mechanisms of action, potential for toxicity and bioavailability in rats. J Women's Cancer 2:11–19

Lampe JW (2003) Isoflavonoid and lignan phytoestrogens as dietary biomarkers. J Nutr 133 (3):956S–964S

Latorraca MQ, da Costa RP, Laux MC, da Rosa CA, Arantes VC, de Barros Reis MA (2019) Effects of the soybean flour diet on insulin secretion and action. In: Preedy VR, Watson RR (eds) Flour and breads and their fortification in health and disease prevention. Academic Press, London, pp 423–434

Lee JS (2006) Effects of soy protein and genistein on blood glucose, antioxidant enzyme activities, and lipid profile in streptozotocin-induced diabetic rats. Life Sci 79(16):1578–1584

Lee SH, Park HJ, Chun HK, Cho SY, Cho SM, Lillehoj HS (2006) Dietary phytic acid lowers the blood glucose level in diabetic KK mice. Nutr Res 26(9):474–479

Lee SH, Park HJ, Chun HK, Cho SY, Jung HJ, Cho SM, Kim DY, Kang MS, Lillehoj HS (2007) Dietary phytic acid improves serum and hepatic lipid levels in aged ICR mice fed a high-cholesterol diet. Nutr Res 27(8):505–510

Lee SJ, Kim JJ, Moon HI, Ahn JK, Chun SC, Jung WS, Lee OK, Chung IM (2008) Analysis of isoflavones and phenolic compounds in Korean soybean [Glycine max (L.) Merrill] seeds of different seed weights. J Agric Food Chem 56(8):2751–2758

Lethaby A, Marjoribanks J, Kronenberg F, Roberts H, Eden J, Brown J (2007) Phytoestrogens for vasomotor menopausal symptoms. Cochrane Database Syst Rev 11:1–82. https://doi.org/10.1002/14651858.CD001395.pub3

Levis S, Strickman-Stein N, Ganjei-Azar P, Xu P, Doerge DR, Krischer J (2011) Soy isoflavones in the prevention of menopausal bone loss and menopausal symptoms: a randomized, double-blind trial. Arch Intern Med 171(15):1363–1369

Li Z, Hong K, Saltsman P, DeShields S, Bellman M, Thames G, Liu Y, Wang HJ, Elashoff R, Heber D (2005) Long-term efficacy of soy-based meal replacements vs an individualized diet plan in obese type II DM patients: relative effects on weight loss, metabolic parameters, and C-reactive protein. Eur J Clin Nutr 59(3):411–418

Li M, Liu RM, Timblin CR, Meyer SG, Mossman BT, Fukagawa NK (2006) Age affects ERK1/2 and NRF2 signaling in the regulation of GCLC expression. J Cell Physiol 206(2):518–525

Li H, Liu H, Han Y, Wu X, Teng W, Liu G, Li W (2010) Identification of QTL underlying vitamin E contents in soybean seed among multiple environments. Theor Appl Genet 120(7):1405–1413

Li N, Wu X, Zhuang W, Xia L, Chen Y, Zhao R, Yi M, Wan Q, Du L, Zhou Y (2019) Soy and isoflavone consumption and multiple health outcomes: umbrella review of systematic reviews and meta-analyses of observational studies and randomised trials in humans. Mol Nutr Food Res. https://doi.org/10.1002/mnfr.201900751

Liang YL, Teede H, Dalais F, McGrath BP (2006) The effects of phytoestrogen on blood pressure and lipids in healthy volunteers. Zhonghua Xin Xue Guan Bing Za Zhi 34(8):726–729

Liao FH, Shieh MJ, Yang SC, Lin SH, Chien YW (2007) Effectiveness of a soy-based compared with a traditional low-calorie diet on weight loss and lipid levels in overweight adults. Nutrition 23(7-8):551–556

Liener IE (1991) From soybeans to lectins: a trail of research revisited. Carbohydr Res 213:1–5

Liener IE (1994) Implications of antinutritional components in soybean foods. Crit Rev Food Sci Nutr 34(1):31–67

Liener IE (2000) Non-nutritive factors and bioactive compounds in soy. In: Drackley JK (ed) Soy in animal nutrition. Federation of Animal Science Societies, Savoy, IL, pp 1–2

Lin J, Wang C (2004) An analytical method for soy saponins by HPLC/ELSD. J Food Sci 69(6): C456–C462

Lin S, Huff HE, Hsieh F (2000) Texture and chemical characteristics of soy protein meat analog extruded at high moisture. J Food Sci 65(2):264–269

Lin TH, Tan TW, Tsai TH, Chen CC, Hsieh TF, Lee SS, Liu HH, Chen WC, Tang CH (2013) D-pinitol inhibits prostate cancer metastasis through inhibition of αVβ3 integrin by modulating FAK, c-Src and NF-κB pathways. Int J Mol Sci 14(5):9790–9802

List GR (2016) Oilseed composition and modification for health and nutrition. In: Functional Dietary Lipids. Woodhead Publishing, Waltham, MA, pp 23–46

Liu K (1997) Chemistry and nutritional value of soybean components. In: Liu K (ed) Soybeans. Springer, Boston, MA, pp 25–113

Liu S, Stampfer MJ, Hu FB, Giovannucci E, Rimm E, Manson JE, Hennekens CH, Willett WC (1999) Whole-grain consumption and risk of coronary heart disease: results from the Nurses' Health Study. Am J Clin Nutr 70(3):412–419

Liu JR, Wang SY, Chen MJ, Chen HL, Yueh PY, Lin CW (2006) Hypocholesterolaemic effects of milk-kefir and soyamilk-kefir in cholesterol-fed hamsters. Br J Nutr 95(5):939–946

Liu ZM, Ho SC, Chen YM, Woo J (2013) Effect of soy protein and isoflavones on blood pressure and endothelial cytokines: a 6-month randomized controlled trial among postmenopausal women. J Hypertens 31(2):384–392

Liu ZM, Ho SC, Chen YM, Liu J, Woo J (2014) Cardiovascular risks in relation to daidzein metabolizing phenotypes among Chinese postmenopausal women. PLoS One 9(2):e87861. https://doi.org/10.1371/journal.pone.0087861

Liu C, Lin XL, Wan Z, Zou Y, Cheng FF, Yang XQ (2016) The physicochemical properties, in vitro binding capacities and in vivo hypocholesterolemic activity of soluble dietary fiber extracted from soy hulls. Food Funct 7(12):4830–4840

Lockley M (1991) Contested meanings of the menopause. Lancet 337(8752):1270–1272

López-González AA, Grases F, Roca P, Mari B, Vicente-Herrero MT, Costa-Bauzá A (2008) Phytate (myo-inositol hexaphosphate) and risk factors for osteoporosis. J Med Food 11 (4):747–752

Lu J, Zeng Y, Hou W, Zhang S, Li L, Luo X, Xi W, Chen Z, Xiang M (2012) The soybean peptide aglycin regulates glucose homeostasis in type 2 diabetic mice via IR/IRS1 pathway. J Nutr Biochem 23(11):1449–1457

Lutzow S (1996) Better, healthier baking with soy. Bakery Prod Market 31:16–22

Ma CY, Liu WS, Kwok KC, Kwok F (1996) Isolation and characterization of proteins from soymilk residue (okara)∗. Food Res Int 29(8):799–805

Macedo-Silva A, Shimokomaki M, Vaz AJ, Yamamoto YY, Tenuta-Filho A (2001) Textured soy protein quantification in commercial hamburger. J Food Compos Anal 14(5):469–478

Machado FP, Queiróz JH, Oliveira MG, Piovesan ND, Peluzio MC, Costa NM, Moreira MA (2008) Effects of heating on protein quality of soybean flour devoid of Kunitz inhibitor and lectin. Food Chem 107(2):649–655

Mahmoud AM, Yang W, Bosland MC (2014) Soy isoflavones and prostate cancer: a review of molecular mechanisms. J Steroid Biochem Mol Biol 140:116–132

Maidala A, Doma UD, Egbo LM (2013) Effects of different processing methods on the chemical composition and antinutritional factors of soybean [Glycine max (L.) Merrill]. Pak J Nutr 12 (12):1057–1060

Mani V, Ming LC (2017) Tempeh and other fermented soybean products rich in isoflavones. In: Frías J, Martínez-Villaluenga C, Peñas E (eds) Fermented foods in health and disease prevention. Academic Press, Oxford, pp 453–474

Marques M, Bora PS, Narain N (2000) Development of some high-protein conventional foods based on wheat and oilseed flours. J Food Sci Technol 37(4):394–399

Masilamani M, Wei J, Bhatt S, Paul M, Yakir S, Sampson HA (2011) Soybean isoflavones regulate dendritic cell function and suppress allergic sensitization to peanut. J Allergy Clin Immunol 128 (6):1242–1250

Massey LK, Roman-Smith H, Sutton RA (1993) Effect of dietary oxalate and calcium on urinary oxalate and risk of formation of calcium oxalate kidney stones. J Am Diet Assoc 93(8):901–906

Massey LK, Palmer RG, Horner HT (2001) Oxalate content of soybean seeds (Glycine max: Leguminosae), soyfoods, and other edible legumes. J Agric Food Chem 49(9):4262–4266
McVeigh BL, Dillingham BL, Lampe JW, Duncan AM (2006) Effect of soy protein varying in isoflavone content on serum lipids in healthy young men. Am J Clin Nutr 83(2):244–251
Medić DR, Ristić V, Arsić A, Poštić M, Ristić G, Mladenović VB, Tepšić J (2006) Effects of soybean D-LeciVita product on serum lipids and fatty acid composition in type 2 diabetic patients with hyperlipidemia. Nutr Metab Cardiovasc Dis 16(6):395–404
Medina-Juarez LA, Gonzalez-Diaz P, Gámez-Meza N, Ortega-Garcia J, Moreno-Varela AB, Bringas-Alvarado L, Angulo-Guerrero O (1998) Effects of processing on the oxidative stability of soybean oil produced in Mexico. J Am Oil Chem Soc 75(12):1729–1733
Meezan E, Meezan EM, Jones K, Moore R, Barnes S, Prasain JK (2005) Contrasting effects of puerarin and daidzin on glucose homeostasis in mice. J Agric Food Chem 53(22):8760–8767
Mentor-Marcel R, Lamartiniere CA, Eltoum IE, Greenberg NM, Elgavish A (2001) Genistein in the diet reduces the incidence of poorly differentiated prostatic adenocarcinoma in transgenic mice (TRAMP). Cancer Res 61(18):6777–6782
Merritt JC (2004) Metabolic syndrome: soybean foods and serum lipids. J Natl Med Assoc 96 (8):1032
Messina MJ (1999) Legumes and soybeans: overview of their nutritional profiles and health effects. Am J Clin Nutr 70(3):439s–450s
Messina M (2014) Soy foods, isoflavones, and the health of postmenopausal women. Am J Clin Nutr 100(suppl_1):423S–430S. https://doi.org/10.3945/ajcn.113.071464
Mikić A, Perić V, Đorđević V, Srebrić M, Mihailović V (2009) Anti-nutritional factors in some grain legumes. Biotechnol Anim Husbandry 25(5-6-2):1181–1188
Monteiro MR, Moreira MA, Costa NM, Oliveira MG, Pires CV (2003) Avaliação da digestibilidade protéica de genótipos de soja com ausência e presença do inibidor de tripsina Kunitz e lipoxigenases. Braz J Food Technol 6(1):99–107
Montonen J, Knekt P, Järvinen R, Aromaa A, Reunanen A (2003) Whole-grain and fiber intake and the incidence of type 2 diabetes. Am J Clin Nutr 77(3):622–629
Mudryj AN, Yu N, Aukema HM (2014) Nutritional and health benefits of pulses. Appl Physiol Nutr Metab 39(11):1197–1204
Murkies AL, Lombard C, Strauss BJ, Wilcox G, Burger HG, Morton MS (1995) Dietary flour supplementation decreases post-menopausal hot flushes: effect of soy and wheat. Maturitas 21 (3):189–195
Muscogiuri G, Palomba S, Laganà AS, Orio F (2016) Inositols in the treatment of insulin-mediated diseases. Int J Endocrinol 2016:1–6. https://doi.org/10.1155/2016/3058393
Nagaoka S, Miwa K, Eto M, Kuzuya Y, Hori G, Yamamoto K (1999) Soy protein peptic hydrolysate with bound phospholipids decreases micellar solubility and cholesterol absorption in rats and Caco-2 cells. J Nutr 129(9):1725–1730
Nagasawa A, Fukui K, Funahashi T, Maeda N, Shimomura I, Kihara S, Waki M, Takamatsu K, Matsuzawa Y (2002) Effects of soy protein diet on the expression of adipose genes and plasma adiponectin. Horm Metab Res 34(11/12):635–639
Nagata C, Shimizu H, Takami R, Hayashi M, Takeda N, Yasuda K (2003) Soy product intake is inversely associated with serum homocysteine level in premenopausal Japanese women. J Nutr 133(3):797–800
Nagata C, Wada K, Tamura T, Konishi K, Goto Y, Koda S, Kawachi T, Tsuji M, Nakamura K (2016) Dietary soy and natto intake and cardiovascular disease mortality in Japanese adults: the Takayama study. Am J Clin Nutr 105(2):426–431
Neacsu M, Fyfe C, Horgan G, Johnstone AM (2014) Appetite control and biomarkers of satiety with vegetarian (soy) and meat-based high-protein diets for weight loss in obese men: a randomized crossover trial. Am J Clin Nutr 100(2):548–558

Neus JD, Fehr WR, Schnebly SR (2005) Agronomic and seed characteristics of soybean with reduced raffinose and stachyose. Crop Sci 45(2):589–592

Newton KM, Grady D (2011) Soy isoflavones for prevention of menopausal bone loss and vasomotor symptoms: comment on "Soy isoflavones in the prevention of menopausal bone loss and menopausal symptoms". Arch Intern Med 171(15):1369–1370

Nguyen CT, Pham NM, Do VV, Binns CW, Hoang VM, Dang DA, Lee AH (2017) Soyfood and isoflavone intake and risk of type 2 diabetes in Vietnamese adults. Eur J Clin Nutr 71(10):1186

Nishi T, Hara H, Asano K, Tomita F (2003) The soybean β-conglycinin β 51–63 fragment suppresses appetite by stimulating cholecystokinin release in rats. J Nutr 133(8):2537–2542

Nishibori N, Kishibuchi R, Morita K (2017) Soy pulp extract inhibits angiotensin I-converting enzyme (ACE) activity in vitro: Evidence for its potential hypertension-improving action. J Diet Suppl 14(3):241–251

Norouzi S, Pourjafar H, Ansari F, Homayouni A (2019) A Survey on the survival of Lactobacillus paracasei in fermented and non-fermented frozen soy dessert. Biocatal Agric Biotechnol 21:101297. https://doi.org/10.1016/j.bcab.2019.101297

Ohr LM (1997) Fortifying for the health of it. Prepared Foods 166(12):55–60

Onayemi O, Lorenz K (1978) Concentrates and isolate in bread making. Bakers Dig 52:18–24

Ørgaard A, Jensen L (2008) The effects of soy isoflavones on obesity. Exp Biol Med 233 (9):1066–1080

Ostlund RE Jr (2004) Phytosterols and cholesterol metabolism. Curr Opin Lipidol 15(1):37–41

Oyama A, Ueno T, Uchiyama S, Aihara T, Miyake A, Kondo S, Matsunaga K (2012) The effects of natural S-equol supplementation on skin aging in postmenopausal women: a pilot randomized placebo-controlled trial. Menopause 19(2):202–210

Patel S, Peretz J, Pan YX, Helferich WG, Flaws JA (2016) Genistein exposure inhibits growth and alters steroidogenesis in adult mouse antral follicles. Toxicol Appl Pharmacol 293:53–62

Patten GS, Abeywardena MY, Bennett LE (2016) Inhibition of angiotensin converting enzyme, angiotensin II receptor blocking, and blood pressure lowering bioactivity across plant families. Crit Rev Food Sci Nutr 56(2):181–214

Pavese JM, Krishna SN, Bergan RC (2014) Genistein inhibits human prostate cancer cell detachment, invasion, and metastasis. Am J Clin Nutr 100(suppl_1):431S–436S

Peisker M (2001) Manufacturing of soy protein concentrate for animal nutrition. Cah Options Mediterr 54:103–107. http://ressources.ciheam.org/om/pdf/c54/01600017.pdf

Pele GI, Ogunsua AO, Adepeju AB, Esan YO, Oladiti EO (2016) Effects of processing methods on the nutritional and anti-nutritional properties of soybeans (glycine max). Afr J Food Sci Technol 7(1):009–012

Petruzziello L, Iacopini F, Bulajic M, Shah S, Costamagna G (2006) Uncomplicated diverticular disease of the colon. Aliment Pharmacol Ther 23(10):1379–1391

Pipe EA, Gobert CP, Capes SE, Darlington GA, Lampe JW, Duncan AM (2009) Soy protein reduces serum LDL cholesterol and the LDL cholesterol: HDL cholesterol and apolipoprotein B: apolipoprotein AI ratios in adults with type 2 diabetes. J Nutr 139(9):1700–1706

Poschner S, Maier-Salamon A, Zehl M, Wackerlig J, Dobusch D, Pachmann B, Sterlini KL, Jäger W (2017) The impacts of genistein and daidzein on estrogen conjugations in human breast cancer cells: a targeted metabolomics approach. Front Pharmacol 8:699

Pryme IF, Bardocz S (2001) Anti-cancer therapy: diversion of polyamines in the gut. Eur J Gastroenterol Hepatol 13(9):1041–1046

Psodorov ĐB, Plavšić D, Nedeljković N, Jambrec D, Psodorov DĐ, Simić S, Banjac V (2015) Soy bran separation process for application of product in the food industry. Qual Life 6(1-2):32–36. https://doi.org/10.7251/QOL1501032P

Pusztai A, Watt WB, Stewart JC (1991) A comprehensive scheme for the isolation of trypsin inhibitors and the agglutinin from soybean seeds. J Agric Food Chem 39(5):862–866

Pusztai A, Grant G, Duguid T, Brown DS, Peumans WJ, van Damme EJ, Bardocz S (1995) Inhibition of starch digestion by α-amylase inhibitor reduces the efficiency of utilization of dietary proteins and lipids and retards the growth of rats. J Nutr 125(6):1554–1562

Qiao N, Liu Q, Meng H, Zhao D (2014) Haemolytic activity and adjuvant effect of soyasaponins and some of their derivatives on the immune responses to ovalbumin in mice. Int Immunopharmacol 18(2):333–339
Qiu L, Chang R (2010) The origin and history of soybean. In: Singh G (ed) The soybean: botany, production and uses. Punjab Agricultural University, India, pp 1–23
Quan J, Yin X, Jin M, Shen M (2003) Study on the inhibition of alpha-glucosidase by soyasaponins. J Chin Med Mater 26(9):654–656
Rackis JJ, Gumbmann MR (1981) Protease inhibitors: physiological properties and nutritional significance. Antinutrients and natural toxicants in foods. In: Ory RL (ed) Antinutrients and natural toxicants in foods. Food & Nutrition Press, Westport, CT, pp 203–237. https://naldc.nal.usda.gov/download/26115/PDF
Radcliffe JD, Czajka-Narins DM (1998) Partial replacement of dietary casein with soy protein isolate can reduce the severity of retinoid-induced hypertriglyceridemia. Plant Foods Hum Nutr 52(2):97–108
Rakosky JJ (1974) Soy grits, flour, concentrates, and isolates in meat products. J Am Oil Chem Soc 51(Part1):123A–127A
Read JW, Haas LW (1938) Studies on the baking quality of flour as affected by certain enzyme actions. V. Further studies concerning potassium bromate and enzyme activity. Cereal Chem 15:59–68
Redondo-Cuenca A, Villanueva-Suárez MJ, Rodríguez-Sevilla MD, Mateos-Aparicio I (2007) Chemical composition and dietary fibre of yellow and green commercial soybeans (Glycine max). Food Chem 101(3):1216–1222
Rezaei R, Khomeiri M, Kashaninejad M, Mazaheri-Tehrani M, Aalami M (2019) Potential of β-d-glucan to enhance physicochemical quality of frozen soy yogurt at different aging conditions. Iran Food Sci Technol Res J 15(3):1–12
Ricketts ML, Moore DD, Banz WJ, Mezei O, Shay NF (2005) Molecular mechanisms of action of the soy isoflavones includes activation of promiscuous nuclear receptors. A review. J Nutr Biochem 16(6):321–330
Ridout CL, Wharf SG, Price KR, Johnson IT, Fenwick GR (1988) UK mean daily intakes of saponins-intestine-permeabilizing factors in legumes. Food Sci Nutr 42(2):111–116
Rijavec T, Zupin Z (2011) Soybean protein fibres (SPF). In: Krezhova D (ed) Recent trends for enhancing the diversity and quality of soybean products. InTech Open, pp 501–522. https://doi.org/10.5772/19614
Rimbach G, Pallauf J (1998) Phytic acid inhibits free radical formation in vitro but does not affect liver oxidant or antioxidant status in growing rats. J Nutr 128(11):1950–1955
Rivas M, Garay RP, Escanero JF, Cia P Jr, Cia P, Alda JO (2002) Soy milk lowers blood pressure in men and women with mild to moderate essential hypertension. J Nutr 132(7):1900–1902
Rizzi C, Galeoto L, Zoccatelli G, Vincenzi S, Chignola R, Peruffo AD (2003) Active soybean lectin in foods: quantitative determination by ELISA using immobilised asialofetuin. Food Res Int 36 (8):815–821
Sacks FM, Lichtenstein A, Van Horn L, Harris W, Kris-Etherton P, Winston M (2006) Soy protein, isoflavones, and cardiovascular health: an American Heart Association Science Advisory for professionals from the Nutrition Committee. Circulation 113(7):1034–1044
Sahebkar A (2013) Fat lowers fat: purified phospholipids as emerging therapies for dyslipidemia. Biochim Biophys Acta 1831(4):887–893
Sakamoto Y, Kanatsu J, Toh M, Naka A, Kondo K, Iida K (2016) The dietary isoflavone daidzein reduces expression of pro-inflammatory genes through PPARα/γ and JNK pathways in adipocyte and macrophage co-cultures. PLoS One 11(2):e0149676. https://doi.org/10.1371/journal.pone.0149676
Schmitt C, Sanchez C, Desobry-Banon S, Hardy J (1998) Structure and technofunctional properties of protein-polysaccharide complexes: a review. Crit Rev Food Sci Nutr 38(8):689–753
Sebastian S, Touchburn SP, Chavez ER (1998) Implications of phytic acid and supplemental microbial phytase in poultry nutrition: a review. Worlds Poult Sci J 54(1):27–47

Sedaghat A, Shahbazian H, Rezazadeh A, Haidari F, Jahanshahi A, Latifi SM, Shirbeigi E (2019) The effect of soy nut on serum total antioxidant, endothelial function and cardiovascular risk factors in patients with type 2 diabetes. Diabetes Metab Syndr Clin Res Rev 13(2):1387–1391
Selle PH, Ravindran V, Caldwell A, Bryden WL (2000) Phytate and phytase: consequences for protein utilisation. Nutr Res Rev 13(2):255–278
Shamsuddin AM (2002) Anti-cancer function of phytic acid. Int J Food Sci Technol 37(7):769–782
Sharma S, Adrogue JV, Golfman L, Uray I, Lemm J, Youker K, Noon GP, Frazier OH, Taegtmeyer H (2004) Intramyocardial lipid accumulation in the failing human heart resembles the lipotoxic rat heart. FASEB J 18(14):1692–1700
Shike M, Doane AS, Russo L, Cabal R, Reis-Filho JS, Gerald W, Cody H, Khanin R, Bromberg J, Norton L (2014) The effects of soy supplementation on gene expression in breast cancer: a randomized placebo-controlled study. J Natl Cancer Inst 106(9):dju189. https://doi.org/10.1093/jnci/dju189
Shimazu T, Kuriyama S, Hozawa A, Ohmori K, Sato Y, Nakaya N, Nishino Y, Tsubono Y, Tsuji I (2007) Dietary patterns and cardiovascular disease mortality in Japan: a prospective cohort study. Int J Epidemiol 36(3):600–609
Shu XO, Jin F, Dai Q, Wen W, Potter JD, Kushi LH, Ruan Z, Gao YT, Zheng W (2001) Soyfood intake during adolescence and subsequent risk of breast cancer among Chinese women. Cancer Epidemiol Prev Biomarkers 10(5):483–488
Shurtleff W, Aoyagi A (2012) History of soy yogurt, soy acidophilus milk and other cultured soymilks (1918–2012). Soyinfo Center. http://www.soyinfocenter.com/books/156
Singh P, Kumar R, Sabapathy SN, Bawa AS (2008) Functional and edible uses of soy protein products. Compr Rev Food Sci Food Safety 7(1):14–28
Sivakumar S, Subramanian SP (2009) D-pinitol attenuates the impaired activities of hepatic key enzymes in carbohydrate metabolism of streptozotocin-induced diabetic rats. Gen Physiol Biophys 28(3):233–241
Sonntag NO (1985) Growth potential for soybean oil products as industrial materials. J Am Oil Chem Soc 62(5):928–933
Spear JD, Fehr WR (2007) Genetic improvement of seedling emergence of soybean lines with low phytate. Crop Sci 47(4):1354–1360
Spilburg CA, Goldberg AC, McGill JB, Stenson WF, Racette SB, Bateman J, McPherson TB, Ostlund RE Jr (2003) Fat-free foods supplemented with soy stanol-lecithin powder reduce cholesterolabsorption and LDL cholesterol. J Am Diet Assoc 103(5):577–581
Stauffer CE (2006) Soy protein in baking. Soya Update 3(11):1–2
Steffen LM, Jacobs DR Jr, Stevens J, Shahar E, Carithers T, Folsom AR (2003) Associations of whole-grain, refined-grain, and fruit and vegetable consumption with risks of all-cause mortality and incident coronary artery disease and ischemic stroke: the Atherosclerosis Risk in Communities (ARIC) Study. Am J Clin Nutr 78(3):383–390
Steggerda FR, Richards EA, Rackis JJ (1966) Effects of various soybean products on flatulence in the adult man. Proc Soc Exp Biol Med 121(4):1235–1239
Stephenson TJ, Setchell KD, Kendall CW, Jenkins DJ, Anderson JW, Fanti P (2005) Effect of soy protein-rich diet on renal function in young adults with insulin-dependent diabetes mellitus. Clin Nephrol 64(1):1–11. https://doi.org/10.5414/cnp64001
Sun T, Yan X, Guo W, Zhao D (2014) Evaluation of cytotoxicity and immune modulatory activities of soyasaponin Ab: an in vitro and in vivo study. Phytomedicine 21(13):1759–1766
Suzuki O, Nozawa Y, Kawaguchi T, Abe M (1999) Phaseolus vulgaris leuko agglutinating lectin binding reactivity in human diffuse large B-cell lymphoma and its relevance to the patient's clinical outcome: Lectin histochemistry and lectin blot analysis. Pathol Int 49(10):874–880
Szuhaj BF (1983) Lecithin production and utilization. J Am Oil Chem Soc 60(2):306–309
Tai TY, Tsai KS, Tu ST, Wu JS, Chang CI, Chen CL, Shaw NS, Peng HY, Wang SY, Wu CH (2012) The effect of soy isoflavone on bone mineral density in postmenopausal Taiwanese women with bone loss: a 2-year randomized double-blind placebo-controlled study. Osteoporos Int 23(5):1571–1580

Takamatsu H, Kawajiri H, Takahashi Y, Ali AM, Yoshimoto T (1999) Continuous antibody production by phytohemagglutinin-L-aggregated hybridoma cells. J Immunol Methods 223 (2):165–170

Taku K, Umegaki K, Ishimi Y, Watanabe S (2008) Effects of extracted soy isoflavones alone on blood total and LDL cholesterol: Meta-analysis of randomized controlled trials. Ther Clin Risk Manag 4(5):1097–1103

Taku K, Melby MK, Takebayashi J, Mizuno S, Ishimi Y, Omori T, Watanabe S (2010) Effect of soy isoflavone extract supplements on bone mineral density in menopausal women: meta-analysis of randomized controlled trials. Asia Pac J Clin Nutr 19(1):33–42

Tan ST, Ismail A, Hamid M, Chong PP, Sun J (2019) Soy husk extract improves physical and biochemical parameters of obese–diabetic rats through the regulation of PPARγ expression. J Food Biochem 43(5):e12843. https://doi.org/10.1111/jfbc.12843

Tatsumi Y, Morimoto A, Deura K, Mizuno S, Ohno Y, Watanabe S (2013) Effects of soybean product intake on fasting and postload hyperglycemia and type 2 diabetes in J apanese men with high body mass index: the Saku Study. J Diabetes Investig 4(6):626–633

Teixeira SR, Tappenden KA, Carson L, Jones R, Prabhudesai M, Marshall WP, Erdman JW Jr (2004) Isolated soy protein consumption reduces urinary albumin excretion and improves the serum lipid profile in men with type 2 diabetes mellitus and nephropathy. J Nutr 134 (8):1874–1880

Tokede OA, Onabanjo TA, Yansane A, Gaziano JM, Djoussé L (2015) Soya products and serum lipids: a meta-analysis of randomised controlled trials. Br J Nutr 114(6):831–843

Toldrà M, Parés D, Saguer E, Carretero C (2019) Utilisation of protein fractions from porcine spleen as technofunctional ingredients in emulsified cooked meat sausages. Int J Food Sci Technol. https://doi.org/10.1111/ijfs.14298

Tonstad S, Smerud K, Høie L (2002) A comparison of the effects of 2 doses of soy protein or casein on serum lipids, serum lipoproteins, and plasma total homocysteine in hypercholesterolemic subjects. Am J Clin Nutr 76(1):78–84

Torres N, Torre-Villalvazo I, Tovar AR (2006) Regulation of lipid metabolism by soy protein and its implication in diseases mediated by lipid disorders. J Nutr Biochem 17(6):365–373

Tovar AR, Murguía F, Cruz C, Hernández-Pando R, Aguilar-Salinas CA, Pedraza-Chaverri J, Correa-Rotter R, Torres N (2002) A soy protein diet alters hepatic lipid metabolism gene expression and reduces serum lipids and renal fibrogenic cytokines in rats with chronic nephrotic syndrome. J Nutr 132(9):2562–2569

Trottier G, Boström PJ, Lawrentschuk N, Fleshner NE (2010) Nutraceuticals and prostate cancer prevention: a current review. Nat Rev Urol 7(1):21–30. https://doi.org/10.1038/nrurol.2009.234

Trujillo J, Ramírez V, Pérez J, Torre-Villalvazo I, Torres N, Tovar AR, Munoz RM, Uribe N, Gamba G, Bobadilla NA (2005) Renal protection by a soy diet in obese Zucker rats is associated with restoration of nitric oxide generation. Am J Physiol 288(1):F108–F116

Tsukamoto C, Shimada S, Igita K, Kudou S, Kokubun M, Okubo K, Kitamura K (1995) Factors affecting isoflavone content in soybean seeds: changes in isoflavones, saponins, and composition of fatty acids at different temperatures during seed development. J Agric Food Chem 43 (5):1184–1192

Unger RH (2003) Lipid overload and overflow: metabolic trauma and the metabolic syndrome. Trends Endocrinol Metab 14(9):398–403

US Department of Agriculture (USDA) (2019) Food composition databases show foods list. https://ndb.nal.usda.gov/ndb/search/list. Accessed 20 Oct 2018

Usui T, Tochiya M, Sasaki Y, Muranaka K, Yamakage H, Himeno A, Shimatsu A, Inaguma A, Ueno T, Uchiyama S, Satoh-Asahara N (2013) Effects of natural S-equol supplements on overweight or obesity and metabolic syndrome in the Japanese, based on sex and equol status. Clin Endocrinol (Oxf) 78(3):365–372

Valachovicova T, Slivova V, Bergman H, Shuherk J, Sliva D (2004) Soy isoflavones suppress invasiveness of breast cancer cells by the inhibition of NF-κB/AP-1-dependent and-independent pathways. Int J Oncol 25(5):1389–1395

Vasconcelos IM, Siebra EA, Maia AA, Moreira RA, Neto AF, Campelo GJ, Oliveira JT (1997) Composition, toxic and antinutritional factors of newly developed cultivars of Brazilian soybean (Glycine max). J Sci Food Agric 75(4):419–426

Vasconcelos IM, Maia AA, Siebra EA, Oliveira JT, Carvalho AF, Melo VM, Carlini CR, Castelar LI (2001) Nutritional study of two Brazilian soybean (Glycine max) cultivars differing in the contents of antinutritional and toxic proteins. J Nutr Biochem 12(1):55–62

Velasquez MT, Bhathena SJ (2007) Role of dietary soy protein in obesity. Int J Med Sci 4(2):72–82

Villares A, Rostagno MA, García-Lafuente A, Guillamón E, Martínez JA (2011) Content and profile of isoflavones in soy-based foods as a function of the production process. Food Bioproc Tech 4(1):27–38

Villegas R, Gao YT, Yang G, Li HL, Elasy TA, Zheng W, Shu XO (2008) Legume and soy food intake and the incidence of type 2 diabetes in the Shanghai Women's Health Study. Am J Clin Nutr 87(1):162–167

Vital R, Bassinello P, Cruz Q, Carvalho R, de Paiva J, Colombo A (2018) Production, quality, and acceptance of tempeh and white bean tempeh burgers. Foods 7(9):1–9

Vucenik I, Shamsuddin AM (2004) IP-6 and Inositol against Breast Cancer: Modulation of PKC delta and p27 (Kip1). Anticancer Res 24(5):3665–3666

Wang B, Wu C (2017) Dietary soy isoflavones alleviate dextran sulfate sodium induced inflammation and oxidative stress in mice. Exp Ther Med 14(1):276–282

Wang Y, Mo X, Sun XS, Wang D (2007) Soy protein adhesion enhanced by glutaraldehyde crosslink. J Appl Polym Sci 104(1):130–136

Wang J, Huang M, Yang J, Ma X, Zheng S, Deng S, Huang Y, Yang X, Zhao P (2017) Anti-diabetic activity of stigmasterol from soybean oil by targeting the GLUT4 glucose transporter. Food Nutr Res 61(1):1364117

Wang L, Yun L, Wang X, Sha L, Wang L, Sui Y, Zhang H (2019) Endoplasmic reticulum stress triggered by Soyasapogenol B promotes apoptosis and autophagy in colorectal cancer. Life Sci 218:16–24

Warle B, Riar C, Gaikwad S, Mane V (2015) Effect of germination on nutritional quality of soybean (Glycine Max). J Environ Sci Toxicol Food Technol 9(4):12–15

Watanabe S, Uehara M (2019) Health effects and safety of soy and isoflavones. In: Singh RB, Watson RR, Takahashi T (eds) The role of functional food security in global health. Academic Press, London, pp 379–394

Welty FK, Lee KS, Lew NS, Zhou JR (2007) Effect of soy nuts on blood pressure and lipid levels in hypertensive, prehypertensive, and normotensive postmenopausal women. Arch Intern Med 167(10):1060–1067

Werner MH, Wemmer DE (1992) Three-dimensional structure of soybean trypsin/chymotrypsin Bowman-Birk inhibitor in solution. Biochemistry 31(4):999–1010

Whelton SP, Hyre AD, Pedersen B, Yi Y, Whelton PK, He J (2005) Effect of dietary fiber intake on blood pressure: a meta-analysis of randomized, controlled clinical trials. J Hypertens 23 (3):475–481

Wiseman H, O'Reilly JD, Adlercreutz H, Mallet AI, Bowey EA, Rowland IR, Sanders TA (2000) Isoflavone phytoestrogens consumed in soy decrease F2-isoprostane concentrations and increase resistance of low-density lipoprotein to oxidation in humans. Am J Clin Nutr 72 (2):395–400

Wolf WJ (1970) Soybean proteins. Their functional, chemical, and physical properties. J Agric Food Chem 18(6):969–976

Wong KM, Corradini MG, Autio W, Kinchla AJ (2019) Sodium reduction strategies through use of meat extenders (white button mushrooms vs. textured soy) in beef patties. Food Sci Nutr 7 (2):506–518

Wu Y, Wang T (2003) Soybean lecithin fractionation and functionality. J Am Oil Chem Soc 80 (4):319–326

Wynn TA, Chawla A, Pollard JW (2013) Macrophage biology in development, homeostasis and disease. Nature 496(7446):445–455

Xiao CW, Wood CM, Weber D, Aziz SA, Mehta R, Griffin P, Cockell KA (2014) Dietary supplementation with soy isoflavones or replacement with soy proteins prevents hepatic lipid droplet accumulation and alters expression of genes involved in lipid metabolism in rats. Genes Nutr 9(1):373

Xu Q, Kanthasamy AG, Reddy MB (2008) Neuroprotective effect of the natural iron chelator, phytic acid in a cell culture model of Parkinson's disease. Toxicology 245(1-2):101–108

Xu Q, Nakajima M, Liu Z, Shiina T (2011) Biosurfactants for microbubble preparation and application. Int J Mol Sci 12(1):462–475

Yang SC, Liu SM, Yang HY, Lin YH, Chen JR (2007) Soybean protein hydrolysate improves plasma and liver lipid profiles in rats fed high-cholesterol diet. J Am Coll Nutr 26(5):416–423

Yasothai R (2016) Antinutritional factors in soybean meal and its deactivation. Int J Sci Environ Technol 5(6):3793–3797

Yin D, Zhu Y, Liu L, Xu H, Huang J, Li Y (2014) Potential detrimental effect of soy isoflavones on testis sertoli cells. Zhong Nan Da Xue Xue Bao Yi Xue Ban 39(6):598–604; Journal of Central South University. Medical Sciences

Yoshiki Y, Kudou S, OKuBo K. (1998) Relationship between chemical structures and biological activities of triterpenoid saponins from soybean. Biosci Biotechnol Biochem 62(12):2291–2299

Yousef MI, Kamel KI, Esmail AM, Baghdadi HH (2004) Antioxidant activities and lipid lowering effects of isoflavone in male rabbits. Food Chem Toxicol 42(9):1497–1503

Yu Y, Jing X, Li H, Zhao X, Wang D (2016) Soy isoflavone consumption and colorectal cancer risk: a systematic review and meta-analysis. Sci Rep 6:25939. https://doi.org/10.1038/srep25939

Zanella I, Marrazzo E, Biasiotto G, Penza M, Romani A, Vignolini P, Caimi L, Di Lorenzo D (2015) Soy and the soy isoflavone genistein promote adipose tissue development in male mice on a low-fat diet. Eur J Nutr 54(7):1095–1107

Zhan S, Ho SC (2005) Meta-analysis of the effects of soy protein containing isoflavones on the lipid profile. Am J Clin Nutr 81(2):397–408

Zhang X, Shu XO, Li H, Yang G, Li Q, Gao YT, Zheng W (2005) Prospective cohort study of soy food consumption and risk of bone fracture among postmenopausal women. Arch Intern Med 165(16):1890–1895

Zhi L, Song D, Ma L, Feng T, Soyasapogenol B (2019) Attenuates laryngeal carcinoma progression through inducing apoptotic and autophagic cell death. Anat Rec. https://doi.org/10.1002/ar.24274

Zhou JR, Gugger ET, Tanaka T, Guo Y, Blackburn GL, Clinton SK (1999) Soybean phytochemicals inhibit the growth of transplantable human prostate carcinoma and tumor angiogenesis in mice. J Nutr 129(9):1628–1635

Zuniga KE, Clinton SK, Erdman JW (2013) The interactions of dietary tomato powder and soy germ on prostate carcinogenesis in the TRAMP model. Cancer Prev Res 6(6):548–557

Chapter 2
Rapeseed/Canola (*Brassica napus*) Seed

Ankit Goyal, Beenu Tanwar, Manvesh Kumar Sihag, Vikas Kumar, Vivek Sharma, and Suman Soni

Abstract Canola (*Brassica napus*), previously known as rapeseed, is one of the most commonly grown oil seeds. Its oil is commonly known for higher amount of monounsaturated fatty acids, moderate amount of polyunsaturated fatty acids, and substantial content of tocopherols, phytosterols, and omega-3 fatty acids. Native rapeseed used to contain higher concentration of toxic erucic acid (22–60%) and glucosinolates (80 μmol/g), which had adverse effect on nutrient's bioavailability and growth performance of the animals. Consequently, canola, a cross-bred of rapeseed cultivar, was developed having <2% erucic acid in oil and <30 μmol glucosinolates per gram of rapeseed meal. Canola oil is probably the only edible vegetable oil by today's standards which is nutritionally well-balanced among all other vegetable oils. Canola oil has been reported to prevent the risks of heart diseases, type II diabetes, hypercholesterolemia, etc. Apart from the cooking oil, canola oil is also used in the preparation of salad oils, salad dressings, margarines and organogels. On the other hand, defatted canola meal offers the possibility to be used as an emulsifier, gelling agents, absorbent, stabilizer, thicker, oleogelator,

---

A. Goyal
Department of Dairy Chemistry, Mansinhbhai Institute of Dairy and Food Technology, Mehsana, Gujarat, India

B. Tanwar (✉)
Department of Dairy Technology, Mansinhbhai Institute of Dairy and Food Technology, Mehsana, Gujarat, India

M. K. Sihag
Department of Dairy Chemistry, College of Dairy Science & Technology, Guru Angad Dev Veterinary and Animal Sciences University, Ludhiana, Punjab, India

V. Kumar
Department of Food Science and Technology, College of Agriculture, Punjab Agricultural University, Ludhiana, Punjab, India

V. Sharma
Dairy Chemistry Division, ICAR-National Dairy Research Institute, Karnal, Haryana, India

S. Soni
Regional Agmark Laboratory, Directorate of Marketing & Inspection, Ministry of Agriculture & Farmers' Welfare, Amritsar, Punjab, India

B. Tanwar, A. Goyal (eds.), *Oilseeds: Health Attributes and Food Applications*,
https://doi.org/10.1007/978-981-15-4194-0_2

texturizer, etc. in a variety of food products. However, long-term human studies are required to claim the health benefits of canola protein.

Keywords Canola (*Brassica napus*) · Rapeseed · Glucosinolates · Erucic acid · Oleogels · Organogels · Hypercholesterolemia · Canola protein

2.1 History and Introduction

Canola, previously known as rapeseed (*Brassica napus*), is one of the most common agriculture crops produced worldwide. Rapeseed is believed to be originated in India over 3000 years ago (Shahidi 1990). Later on, it was introduced into China and Japan. It is reported that the seeds of rapeseed were carried out to Europe for the first time in the thirteenth century. Afterward, in 1936, rapeseed was introduced in Canada from a traveler from Poland, where it grew well due to adaptable agro-climatic conditions. Though its cultivation was continued, its industrial use was not exploited until the World War II (1939–1945), where it was used as lubricant oil for the steam marine engines. Rapeseed oil adheres well to the metal surfaces washed with water and steam in highly heated engines.

Initially, rapeseed oil was utilized for the illumination as well as lubrication purposes. However, it was adopted as edible oil after the end of World War II, when shortage for cooking and edible oils was experienced. Initially, the food applications of rapeseed oil were very limited due to the presence of high amount (22–60%) of erucic acid (C22:1: ω-9). Erucic acid has been reported to have adverse effects such as cardiac injury and fatty deposits in heart muscles and adrenals in rodents along with impaired growth in animals (Gunstone 2004). Another major problem with rapeseed was the presence of goitrogenic glucosinolates (sulfur-containing glucosides) in rapeseed meal (80 μmol/g) which interferes with the absorption of iodine by thyroid gland and affects normal growth and development (Gunstone 2004). After the World War II, shortage of edible oil suggested to grow such oilseed crops suitable for human consumption. Therefore, in Canada, plant-breeding programs were initiated to reduce the level of erucic acid and glucosinolates in rapeseed oil and meal, respectively. In 1968, the first low erucic acid cultivar of *Brassica napus* containing less than 5% erucic acid and named as low erucic acid rapeseed (LEAR) oil variety was produced in Canada. Such varieties were also known as "single-low" or "zero-low" rapeseed varieties. Through further improvements these "single-low" varieties were converted into "double-low" or "double-zero" varieties using genetic modifications, which contained low erucic acid (<5%) in oil and low glucosinolates in rapeseed meal. In 1979, the term "canola" was adopted for all the rapeseed cultivars containing lower amount of erucic acid (<5%) in oil and glucosinolates (3 mg/g) in air-dried defatted meal. Later on, in 1986, the definition of canola was amended by Canola Council of Canada, which referred "canola" to all those rapeseed cultivars having <2% erucic acid in oil and <30 μmol glucosinolates per gram of rapeseed meal. Canola was also granted the GRAS (generally recognized as safe) status by Food and Agriculture Administration (FAO) in 1986. It is important to note that traditional high erucic acid rapeseed (HEAR) varieties are still produced for the industrial use in plastics, lubricants, lacquers, and detergents. In the recent times, HEAR varieties with low glucosinolate content are being produced by plant breeders so that de-oiled meal can be marketed as animal feed.

2.2 Production

Canola/rapeseed grows well under low temperature and moderate humidity conditions. Therefore, temperate zone is most appropriate environment for the growth of canola crop, where soybean and sunflower may not be able to survive. It can grow on a wide range of soil types, but clay loam soils are best suited.

Canola is one of the most important oilseed crops all over the world as its production has increased at much faster rate [approximately 50% in the last 10 years (2007–2017)] than any other oilseed crop. According to the Food and Agricultural Organization, Canada is the largest producer of canola/rapeseed followed by China, India, France, and Australia (FAO 2019). The total world and country-wise production of rapeseed for the years 2013–2017 is shown in Fig. 2.1. In India, canola accounts approximately 30% of the total oilseeds produced. According to a recent report, rapeseed/canola ranks second in terms of total oilseeds production in India after soybean (GAIN 2018). In India, major canola producing states are Rajasthan, Uttar Pradesh, Haryana, Madhya Pradesh, and Gujarat. A trend in the production behavior of most common oilseeds in India is shown in Fig. 2.2.

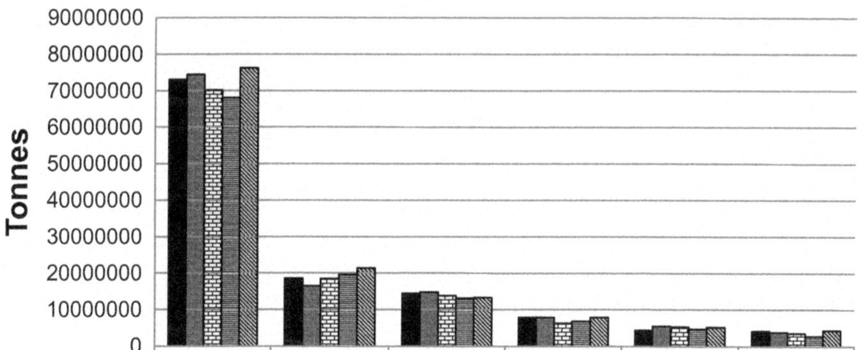

	World	Canada	China	India	France	Australia
2013	73096861	18551000	14458015	7820000	4366200	4141731
2014	74461405	16410100	14772200	7877000	5524094	3832049
2015	70196634	18376500	13859000	6282000	5334404	3540021
2016	68085491	19599200	13128000	6797000	4727961	2775479
2017	76238340	21328000	13274000	7917000	5200000	4313230

Fig. 2.1 Global canola/rapeseed production (2013–2017) (FAO 2019)

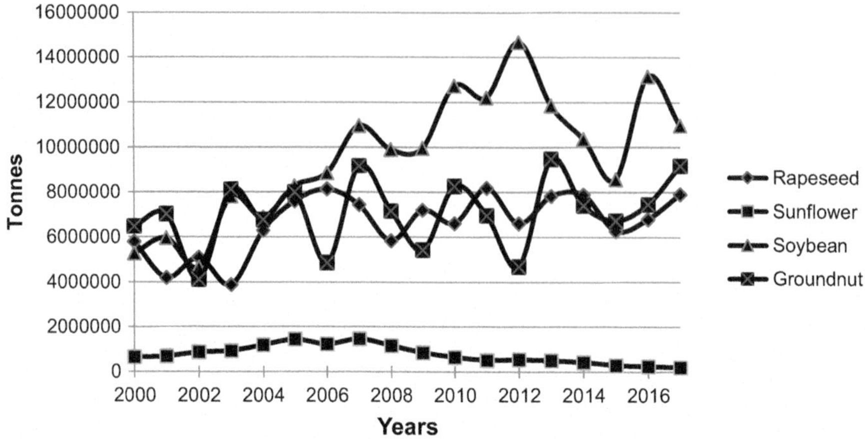

Fig. 2.2 Total production of common oilseeds in India (2000–2017)

2.3 Chemical Composition

Canola/rapeseed belongs to the genus *Brassica* from Cruciferae family and, thus, is close relative of mustard, turnip, cabbage, etc. Canola is a standard of rapeseed containing low erucic acid in oil and low glucosinolates in meal. A wide range of literature is available on the composition of rapeseed/canola oil and its meal. Henceforth, to avoid confusion and to make it simpler, instead of "rapeseed," "canola" term would be used throughout the chapter. Canola is grown majorly for its oil content, which has become third largest consumed vegetable oil (27.83) in the world after palm (69.57) and soybean oil (57.05 million metric tons) (Statista 2019).

Proximate composition of any crop or oilseed depends upon the agro-climatic conditions. Different cultivars of canola have been shown a wide variation in oil content as well as its composition. Genetic modifications and plant-breeding techniques have produced several cultivars or varieties of canola containing varied amount of glucosinolates, antinutritional factors, and so on. Canola seed contains approximately 3–4.5% moisture, 40–45% oil, 19–28% crude proteins, 5–6% ash content, 8–10% soluble sugars, and 14–15% crude fibers (Cakmakli and Ünal 1988; Assadi et al. 2011; Carré et al. 2016). The proximate composition of canola seed is given in Table 2.1, and the different constituents are discussed below:

2.3.1 *Lipids*

Oil is the most important constituent of canola seed for which it is grown worldwide. Though, traditional high erucic acid rapeseed (HEAR) varieties are still grown for industrial use, LEAR (low erucic acid rapeseed) cultivars are mainly produced for

Table 2.1 Proximate composition of whole canola seed (rapeseed)

Parameter	Range (g/100 g)	Reference/s
Moisture content	3.75–4.72	Cakmakli and Ünal (1988)
Lipids	30.60–50.40	Cakmakli and Ünal (1988), Carré et al. (2016), Assadi et al. (2011), Matthaus et al. (2016)
Crude protein	19.50–28.17	Cakmakli and Ünal (1988), Assadi et al. (2011), Carré et al. (2016)
Crude fiber	15.72[a]	Carré et al. (2016)
Ash	5.69–6.93[a]	Carré et al. (2016)

[a]On dry matter basis

Table 2.2 Major fatty acid profile of rape seed oil [HEAR and LEAR (canola)] and other commonly used edible oils

Fatty acid (% of total fatty acids)	Rape seed oil		Soybean oil	Sunflower oil	Peanut oil
	HEAR oil	LEAR oil (canola)			
Oleic acid (C18:1)	15.0	56.80–64.92	23.80–24.00	20.40–28.00	47.10–71.1
Linoleic acid (C18:2) (ω-6)	14.0	17.11–20.92	53.30–54.00	62.20–68.80	18.20–33.6
α-Linolenic acid (C18:3) (ω-3)	9.0	2.63–12.99	7.10–8.00	0–0.16	N.D.
Erucic acid (C22:1)	45.0	0.08–1.63	N.D.	N.D.	N.D.

HEAR high erucic acid rapeseed, *LEAR* low erucic acid rapeseed (Shahidi 1990; Daun et al. 2015; Orsavova et al. 2015)
N.D. Not detected

food applications. Oil content in different cultivars of canola ranges from 30.60% to 50.40% on dry weight basis (Assadi et al. 2011; Matthaus et al. 2016). Canola oil is rich in oleic acid (C18:1), a monounsaturated fatty acid, which has been associated with several positive health effects, which will be discussed in further sections. Apart from that, canola oil has a well-balanced ratio of omega-6 and omega-3 fatty acid (2:1) unlike other vegetable oils such as soybean and sunflower oil (5:1). Moreover, α-linolenic acid (ALA, ω-3) rich varieties were also developed through plant-breeding techniques for the enhanced health benefits. The fatty acid profile of canola oil and other vegetable oils is shown in Table 2.2. Recently, Matthaus et al. (2016) estimated the oil content and fatty acid profile of 14 varieties of rapeseed and canola in Turkey. It was observed that the oil content was comparatively higher (41.8–48.3%) in rapeseed varieties (Orlan, Forza, Mavi, Hyola) than that of canola cultivars (30.6–36.2%) such as Forte, Juna, Ras-5, and Ras-6 (Matthaus et al. 2016). Efforts have also been done to enhance the level of oleic acid (MUFA) from normal 60% to 85% through genetic engineering to enhance the oxidative stability and shelf life of the oil.

2.3.2 Carbohydrates

Carbohydrates constitute relatively small portion of canola seed and are mainly present in hull and cotyledons. As discussed above, carbohydrates range 14–15% (on dry basis) in canola seed. Low-molecular-weight carbohydrates such as sugars (glucose, fructose, and sucrose), stachyose, raffinose, etc. are majorly present in cotyledons; while high-molecular-weight carbohydrates such as cellulose, hemicellulose, and pectin are majorly present in hulls.

Carbohydrates and its fractions in canola/rapeseed have been comprehensively studied and reviewed by Siddiqui and Wood (1972, 1974, 1977) and Shahidi (1990). It is reported that dehulled and oil-free canola meal contained approximately 52% proteins and 48% carbohydrates (Siddiqui and Wood 1977). There are several glycosides of sugars particularly raffinose and stachyose, present in hull of the seed, which cause the flatulence in human and animals. Due to presence of high amount of hull and indigestible carbohydrates, canola meal is limited to be used as animal feed. However, removing of hull by processing techniques could improve the nutritional value of canola meal for human use.

2.3.3 Proteins

Like other members of the *Brassica* genus, whole canola seed contains approximately 17–26% proteins. However, de-oiled canola meal contains up to 50% protein content (on dry basis). Canola protein consists of majorly storage proteins, namely, cruciferin and napin, which constitute approximately 60% and 20% of the total proteins, respectively (Ghodsvali et al. 2005). Another canola protein is oleosin, which is a structural protein associated with oil bodies in the seed (Wu and Muir 2008). There are several studies indicating the superior quality and well-balanced amino acid profile of canola protein as compared to other oilseeds' protein such as soybean, sunflower, safflower, etc. (Sosulski and Sarwar 1973; Ohlson and Anjou 1979; Ghodsvali et al. 2005; Yoshie-Stark et al. 2006; Wanasundara et al. 2016). In terms of essential amino acids, canola protein has a very high amount of sulfur-containing amino acids, i.e., lysine, cysteine, and methionine (3.0–4.0% of total protein), which is closer to FAO requirements for humans than any other quality vegetable protein available (Ohlson and Anjou 1979; FAO/WHO/UNU 1985).

Canola protein has high content of essential amino acids (>400 mg/g protein), particularly sulfur-containing amino acids (>40–49 mg/g protein) (Grala et al. 1997). Earlier, Sosulski and Sarwar (1973) evaluated amino acid profile of various oilseed meals and protein isolates and concluded that canola (rapeseed) and soybean contained high proportions of essential amino acids, while sunflower, safflower, and flaxseed contained high amount of nonessential amino acids. The protein scores were found to be higher for rapeseed meal (70) and flax meal (82) than that of casein (57), soybean (67), sunflower (63), and safflower (63) meal (Sosulski and Sarwar

Table 2.3 Amino acid content of canola, soybean, sunflower, safflower, and casein protein (g/100 g protein)

Amino acid	Rapeseed/canola meal (g/100 g protein)[a]	Soybean meal (g/100 g protein)[b]	Sunflower meal (g/100 g protein)[c]	Safflower meal (g/100 g protein)[d]	Casein (milk protein) (g/100 g protein)[e]
Cysteine	2.29	1.20	0.31	NR	0.10
Histidine	3.39	2.53	2.79	3.13	1.70
Isoleucine	3.47	5.04	4.40	2.24	2.30
Leucine	6.19	8.04	6.92	5.78	4.60
Lysine	5.92	6.39	3.87	2.16	1.60
Methionine	1.94	1.17	2.26	1.11	3.10
Phenylalanine	4.06	5.05	5.32	4.63	3.10
Threonine	4.27	3.93	3.70	1.94	2.60
Tryptophan	1.33	NR	1.57	1.71	NR
Tyrosine	2.50	3.64	3.21	3.35	3.40
Valine	4.97	5.01	5.29	2.84	3.00
Arginine	6.62	7.48	10.12	11.52	2.10
Glutamine and/or glutamate	18.14	19.30	22.45	21.89	13.9
Glycine	4.92	4.25	5.35	4.82	1.20
Proline	5.97	5.51	3.61	3.62	6.50
Alanine	4.36	4.48	4.59	3.74	2.00
Aspartic acid + aspartate	7.25	12.1	10.02	9.94	NR
Serine	4.00	4.85	4.05	4.15	3.40

NR Not reported
[a]Canola Council of Canada (2015)
[b]Cavins et al. (1972)
[c]Robinson (1975)
[d]Paredes-Lopez and Ordorica-Falomir (1986)
[e]Gorissen et al. (2018)

1973). Similarly, Delisle et al. (1984) conducted a study to evaluated the nutritive value of soybean, rapeseed, and wheat flours in Sprague-Dawley male rats and reported higher protein efficiency ratio and apparent digestibility coefficient for rapeseed meal (2.64 ± 0.10 and 82.5 ± 0.57) when compared to soybean (2.19 ± 0.10 and 76.1 ± 0.63). In another study conducted by Bos et al. (2007) in humans, post-prandial biological value was found to be higher for rapeseed protein (84.0) than that of milk (82.0), soybean (80), wheat (73), pea (79), and lupin (81.5) proteins. The amino acid content of canola, soybean, sunflower, safflower, and casein proteins is shown in Table 2.3.

2.3.4 Minerals and Vitamins

Canola seed/meal is a rich source of minerals particularly calcium, phosphorus, potassium, iron, zinc, and selenium (Assadi et al. 2011; Shahidi 1990; Beyzi et al. 2019). Reports have suggested that canola meal contained comparatively higher amount of essential minerals (calcium and phosphorus) than that of soybean meal (Wickramasuriya et al. 2015; Summers and Leeson 1985). Though a major proportion of phosphorus is present in the form of phytates in canola meal, which is not bioavailable in monogastric animals; nonetheless, it is a quality source of Ca and P compared to soybean meal (Wickramasuriya et al. 2015). A comparison of minerals content of canola meal with soybean meal is shown in Table 2.4.

In terms of vitamin content, canola meal has higher proportions of all the vitamins except pantothenic acid when compared with soybean meal (Wickramasuriya et al. 2015). As canola meal is used as animal feed, little information is available on the vitamin content of canola meal. The content of various vitamins is shown in Table 2.5.

2.3.5 Phytonutrients and Other Minor Components

In crude vegetable oil, several nutritionally important components such as tocopherols, phytosterols, carotenoids, ubiquinones (coenzyme Q), squalene, polyphenols, and phospholipids are found at lower concentrations, which make the part of unsaponifiable matter. As compared to other regular vegetable oils, canola oil has higher amount of brassicasterol, campesterol, and β-sitosterol. The range of different

Table 2.4 Mineral composition of canola meal

	Canola meal		Soybean meal
Mineral	Shahidi (1990)	Canola Council of Canada (2015)	Shahidi (1990)
Calcium (g/100 g)	0.63	0.65	0.30
Phosphorus (g/100 g)	1.02	0.99	0.63
Sodium (g/100 g)	NR	0.07	NR
Chlorine (g/100 g)	NR	0.10	NR
Potassium (g/100 g)	1.21	1.13	1.97
Sulfur (g/100 g)	0.85	0.63	0.43
Magnesium (g/100 g)	0.51	0.54	0.27
Copper (mg/kg)	5.70	4.70	23.00
Iron (mg/kg)	141.00	162.00	119.00
Manganese (mg/kg)	49.20	58.00	29.00
Molybdenum (mg/kg)	1.30	1.40	–
Zinc (mg/kg)	68.20	47.00	43.00
Selenium (mg/kg)	1.10	1.10	0.30

NR not reported

Table 2.5 Vitamin composition of canola meal

	Canola meal		Soybean meal
Vitamin (mg/kg)	Liang (2000)	Canola Council of Canada (2015)	National Research Council (1994)
Vitamin E	14.00	13.00	4.47
Pantothenic acid	9.50	9.30	15.00
Niacin	160.00	156.00	22.00
Choline	6700.00	6500.00	2700.00
Riboflavin	5.80	5.70	2.90
Biotin	1.10	0.96	0.32
Folic acid	2.30	0.80	1.30
Pyridoxine	7.20	7.00	5.00
Thiamin	5.20	5.10	3.20

Table 2.6 Concentration of minor components and phytonutrients in canola/rapeseed oil

Constituent	Canola oil	Soybean oil	Sunflower oil
Total phytosterols (mg/kg oil)	3459.40–6900.00	4600	4100
1. Brassicasterol (mg/kg oil)	455.00–986.70	Traces	Traces
2. Stigmasterol (mg/kg oil)	ND	772.80	319.80
3. Campesterol (mg/kg oil)	1331.50–1904.40	1191.40	266.50
4. β-Sitosterol (mg/kg oil)	1590.10–3608.00	2060.80	2300.00
5. Avenasterol	82.80–234.60	427.80	848.70
Total tocopherols (mg/100 g)	56.40–94.28	114.64	89.88
1. α-Tocopherol (mg/100 g)	19.70–42.21	17.93	20.65
2. β-Tocopherol (mg/100 g)	ND-0.18	ND	ND
3. γ-Tocopherol (mg/100 g)	31.00–50.76	63.68	2.39
4. δ-Tocopherol (mg/100 g)	ND-1.13	33.03	1.94
coenzyme Q10/ubiquinones (mg/100 g)	7.3	9.2	0.4

Source: Siger et al. (2015), Rekas et al. (2016), Ghazani and Marangoni (2013), Shahidi (2002)
ND Not detected

phytosterols and other minor components is shown in Table 2.6. Phytosterols have been reported to show cholesterol-lowering effects and improve oxidative stability of oil during heating (Gertz and Kochhar 2001).

Siger et al. (2015) studied the effect of roasting of canola seeds followed by cold extraction of oil and observed majorly brassicasterol, campesterol, sitosterol, and avenasterol to the value 455.0 ± 7.1, 1331.5 ± 20.8, 1590.1 ± 24.8, and 82.8 ± 1.3 μg/g oil, respectively, in unroasted seed oil. Moreover, the authors did not observe any significant effect ($p < 0.05$) on the concentration of total as well as individual phytosterol/s. In another study, Rekas et al. (2016) characterized the chemical composition of cold-pressed oils extracted from different rapeseed varieties and reported α-, γ-, δ-, and total tocopherols in the range of 21.3–26.8, 31.0–42.4, 0.1–1.3 and 56.4–65.1 mg/100 g, respectively. In the study, carotenoid pigments and chlorophyll have also been reported in crude canola oil obtained from different varieties ranging from 6.66 to 17.39 and 2.6 to 2.92 mg/kg, respectively (Rekas et al. 2016).

2.3.6 Antinutritional Factors

Rapeseed contains several antinutritional factors such as glucosinolates, tannins, phytates, erucic acid, etc., which affect the nutrient's bioavailability adversely and reduce the growth and performance of the animals.

2.3.6.1 Glucosinolates

Glucosinolates, chemically, are a class of water-soluble, sulfur- or nitrogen-containing glucosides that occur as secondary metabolites in virtually all species of *Brassica* (Boot 2016). Traditional rapeseed is reported to contain 120–150 μmol glucosinolates/g of rapeseed meal. Through breeding techniques, the amount of glucosinolates has been reduced to 4–10 μmol/g (approximately 10–12% of original level) in canola meal. The most common glucosinolates in canola meal are progoitrin (2(R)-2-hydroxy-3-butenyl glucosinolate), gluconapin (3-butenylglucosinolate), gluconapoleiferin (2-hydroxy-4-pentenyl glucosinolate), glucobrassicanapin (4-pentenyl glucosinolate) (Konkol et al. 2019). In their natural and non-hydrolyzed form, glucosinolates are not harmful to animals or humans. However, in presence of myrosinase (naturally present in seed), glucosinolates get hydrolyze and produce glucosinolate derivatives including isothiocyanates, nitriles, thiocyanates, and 5-vinyloxazolidine-2-thione (VOT) (Liang 2000). Isothiocyanates, thiocyanates, and VOT have been reported to interfere with iodine uptake by thyroid gland and induce goitrogenic effects in humans. On the other hand, nitriles are toxic and have shown to interfering with liver and kidney functions. Glucosinolate levels of 18–30 μmol/g canola meal have been reported to have antinutritional or toxic effects in animal studies (Boot 2016) (Fig. 2.3).

2.3.6.2 Phytic Acid/Phytates

Phytate (myoinositol 1,2,3,4,5,6-hexakis dihydrogen phosphate or inositol hexaphosphate, InsP6) represents the major storage form of phosphorus and inositol in plant species. Whole canola/rapeseed contains about 2–4 g/100 g phosphorus, of which approximately 65% is present in the form of phytates, which are salts of phytic acid with divalent minerals) (Thompson 1990; Liang 2000). Phytates act as antinutrients and bind with divalent cations such as Zn, Ca, Fe, Cu, Mn, etc. in gut of monogastric animals including poultry and humans. Monogastric animals are not able to digest phytates due to deficiency of hydrolyzing enzyme (phytase). Thus, phytates reduce the bioavailability of essential ions and adversely affect the

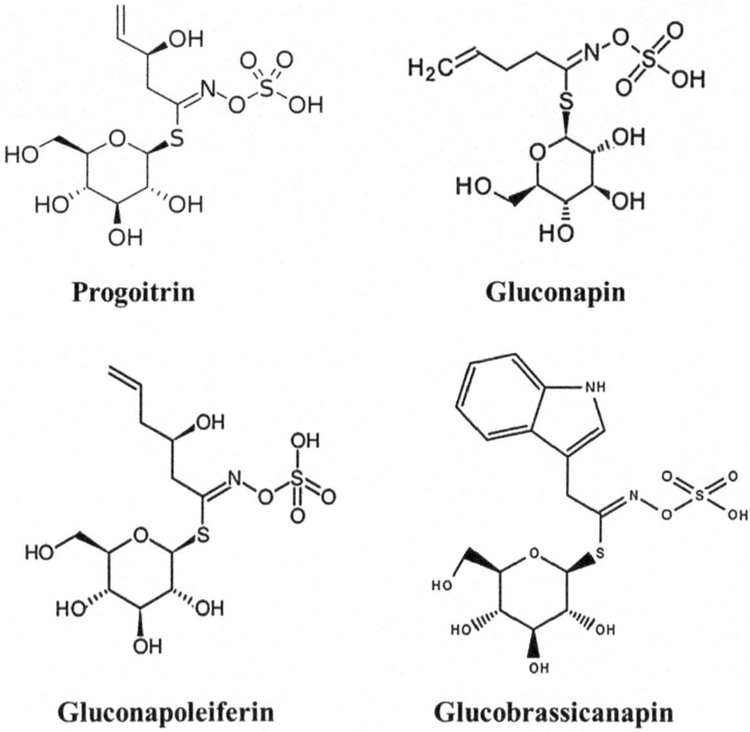

Fig. 2.3 The structures of most common glucosinolates present in rapeseed/canola

digestibility and bioavailability of other nutrients also. It is also reported that phytates may bind with the digestive enzymes such as trypsin, chymotrypsin, etc. and inhibit their enzymatic action. As canola meal is generally used in animal feed, sometimes, phytase enzyme is already added in diet formulation to break phytic acid in order to improve the digestibility and biological quality of feed. A wide range of literature is available on the negative effects of phytates on the growth performance and bioavailability of nutrients in different animals (Thompson 1990; Kumar et al. 2012; Gupta et al. 2015; Dersjant-Li and Dusel 2019).

2.3.6.3 Phenolic Acids (Expressed as Sinapin) and Tannins

Phenolic acids are naturally present in plants and are responsible for bitter and astringent taste depending on the concentration. In canola, phenolic acids constitute approximately 1.5–1.8 g/100 g of defatted meal. It is reported that canola flour contains the highest amount of phenolic acids (639.9 mg/100 g) among other oilseed flours such as soybean (23.4 mg/100 g), cottonseed (56.7 mg/100 g) or peanut (63.6 mg/100 g) (Shahidi 1990). In canola meal, the most common phenolic acid is sinapin (sinapine), which is a choline ester of 3,5-dimethoxy-4-hydroxycinnamic

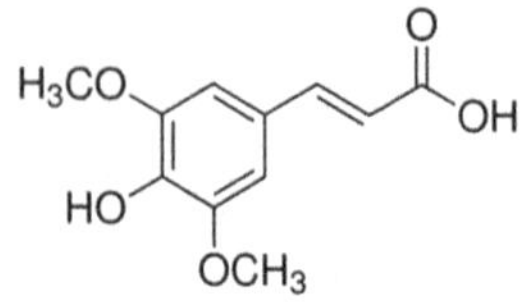

Fig. 2.4 Structure of sinapic acid

Table 2.7 Concentration of free phenolic acids in defatted rapeseed meal (on dry basis) (Malgorzata and Aleksander 2010)

Free phenolic acid (15% of total phenolic acids)	Concentration (mg/100 g dry matter of defatted rapeseed meal)
Protocatechuic acid	0.15
p-Hydroxybenzoic acid	0.04
Vanillic acid	0.45
Caffeic acid	1.97
p-Coumaric acid	0.18
Ferulic acid	0.37
Sinapic acid	25.20
Sinapic acid derivatives	46.62

acid (Malgorzata and Aleksander 2010) (Fig. 2.4)). Sinapin is present in free (15%), esterified (80%), as well as bound-insoluble form (<5%) in canola seed (Shahidi 1990). The concentration of free form of sinapin and other minor phenolic acids present in canola meal is shown in Table 2.7.

It is reported that sinapin is associated with the biosynthesis of lignin and flavonoids (Neish 1960). It is noteworthy that sinapin imparts bitter taste to feed and thus affecting the feed intake and growth performance of poultry animals (Liang 2000). Sinapin has also been shown to impart fishy and off-flavor in chicken eggs (Khajali and Slominski 2012). Because of their association/binding with digestive enzymes, amino acid, proteins, and other nutrients, the amount of phenolic acids should be given importance when canola meal is used as animal feed (Shahidi 1990).

Tannins are polyphenolic compounds present in almost all foods and feed of plant origin. In canola, tannins are majorly found in hull of the seed. The amount of tannins can vary from 1.5% to 3.0% in canola meal (Bell 1993), with higher amount in brown-seeded variety than that of yellow-seeded variety. Tannins can be classified as condensed tannins and hydrolysable tannins. Tannins can form soluble and/or insoluble complexes with proteins and digestive enzyme and, thus, reduce the bioavailability of the nutrients in monogastric animals and humans (Tanwar et al. 2018; Tanwar et al. 2019). Though tannins have been reported to affect palatability, feed intake, and digestibility of food and feed in several cereals and legumes; they do not appear to have similar negative effects in canola feed-based diets (Khajali and Slominski 2012). Khattab et al. (2010) studied 27 varieties of canola and reported tannin content in the range of 0.68–1.53 g/100 g in the defatted canola seeds. However, Naczk et al. (2000) reported tannins content varying from 1913 to 6213 mg/100 g canola/rapeseed hull.

2.4 Health Effects

As discussed previously, canola oil is used as a cooking and/ingredient oil for food consumption, while canola meal is generally used as animal feed; therefore, under this section, health benefits of canola oil will only be discussed. The negative effects associated with canola meal and its antinutrients have already been discussed in brief in above section.

Canola oil is characterized by having low amount of saturated fatty acids (4–7%), high amount of monounsaturated fatty acid (MUFA) (oleic acid: ~60%), and appreciable amount of polyunsaturated fatty acids (PUFAs) (linoleic and α-linolenic acid). The intake of saturated fatty acids has been positively linked with increased risks of atherosclerosis and oxidation of cholesterol. Canola oil contains approximately half of the saturated fatty acids than that of soybean oil, olive oil, or corn oil. Hence, canola oil fits well with the recommendations of several health regulatory agencies to decrease the dietary intake of saturated fatty acids. A wide range of literature is available on the health benefits of consuming MUFA and PUFAs individually and/or in the form of canola oil, which will be discussed in further sections:

2.4.1 In Hypercholesterolemia

Appreciable research has been conducted on the effects of canola oil along with its MUFA and PUFAs on the plasma and total cholesterol level in animals and humans. In 2006, a randomized, crossover study was carried on 23 overweight, hyperlipidemic men consuming canola and olive oil (control) for 6 weeks, and the levels of triacylglycerides, total cholesterol (TC), LDL-C, and HDL-C were measured (Rudkowska et al. 2006). Canola group showed reduced levels of TC (4.71 vs. 5.14 mmol/L), LDL-C (3.12 vs. 3.54 mmol/L), HDL-C (0.89 vs. 0.93 mmol/L) and triglycerides (1.53 vs. 1.48 mmol/L) when compared with olive oil group. In another study, the effect of canola oil consumption was evaluated on 36 hypercholesterolemic and/or hypertriglyceridemic subjects for a longer duration (4 months) (Bierenbaum et al. 1991). The subjects consumed 30 ml/d canola oil as a replacement of edible oil taken in usual diet. Results showed significantly ($p < 0.025$) reduced level of LDL-C from 173 ± 9.0 from 160 ± 10 mg/dL in blood serum. However, no significant changes were observed in total cholesterol, HDL-C and triglycerides level (Bierenbaum et al. 1991). Similarly, Gulesserian and Widhalm (2002) evaluated the effect of rapeseed oil-based diets in 17 children and young adolescents (male = 6, female = 11) with familial hypercholesterolemia. The patients received low-fat/low-cholesterol diet having 15 g/d (8–23 g/d) rapeseed oil for first 2 months following 22 g/d (15–30 g/d) for further 3 months. Expected results were obtained with a decrease of 28% in serum triglycerides level (from 181 ± 61 to 85 ± 40 mg/dL), 6% in LDL-C level (from

151 ± 31 to 142 ± 31 mg/dL), 27% in VLDL-C level (from 23 ± 12 to 17 ± 8 mg/dL), and 9% in total cholesterol level (from 233 ± 35 to 213 ± 36 mg/dL). However, no significant ($P > 0.05$) change was observed in the levels of HDL-C (Gulesserian and Widhalm 2002). A lot of literature is available on the cholesterol and triglycerides lowering effects of rapeseed or canola oil (Gylling et al. 1999; Chisholm et al. 2005; Ferguson et al. 2016; Kruse et al. 2015), but the mechanism behind hypotriacyglycerolemic and/or hypocholesterolemic effect is still not well understood (Gulesserian and Widhalm 2002). It is suggested that there are some compositional changes in VLDL or in the expressed activities of the enzymes and proteins involved in intravascular processing and catabolism of VLDL, which could play an important role in lowering serum triacylglycerides level (McNamara 1992; Ruiz-Gutierrez et al. 1998; Campos et al. 1996; Montalto and Bensadoun 1993). Fumeron et al. (2017) reviewed the possible mechanisms for cholesterol-lowering effects of plant stanols/phytosterols. It was suggested that phytosterols (which are present in appreciable amount in canola oil) compete with cholesterol in the micelles of lecithin and bile salts, which causes a reduction in cholesterol solubilization. When the concentration of phytosterols is high in diet, cholesterol almost loses its solubility for intestinal absorption and excreted out through feces (De Smet et al. 2012; Ikeda et al. 1989).

2.4.2 *In Cardiovascular Diseases*

There are evidences suggesting that consumption of MUFA-rich oils such as canola can reduce the risk factors related to cardiovascular diseases, which have comprehensively reviewed by Hammad et al. (2016) and Baum et al. (2012). American Heart Association has recommended a diet that provides <10% of calories from SFA, up to 10% from PUFA, and as much as 15% from MUFA for cardiovascular health (Kris-Etherton 1999). The serum levels of LDL and HDL are directly correlated with the risk of heart diseases, which have already been discussed in previous section. Currently, there is an increased interest in the relation of thrombogenesis and MUFA-rich diet. A few studies have suggested that MUFA decreases the platelet aggregation (Sirtori et al. 1986), increases bleeding time (McDonald et al. 1989), and increases fibrinolysis (Lopez-Segura et al. 1996), thereby protecting against thrombogenesis. Several researchers observed a slight protective effect against CHD on replacing energy from complex carbohydrates with MUFAs (Hu et al. 1997; Kris-Etherton 1999). In 1997, a prospective study was conducted on 80,082 women (age 34–59 years), who had no known coronary disease, stroke, cancer, hypercholesterolemia, or diabetes in 1980. During 14 years of follow-up, it was observed that for a 5% increment in energy from monounsaturated fats, the risk of coronary disease was reduced to 0.81 (95% confidence interval, 0.65–1.00; $P = 0.05$) (Hu et al. 1997). A wide range of literature has suggested that MUFA reduced the level of LDL-cholesterol, triglycerides, and total cholesterol, whereby increased the HDL-cholesterol and, thus, played a positive role

Table 2.8 Possible mechanisms behind reducing the cardiovascular diseases risk factors by higher consumption of MUFA

S. No.	Mechanism	References
1	Enhanced catabolism of triacylglycerol-rich lipoproteins (carrier of triglycerides in serum)	Schoonjans et al. (1996)
2	Reduced hepatic production and/or secretion of VLDL	Schoonjans et al. (1996)
3	Change in VLDL composition due to change in dietary fat	McNamara (1992)
4	Stimulation of acyl-CoA: cholesterol acyltransferase in the liver → increased cholesterol ester formation → decreased sterol pool → higher expression of the LDL receptor in the liver → decreased LDL levels in serum	Kien et al. (2014)
5	Degradation of insulin-induced gene-1 protein → inactivation of the transcription factor sterol regulatory element binding protein → reduced cholesterol synthesis, cellular LDL uptake, and fat oxidation → reduced biosynthesis and cellular uptake of cholesterol	Kien et al. (2014)
6	By lowering LDL-proteoglycan binding	Pu et al. (2015)
7	By elevating post-prandial oleoylethanolamide (OEA) levels (OEA controls appetite sensation) → reduced energy intake	Mennella et al. (2015)
8	Increased hepatic LDL-receptors	Fernandez and West (2005)

in reducing the risk of cardiovascular diseases (Sanders 2009; Venturini et al. 2015; Mozaffarian and Clarke 2009; Hammad et al. 2016). The possible mechanisms of MUFA for reducing the risk factors of cardiovascular diseases are summarized in Table 2.8.

Contrary to the positive association, several studies have demonstrated no correlation between the consumption of MUFA-rich oils and CHD (Kromhout and Coulander 1984; Kromhout et al. 1995; McGee et al. 1984; Garcia-Palmieri et al. 1980). Similarly, in another study conducted by Rudel et al. (1998), it was observed that dietary cis or trans monounsaturated fat did not protect against atherosclerosis development in mouse model.

2.4.3 *In Diabetes Mellitus*

Type II diabetes is directly associated with risk of cardiovascular diseases. A meta-analysis of randomized controlled trials suggested that metabolic risk factors can be improved among diabetes patients by increasing MUFA content in diet (Qian et al. 2016). In the study, when high-MUFA diet was compared with high-carbohydrate and high-PUFA diet (individually), a significant ($p < 0.05$) reduction in fasting plasma glucose level, viz., −0.57 and −0.87 mmol/L, respectively, was observed. Bozzetto et al. (2012) studied the effect of isoenergetic MUFA-rich diet over high-carbohydrate/high-fiber diet in type II diabetic patients ($n = 45$, 37 men and

8 women, aged 35–70 years) and found significant change in HbA_{1c} (from 6.6% to 6.2%) and liver fat content (from 7.4 ± 2.8 to 5.2 ± 2.7) after 8-week intervention period. Similarly, Garg (1998) conducted a meta-analysis of ten randomized cross-over trials comparing isoenergetic high-MUFA and high-carbohydrate diets in type II diabetic patients and observed improved fasting and post-prandial glucose level with no significant difference on fasting insulin, HbA_{1c}, or insulin sensitivity in high-MUFA group when compared with other group.

2.5 Food Applications

2.5.1 As Cooking Oil, Salad Oils, and Margarines

The most common food applications of canola oil are in cooking and liquid oils. Currently, about 90% of canola oil is used in the preparation of salad dressing, salad oil, soft and hard margarines, mayonnaises, and shortenings. Approximately 65% of produced oil is exported from Canada and most of it is used as a liquid oil. There is a growing interest in cold-pressed or virgin canola oil due to the retention of higher amount of phytosterols and bioactive components. Although there is problem of flavor with cold-pressed canola oil, still, the global market is growing rapidly in this area. Germany is the most advanced producer of virgin canola oil, and a lot of research has been done to establish its health benefits.

Canola oil is the principal oil used in salad dressings. The advantage of canola oil for the use in salad dressings/salad oils is the lowest amount of saturated fatty acids among all common vegetable oils. Therefore, canola oil remains transparent and/or liquid even at refrigeration temperature without producing haziness. Other oils need fractionation and/or winterization before use as a salad oil. Another reason to use canola oil in salad dressing is its superior nutritional value. Canola oil has been reported to contain the highest amount of monounsaturated fatty acids (oleic acid), less amount of linoleic acid (omega-6), and good amount of α-linolenic acid (omega-3) among commonly used vegetable oils such as soybean, sunflower oil, and peanut oil. For frying applications, canola oil, again, shows some chemical properties which are important for improved oxidative stability and, thus, enhanced shelf life during processing conditions. As previously discussed, canola oil has lower amount of linoleic acid (PUFA) as compared to other oils, it is more oxidative stable during frying operations. Though canola oil has good amount of highly oxidative α-linolenic acid; nevertheless, it is more stable to high temperatures probably due to presence of high amount of natural antioxidants and high amount of MUFA.

2.5.2 In the Preparation of Organogels/Oleogels

Organogels are the novel modified lipids prepared by entrapping the liquid vegetable oil into a three-dimensional, stand-alone, thermo-reversible, viscoelastic, and anhydrous network of organogelators such as mono- and diacylglycerides, fatty acid's alcohols and esters, phytosterols, phospholipids, higher chain fatty alcohols, waxes, etc. (Rogers et al. 2009; Pehlivanoğlu et al. 2018). In organogels, vegetable oil remains in the continuous phase without any change in fatty acids composition. When vegetable oil is mixed with organogelator/s, there is the only change in physical characteristics of the oil in consequently formed organogel, thus, offering the physical and sensorial characteristics similar to partially hydrogenated vegetable oils (PHVOs) without generation of saturated fatty acids and/or *trans* fats (Chaves et al. 2018). Another advantage of using oleogels is that it can also be used to minimize some quality defects such as fat bloom, change in melting temperature in chocolates/chocolate-based products, etc. (Hughes et al. 2009). Therefore, such structured lipids offer potential alternatives to the saturated fats and PHVOs in a number of food applications particularly in bakery products (Jang et al. 2015; Mert and Demirkesen 2016), frozen desserts, ice cream (Zulim Botega et al. 2013), meat products (Zetzl et al. 2012), etc.

Canola oil is one of the richest sources of unsaturated fatty acids and confers several health benefits as discussed in previous sections. Recently, canola oil has been reported to be used for the preparation of organogels using ethyl cellulose (Zetzl et al. 2012), candelilla wax (Mert and Demirkesen 2016; Jang et al. 2015; Lim et al. 2017), carnauba wax,and beeswax (Lim et al. 2017) as organogelators. Previously, Jang et al. (2015) prepared canola oil oleogels with candelilla wax and utilized it as a shortening replacer to produce cookies with a high level of unsaturated fatty acids (UFAs). In the study, replacement of shortening with canola oil-based oleogel reduced the viscoelastic parameters of dough, increased the UFAs content [92% vs. 47.2% (control) of total fatty acids], showed desirable spreadable properties, and produced cookies with soft eating characteristics (Jang et al. 2015). Similarly, in another study conducted by Mert and Demirkesen (2016), canola oil oleogel prepared with candelilla wax (3 and 6%) was partially replaced with commercial shortening (oleogel/shortening: 30–40:60–70) in cookies and observed increased extensibility, lowered hardness, and improved physical properties. Previously, Zetzl et al. (2012) studied the mechanical properties of frankfurters prepared with oleogels based on vegetable oils (canola, soybean, and flaxseed) (90%) and ethylcellulose (10%). Texture profile analysis indicated no significant difference between the chewiness or hardness of cooked frankfurters made with oleogels and beef fat and concluded that ethylcellulose oleogels could be used to replace saturated fats and provide desired textural properties in a variety of food products.

2.5.3 As a Protein Supplement and Functional Ingredient

The applications of canola/rapeseed meal as protein supplement are very limited to animal feed due to the presence of non-protein components such as phytic acid, glucosinolates, erucic acid, phenolic compounds, etc., which, apart from being toxic (as discussed previously), have been reported to reduce the digestibility and bioavailability of macro- and micronutrients in the human gut. Canola protein is reported to contain well-balanced amino acids profile and comparable to milk and egg proteins in terms of quality (Bos et al. 2007; Fleddermann et al. 2013). Therefore, efforts were made to reduce the content of non-protein components from canola meal and to produce canola protein isolates (CPI) and concentrates (CPC) with little amount of glucosinolates (<0.1 μmol/g), erucic acid (<0.005%), phytic acid (<0.14%), and phenolic components (0.07%) (GRAS 2016). CPC and CPI are expected to provide several desired techno-functional properties such as thickener, superabsorbent hydrogels (Shi et al. 2014), oleogelators, emulsifier (Wu and Muir 2008), gelling agent (Pinterits and Arntfield 2008), foaming agent, water binder, stabilizer (Wu and Muir 2008), texturizer, antioxidants (Cumby et al. 2008), bioactive components (Wu and Muir 2008), etc. in a variety of food applications (GRAS 2010; GRAS 2016; Von Der Haar et al. 2014; Aider and Barbana 2011; Wanasundara et al. 2016). A few studies on the addition of CPC and CPI and the effect on physico-chemical and sensory properties of the final food products are shown in Table 2.9.

2.6 Future Scopes

Canola oil contains high amount of MUFA, balanced ratio of omega-3 and omega-6 fatty acids and sufficient quantity of bioactive components. A wide range of literature is available on the health-promoting effects and food applications of canola oil and meal. However, more research is required on the potential applications of minor/bioactive components of canola oil and protein. In addition to high amount of quality protein, canola meal, and protein concentrates have the potential to be used to as an emulsifier, gelling agents, absorbent, stabilizer, thickener, oleogelator, texturizer, antioxidant, etc. However, long-term human studies are required to claim the health benefits of canola protein.

Table 2.9 Food applications of canola/rapeseed protein

Food product	Form of canola/rapeseed protein added	Major findings	References
Bread	Canola and pumpkin protein concentrate and isolate (individually and blends)	Canola and pumpkin protein concentrates could be supplemented in wheat flour up to 18% without detrimental effects on dough and loaf quality. Improved protein (11–38%), lysine (90–200%), and mineral content (70–135%) in supplemented bread as compared to control	Mansour et al. (1999)
Bologna type sausage	Rapeseed and pumpkin seed (protein concentrates/isolates)	No appreciable effect on viscosity, pH, crude protein, fat content or amino acids composition. Cooking yield and water-holding capacity significantly improved, while sensory scores significantly decreased in case of rapeseed added sausages than that of control	Mansour et al. (1996)
Weiner (sausage)	Rapeseed protein concentrate	Rapeseed protein improved emulsion stability, firmness, and protein content and decreased cooking yield, water absorption, sensory acceptability of weiners as compared to control	Thompson et al. (1982)
Meat patties	Rapeseed protein concentrate	Rapeseed protein improved protein content, cooking yield, moisture and fat retention, and firmness and decreased percent shrinkage and flavor and overall acceptability in patties when compared to control	Thompson et al. (1982)
Sausages	Rapeseed protein concentrate	Instead of casein, when rapeseed protein concentrate (steamed) was added in sausages, sensory analysis showed improved taste, good texture, and characteristic aroma	Yoshie-Stark et al. (2006)
Food products (dairy, meat, bakery, beverages, nutritional bars, etc.))	Rapeseed/canola protein isolates	Cruciferin-rich canola/rapeseed protein isolate (≥80% cruciferin) under trade name of "Puratein" and the napin-rich canola/rapeseed protein isolate (≥80% napin) under the trade name of "Supertein" could be added up to 60%, 5%, 2%, 10%, 95%, and 50% in powered egg/egg substitutes, dairy products, processed meat and grain products, fruit and vegetable beverages, protein supplement powders, and nutritional bars, respectively	GRAS (2010)

References

Aider M, Barbana C (2011) Canola proteins: composition, extraction, functional properties, bioactivity, applications as a food ingredient and allergenicity–a practical and critical review. Trends Food Sci Technol 22(1):21–39

Assadi E, Janmohammadi H, Taghizadeh A, Alijani S (2011) Nutrient composition of different varieties of full-fat canola seed and nitrogen-corrected true metabolizable energy of full-fat canola seed with or without enzyme addition and thermal processing. J Appl Poult Res 20 (1):95–101

Baum SJ, Kris-Etherton PM, Willett WC, Lichtenstein AH, Rudel LL, Maki KC, Whelan J, Ramsden CE, Block RC (2012) Fatty acids in cardiovascular health and disease: a comprehensive update. J Clin Lipidol 6(3):216–234

Bell JM (1993) Factors affecting the nutritional value of canola meal: a review. Can J Anim Sci 73:679–697

Beyzi E, Gunes A, Beyzi SB, Konca Y (2019) Changes in fatty acid and mineral composition of rapeseed (Brassica napus ssp. oleifera L.) oil with seed sizes. Ind Crop Prod 129:10–14

Bierenbaum ML, Reichstein RP, Watkins TR, Maginnis WP, Geller M (1991) Effects of canola oil on serum lipids in humans. J Am Coll Nutr 10(3):228–233

Boot A (2016) The safety and the generally recognized as safe (GRAS) status of the proposed uses of canola /rapeseed protein isolate in human food. https://www.fda.gov/media/103212/download. Accessed 22 Feb 2019

Bos C, Airinei G, Mariotti F, Benamouzig R, Bérot S, Evrard J, Fénart E, Tomé D, Gaudichon C (2007) The poor digestibility of rapeseed protein is balanced by its very high metabolic utilization in humans. J Nutr 137(3):594–600

Bozzetto L, Prinster A, Annuzzi G, Costagliola L, Mangione A, Vitelli A, Mazzarella R, Longobardo M, Mancini M, Vigorito C, Riccardi G (2012) Liver fat is reduced by an isoenergetic MUFA diet in a controlled randomized study in type 2 diabetic patients. Diabetes Care 35(7):1429–1435

Cakmakli Ü, Ünal MK (1988) Composition and mineral distribution of rapeseed varieties tested for adaptation in Turkey. Lipid/Fett 90(10):386–389

Campos H, Dreon DM, Krauss RM (1996) Associations of hepatic and lipoprotein lipase activities with changes in dietary composition and low-density lipoprotein subclasses. J Lipid Res 36:462–472

Canola Council of Canada (2015) Canola meal feeding guide. https://wwwcanolacouncilorg/media/516716/2015_canola_meal_feed_industry_guidepdf. Accessed 19 July 2019

Carré P, Citeau M, Robin G, Estorges M (2016) Hull content and chemical composition of whole seeds, hulls and germs in cultivars of rapeseed (Brassica napus). OCL 23(3):1–8. https://doi.org/10.1051/ocl/2016013

Cavins JF, Kwolek WF, Inglett GE, Cowan JC (1972) Amino acid analysis of soybean meal: interlaboratory study. J Assoc Off Anal Chem 55(4):686–691

Chaves KF, Barrera-Arellano D, Ribeiro AP (2018) Potential application of lipid organogels for food industry. Food Res Int 105:863–872

Chisholm A, Mc Auley K, Mann J, Williams S, Skeaff M (2005) Cholesterol lowering effects of nuts compared with a canola oil enriched cereal of similar fat composition. Nutr Metab Cardiovasc Dis 15(4):284–292

Cumby N, Zhong Y, Naczk M, Shahidi F (2008) Antioxidant activity and water-holding capacity of canola protein hydrolysates. Food Chem 109(1):144–148

Daun JK, Eskin MN, Hickling D (eds) (2015) Canola: chemistry, production, processing, and utilization. AOCS Press, Urabana, IL

De Smet E, Mensink RP, Plat J (2012) Effects of plant sterols and stanols on intestinal cholesterol metabolism: suggested mechanisms from past to present. Mol Nutr Food Res 56:1058–1072

Delisle J, Amiot J, Goulet G, Simard C, Brisson GJ, Jones JD (1984) Nutritive value of protein fractions extracted from soybean, rapeseed, and wheat flours in the rat. Qual Plant Plant Foods Hum Nutr 34:243–251

Dersjant-Li Y, Dusel G (2019) Increasing the dosing of a Buttiauxella phytase improves phytate degradation, mineral, energy, and amino acid digestibility in weaned pigs fed a complex diet based on wheat, corn, soybean meal, barley, and rapeseed meal. J Anim Sci 97(6):2524–2533

FAO (2019) FAOSTAT. http://www.fao.org/faostat/en/#data/QC. Accessed 26 June 2019

FAO/WHO/UNU (1985) Energy and protein requirements. Report of a joint FAO/WHO/UNU meeting. WHO, Geneva, Technical report series No. 724

Ferguson JJ, Stojanovski E, MacDonald-Wicks L, Garg ML (2016) Fat type in phytosterol products influence their cholesterol-lowering potential: a systematic review and meta-analysis of RCTs. Prog Lipid Res 64:16–29

Fernandez ML, West KL (2005) Mechanisms by which dietary fatty acids modulate plasma lipids. J Nutr 135(9):2075–2078

Fleddermann M, Fechner A, Rößler A et al (2013) Nutritional evaluation of rapeseed protein compared to soy protein for quality, plasma amino acids, and nitrogen balance—a randomized crossover intervention study in humans. Clin Nutr 32:519–526

Fumeron F, Bard JM, Lecerf JM (2017) Interindividual variability in the cholesterol-lowering effect of supplementation with plant sterols or stanols. Nutr Rev 75(2):134–145

GAIN (2018) India: oilseeds and products annual. GAIN report number: IN8040. https://apps.fas.usda.gov/newgainapi/api/report/downloadreportbyfilename?filename=Oilseeds%20and%20Products%20Annual_New%20Delhi_India_4-13-2018.pdf. Accessed 26 June 2019

Garcia-Palmieri MR, Sorlie P, Tillotson J, Costas R Jr, Cordero E, Rodriguez M (1980) Relationship of dietary intake to subsequent coronary heart disease incidence: the Puerto Rico heart health program. Am J Clin Nutr 33:1818–1827

Garg A (1998) High-monounsaturated-fat diets for patients with diabetes mellitus: a meta-analysis. Am J Clin Nutr 67(suppl):577S–582S

Gertz C, Kochhar P (2001) A new method to determine oxidative stability of vegetable fats and oils at simulated frying temperature. Oilseeds Fats Corps Lipids 8:82–85

Ghazani SM, Marangoni AG (2013) Minor components in canola oil and effects of refining on these constituents: a review. J Am Oil Chem Soc 90(7):923–932

Ghodsvali A, Khodaparast MH, Vosoughi M, Diosady LL (2005) Preparation of canola protein materials using membrane technology and evaluation of meals functional properties. Food Res Int 38(2):223–231

Gorissen SH, Crombag JJ, Senden JM, Waterval WH, Bierau J, Verdijk LB, van Loon LJ (2018) Protein content and amino acid composition of commercially available plant-based protein isolates. Amino Acids 50(12):1685–1695

Grala W, Verstegen MWA, Van Leeuwen P, Huisman J, Jansman AJM, Tamminga S (1997) Nitrogen balance of pigs as affected by feedstuffs causing different endogenous nitrogen flow at the terminal ileum. Livest Prod Sci 48:143–155

GRAS (2010) Cruciferin-rich canola/rapeseed protein isolate and napin-rich canola/rapeseed protein isolate. http://wayback.archive-it.org/7993/20171031043650/https://www.fda.gov/downloads/Food/IngredientsPackagingLabeling/GRAS/NoticeInventory/UCM269508.pdf.Accessed 10 Oct 2019

GRAS (2016) The safety and the generally recognized as safe (GRAS) status of the proposed uses of canola /rapeseed protein isolate in human food. https://www.fda.gov/downloads/Food/IngredientsPackagingLabeling/GRAS/NoticeInventory/ucm542217.pdf. Accessed 10 Oct 2019

Gulesserian T, Widhalm K (2002) Effect of a rapeseed oil substituting diet on serum lipids and lipoproteins in children and adolescents with familial hypercholesterolemia. J Am Coll Nutr 21 (2):103–108

Gunstone FD (ed) (2004) Rapeseed and canola oil: production, processing, properties and uses. CRC Press, Boca Raton, FL

Gupta RK, Gangoliya SS, Singh NK (2015) Reduction of phytic acid and enhancement of bioavailable micronutrients in food grains. J Food Sci Technol 52(2):676–684

Gylling H, Puska P, Vartiainen E, Miettinen TA (1999) Serum sterols during stanol ester feeding in a mildly hypercholesterolemic population. J Lipid Res 40(4):593–600

Hammad S, Pu S, Jones PJ (2016) Current evidence supporting the link between dietary fatty acids and cardiovascular disease. Lipids 51(5):507–517

Hu FB, Stampfer MJ, Manson JE, Rimm E, Colditz GA, Rosner BA et al (1997) Dietary fat intake and the risk of coronary heart disease in women. New Engl J Med 337:1491–1499

Hughes NE, Marangoni AG, Wright AJ, Rogers MA, Rush JWE (2009) Potential applications of edible oil organogels. Trends Food Sci Tech 20:470–480

Ikeda I, Tanabe Y, Sugano M (1989) Effects of sitosterol and sitostanol on micellar solubility of cholesterol. J Nutr Sci Vitaminol 35:361–369

Jang A, Bae W, Hwang HS, Lee HG, Lee S (2015) Evaluation of canola oil oleogels with candelilla wax as an alternative to shortening in baked goods. Food Chem 187:525–529

Khajali F, Slominski BA (2012) Factors that affect the nutritive value of canola meal for poultry. Poult Sci 91:2564–2575

Khattab R, Goldberg E, Lin L, Thiyam U (2010) Quantitative analysis and free-radical-scavenging activity of chlorophyll, phytic acid, and condensed tannins in canola. Food Chem 122 (4):1266–1272

Kien CL, Bunn JY, Stevens R, Bain J, Ikayeva O, Crain K, Koves TR, Muoio DM (2014) Dietary intake of palmitate and oleate has broad impact on systemic and tissue lipid profiles in humans. Am J Clin Nutr 99(3):436–445

Konkol D, Szmigiel I, Domżał-Kędzia M, Kułażyński M, Krasowska A, Opaliński S, Korczyński M, Łukaszewicz M (2019) Biotransformation of rapeseed meal leading to production of polymers, biosurfactants, and fodder. Bioorg Chem 93:102,865. https://doi.org/10.1016/j.bioorg.2019.03.039

Kris-Etherton PM (1999) Monounsaturated fatty acids and risk of cardiovascular disease. Circulation 100(11):1253–1258

Kromhout D, Coulander CDL (1984) Diet, prevalence and 10-year mortality from coronary heart disease in 871 middle-aged men: the Zutphen study. Am J Epidemiol 119:733–741

Kromhout D, Menotti A, Bloemberg B, Aravanis C, Blackburn H, Buzina R, Dontas AS, Fidanza F, Giampaoli S, Jansen A, Karvonen M, Katan M, Nissinen A, Nedeljkovic S, Pekkanen G, Pekkarinen M, Punsar S, Rasanen L, Simic B, Toshima H (1995) Dietary saturated and trans fatty acids and cholesterol and 25-year mortality from coronary heart disease: the seven countries study. Prev Med 24:308–315

Kruse M, von Loeffelholz C, Hoffmann D, Pohlmann A, Seltmann AC, Osterhoff M, Hornemann S, Pivovarova O, Rohn S, Jahreis G, Pfeiffer AF (2015) Dietary rapeseed/canola-oil supplementation reduces serum lipids and liver enzymes and alters postprandial inflammatory responses in adipose tissue compared to olive-oil supplementation in obese men. Mol Nutr Food Res 59 (3):507–519

Kumar V, Sinha AK, Makkar HP, De Boeck G, Becker K (2012) Phytate and phytase in fish nutrition. J Anim Physiol Anim Nutr 96(3):335–364

Liang D (2000) Effect of enzyme supplementation on the nutritive value of canola meal for broiler chickens. M.Sc. dissertation, Faculty of graduate studies, University of Manitoba. http://www.collectionscanada.gc.ca/obj/s4/f2/dsk1/tape2/PQDD_0012/MQ53178.pdf. 21 May 2014

Lim J, Hwang HS, Lee S (2017) Oil-structuring characterization of natural waxes in canola oil oleogels: rheological, thermal, and oxidative properties. Appl Biol Chem 60(1):17–22

Lopez-Segura F, Velasco F, Lopez-Miranda J, Castro P, Lopez-Pedrera R, Blanco A, Jimenez-Pereperez J, Torres A, Trujillo J, Ordovas JM, Perez-Jimenez F (1996) Monounsaturated fatty acid-enriched diet decreases plasma plasminogen activator inhibitor type 1. Arterioscler Thromb Vasc Biol 16:82–88

Malgorzata NK, Aleksander S (2010) Changes of phenolic content in rapeseed during preliminary drying. J Oilseed Brassica 1(1):33–38

Mansour EH, Dworschak E, Huszka T, Hovari J, Gergely A (1996) Utilization of pumpkin seed and rapeseed proteins in the preparation of bologna type sausages. Acta Aliment 25(1):25–36

Mansour EH, Dworschak E, Pollhamer ZS, Gergely A, Hovari J (1999) Pumpkin and canola seed proteins and bread quality. Acta Aliment 28(1):59–70

Matthaus B, Özcan MM, Al JF (2016) Some rape/canola seed oils: fatty acid composition and tocopherols. Z Naturforsch C J Biosci 71(3–4):73–77

McDonald BE, Gerrard JM, Bruce VM, Corner EJ (1989) Comparison of the effect of canola oil and sunflower oil on plasma lipids and lipoproteins and on in vivo thromboxane A2 and prostacyclin production in healthy young men. Am J Clin Nutr 50:1382–1388

McGee DL, Reed DM, Yano K, Kagan A, Tillotson J (1984) Ten-year incidence of coronary heart disease in the Honolulu heart program. Am J Epidemiol 119:667–676

McNamara DJ (1992) Dietary fatty acids, lipoproteins, and cardiovascular disease. Adv Food Nutr Res 36:253–351

Mennella I, Savarese M, Ferracane R, Sacchi R, Vitaglione P (2015) Oleic acid content of a meal promotes oleoylethanolamide response and reduces subsequent energy intake in humans. Food Funct 6(1):204–210

Mert B, Demirkesen I (2016) Reducing saturated fat with oleogel/shortening blends in a baked product. Food Chem 199:809–816

Montalto MB, Bensadoun A (1993) Lipoprotein lipase synthesis and secretion: effects of concentration and type of fatty acids in adipocyte cell culture. J Lipid Res 34:397–407

Mozaffarian D, Clarke R (2009) Quantitative effects on cardiovascular risk factors and coronary heart disease risk of replacing partially hydrogenated vegetable oils with other fats and oils. Eur J Clin Nutr 63(S2):S22

Naczk M, Amarowicz R, Pink D, Shahidi F (2000) Insoluble condensed tannins of canola/rapeseed. J Agric Food Chem 48(5):1758–1762

National Research Council (1994) Nutrient requirements of poultry, 9th edn. The National Academies Press, Washington, DC. https://doi.org/10.17226/2114

Neish AC (1960) Biosynthetic pathways of aromatic compounds. Annu Rev Plant Physiol 11 (1):55–80

Ohlson R, Anjou K (1979) Rapeseed protein products. J Am Oil Chem Soc 56(3Part3):431–437

Orsavova J, Misurcova L, Ambrozova J, Vicha R, Mlcek J (2015) Fatty acids composition of vegetable oils and its contribution to dietary energy intake and dependence of cardiovascular mortality on dietary intake of fatty acids. Int J Mol Sci 16(6):12871–12890

Paredes-Lopez O, Ordorica-Falomir C (1986) Production of safflower protein isolates: composition, yield and protein quality. J Sci Food Agric 37(11):1097–1103

Pehlivanoğlu H, Demirci M, Toker OS, Konar N, Karasu S, Sagdic O (2018) Oleogels, a promising structured oil for decreasing saturated fatty acid concentrations: production and food-based applications. Crit Rev Food Sci Nutr 58(8):1330–1341

Pinterits A, Arntfield SD (2008) Improvement of canola protein gelation properties through enzymatic modification with transglutaminase. LWT Food Sci Technol 41(1):128–138

Pu S, Rodríguez-Pérez C, Ramprasath V, Segura-Carretero A (2015) Dietary high oleic canola oil supplemented with DHA oil attenuates plasma PCSK9 levels in humans. FASEB 29 (1 Suppl):743.12S

Qian F, Korat AA, Malik V, Hu FB (2016) Metabolic effects of monounsaturated fatty acid–enriched diets compared with carbohydrate or polyunsaturated fatty acid–enriched diets in patients with type 2 diabetes: a systematic review and meta-analysis of randomized controlled trials. Diabetes Care 39(8):1448–1457

Rekas A, Wroniak M, Szterk A (2016) Characterization of some quality properties and chemical composition of cold-pressed oils obtained from different rapeseed varieties cultivated in Poland. Polish J Nat Sci 31(2):249–261

Robinson RG (1975) Amino acid and elemental composition of sunflower and pumpkin seeds 1. Agron J 67(4):541–544

Rogers MA, Wrightb AJ, Marangoni AG (2009) Oil organogels: the fat of the future? Soft Matter 5 (8):1594–1596. https://doi.org/10.1039/B822008P
Rudel LL, Kelley K, Sawyer JK, Shah R, Wilson MD (1998) Dietary monounsaturated fatty acids promote aortic atherosclerosis in LDL receptor–null, human apoB100–overexpressing transgenic mice. Arterioscler Thromb Vasc Biol 18(11):1818–1827
Rudkowska I, Roynette CE, Nakhasi DK, Jones PJ (2006) Phytosterols mixed with medium-chain triglycerides and high-oleic canola oil decrease plasma lipids in overweight men. Metabolism 55 (3):391–395
Ruiz-Gutierrez V, Morgado N, Prada JL, Perez-Jimenez F, Mutiana FJG (1998) Composition of human VLDL triacylglycerols after ingestion of olive oil and high oleic sunflower oil. J Nutr 128:570–576
Sanders TA (2009) Fat and fatty acid intake and metabolic effects in the human body. Ann Nutr Metab 55(1–3):162–172
Schoonjans K, Staels B, Auwerx J (1996) Role of the peroxisome proliferatoractivated receptor (PPAR) in mediating the effects of fibrates and fatty acids on gene expression. J Lipid Res 37:907–925
Shahidi F (ed) (1990) Canola and rapeseed: production, chemistry, nutrition, and processing technology. Springer Science & Business Media, Boston, MA, pp 80–85
Shahidi F (2002) Phytochemicals in oilseeds. In: Meskin MS, Bidlack WR, Davies AJ, Omaye ST (eds) Phytochemicals in nutrition and health. CRC Press, Boca Raton, FL, pp 155–172
Shi W, Dumont MJ, Ly EB (2014) Synthesis and properties of canola protein-based superabsorbent hydrogels. Eur Polym J 54:172–180
Siddiqui IR, Wood PJ (1972) Structural investigation of water-soluble rapeseed (B. campestris) polysaccharides. Part n. an acidic arabinogalactan. Carbohydr Res 24:1–9
Siddiqui IR, Wood PJ (1974) Structural investigation of oxalate-soluble rapeseed (B. campestris) polysaccharides. Part m. an arabinan. Carbohydr Res 36:35–44
Siddiqui IR, Wood PJ (1977) Carbohydrates of rapeseed: a review. J Sci Food Agric 28(6):530–538
Siger A, Kaczmarek A, Rudzińska M (2015) Antioxidant activity and phytochemical content of cold-pressed rapeseed oil obtained from roasted seeds. Eur J Lipid Sci Technol 117 (8):1225–1237
Sirtori CR, Tremoli E, Gatti E, Montanari G, Sirtori M, Colli S, Gianfranceschi G, Maderna P, Dentone CZ, Testolin G, Galli C (1986) Controlled evaluation of fat intake in the Mediterranean diet: comparative activities of olive oil and corn oil on plasma lipids and platelets in high-risk patients. Am J Clin Nutr 44:635–642
Sosulski FW, Sarwar G (1973) Amino acid composition of oilseed meals and protein isolates. Can Inst Food Sci Technol J 6(1):1–5
Statista (2019) Consumption of vegetable oils worldwide from 2013/14 to 2018/2019, by oil type (in million metric tons). https://www.statista.com/statistics/263937/vegetable-oils-global-consumption/. Accessed 26 June 2019
Summers JD, Leeson S (1985) Mineral profile of canola and soybean meal. Can J Anim Sci 65 (4):913–919
Tanwar B, Modgil R, Goyal A (2018) Antinutritional factors and hypocholesterolemic effect of wild apricot kernel (Prunus armeniaca L.) as affected by detoxification. Food Funct 9 (4):2121–2135
Tanwar B, Modgil R, Goyal A (2019) Effect of detoxification on biological quality of wild apricot (Prunus armeniaca L.) kernel. J Sci Food Agric 99(2):517–528
Thompson LU (1990) Phytates in canola/rapeseed. In: Shahidi F (ed) Canola and rapeseed. Springer, Boston, MA, pp 173–192
Thompson LU, Liu RF, Jones JD (1982) Functional properties and food applications of rapeseed protein concentrate. J Food Sci 47(4):1175–1180
Venturini D, Simão AN, Urbano MR, Dichi I (2015) Effects of extra virgin olive oil and fish oil on lipid profile and oxidative stress in patients with metabolic syndrome. Nutrition 31(6):834–840

Von Der Haar D, Müller K, Bader-Mittermaier S, Eisner P (2014) Rapeseed proteins–production methods and possible application ranges. OCL 21(1):D104

Wanasundara JP, McIntosh TC, Perera SP, Withana-Gamage TS, Mitra P (2016) Canola/rapeseed protein-functionality and nutrition. OCL 23(4):D407

Wickramasuriya SS, Yi YJ, Yoo J, Kang NK, Heo JM (2015) A review of canola meal as an alternative feed ingredient for ducks. J Anim Sci Technol 57(1):29

Wu J, Muir AD (2008) Comparative structural, emulsifying, and biological properties of 2 major canola proteins, cruciferin and napin. J Food Sci 73(3):C210–C216

Yoshie-Stark Y, Wada Y, Schott M, Wäsche A (2006) Functional and bioactive properties of rapeseed protein concentrates and sensory analysis of food application with rapeseed protein concentrates. LWT Food Sci Technol 39(5):503–512

Zetzl AK, Marangoni AG, Barbut S (2012) Mechanical properties of ethylcellulose oleogels and their potential for saturated fat reduction in frankfurters. Food Funct 3(3):327–337

Zulim Botega DC, Marangoni AG, Smith AK, Goff HD (2013) Development of formulations and processes to incorporate wax oleogels in ice cream. J Food Sci 78(12):C1845. https://doi.org/10.1111/1750-3841.12248

Chapter 3
Cottonseed (*Gossypium hirsutum*)

Manvesh Kumar Sihag, Ami Patel, and Vikas Kumar

Abstract Cotton is both a food and a fiber crop. The cottonseed fibers/linters are mainly used for textile industry, while, as a food, its oil is used worldwide in various food applications such as cooking oil, frying oil, salad dressings, shortenings, and margarines. Unlike other oilseeds/vegetable oils, cottonseed has an antinutrient known as gossypol, which is highly toxic to monogastric animals and causes cardiotoxic and hemolytic effects and interferes with fertility and reproductive functions. Due to the oil refining technology, most of the gossypol goes to cottonseed meal/cake, which, further, is used as a feed ingredient for ruminants. Ruminants are more resistant to gossypol toxicity than monogastric species, because it remains in bound form with proteins and thus unavailable to the body. Cottonseed oil is a rich source of essential fatty acid (linoleic acid, ω-6), which has been reported to show positive effects on cardiovascular health and immunomodulatory functions, if consumed at recommended levels.

Keywords Cottonseed · Gossypol · Antinutrients · Phytonutrients · Cardiovascular diseases · Salad dressing

M. K. Sihag (✉)
Department of Dairy Chemistry, College of Dairy Science & Technology, Guru Angad Dev Veterinary and Animal Sciences University, Ludhiana, Punjab, India

A. Patel
Dairy Microbiology Department, Mansinhbhai Institute of Dairy & Food Technology (MIDFT), Mehsana, Gujarat, India
e-mail: ami@midft.com

V. Kumar
Department of Food Science and Technology, College of Agriculture, Punjab Agricultural University, Ludhiana, Punjab, India

B. Tanwar, A. Goyal (eds.), *Oilseeds: Health Attributes and Food Applications*,
https://doi.org/10.1007/978-981-15-4194-0_3

3.1 Origin and History

Cotton, the single and largest natural source of fiber, is an important crop of global significance. Cotton plant (*Gossypium hirsutum*) is an herb, grown annually and classified under *Gossypium* genus of Malvaceae family. The uniqueness of the cottonseed plant is that it is grown for both the purpose of seed (for both food and feed) and lint (as fiber in textile industry). This is not known exactly from where the history of cotton agriculture started. However, the plant is native of tropical and subtropical America, Carabbia, and some Pacific Islands. Cotton was brought to Europe in 800 A.D. by Arab merchants. The plant height varies from approximately 1.5 to 2.0 m. Moderate rainfall, adequate sunshine, and soil with moisture holding capability are ideal conditions for optimum growth of the plant. The plant produces cup-shaped flowers along with big, fleshy petals and a purple or reddish base, which after pollination gets converted into the shape of a leathery capsule called boll. These bolls are usually 4–6 cm long and beaked at the tip. They may contain either three or five compartments. Each compartment may contain up to 11 hairy and fuzzy cottonseeds covered with long wooly hair-like fiber. The seeds are removed from these fibers by the process known as "ginning." The oil can be extracted from fiber-free cottonseeds either through mechanical extraction or direct solvent extraction process or pre-press solvent extraction method. This cottonseed oil has wide applications in food, cosmetic, and pharmaceutical industry. The dry cake left behind after pressing is known as cottonseed meal (CSM). This meal is rich in protein, therefore, used as a feed for ruminants in countries where cotton is a regular crop such as India, China, and the USA. Additionally, CSM is also used for production of potential value-added products such as wood adhesives (Cheng et al. 2013), bioplastics (Yue et al. 2012), films, super adsorbent hydrogels (Zhang et al. 2010), antioxidant meal hydrolysates (Gao et al. 2010), and bio oil/diesel (Singh et al. 2014).

3.2 Production

As per FAOSTAT, total production of cottonseed and cotton lint worldwide in 2014 was 4698.80 and 2615.67 MMT, respectively. China is the leading producer of cottonseed with total production of 1232 MMT in 2014, followed by India (1230 MMT), the USA (464.93 MMT), and Pakistan (444.27 MMT). The cottonseed production (1965–2014) of the major producing countries is shown in Fig. 3.1.

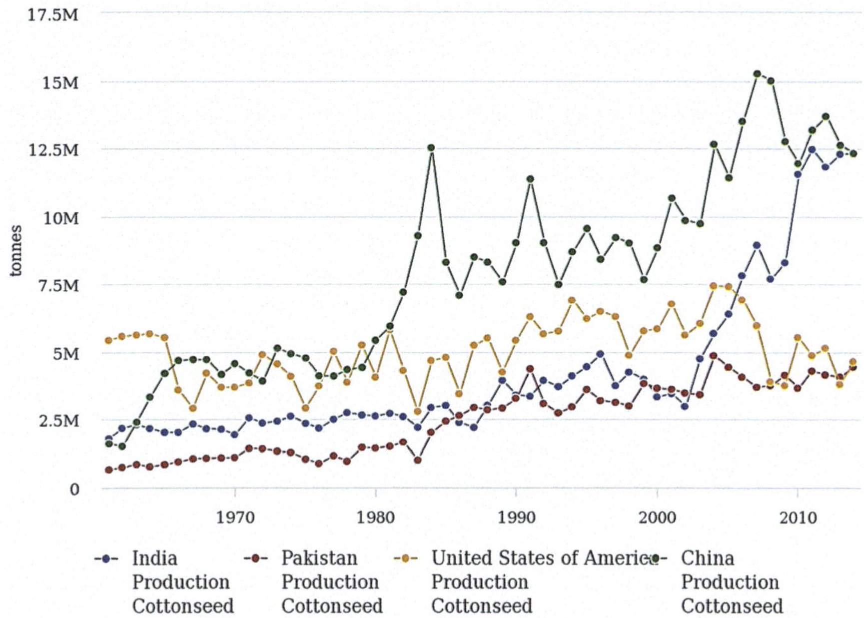

Fig. 3.1 Countries leading in cottonseed production (FAOSTAT 2018)

3.3 Chemical Composition

Basically, the cottonseed consists of two parts: hull and kernel. Fiber or linters used in textile industry are obtained from hull part, while the kernel part is composed of carbohydrates, protein, oil, dietary fibers, vitamin, mineral, etc. The whole cottonseed can be separated through crushing into 16% oil, 26% hulls, 45.5% meal, and 8.5% linters (O'Brien et al. 2005).

3.3.1 Oil

Cottonseed oil is light golden in color and possesses a mild but pleasant taste. The color degrades with the degree of refining. The oil is also known as "heart oil" as it is one of the most unsaturated vegetable oils available in the market. It is interesting to note that cottonseed oil also contains highest amount of saturated fatty acids (25%) among all vegetable oils; therefore, it is either not hydrogenated or only partially hydrogenated, where improved techno-functionality is required. On the other hand, cottonseed oil also has large amount of linoleic acid (PUFA) (50–55%). Consequently, some workers have reported lower oxidative stability of cottonseed oil as compared to other vegetable oils, especially during the heat processing treatments such as deep frying (Dowd et al. 2010). This results in production of off-flavors and reduces the shelf life of the oil. Therefore, sometimes, partial hydrogenation is advised to avoid this situation, which also often leads to production of *trans*-fatty acids (Sacks and Katan 2002).

In terms of fatty acids, oleic and linoleic are the dominant unsaturated fatty acids contributing 18–24% and 42–52% of the total share of unsaturated fatty acids, respectively, while palmitic and stearic acids are the major saturated fatty acids (26–35%) present in cottonseed oil. It is cholesterol-free and quite high in tocopherols content, ascertaining both its nutritional content and natural antioxidant properties. During frying, these antioxidants are transferred to the food fried, imparting them a comparatively longer shelf life and preserving their freshness. Average tocopherol content of cottonseed oil is 65 mg/100 g of oil, which is further equivalent to 38 mg of α-tocopherol. Cottonseed oil has wide applications in deep frying, baking, margarine, icing, whipped toppings, and salad dressings. Additionally, the refined cottonseed oil also has some non-food applications such as biodiesel source, in paint industry, and as environment friendly lubricant additive to base oils (Durak and Karaosmanoglu 2004), which will be discussed in brief in the last section of this chapter.

3.3.2 Protein

Cottonseed proteins are considered as potential source of nutrition for both animals and plants. Along with topmost fiber crop worldwide, cottonseed is also regarded as the best source for plant proteins after soybean (Ory and Flick 1994). After oil extraction from cottonseed, whatever dry matter remains behind is known as CSM. High concentration of toxic component known as gossypol restricts its use for human consumption. However, the animals are capable of tolerating its presence. Therefore, CSM (by-product of oil industry) is mainly used as an animal feed. However, despite of much compositional variability, CSM is a rich source of protein with 30–50% protein of dry matter (He et al. 2015). As per the findings of Cheng and Hardy (2002), crude protein, fat, and phosphorus content of CSM from four different locations of Southern USA varied from 34.9% to 41.3%, 1.43% to 3.70%, and 9.95% to 1.10%, respectively. Similarly, lysine, methionine, threonine, and tryptophan content of CSM also varies from 1.68% to 1.90%, 0.57% to 0.63%, 1.17% to 1.30%, and 0.42% to 0.47%, respectively. Along with variation in composition, protein digestibility is also an issue of major concern in CSM. The main limiting amino acid in cottonseed protein is lysine. During the process of oil extraction, gossypol binds with epsilon (ε)-amino group of the lysine amino acid, resulting in non-availability of this essential amino acid (Nagalakshmi et al. 2007). Huang et al. 2007 reported 0.72 as the ideal digestibility coefficient for crude protein present in CSM.

3.3.3 Crude Fiber

Crude fiber content ranges from 25.00% to 26.20% in CSM (Zotte et al. 2013; Sahin et al. 2006). Acid detergent fiber (ADF) and neutral detergent fiber (NDF) are the two parameters to judge digestibility of the plant material present in animal's food. Cellulose and lignin present in the plant material is termed as ADF, while NDF also includes hemicellulose. Basically, ADF is used to calculate the energy derived from the feed and could be used by the animal, while NDF is used to calculate the amount of feed that can be hold by animal. As per the findings of Calhoun et al. (1995), whole glanded cottonseed meal contains 38.8% ADF and 47.2% NDF on dry weight basis. He et al. (2015) reported 5.47%, 11.58%, and 0.95% ADF, NDF, and acid detergent lignin (ADL) content, respectively. Water-soluble fractions of the cottonseed meal contained no fiber component indicating their highly insoluble nature. The water-insoluble fraction consisted of cellulose and hemicellulose fractions. In addition to these two fractions, other carbohydrates reported by He et al. (2015) were galactose, arabinose, glucose, xylose, rhamnose, mannose, and fucose, representing 16.6% of the total carbohydrate content on dry weight basis. Out of these seven carbohydrates, glucose and galactose are the main water-soluble components.

3.3.4 Minerals and Vitamins

Cottonseed meal is a rich source of minerals. Thirumalaisamy et al. (2015) have reported that calcium (16%), phosphorus (0.75%), potassium (1.21%), magnesium (0.35%), sodium (0.31%), and sulfur (0.26%) represent the major minerals in CSM, while iron (120 mg/kg), zinc (61 mg/kg), copper (9.88 mg/kg), manganese (58.30 mg/kg), aluminum, and boron represent the minor minerals. The ratio of calcium (0.15–0.25%) and phosphorus (0.95–1.71%) content in CSM is 1:6 (Thirumalaisamy et al. 2015). Sulfur is present and derived from the two sulfur-rich amino acids, viz., methionine and cysteine, while phosphorus may be present in phytate form in association with seed proteins (Han 1988).

Much literature is not available on the vitamin content in cottonseed. However, cottonseed oil is rich source of natural antioxidant tocopherol which possesses vitamin E activity. As per the findings of Wassef et al. (2015), the cottonseed oil contains 65 mg of tocopherol per 100 g oil.

3.4 Antinutrient Content

Cotton plants possess some pigment glands in stems, leaves, seeds, and flower buds. Although these glands are distributed all over the cotton plant, their maximum concentration is found in seed part. These glands are meant for secretion of a toxic

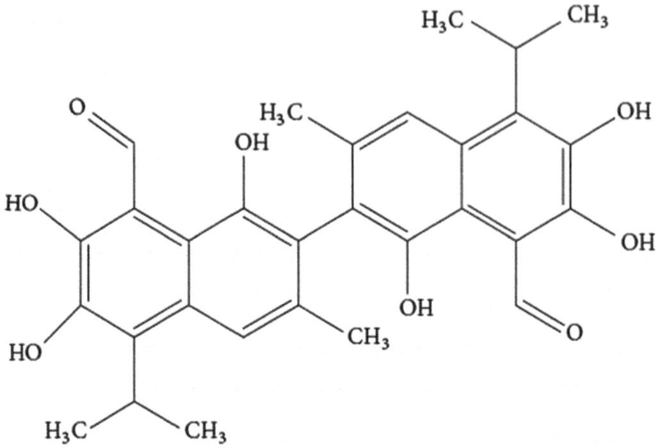

Fig. 3.2 Structure of gossypol

phenolic compound known as gossypol which provides resistance/protection to cotton plant from pests (Soto-Blanco 2008; Rogers et al. 2002; EFSA 2009). The gossypol content of cottonseeds varies from 0.02% to 6.64% among the various cotton varieties (Price et al. 1993). Presence of this compound in cottonseed and CSM hinders their application for human use and limits its applications in animal feed only. Gossypol, a crystalline yellow pigment with molecular weight of 518.55 Dalton (Fig. 3.2), was first isolated from cottonseed in 1899. It is completely soluble in acetone, chloroform, butanone, and ether, partially soluble in crude vegetable oils and insoluble in water and hexane (Gadelha et al. 2014). Its structural formula and scientific name are $C_{30}H_{30}O_8$ and 2,2′-bis(8-formyl-1,6,7 trihydroxy-5-isopropyl-3-methylnaphthalene), respectively. However, other phenolic compounds are also secreted by pigment glands, but due to their low concentration, they have non-significant toxicological importance (Soto-Blanco 2008).

Gossypol exists in two enantiomeric forms in cottonseeds, viz., (−) gossypol and (+) gossypol (Soto-Blanco 2008; Hron et al. 1999; Kakani et al. 2010). The (−) form is more biologically active, more toxic, and more slowly removed form cottonseeds by processing operations compared to latter (Kakani et al. 2010; Bailey et al. 2000). The plant produces both the enantiomeric forms in varying proportion depending upon the genetics of the plant. Further, gossypol also exists in bound and free forms, the later form being more toxic. The bound form comes in existence during the browning or Maillard reaction while heating/processing operations. Gossypol gets covalently bonded with the epsilon (ε) group of basic amino acids such as lysine and arginine. This binding reduces its toxic effects but also results in loss of these essential amino acids (Mahmood et al. 2011).

3.4.1 Adverse Health Effects of Gossypol

During ruminal fermentation, gossypol gets converted into the bound form, leading to its reduced toxicity. Therefore, susceptibility to gossypol poisoning is more in monogastric animals (with single-chambered stomach) such as pigs, rodents, birds, and fish compared to ruminants (Kenar 2006; Alexander et al. 2008; Randel et al. 1992; Zhang et al. 2007). However, few workers have also reported its poisoning in ruminants such as sheep and goats (Morgan et al. 1988; East et al. 1994).

Gossypol forms a complex with iron, leading to inhibition in iron absorption and finally erythrocyte fragility and erythropoiesis (Randel et al. 1996; Lindsey et al. 1980; Zhang et al. 2007; Mena et al. 2004). Several studies have also reported decreased concentration of T4 and T3 hormones in blood of experimental rats fed with gossypol thereby indicating adverse effect on thyroid metabolism (El-Mokadem et al. 2012; Tang and Wong 1984; Rikihisa and Lin 1989; Lin et al. 1990; Udoh et al. 1992). Similarly, hepatotoxic effects of gossypol intake have also been observed in experimental rats, characterized by vacuolation in mitochondria, enlarged endoplasmic reticulum, proliferation of collagen fiber, and expanded perinuclear space in morphology of liver cells (Kakani et al. 2010; Blevins et al. 2010; El-Sharaky et al. 2010; Haschek et al. 1989; Gadelha et al. 2014; Manabe et al. 1991). Adverse effects of gossypol on male fertility through inhibiting spermatogenesis, decreasing sperm counts, specific mitochondrial injuries to the sperm tail cells, damage to germinal epithelium, low sperm volume, interference with the system of utilization of ATP by sperm cells, calcium influx inhibition, reduction in Mg-ATPase and Ca-Mg-ATPase activity in plasmid membranes of spermatozoa, and reducing sperm motility have also been reported and discussed by several workers (Randel et al. 1992; El-Sharaky et al. 2010; Chenoweth et al. 1994; Gu and Anderson 1985; Fornes et al. 1993; Chongthammakun et al. 1986; Yuan and Shi 2000). Gossypol consumption has also been linked with poor immunocompetence of organisms (Braga et al. 2012). Its immunosuppressive behavior is characterized by reduction in levels of leukocytes and lymphocytes, apoptosis induction, decreased $CD4^+$ thymocyte population, and increased $CD8^+$ lymphocyte population (Xu et al. 2009; Quintana et al. 2000).

3.5 Bioactive Components

Cottonseed meal (CSM) is one of the most important by-products of oil industry and is used as a protein supplement in animal feed, for biodiesel production, and as a fertilizer. CSM also contains some bioactive components such as lignin, cellulose, amino acids, proteins, and some polyphenolic and flavonoid components. This bioactivity of these components is due to presence of some functional groups such as carboxyl groups, methyl groups, hydroxyl groups, and fixed ionic moieties (both anionic and cationic). These groups are responsible for therapeutic effects such as anti-microbial, anti-inflammatory, anti-cancer, and antioxidant activities (Henry et al. 2002; Oskoueian et al. 2011b; Namuli et al. 2011; Hendra et al. 2011). Cottonseed oil has reported to possesses antioxidant and anti-inflammatory activities due to the presence of various saturated and unsaturated fatty acids also (Henry et al. 2002). Oskoueian et al. (2011a) reported higher antioxidant activity of CSM due to presence of flavonoid compounds, compared to rapeseed meal. Rutin, naringenin, and kaempferol were the major flavonoids present in CSM. These flavonoids possess anti-inflammatory, antioxidant, free radical scavenging, and anti-hypertensive effects (Xiao et al. 2011). The flavonoid compounds present in CSM can react with the reactive sites of xanthine oxidase resulting in its inhibition (Oskoueian et al. 2011a; Umamaheswari et al. 2009).

Gossypol, the major antinutrient component of cottonseed, has been shown to exhibit anti-cancer activities against breast, colon, pancreatic, and prostate cancer (Zhong et al. 2013; Liu et al. 2002; Chien et al. 2012; Yuan et al. 2013; Thakur et al. 2012; Pang et al. 2011; Huang et al. 2009). The gossypol inhibits DNA synthesis by suppressing expressions of Bcl-2 and Bcl-xL cells of breast cancer cells (Hu et al. 1993; Li et al. 2011) and Bcl-2 and Mcl-1 cells of pancreatic cancer cells (Banerjee et al. 2010). Gossypol has also been reported to show anti-obesity (Zhong et al. 2013; Zhong et al. 2010), anti-inflammatory (Huo et al. 2013; Oskoueian et al. 2011a; b), anti-fungal (Mellon et al. 2012; Puckhaber et al. 2002), and antidepressant activity (Zhang et al. 2001). Cao et al. (2018) isolated the bioactive components form the kernels and coat extracts of glanded and glandless cottonseeds in ethanol for determining the anti-cancer activity in terms of mitochondrial activity of pancreatic and breast cancer cells. As per their findings, both anti-cancer and pro-cancer compounds were present in cottonseeds. Anti-cancer bioactive components were present in cottonseed kernel extract, while the pro-cancer components were present in coat extract of glanded cottonseed.

3.6 Health Attributes

Cottonseed oil contains about 50% essential polyunsaturated fatty acid (PUFA) especially linoleic acid, which is required in human diet as it is not synthesized in the human body. Cottonseed oil is considered as "heart oil" and one of the few oil types which are in the *ok food* list of American Heart Association (AHA). Furthermore, the estimated glycemic index of cottonseed oil is zero (Sekhar and Rao 2011).

Cottonseed oil is high in vitamin E (alpha-tocopherol), which is an antioxidant. Antioxidants present in cottonseed oil work against the free radicals that cause cell damage, aging and plays an important role in cell metabolism. This important vitamin may also help to protect against certain diseases including cancer, cardiovascular disease, diabetes, and Alzheimer's disease (Frey 2018; Sekhar and Rao 2011). Contrary to the other vegetable oils such as flaxseed, chia seed, and canola, it has low amounts of heart-healthy omega-3 and monounsaturated fatty acids. Regular intake of cottonseed oil or its component has been proven to cure a number of health problems. Some of these studies have been discussed in further sections.

3.6.1 In Cardiovascular Diseases

Hypercholesterolemia, in which low-density lipoprotein (LDL)-cholesterol increases, has been associated with enhanced risk for cardiovascular disease (Carson et al. 2005; Roger et al. 2011). Modification of diet, body mass index (BMI), and lifestyle have vital roles to play in strategies for hypercholesterolemia prevention and subsequently to manage cardiovascular disease. The literature suggests a differential role of kinds of cooking oil and fat in diet on total cholesterol and its components including high-density lipoprotein (HDL) and LDL (Carson et al. 2005; Harris et al. 2011; Roger et al. 2011). The effects of a cottonseed oil (CSO)-rich diet have not been evaluated in humans, but animal data indicates that dietary cottonseed oil may lower cholesterol. In context to this, Radcliffe and coworker(s) (Radcliffe et al. 2001; Radcliffe and Czajka-Narins 2006) have reported that replacement of corn oil with cottonseed oil (CSO) for 4 weeks decreased total cholesterol along with HDL-C in rats. In spite of relatively high concentrations of saturated fatty acids compared to corn oil (26% saturated fat compared to 13%) and lower concentrations of mono-unsaturated fats (20% compared to 28%), cottonseed oil has been shown the lipid lowering effects in animal experiments. However, in another study, same authors (Radcliffe et al. 2004) showed that rats fed with CSO had increased adipose tissue levels of saturated fatty acids. Saturated fats are well-known to be associated with atherogenic lipid profiles (Siri-Tarino et al. 2010); hence, these separate findings are somewhat mystifying.

A group of researchers evaluated the effect of consumption of a diet rich in cottonseed oil (95 g/d) in 38 healthy adults (aged 18–40; 12 males, 26 females) normo-cholesterolemic subjects for a week (Davis et al. 2012). It is reported that there was no notable change in weight or waist circumference as well as HDL or triglycerides (TG) levels among participants. Fascinatingly, total cholesterol (pre, 4.39 ± 0.9 mmol/L; post, 4.16 ± 0.8 mmol/L) and LDL (pre, 2.70 ± 0.8 mmol/L; post, 2.47 ± 0.6 mmol/L) were reduced in all participants. Moreover, reduction of total cholesterol was significantly high (pre, 4.34 ± 0.9 mmol/L; post, 4.09 ± 0.8 mmol/L) in female participants. In a recent clinical trial, Polley et al. (2018) observed no changes in blood lipids following olive oil diet; interestingly, a CSO-rich diet led to improvements in cholesterol and TGs. (total cholesterol, 148.40 ± 6.39 to 135.93 ± 6.31 mg/dL; LDL cholesterol, 92.20 ± 5.57 to 78.13 ± 5.60 mg/dL; TG, 80.11 ± 4.91 to 56.37 ± 5.46 mg/dL for pre- to post-diet, respectively; $P < 0.05$). However, the complete mechanism behind the decrease of blood lipids is not known yet. Further, replacement of saturated fat with mono-unsaturated fat reduces the risk for cardiovascular events or cardiovascular death and an increased intake of monounsaturated fat reduces the risk for all-cause mortality and stroke (Schwingshackl and Hoffmann 2012). Presence of unsaturated fatty acids (especially, omega-3 and omega-6) and monounsaturated fatty acids is also believed to reduce the risk of cardiovascular diseases. One of the assumed possibilities is the non-saponifiable portion of the cottonseed oil, which contains tocopherols and beta-sitosterol and may be responsible for modifying the blood lipid response together with overall alteration of the dietary pattern due to the dietary intervention (Davis et al. 2012; Radcliffe et al. 2004).

3.6.2 Risk of Type 2 Diabetes

Type 2 diabetes, the most common form of *diabetes mellitus*, is a disorder of carbohydrate metabolism characterized by impaired ability of the body to produce or respond to insulin hormone and thereby maintain proper sugar (glucose) levels in the blood. Several animal studies triggered the role of cottonseed aqueous extract in the diabetes mellitus; conversely, no study is documented with CSO. In one of the earlier approaches, the aqueous extract of cottonseed was able to reduce blood sugar in alloxan-induced diabetes mellitus in rats where a dose of 1000 mg/kg was found to be an effective dose (Sadique et al. 1987). Cottonseed extract was able to enhance the liver glycogen (such as glibenclamid) and was also able to reduce blood cholesterol which was found raised in the diabetic state. Additionally, it was able to normalize the altered level in the liver lipid peroxide content.

In a separate study, the effect of gland cottonseed dietary fiber (CSDF) containing 86% dietary fiber (mainly cellulose) was studied on serum glucose levels in diabetic rats and in non-insulin-dependent diabetes mellitus (NIDDM) patients (Madar et al. 1988). Diet containing 15% CSDF given to streptozotocin-induced diabetic rats for 30 days tended to diminish the postprandial plasma glucose level curve but had no effect on body weight and serum lipid levels. Twelve NIDDM subjects were given a meal tolerance test (MTT) with or without CSDF before and after daily supplementation of CSDF (16.5 g) in pita twice a day for a month. Incremental glucose levels were significantly ($P < 0.05$) lower at 30, 60, and 180 min after the MTT containing CSDF than in subjects consuming a meal without CSDF. The insulin levels also tended to be lower. The NIDDM subjects tolerated the CSDF well; no flatulence or other side effects were noticed by the researchers (Madar et al. 1988).

3.6.3 Insulin Resistance and Glycemic Control

For the regulation of glucose homeostasis, absorption of glucose from the gut and subsequent glucose uptake into muscles is greatly essential. In one of the recent studies, a group of researchers investigated if gossypol has the potential in managing and preventing diabetes by ameliorating glucose uptake and improving glucose homeostasis (Alam et al. 2018). Further, they also determined the molecular mechanisms underlying those processes in vitro and in vivo. According to the report, gossypol strongly inhibited α-glucosidase on concentration dependent manner by functioning as a competitive inhibitor. Gossypol activated the insulin receptor substrate 1 (IRS-1)/protein kinase B (Akt) signaling pathways and increased uptake of glucose through the translocation of glucose transporter 4 (GLUT4) into plasma membrane in C2C12 myotubes. Consistent with the in vitro study, a higher dose of gossypol (2.5 mg/kg^{-1}) dramatically decreased the postprandial blood glucose levels associated with the upregulated expressions of GLUT4 and the IRS-1/Akt-mediated signaling cascade in skeletal muscle of streptozotocin-induced diabetic mouse. Further, gossypol treatment considerably boosted antioxidant enzyme expression and mitigated gluconeogenesis in the liver (Alam et al. 2018).

3.6.4 Antitumor and Anti-cancer Activity

Gossypol, a complex polyphenol with a highly yellow-colored pigment is found in the small intercellular pigment glands in seeds, roots, stems, and leaves of cotton. As discussed earlier, it is a racemic mixture of two enantiomers, (+) and (−). Gossypol is basically considered as a toxic compound. Conversely, recent investigations have demonstrated that gossypol and related compounds have anti-cancer activities, including breast cancer (Liu et al. 2002; Zhong et al. 2013), colon cancer (Chien et al. 2012), pancreatic cancer (Thakur et al. 2012; Yuan et al. 2013), and prostate cancer (Huang et al. 2009; Pang et al. 2011). These novel findings have generated great interest in biomedical field to understand and investigate potential of gossypol and related compounds. In a recent investigation, Cao et al. (2018) reported the anti-cancer activity of glanded cottonseeds in which they isolated ethanol extracts from cottonseeds and investigated their effects on human cancer cells derived from breast and pancreas followed by MTT assay for cell viability. Ethanol extracts from glanded and glandless cottonseed kernels and gossypol notably decreased breast cancer cell mitochondrial activity. On the other hand, ethanol extract from glanded cottonseed kernel and gossypol also significantly decreased pancreas cancer cell mitochondrial activity. Such fascinating outcomes propose anti-cancer effect of gossypol from cottonseeds.

In a previous study, Ko et al. (2007) showed that gossypol exhibits potent cytotoxic effects via apoptosis induction against human colorectal carcinoma cells. Hence in further approach, they reported that addition of 12-O-tetradecanoylphorbol-13-acetate (TPA) significantly inhibited GOS-induced apoptosis in human colorectal carcinoma HT-29 cells in accordance with inducing COX-2 protein/PGE(2) production (Chien et al. 2012). The results of one of the studies on anti-cancer activity of gossypol showed that (−) gossypol was the major inhibitory component of cancerous breast cell growth. The inhibitory activity of (−) gossypol was associated with the reduction of the cell cycle regulator, cyclin D1, and the induction of the cell proliferation inhibitor, TGFbeta (Liu et al. 2002). Another findings illustrated that (−) gossypol reduced in vitro invasion of both the parental

MAT-LyLu cells and the isolated MLL cells, suggesting that (−) gossypol might serve as a chemotherapeutic and/or chemopreventive agent (Huang et al. 2009).

Zhong et al. (2013) reported that (−) gossypol-enriched cottonseed oil [(−)-GPCSO] inhibited proliferation of pre-adipocytes and downregulated the expression of cyclin D1 and B-cell lymphoma-2 (BCL-2). Further, it significantly decreased adipogenesis and, hence, suggested that (−)-GPCSO has the potential as a food supplement to reduce obesity. Mouse double minute 2 (MDM2) and vascular endothelial growth factor (VEGF) are important molecules involved in tumor progression. An investigation by Pang et al. (2011) showed that (−) gossypol potently inhibits human prostate tumor growth on dose-dependent manner inhibited through modulating VEGF signaling pathway, which further validates its great potential in clinical practice. More recently, Xiong et al. (2017) reported that the anti-cancer effects of gossypol are by dual-targeting MDM2 and VEGF in human breast cancer.

3.7 Adverse Effects and Individual Concerns

Cottonseed oil contains both naturally occurring toxins and pesticide contaminants. Cottonseed oil is generally extracted by using harsh chemical solvents and heat which may alter the chemistry of the oil. Most nutritionists are still uncertain about the long-term implications of these changes. In addition, cottonseed oil contains gossypol, which reacts with protein and reduces the nutritional value. Gossypol remains in the free state in the whole seed and on cooking; a bound form is formed in which gossypol combines with free amino or free carboxyl groups of cottonseed protein. Bound gossypol decreases the nutritive value of protein and availability of lysine, an essential amino acid (Sekhar and Rao 2011). Gossypol also has an inhibitory action on the proteolytic enzymes such as pepsin and trypsin, present in the gastrointestinal tract of birds, which negatively interferes with the protein digestion reducing its digestibility (Ryan et al. 1986). Gossypol can trigger alterations in the hematological and biochemical parameters of birds as reported by Aletor (1989).

Moreover, gossypol is toxic to non-ruminant animals including humans, swine, equines, reptiles, avian, fish, canines, and felines (Coutinho 2002). It has been reported to cause reproductive problems in ruminants like cattle, sheep, and goats (Gadelha et al. 2016). It has been known that long-term consumption of gossypol-containing cottonseed oil contributes to its toxicity resulting in male infertility (Coutinho 2002; Cope 2018). Gossypol decreases spermatogenesis and sperm motility in men. Conversely, in a study done in 2006 at the University of Lecce, Italy showed that gossypol in cottonseed oil is not an effective contraceptive, because if combined with most proteins, gossypol no longer causes infertility (Akinola et al. 2009).

Gossypol was found to induce reproductive problems in females by disrupting estrous cycles, pregnancy, and early embryo development in non-ruminant species (Randel et al. 1992). The probable mechanisms comprise an endocrine effect on the ovary as well as a cytotoxic effect on the uterus or embryo. The female ruminants appear to be relatively insensitive to the antifertility effect of gossypol; nevertheless, in vitro data indicate embryonic development inhibition and ovarian steroidogenesis. The same report also mentioned the negative impacts of gossypol on male reproductive health, such as sperm immotility, reduced sperm counts, and infertility (Coutinho 2002; Cope 2018). Other symptoms of gossypol toxicity include breathing problems and anorexia. Recent study demonstrated that gossypol promotes the degeneration of chicken ovarian follicles in vitro. In laying hens fed with cottonseed, gossypol was found to reduce egg production and weight and also associated with discoloration of the egg yolk and/or albumen (Gadelha et al. 2016). Increased proportion of atretic follicles at all stages of development in cultivated chicken ovaries was observed in the presence of gossypol. Hence, authors concluded that gossypol may affect ovarian follicular viability and maturation, which might interfere with female fertility. Similar results were also obtained by Câmara et al. (2015) while studying the effect of gossypol present in cottonseeds on ovarian follicles in sheep to increase atresia. Significant efforts have been directed to reduce gossypol content in cottonseeds by selecting glandless cotton varieties and genetic engineering of gossypol-free seeds of cotton plants to reduce such adverse health effects.

3.8 Food Applications

The utilization of cottonseed oil can be divided into four broad categories: (1) liquid/cooking oils, (2) shortenings, (3) margarines, and (4) specialty products (Fig. 3.3). Majorly cottonseed oil is used as a liquid oil in the form of cooking oil and salad dressings. Until the World War II, cottonseed oil was the most commonly used oil in the USA. Later on, with the advances of refining technology, the beany flavor of soybean oil was removed, and, thus, soybean oil dominated cottonseed oil probably due to easy and wide availability and better economic value.

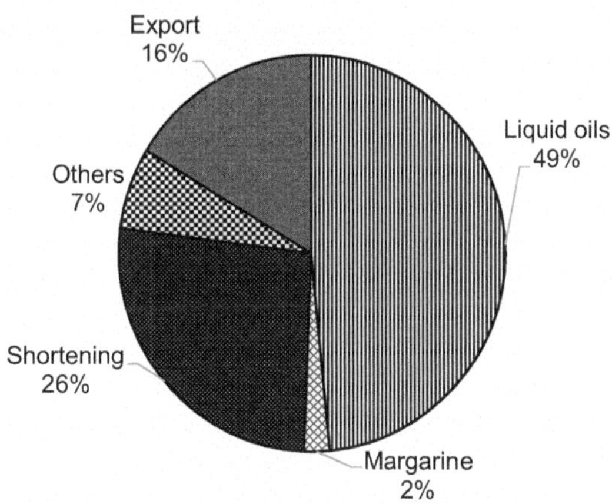

Fig. 3.3 Utilization of cottonseed oil in the USA. (O'Brien et al. 2005)

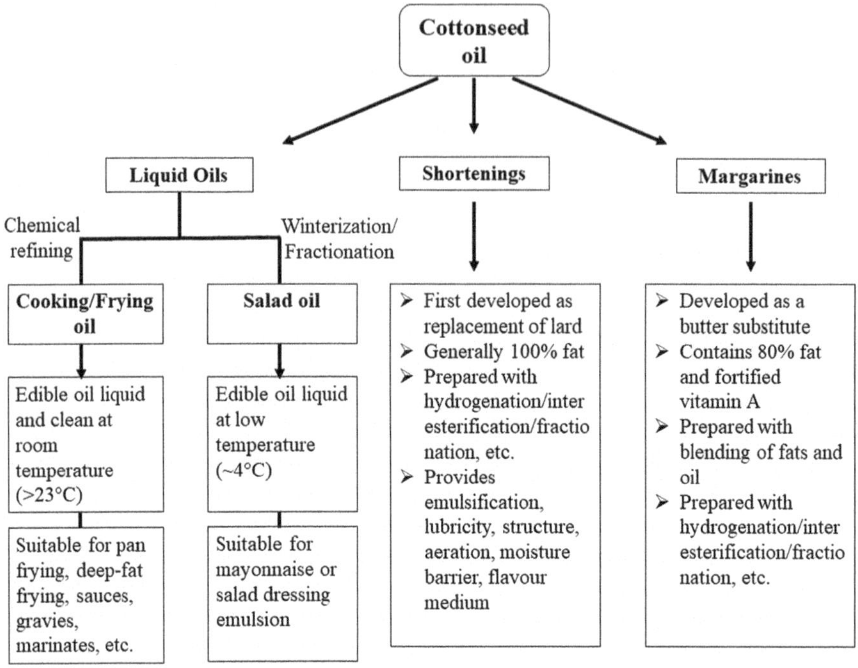

Fig. 3.4 Food applications of cottonseed oil

Cottonseed oil is the only vegetable oil which has the maximum saturated fatty acids content (25–26%) among all the vegetable oils, which means there is no need of hydrogenation to improve the functional properties (such as plasticity, texture, mouthfeel, etc.) of the food products prepared in it. Nowadays, in most cases, cottonseed oil is used as a blend oil mixed with soybean, sunflower, canola, mustard oil, etc. Food applications of cottonseed oil are summarized in Fig. 3.4.

Cottonseed oil is a good source of essential fatty acids and contains approximately 55% linoleic acid. As previously discussed, it has highest amount of saturated fatty acids and, thus, is maximum oxidative stable when heated on high temperatures during processing conditions. Another advantage of high amount of saturated fatty acids in cottonseed oil is that it can be used directly as a blend oil without going through hydrogenation and, thus, product remains free from *trans*-isomers. Unlike other vegetable oils, cottonseed oil still is a preferable oil in snacks industry as it imparts a nutty and pleasant flavor to salted snacks.

In India, majority of cottonseed oil is used as a cooking/frying oil and in the preparation of *Vanaspati ghee* (partially hydrogenated vegetable oil) to provide the consistency, texture, and mouthfeel similar to milk fat (*ghee*).

3.9 Alternative Applications

Cottonseed meal is a rich source of proteins (60–65%) and used as a feed ingredient for ruminants and other animals. In spite of being rich in amino acids and proteins, cottonseed meal is not used as a food ingredient intended for human consumption due to presence of gossypol. As discussed previously, gossypol is a binaphthyl, dialdehyde, and polyhydroxyl pigment which is highly reactive and deleterious to monogastric animals, including humans. According to USFDA, edible cottonseed protein products can have not more than 0.045% "free" gossypol (Gilani and Lee 2003). Cottonseed hulls are used primarily as roughage in cattle and dairy feeds, where they are employed as a bulking agent to reduce digestive disturbances. In India, a small proportion (5–10%) of cottonseed oil is also used in the preparation of soap. Cottonseed linters are used in a variety of applications such as for the manufacturing of furniture's cushions, paper currency, and absorbent cotton products (such as pads). After chemical treatment, cottonseed oil and linters can also be used as bulking agents in ice creams and toothpaste and for the photographic films, toys, flexible pipes, polish, membranes, synthetic plastics and leathers, varnishes, etc. (Gunstone 2001; O'Brien et al. 2005). Recently, cottonseed oil was also used for the preparation of fat replacers/fat substitutes to provide zero calories under the trade name of olestra.

3.10 Conclusion

Cottonseed is a rich source of oil, protein, fibers/linters, and other bioactive components. Cottonseed oil contains highest amount of saturated fatty acids among all the vegetable oils, which eliminates the need of hydrogenation and trans fat formation for further food applications. Cottonseed meal is rich in proteins having the protein efficiency ratio equivalent to casein. However, due to the presence of toxic gossypol, cottonseed meal is not used for human consumption, and after processing, it is used as a feed ingredient. Majorly, cottonseed oil is used as liquid oil in the form of cooking oil and salad oil worldwide. Another food application of cottonseed oil lies in the formulation of margarine and shortenings.

References

Akinola OB, Dosunmu OO, Dini L, Ajayi S (2009) Proteinaceous diet inhibits gossypol-induced spermatotoxicity. Eur J Histochem 50(3):205–208

Alam MB, An H, Ra JS, Lim JY, Lee SH, Yoo CY, Lee SH (2018) Gossypol from cottonseeds ameliorates glucose uptake by mimicking insulin signaling and improves glucose homeostasis in mice with streptozotocin-induced diabetes. Oxid Med Cell Longev 2018:1–11. https://doi.org/10.1155/2018/5796102

Aletor VA (1989) Effect of varying levels of oyster shell on serum constituents and erythrocyte indices in growing chicken fed gossypol-containing Nigerian cottonseed cake. Nahrung 33:905
Alexander J, Andersson HC, Bernhoft A, Brimer L, Cottrill B, Fink-Gremmels J, Jaroszewski J, Sørensen H (2008) Gossypol as undesirable substance in animal feed: scientific opinion of the panel on contaminants in the food chain. EFSA J 908. https://doi.org/10.2903/j.efsa.2009.908
Bailey CA, Stipanovic RD, Ziehr MS, Haq AU, Sattar M, Kubena LF, Kim HL, Vieira R d M (2000) Cottonseed with a high (+)-to (−)-gossypol enantiomer ratio favorable to broiler production. J Agric Food Chem 48(11):5692–5695
Banerjee S, Choi M, Aboukameel A, Wang Z, Mohammad M, Chen J, Yang D, Sarkar FH, Mohammad RM (2010) Preclinical studies of apogossypolone, a novel pan inhibitor of bcl-2 and mcl-1, synergistically potentiates cytotoxic effect of gemcitabine in pancreatic cancer cells. Pancreas 39(3):323
Blevins S, Siegel PB, Blodgett DJ, Ehrich M, Saunders GK, Lewis RM (2010) Effects of silymarin on gossypol toxicosis in divergent lines of chickens. Poult Sci 89(9):1878–1886
Braga AP, MacIel MV, Guerra DGF, Maia ISAS, Oloris SCS, Soto-Blanco B (2012) Extruded-expelled cottonseed meal decreases lymphocyte counts in male sheep. Rev Med Vet 163 (3):147–152
Calhoun MC, Kuhlmann SW, Baldwin BC (1995) September. Cotton feed product composition and gossypol availability and toxicity. In: Proc. National Invitational Symposium on Alternative Feeds for Dairy and Beef Cattle, St. Louis, MO, p 125–145
Câmara ACL, Gadelha ICN, Borges PAC, de Paiva SA, Melo MM, Soto-Blanco B (2015) Toxicity of gossypol from cottonseed cake to sheep ovarian follicles. PLoS One 10(11):e0143708
Cao H, Sethumadhavan K, Bland JM (2018) Isolation of cottonseed extracts that affect human cancer cell growth. Sci Rep 8(1):10,458
Carson JA, Burke FM, Hark LA (2005) Cardiovascular nutrition: disease management and prevention. Chicago, IL, American Dietetic Association
Cheng ZJ, Hardy RW (2002) Apparent digestibility coefficients and nutritional value of cottonseed meal for rainbow trout (Oncorhynchus mykiss). Aquaculture 212(1–4):361–372
Cheng HN, Dowd MK, He Z (2013) Investigation of modified cottonseed protein adhesives for wood composites. Ind Crop Prod 46:399–403
Chenoweth PJ, Risco CA, Larsen R, Vele J, Tran T, Chase CC Jr (1994) Effects of dietary gossypol on aspects of semen quality, sperm morphology and sperm production in young Brahman bulls. Theriogenology 42(1):1–13
Chien CC, Ko CH, Shen SC, Yang LY, Chen YC (2012) The role of COX-2/PGE2 in gossypol-induced apoptosis of colorectal carcinoma cells. J Cell Physiol 227(8):3128–3137
Chongthammakun S, Ekavipat C, Sanitwongse B, Pavasuthipaisit K (1986) Effects of gossypol on human and monkey sperm motility in vitro. Contraception 34(3):323–331
Cope RB (2018) Cottonseed Toxicity. In Veterinary Toxicology (Basic and Clinical Principles), Ed. RC Gupta, Academic Press, pp. 967–980
Coutinho EM (2002) Gossypol: a contraceptive for men. Contraception 65:259–263
Davis KE, Prasad C, Imrhan V (2012) Consumption of a diet rich in cottonseed oil (CSO) lowers total and LDL cholesterol in normo-cholesterolemic subjects. Nutrition 4(7):602–610
Dowd MK, Boykin DL, Meredith WR Jr, Campbell BT, Bourland FM, Gannaway JR, Glass KM, Zhang J (2010) Fatty acid profiles of cottonseed genotypes from the national cotton variety trials. J Cotton Sci 14:64–73
Durak E, Karaosmanoglu F (2004) Using of cottonseed oil as an environmentally accepted lubricant additive. Energy Source 26(7):611–625
East NE, Anderson M, Lowenstine LJ (1994) Apparent gossypol-induced toxicosis in adult dairy goats. J Am Vet Med Assoc 204(4):642–643
El-Mokadem MY, Taha TA, Samak MA, Yassen AM (2012) Alleviation of reproductive toxicity of gossypol using selenium supplementation in rams. J Anim Sci 90(9):3274–3285

El-Sharaky AS, Newairy AA, Elguindy NM, Elwafa AA (2010) Spermatotoxicity, biochemical changes and histological alteration induced by gossypol in testicular and hepatic tissues of male rats. Food Chem Toxicol 48(12):3354–3361

European Food Safety Authority (EFSA) (2009) Gossypol as undesirable substance in animal feed-Scientific Opinion of the Panel on Contaminants in the Food Chain. EFSA J 7(1):908

FAOSTAT (2018). Data available from: http://www.fao.org/faostat/en/#search/Cottonseed. Accessed 2 Mar 2019

Fornes MW, Barbieri AM, Burgos MH (1993) Sperm motility loss induced by gossypol: relation with OH. scavengers, motile stimulators and malondialdehyde production. Biochem Biophys Res Commun 195(3):1289–1293

Frey M (2018) Cottonseed oil nutrition facts-calories, carbs, and health benefits of cottonseed oil. https://www.verywellfit.com/cottonseed-oil-nutrition-facts-calories-carbs-and-health-benefits-4165691. Accessed 22 Feb 2019

Gadelha ICN, Fonseca NBS, Oloris SCS, Melo MM, Soto-Blanco B (2014) Gossypol toxicity from cottonseed products. Sci World J 2014:1–11. https://doi.org/10.1155/2014/231635

Gadelha ICN, Lima MA, Melo MM, Soto-Blanco B (2016) Gossypol promotes the degeneration of chicken ovarian follicles in vitro. Braz J Poult Sci 18(3):505–510

Gao D, Cao Y, Li H (2010) Antioxidant activity of peptide fractions derived from cottonseed protein hydrolysate. J Sci Food Agric 90:1855–1860. https://doi.org/10.1002/jsfa.4024

Gilani GS, Lee N (2003) Protein: sources of food-grade protein. In: Caballero B (ed) Encyclopedia of food sciences and nutrition, 2nd edn. Academic Press, Cambridge, MA, pp 4873–4879. https://doi.org/10.1016/B0-12-227055-X/00834-8

Gu Y, Anderson NO (1985) Effects of gossypol on the estrous cycle and ovarian weight in the rat. Contraception 32(5):491–496

Gunstone F (2001) Vegetable oils in food technology: composition, properties and uses. Wiley, Chichester, West Sussex

Han YW (1988) Removal of phytic acid from soybean and cottonseed meals. J Agric Food Chem 36(6):1181–1183

Harris WS, Mozafarian D, Rimm E, Kris-Etherton P, Rudel LL, Appel LJ, Engler MM, Sacks F (2011) Omega-6 fatty acids and risk for cardiovascular disease: a science advisory from the American Heart Association Nutrition Subcommittee of the Council on Nutrition, Physical Activity, and Metabolism. Circulation 119:902–907

Haschek WM, Beasley V, Buck WB, Finnell JH (1989) Cottonseed meal (gossypol) toxicosis in a swine herd. J Am Vet Med Assoc 195(5):613–615

He Z, Zhang H, Olk DC (2015) Chemical composition of defatted cottonseed and soy meal products. PLoS One 10(6):e0129933

Hendra R, Ahmad S, Sukari A, Shukor MY, Oskoueian E (2011) Flavonoid analyses and antimicrobial activity of various parts of Phaleria macrocarpa (Scheff.) Boerl fruit. Int J Mol Sci 12 (6):3422–3431

Henry GE, Momin RA, Nair MG, Dewitt DL (2002) Antioxidant and cyclooxygenase activities of fatty acids found in food. J Agric Food Chem 50(8):2231–2234

Hron RJ, Kim HL, Calhoun MC, Fisher GS (1999) Determination of (+)-,(−)-, and total gossypol in cottonseed by high-performance liquid chromatography. J Am Oil Chem Soc 76 (11):1351–1355

Hu YF, Chang CJG, Brueggemeier RW, Lin YC (1993) Gossypol inhibits basal and estrogen-stimulated DNA synthesis in human breast carcinoma cells. Life Sci 53(25):PL433–PL438

Huang KH, Ravindran V, Li X, Ravindran G, Bryden WL (2007) Apparent ileal digestibility of amino acids in feed ingredients determined with broilers and layers. J Sci Food Agric 87 (1):47–53

Huang YW, Wan LS, Dowd MK, Wan PJ, Lin YC (2009) (-)-Gossypol reduces invasiveness in metastatic prostate cancer cells. Anticancer Res 29(6):2179–2188

Huo M, Gao R, Jiang L, Cui X, Duan L, Deng X, Guan S, Wei J, Soromou LW, Feng H, Chi G (2013) Suppression of LPS-induced inflammatory responses by gossypol in RAW 264.7 cells and mouse models. Int Immunopharmacol 15(2):442–449

Kakani R, Gamboa DA, Calhoun MC, Haq AU, Bailey CA (2010) Relative toxicity of cottonseed gossypol enantiomers in broilers. Open Toxicol J 4:26–31

Kenar JA (2006) Reaction chemistry of gossypol and its derivatives. J Am Oil Chem Soc 83 (4):269–302

Ko CH, Shen SC, Yang LY, Lin CW, Chen YC (2007) Gossypol reduction of tumor growth through ROS-dependent mitochondria pathway in human colorectal carcinoma cells. Int J Cancer 121(8):1670–1679

Li H, Piao L, Xu P, Ye W, Zhong S, Lin SH, Kulp SK, Mao Y, Cho Y, Lee LJ, Lee RJ (2011) Liposomes containing (-)-gossypol-enriched cottonseed oil suppress Bcl-2 and Bcl-xL expression in breast cancer cells. Pharm Res 28(12):3256–3264

Lin YC, Chitcharoenthum M, Rikihisa Y (1990) Effect of gossypol on thyroid hormones in young female rats. Contraception 41(4):431–440

Lindsey TO, Hawkins GE, Guthrie LD (1980) Physiological responses of lactating cows to gossypol from cottonseed meal rations. J Dairy Sci 63(4):562–573

Liu S, Kulp SK, Sugimoto Y, Jiang J, Chang HL, Dowd MK, Wan P, Lin YC (2002) The (-)-enantiomer of gossypol possesses higher anticancer potency than racemic gossypol in human breast cancer. Anticancer Res 22(1A):33–38

Madar Z, Nir M, Trostler N, Norenberg C (1988) Effects of cottonseed dietary fiber on metabolic parameters in diabetic rats and non-insulin-dependent diabetic humans. J Nutr 118 (9):1143–1148

Mahmood F, Khan MZ, Khan A, Muhammad G, Javed I (2011) Lysine induced modulation of toxico-pathological effects of cottonseed meal in broiler breeder males. Pak J Zool 43 (2):357–365

Manabe S, Nuber DC, Lin YC (1991) Zone-specific hepatotoxicity of gossypol in perfused rat liver. Toxicon 29(6):787–790

Mellon JE, Zelaya CA, Dowd MK, Beltz SB, Klich MA (2012) Inhibitory effects of gossypol, gossypolone, and apogossypolone on a collection of economically important filamentous fungi. J Agric Food Chem 60(10):2740–2745

Mena H, Santos JEP, Huber JT, Tarazon M, Calhoun MC (2004) The effects of varying gossypol intake from whole cottonseed and cottonseed meal on lactation and blood parameters in lactating dairy cows. J Dairy Sci 87(8):2506–2518

Morgan S, Stair EL, Martin T, Edwards WC, Morgan GL (1988) Clinical, clinicopathologic, pathologic, and toxicologic alterations associated with gossypol toxicosis in feeder lambs. Am J Vet Res 49(4):493–499

Nagalakshmi D, Rao SVR, Panda AK, Sastry VR (2007) Cottonseed meal in poultry diets: a review. J Poult Sci 44(2):119–134

Namuli A, Abdullah N, Sieo CC, Zuhainis SW, Oskoueian E (2011) Phytochemical compounds and antibacterial activity of Jatropha curcas Linn. extracts. J Med Plant Res 5(16):3982–3990

O'Brien RD, Jones, LA, King CC, Wakelyn PJ, Wan PJ (2005) Cottonseed oil, In: Shahidi F (ed) Bailey's industrial oil and fat products, vol. 2, 6th edn. John Wiley & Sons, Inc., Hoboken, NJ, pp. 173–279

Ory RL, Flick GJ (1994) Peanut and cottonseed proteins for food uses. In: Hudson BJF (ed) New and developing sources of food proteins. Springer, Boston, MA, pp 195–240

Oskoueian E, Abdullah N, Hendra R, Karimi E (2011a) Bioactive compounds, antioxidant, xanthine oxidase inhibitory, tyrosinase inhibitory and anti-inflammatory activities of selected agro-industrial by-products. Int J Mol Sci 12(12):8610–8625

Oskoueian E, Abdullah N, Ahmad S, Saad WZ, Omar AR, Ho YW (2011b) Bioactive compounds and biological activities of Jatropha curcas L. kernel meal extract. Int J Mol Sci 12 (9):5955–5970

Pang X, Wu Y, Wu Y, Lu B, Chen J, Wang J, Yi Z, Qu W, Liu M (2011) (−)-Gossypol suppresses the growth of human prostate cancer xenografts via modulating VEGF signaling–mediated angiogenesis. Mol Cancer Ther 10(5):795–805

Polley KR, Oswell NJ, Pegg RB, Paton CM, Cooper JA (2018) A 5-day high-fat diet rich in cottonseed oil improves cholesterol profiles and triglycerides compared to olive oil in healthy men. Nutr Res 60:43–53

Price WD, Lovell RA, McChesney DG (1993) Naturally occurring toxins in feedstuffs: center for veterinary medicine perspective. J Anim Sci 71(9):2556–2562

Puckhaber LS, Dowd MK, Stipanovic RD, Howell CR (2002) Toxicity of (+)-and (−)-gossypol to the plant pathogen, Rhizoctonia solani. J Agric Food Chem 50(24):7017–7021

Quintana PJ, de Peyster A, Klatzke S, Park HJ (2000) Gossypol-induced DNA breaks in rat lymphocytes are secondary to cytotoxicity. Toxicol Lett 117(1–2):85–94

Radcliffe JD, Czajka-Narins DM (2006) Lipids and tocopherols in serum and liver of female rats fed diets containing corn oil or cottonseed oil. Plant Foods Hum Nutr 61:35–38

Radcliffe JD, King CC, Czajka-Narins DM, Imrhan V (2001) Serum and liver lipids in rats fed diets containing corn oil, cottonseed oil, or a mixture of corn and cottonseed oils. Plant Foods Hum Nutr 56:51–60

Radcliffe JD, Czajka-Narins DM, Imrhan V (2004) Fatty acid composition of serum, adipose tissue and liver in rats fed diets containing corn oil or cottonseed oil. Plant Foods Hum Nutr 69:73–77

Randel RD, Chase CC Jr, Wyse SJ (1992) Effects of gossypol and cottonseed products on reproduction of mammals. J Anim Sci 70(5):1628–1638

Randel RD, Willard ST, Wyse SJ, French LN (1996) Effects of diets containing free gossypol on follicular development, embryo recovery and corpus luteum function in Brangus heifers treated with bFSH. Theriogenology 45(5):911–922

Rikihisa Y, Lin YC (1989) Effect of gossypol on the thyroid in young rats. J Comp Pathol 100 (4):411–417

Roger V, Go AS, Lloyd-Jones DM, Adams RJ, Berry JD, Brown TM, Carnethon MR, Dai S, Simone G, Ford ES (2011) Heart disease and stroke statistics: 2011 update. Circulation 123: e18–e209

Rogers GM, Poore MH, Paschal JC (2002) Feeding cotton products to cattle. Vet Clin Food Anim Pract 18(2):267–294

Ryan JR, Kratzer FA, Grats GR, Vohrra P (1986) Glandless cottonseed cake for laying and breeding hens and broiler chicks. Poult Sci 65:549–955

Sacks FM, Katan M (2002) Randomized clinical trials on the effects of dietary fat and carbohydrate on plasma lipoproteins and cardiovascular disease. Am J Med 113(9):13–24

Sadique J, Begum VH, Thenmozhi V, Elango V (1987) Hypoglycemic effect of cotton seed aqueous extract in alloxan-induced diabetes mellitus in rats. Biochem Med Metab Biol 38 (1):104–110

Sahin A, Duru M, Kaya S, Camci O (2006) Effects of raw material in finisher diet on broiler performance I choice feeding system. Arch Geflugelk 70(1):8–13

Schwingshackl L, Hoffmann G (2012) Monounsaturated fatty acids and risk of cardiovascular disease: synopsis of the evidence available from systematic reviews and meta-analyses. Nutrition 4(12):1989–2007

Sekhar SC, Rao BVK (2011) Cottonseed oil as health oil. Pertanika J Trop Agric Sci 34(1):17–24

Singh V, Soni A, Kumar S, Singh R (2014) Characterization of liquid product obtained by pyrolysis of cottonseed de-oiled cake. J Biobased Mater Bio 8:338–343

Siri-Tarino PW, Sun Q, Hu FB, Krauss RM (2010) Saturated fat, carbohydrate and cardiovascular disease. Am J Clin Nutr 91(3):502–509

Soto-Blanco B (2008) Gossipol e fatores antinutricionais da soja. Toxicologia Aplicada à Veterinária. Manole, São Paulo, pp 531–545

Tang F, Wong PYD (1984) Serum potassium and aldosterone levels in gossypol-treated rats. Int J Androl 7(2):149–153

Thakur A, Lum LG, Schalk D, Azmi A, Banerjee S, Sarkar FH, Mohommad R (2012) Pan-Bcl-2 inhibitor AT-101 enhances tumor cell killing by EGFR targeted T cells. PLoS One 7(11):e47520
Thirumalaisamy G, Purushothaman MR, Vasantha Kumar P, Selvaraj P (2015) Nutritive and feeding value of cottonseed meal for broilers. M.V.Sc. thesis. Tamil Nadu Veterinary and Animal Sciences University Namakkal-637002
Udoh P, Patil DR, Deshpande MK (1992) Histopathological and biochemical effects of gossypol acetate on pituitary-gonadal axis of male albino rats. Contraception 45(5):493–509
Umamaheswari M, Asokkumar K, Sivashanmugam AT, Remyaraju A, Subhadradevi V, Ravi TK (2009) In vitro xanthine oxidase inhibitory activity of the fractions of Erythrina stricta Roxb. J Ethnopharmacol 124(3):646–648
Wassef EA, Shalaby SH, Saleh NE (2015) Cottonseed oil as a complementary lipid source in diets for gilthead seabream S parus aurata juveniles. Aquacult Res 46(10):2469–2480
Xiao ZP, Peng ZY, Peng MJ, Yan WB, Ouyang YZ, Zhu HL (2011) Flavonoids health benefits and their molecular mechanism. Mini Rev Med Chem 11(2):169–177
Xiong J, Li J, Yang Q, Wang J, Su T, Zhou S (2017) Gossypol has anti-cancer effects by dual-targeting MDM2 and VEGF in human breast cancer. Breast Cancer Res 19(1)
Xu WB, Xu LH, Lu HS, Ou-Yang DY, Shi HJ, Di JF, He XH (2009) The immunosuppressive effect of gossypol in mice is mediated by inhibition of lymphocyte proliferation and by induction of cell apoptosis. Acta Pharmacol Sin 30(5):597
Yuan YY, Shi QX (2000) Inhibition of hamster sperm acrosomal enzyme by gossypol is closely associated with the decrease in fertilization capacity. Contraception 62(4):203–209
Yuan Y, Tang AJ, Castoreno AB, Kuo SY, Wang Q, Kuballa P, Xavier R, Shamji AF, Schreiber SL, Wagner BK (2013) Gossypol and an HMT G9a inhibitor act in synergy to induce cell death in pancreatic cancer cells. Cell Death Dis 4(6):e690
Yue HB, Cui YD, Shuttleworth PS, Clark JH (2012) Preparation and characterization of bioplastics made from cottonseed protein. Green Chem 14:2009–2016
Zhang Q, Yang M, Zhao Y, Luan X, Ke Y (2001) Isolation and structure identification of flavonol glycosides from glandless cotton seeds. Acta Pharm Sin 36(11):827–831
Zhang WJ, Xu ZR, Pan XL, Yan XH, Wang YB (2007) Advances in gossypol toxicity and processing effects of whole cottonseed in dairy cows feeding. Livest Sci 111(1-2):1–9
Zhang B, Cui Y, Yin G, Li X, You Y (2010) Synthesis and swelling properties of hydrolyzed cottonseed protein composite superabsorbent hydrogel. Int J Polym Mater Polym Biomater 59:1018–1032
Zhong S, Ye WP, Xu PP, Feng E, Li H, Lin SH, Liu JY, Ma C, Lin YC (2010) Aromatase expression in leptin-pretreated human breast pre-adipocytes is enhanced by zeranol and suppressed by (−)-gossypol. Anticancer Res 30(12):5077–5084
Zhong S, Leong J, Ye W, Xu P, Lin SH, Liu JY, Lin YC (2013) (−)-Gossypol-enriched cottonseed oil inhibits proliferation and adipogenesis of human breast pre-adipocytes. Anticancer Res 33 (3):949–955
Zotte AD, Brand TS, Hoffman LC, Schoon K, Cullere M, Swart R (2013) Effect of cottonseed oilcake inclusion on ostrich growth performance and meat chemical composition. Meat Sci 93:194–200

Chapter 4
Groundnut (Peanut) (*Arachis hypogaea*)

Faiza Syed, Sania Arif, Iftikhar Ahmed, and Nauman Khalid

Abstract *Arachis hypogaea,* known as the groundnut or peanut, is an annual herbaceous legume grown in tropical and temperate areas of the world. Peanuts are a composite food consisting of a wide variety of nutrients, such as carbohydrates, proteins, lipids, vitamins, minerals, and a good dose of fiber. Bioactive compounds have also been isolated from peanuts that include flavonoids, phytosterols, amino acids, and stilbenes. Large-scale clinical studies have shown that regular peanut consumption has a positive effect on cardiovascular diseases, type 2 diabetes, and Alzheimer's. These bioactive compounds also have anti-inflammatory, antioxidant, anticancer, and antitumor activities. Potential health concerns regarding peanuts include allergies and contamination with aflatoxins. Peanuts are widely used in the food industry for the production of flour, protein concentrates and isolates, confectionaries, oils, and beverages.

Keywords Peanuts · Groundnut · Phytochemicals · Flavonoids · Health benefits

F. Syed
Department of Microbiology, Quaid-i-Azam, University, Islamabad, Pakistan

S. Arif
Institute of Microbiology and Genetics, Georg-August-Universität, Göttingen, Germany

I. Ahmed
National Culture Collection Center of Pakistan (NCCP), Bioresources Conservation Institute (BCI), National Agricultural Research Centre, Islamabad, Pakistan

N. Khalid (✉)
School of Food and Agricultural Sciences, University of Management and Technology, Lahore, Pakistan
e-mail: nauman.khalid@umt.edu.pk

B. Tanwar, A. Goyal (eds.), *Oilseeds: Health Attributes and Food Applications*,
https://doi.org/10.1007/978-981-15-4194-0_4

4.1 Ground Nut (*Arachis hypogaea*)

Arachis hypogaea, commonly known as the groundnut or peanut, is an annual herbaceous legume (belongs to the Fabaceae family). It was originally farmed in Central and South America 3500 years ago, but now owing to its high demand for oil and as a food product, its cultivation has spread to temperate and tropical regions around the globe. *A. hypogea* seeds are nutritionally rich in fats and proteins, and a single cup (146 g) may contain up to 800 calories (Arya et al. 2016). The name of the peanut can be misleading as it is not a nut. However, technically it is a legume just like lentils, chickpeas, and pulses as it grows in a pod below the ground. As *A. hypogea* is a legume, its roots harbor nitrogen-fixing bacteria in root nodules and add valuable nitrates to the soil (Sharma and Bhatnagar-Mathur 2006). This chapter covers the history of peanut cultivation, health, and nutritional benefits.

4.1.1 Origin and History

A. hypogaea was originally cultivated in the New World, where early explorers discovered archeological evidence of its agriculture in South and Central America. Carbon dating of the peanut fruit hull (known as remnant pericarp) from archeological sites in Peru places its agricultural activity between 3700 and 3900 years old. It has been difficult to trace the exact origin of the major peanut plant cultigen, *A. hypogaea* L. (Kochert et al. 1996). RFLP (restriction fragment length polymorphism) and cytogenetic analyses revealed that wild-type *Arachis duranensis* and *Arachis ipaensis* hybridized their genomes to constitute the domesticated tetraploid *A. hypogaea* L. This genome hybridization event took place in Argentina or Southern Bolivia according to RFLP and cytogenetic data (Kochert et al. 1996).

Portuguese, Spanish, German, and Dutch explorers revealed that peanuts were being cultivated in the Antilles region, coastal areas of Brazil, and temperate areas of Rio de la Plata basin (Smith 2002). During the colonial period, *A. hypogaea* was considered as an ornamental garden plant and was cultivated in farmlands for animal feed up till the 1930s. Large-scale cultivation of *A. hypogaea* for human consumption began in the early nineteenth century in America and was later overtaken by Asia, which is currently the largest producer of peanuts (Arya et al. 2016; Shu-bo 2009).

4.1.2 Production

Peanuts are largely cultivated in tropical and temperate regions around the globe (Table 4.1). Ideal conditions for peanut cultivation are warmer temperatures,

Table 4.1 List of countries with respect to peanut production

Global peanut producers		
Rank	Country	Annual production (metric tons)
1	China	13,336,860
2	India	7,156,448
3	Nigeria	2,755,649
4	United States	1,837,519
5	Sudan	1,399,500

500 mm of water, and loamy-sandy soil that has a pH of 5.9–7. Additionally, potassium, phosphorus, magnesium, and other micronutrients are necessary for a higher peanut yield. Moreover, peanut root nodules contain nitrogen-fixing bacteria that enrich the nitrogen content of the soil and in turn reduce the consumption of nitrogen fertilizers.

A. hypogea continues to flower even after an initial round of pollination and fertilization; this results in mature as well as developing pods, and hence, the harvest must be carefully timed to avoid unripe pods (if removed too early) or overly ripe pods that tend to remain in the ground (if removed too late) (Vellidis et al. 2014). During harvesting, the entire plant is uprooted, and the pods are used for human consumption or oil production, whereas the leaves and stalks are used as animal feed. Conventionally, the *A. hypogea* plant was pulled by hand, inverted, and left on the ground to dry with the pods exposed to the air. After a period of 3–4 days, the pods loose approximately 75% of their moisture, and at this point the peanuts are shaken to remove the dry pods from the bush. Currently, the harvesting process is accomplished through machines that in the first step uproot the plant and invert it on the soil; the pods are dried and threshed to separate the pods from the bush (Vellidis et al. 2014).

Storage is also crucial for the peanuts where moisture content is of utmost importance. After the harvest, the peanuts must be stored in a dry place; otherwise there is a risk of development of mold *Aspergillus flavus* which produces aflatoxin, a carcinogenic and toxic substance (Dorner 2008). The largest producer of peanuts is China with an overall production of 16.5 million tons, followed by India (6.6 million tons). Besides China and India, America (2.6 million tons), Nigeria (3 million tons), and Sudan (1.8 million tons) also produce significant levels of peanuts. The main exporters of peanuts include India (exports 541,337 tons) and the United States (exports 250,000 tons) (FAOSTAT 2016).

4.1.3 Chemical Composition

Preliminary studies on the chemical composition of peanuts revealed that they are composed of carbohydrates, proteins, lipids, dietary fibers, minerals, and vitamins (Akhtar et al. 2014) In addition to nutritional elements, they also contain small amounts of moisture (3.26%).

4.1.3.1 Carbohydrates

In 100 g of roasted peanuts, there are 21.51 g of carbohydrates, whereas in raw peanuts there is 25 g.

4.1.3.2 Proteins

Peanuts are an excellent source of proteins, as 100 g of peanuts contain 25 g of protein. Amino acid composition of peanut protein is similar to that of beef protein. The ratios of histidine and arginine in *A. hypogea* are similar to the protein content of an egg. Tryptophan, cysteine, and methionine are found in the lowest quantity out of the 18 amino acids (Akhtar et al. 2014). Out of a total 25 g protein content, the various ratios of amino acids are given in Table 4.2.

Table 4.2 Amino acid composition and content of *A. hypogea*

Protein content of peanuts	
Total protein content	25 g
Tryptophan	0.2445 g
Threonine	0.859 g
Isoleucine	0.882 g
Leucine	1.627 g
Lysine	0.901 g
Methionine	0.308 g
Cysteine	0.322 g
Phenylalanine	1.300 g
Tyrosine	1.020 g
Valine	1.052 g
Arginine	3.001 g
Histidine	0.634 g
Alanine	0.997 g
Aspartic acid	3.060 g
Glutamic acid	5.243 g
Glycine	1.512 g
Proline	1.107 g
Serine	1.236 g

The table shows the variety and relative amount of amino acids present in 100 g of peanut. *A. hypogea* contains 18 essential amino acids out of a 20. Amino acids that are abundant in peanuts include (in descending order) glutamic acid, arginine, and aspartic acid (Conde Nast version SR-21 2014)

4.1.3.3 Lipids

A. hypogea has sufficient quantities of fatty acids including saturated fatty acids (6.89 g), mono-unsaturated fatty acids (24.6 g), and PUFAs (15.69 g) which are contained within 100 g of roasted peanuts (Francisco and Resurreccion 2008). 100 g of unroasted peanuts contain a total of 48 g of lipids, of which 7 g are saturated, 24 g mono-unsaturated, and 16 g PUFAs (Campos-Mondragón et al. 2009).

4.1.3.4 Vitamins

A. hypogea contains a variety of vitamins (Table 4.3) including riboflavin, thiamin, niacin, B6, E, pantothenic acid, folate, and choline. Peanuts contain most of the necessary water-soluble vitamins as well as a fat-soluble vitamin E (He and Hekmat 2014).

Table 4.3 Types and quantity of vitamins present in peanuts

Vitamin content in peanuts		
Vitamin	Function	Quantity in peanuts (g)
Vitamin E	Potent antioxidant. (Rizvi et al. 2014)	69.3×10^{-4}
Riboflavin	Co-enzyme in metabolism, aids in red blood cell production, maintains proper function of neurons, eyes, skin, and liver (J Henriques et al. 2010)	0.98×10^{-4}
Thiamin	Maintains electrolyte balance in neurons and muscles, aids in digestions, supports carbohydrate metabolism (Manzetti et al. 2014)	4.38×10^{-4}
Niacin	Glucose assimilation, fatty acid, and cholesterol synthesis (Jacobson et al. 2012)	13.52×10^{-3}
Pantothenic acid	Carbohydrate and fat metabolism, sex and stress hormone production, and red blood cell production (Byrd-Bredbenner et al. 2014)	1.395×10^{-3}
Folate	Maintains neuronal health (Fenech 2012)	1450×10^{-6}
B6	Neurotransmitter synthesis, norepinephrine and serotonin production and maintains neuronal health (Shils and Shike 2006)	2.56×10^{-4}
Choline	Neurotransmitter synthesis, transport of lipids across cell membranes, and cell membrane signaling (Zeisel and Da Costa 2009)	55.3×10^{-3}

Detailed description of the types of vitamins present in peanuts along with their relative weight per 25 g of peanut. Vitamin E, folate, and niacin are relatively abundant in peanuts as compared to other types of vitamins. Function of each type of vitamin, along with their quantity is also described (Settaluri et al. 2012)

4.1.3.5 Minerals

Peanuts contain adequate levels of various minerals that play an important role in body functions. Peanuts can be a good source of magnesium, and regular consumption of peanuts ensures adequate levels in the body. Magnesium maintains proper functioning of nerves and muscle cells and supports a good immune system. It regulates blood sugar, maintains bone strength, and balances the blood sugar levels (Jahnen-Dechent and Ketteler 2012). Adequate levels of calcium are also found in peanuts. Peanuts contain 50% of the required daily allowance of phosphorus. Peanuts also contain potassium which is an electrolyte and supports neuronal health. Additionally, it is also required for proper muscle development (He and MacGregor 2008). Peanuts contain adequate levels of this mineral and, thus, can be used as a supplement for potassium. Zinc is present in sufficient quantities in peanuts (31 mg per 100 g). Another essential mineral found in peanuts (2.26 mg per 100 g) is iron and is necessary for red blood cell production. It is also a cofactor for various enzymes (catalase, oxidase, and ferrochelatase) involved in metabolism. Iron also maintains cell differentiation and proliferation. Copper is also present in peanuts (0.671 mg per 100 g) and helps in synthesis of proteins including hemoglobin and collagen. Selenium is a micronutrient found in peanuts (7.5 μg per 100 g) and is a potent antioxidant, boosts metabolism, and is also known to prevent cancer. Table 4.4 depicts the levels of minerals per 100 g of peanuts (Rebello et al. 2014).

4.1.3.6 Dietary Fiber

Peanuts contain a healthy amount of fiber (8 g per 100 g) which has several health benefits including regulation of bowel movements, lowers the level of harmful LDC cholesterol, and consequently reduces the risk of heart disease (Atasie et al. 2009).

Table 4.4 Types and quantity of minerals present in peanuts

Mineral	Weight in g
Iron	2.26×10^{-3}
Calcium	54×10^{-3}
Magnesium	176×10^{-3}
Potassium	658×10^{-3}
Phosphorus	358×10^{-3}
Sodium	6×10^{-3}
Copper	0.671×10^{-3}
Zinc	3.31×10^{-3}
Selenium	7.5×10^{-6}
Manganese	2.08×10^{-3}

Detailed descriptions of the types of minerals present in peanuts along with their relative weight per 100 g of peanut are given. Peanuts are rich in calcium, magnesium, potassium, and phosphorus (Settaluri et al. 2012)

4.1.4 Antinutritional Factors

Despite being a rich source of complex carbohydrates, proteins, some minerals, and energy due to higher fat content (approximately 50%), the antinutritional factors such as phytates, condensed tannins, trypsin, and α-amylase inhibitors limit the consumption and nutritional value of oilseeds to a great extent (Tanwar et al. 2019). Raw peanuts contain 852 mg/100 g of phytates, 0.09 mg/100 g of tannins, 16.31 units/mg of trypsin inhibitors, and 62.5 units/mg α-amylase (Rackis et al. 1986; Elkowicz and Sosulski 1982; Stansbury et al. 1950). The growth depression and altered pancreatic function in rats fed with raw or dry heat-treated legumes have been attributed to their proteinase inhibitors and lectins (Herzig et al. 1997). Sitren et al. (1985) studied the in vivo and in vitro assessment of antinutritional factors in peanut and soy. The rats fed with raw peanuts had lower liver weight, while no such change in weight was observed in the moist heated peanut-fed group. Moreover, it was found that the trypsin inhibitor activity was modestly reduced (76%) when dry heat was applied, whereas moist heat effectively decreased it by approximately 82% (Sitren et al. 1985). Ejigui et al. (2005) compared the traditional processing methods such as germination and roasting to reduce the antinutritional factors in peanuts. The results showed that the combination of germination followed by roasting was effective in decreasing the antinutritional contents significantly ($p < 0.05$) except for trypsin inhibitors (Ejigui et al. 2005). The recommendation is to use the wet heat instead of dry heat to effectively reduce the trypsin and α-amylase inhibitors, since the moist heat is more effective in reducing digestive enzymes inhibitors.

4.1.5 Bioactive Compounds

Bioactive compounds are potent phytochemicals found in small quantities in various plant-based foods; however they have tremendous potential to positively affect metabolism. The beneficial effects of bioactive compounds include antioxidant activity, altering gene expression, inducing/inhibiting enzyme activity, and receptor induction/inhibition. Peanuts contain certain bioactive compounds that include flavonoids, phytosterols, essential amino acids, and stilbenes (Davis et al. 2015). This portion pertains to the various bioactive constituents found in peanuts which have important health functions which are described in the following sections.

4.1.5.1 Flavonoids

Flavonoids are aromatic, phenolic, organic compounds produced by plants and are present in leaves, bark, flowers, and seeds. These compounds are derived from phenylalanine and give rise to a wide family of aromatic compounds (Havsteen 2002). More than 4000 flavonoids have been discovered till date and have various

Table 4.5 Biological functions of flavonoids

Function of flavonoids	
Type	Function
Naturally occurring Pesticides	Acts an immune-protectant molecule against predators and parasites
Anti-microbials	Kills invading bacteria and fungi
Antioxidants	Binds and neutralizes reactive oxygen and nitrogen species
Herbivore repellant	Astringent properties of flavonoids discourage pests
Safety from UV	Acts as a filter against harmful UV
Growth inhibitor	Involved in ion-channel regulation and light phase of photosynthesis

Flavonoids are important bioactive compounds present in peanuts. As their structures are quite distinct, their functions also affect multiple cellular and biological processes. The table details the functions of flavonoids (Mierziak et al. 2014; Agati et al. 2013; Brunetti et al. 2013)

Table 4.6 Polyphenolic composition of peanuts

Flavonoid subgroup	Subtypes	Source
Flavanols (A type pro-anthocyanidins)	Procyanidin A_1 Procyanidin A_2 Epicatechin-(2-O-7,4–8)-ent-epicatechin Epicatechin-(2-O-7,4–6)-ent-catechin Epicatechin-(2-O-7,4–6)-catechin Epicatechin-(2-O-7,4–6)-ent-epicatechin	Peanut skins, kernels
Isoflavones	2-Methoxyl-3(3-indoxyl)-propanoic acid 2-Hydroxyl-3[3-(1-N-methyl)-indolyl] propanoic acid Isoflavone glucoside2-(4-hydroxy phenyl)-ethanol	Peanut skins, kernels
Flavanols (Quercetin)	Quercetin-3-O-[2-O-β-xylopyranosyl-6-O-α-rhamnopyranosyl]-β-glucopyranoside Rutin Isohamnetin triglycoside	Peanut skins, kernels

Polyphenols include different types of flavonoids. Peanut kernels as well as the skins and kernels can contain up to six different types of pro-anthocyanidins and can account for 17% of the weight of the peanut skin. Peanuts also contain three types of isoflavones and four different types of flavonols

functions in plant ecosystems. Some flavonoids are brightly colored, imparting color to the flower which in turn attract insects for pollination. Another property of flavonoids is their astringency (Tanwar et al. 2018) which also indicates their role in plant immunity (Brunetti et al. 2013). Overview of plant flavonoids and their functions are given in Table 4.5.

Multiple studies conducted on the chemical composition of peanuts have revealed that they contain multiple flavonoids including β-sitosterol, Δ(5)-avenasterol, campesterol, stigmasterol, clerosterol, Δ(5,24(25))-stigmastadienol, and Δ(7)-sitosterol (Table 4.6). The peanut fruit, peel, hull, and shell all contain significant amounts of polyphenolic flavonoids. Initial studies investigating the presence of flavonoids in peanut shells were conducted by Pendse et al. (1973) where eriodictyol, 5,7-dihydroxychromone, and luteolin were isolated from immature peanut shells. The mature peanut shells had higher concentrations of luteolin; also with maturation there was an increase in the flavonoid concentration in the hull (Pendse et al. 1973). The effectiveness of flavonoids, eriodictyol, 5,7-dihydroxychromone, and luteolin isolated from peanuts, was later confirmed in a separate study (Duh and Yen 1995, 1997; Duh et al. 1992). Peanut skins (testae), a low-value by-product having dark pink color and tangy taste are obtained after removal from the seed and are mostly used as a constituent of cattle feed (Stalker et al. 2016). In the mid-1900s, they were considered poisonous; however research into their constituents proved that it contained useful oligomer pro-anthocyanidins in addition to flavanols (Yu et al. 2005). Furthermore, studies revealed that the skins of mature peanuts contain up to six different types of pro-anthocyanidins and account for 17% of the weight. Isolated pro-anthocyanidins include pro-cyanidins A1 and A2 and 3 epi-catechin oligomers. Furthermore, eight types of flavonoids; two types of indole-based-alkaloids; pro-anthocyanidins, B2, B3, and B4; and epicatechin-4-β were also isolated from the peanut skins (Lou et al. 2001). In another study, flavanols known as catechins (type A and B procyanidin dimers, trimers, and tetramers) were also detected in peanut skins (Yu et al. 2006). Flavanols and isoflavones, due to their antioxidant potential, have been implicated in the prevention of cardiovascular diseases and cancer (Fraser et al. 1992; Ros and Hu 2013; Ellsworth et al. 2001; Pfeffer et al. 1995; Bao et al. 2016).

4.1.5.2 Phytosterols

Chemically, sterols are alcohols of steroids and are an important constituent of cell membranes in eukaryotes. Phytosterols are specifically synthesized by plants which include sterols and stanols. Their generic structure is the same as cholesterol with an additional alcohol molecule as a side chain and can exist either as free molecules, as steryl glycosides, or as esterified to fatty acids. The most commonly occurring phytosterols include campesterol, β-sitosterol, and stigmasterol. Phytosterols can impede the absorption of cholesterol from the gut and are also poorly absorbed through the gut (Ostlund Jr 2004; Racette et al. 2015; Moreau et al. 2018; Gylling and Simonen 2015).

Peanuts and peanut-derived products both contain phytosterols (Table 4.7). The Valencia variety has the highest amount of phytosterols whether raw (0.023–1.792 μg/g), roasted (dry and fried) (0.018–0.080 μg/g), and in the form of peanut butter (0.148–0.504 μg/g) (Sobolev and Cole 1999). Aboriginal cultivars from Bolivia, Peru, and South America have seven types of phytosterols (4-desmethylsterols): campesterol, stigmasterol, Δ 7-stigmasterol, cholesterol, β-sitosterol, Δ 4-avenasterol, and Δ 7-avenasterol. In commercial species (*A. hypogea* L), the concentrations of stigmasterol and β-sitosterol increase as the peanut matures, whereas the campesterol levels remain constant throughout the ripening process (Shin et al. 2010; Mora-Escobedo et al. 2015; Li et al. 2009). Peanut oil, however, has slightly lower concentration of phytosterols (2.7–9.6 ppm) due to the steps involved in refining process (such as filtration, heating, and

Table 4.7 Different types of phytosterols found in various peanut cultivars

Phytosterols in different cultivars of peanuts	
Cultivar	Types of phytosterols
Arachis hypogea L.	β-Sitosterol, Δ(5)-avenasterol, campesterol, stigmasterol, clerosterol, Δ(5,24(25))-stigmastadienol, Δ(7)-sitosterol
Aboriginal cultivars (Bolivia, Peru, South America)	Campesterol, stigmasterol, Δ 7-stigmasterol, cholesterol, β-sitosterol, Δ 4-avenasterol and Δ 7-avenasterol

Each type of cultivar has different concentrations and types of phytosterols. The commercially cultivated *A. hypogea* has six types of phytosterols, whereas aboriginal cultivars contain five types of phytosterols (Mora-Escobedo et al. 2015)

Table 4.8 Different concentrations of various amino acids present in peanuts

Amino acid	Concentration per 25 g of total peanut protein
Glutamic acid	5.243 g
Arginine	3.00 g
Aspartic acid	3.06 g
Phenylalanine	1.3 g
Tyrosine	1.02 g
Glycine	1.5 g
Serine	1.23 g
Proline	1.1 g
Valine	1.05 g

Peanuts are known to contain high levels of proteins and amino acids. Out of the 20 essential amino acids, 18 are found in peanuts. Amino acids that are abundant in peanuts include (in descending order) glutamic acid, arginine, and aspartic acid (Settaluri et al. 2012)

bleaching). The phytosterols being sensitive to degradation because of the refining process (Bortolomeazzi et al. 2003). Commercially available peanut oil contains 20 mg of sterols per 100 g of oil. When compared to pure olive oil, peanut oil has more phytosterols (38%) that make peanut oil more attracted toward health-conscious people (Bortolomeazzi et al. 2003).

4.1.5.3 Essential Amino Acids

Amino acids are building blocks for proteins necessary for healthy development of skin, bones, and joints. In addition to this, amino acids regulate metabolism and are required for a healthy immune system. A total of 18 essential amino acids are found in peanuts, of which 9 are abundant than the others (Settaluri et al. 2012). Details of abundant amino acids and their functions are given in Table 4.8.

4.1.5.4 Stilbenes

Stilbenes are phenolic compounds consisting of two phenyl molecules connected via a methylene bridge at carbon number 2 (Likhtenshtein 2009). These compounds are synthesized by a limited number of plant species; therefore, they are somewhat rare in nature. Their structure and biological activity are similar to isoflavonoids and are, thus, classified among phytoestrogens. Stilbenes are part of an immune response and produced as a response to infection or injury and are also known as phytoalexins due to their antifungal properties especially against *Penicillium*, *Aspergillus*, and *Cladosporium* species (Strange and Rao 2017). The most commonly studied stilbene is resveratrol (3,5,4′-tri-hydroxy-stilbene). It occurs as two isoforms, *cis* and *trans*-resveratrol; however only the *trans*-isomer has estrogenic potential (Aires et al. 2014; Baur and Sinclair 2006).

Phytoalexins are produced in peanut as a response to fungal infection of the plant tissue. Peanut contain three major types of phytoalexins, including resveratrol, isopentadyl resveratrol, and arachidins (Sanders et al. 2000). Resveratrol was detected in significant quantities in the roots, leaves, and shells (1.2–2.6 μg/g). The developing seeds and testa had low levels of resveratrol (0.05–0.06 μg/g). However, as the seed develops, resveratrol levels increase, especially during germination when there is chance of an infection (Chung et al. 2003). Mechanical processing such as harvesting, threshing, drying, and UV exposure can also contribute to phytoalexin production in peanuts. This is because phytoalexins are produced as a response to stress (Tang et al. 2010). Similarly, resveratrol production can artificially be induced in peanuts by soaking them for approximately 20 h, followed by drying for 66 h. This process increases resveratrol levels up to 45- to 65-fold (Tang et al. 2010; Chung et al. 2003). Piceatannol (3,4,3′,5′-tetrahydroxy-trans-stilbene) is another stilbene that occurs in peanuts; however its level is lower than resveratrol. Similar to resveratrol induction, piceatannol can be induced via UV exposure in peanut calluses. One gram of plant calluses produced 2.17–5.31 μg piceatannol and 0.25–11.97 μg resveratrol, upon UV exposure (Ku et al. 2005). Resveratrol concentrations in different parts of the peanut plant are given in Table 4.9.

Table 4.9 Variable concentrations of resveratrol in different peanut products

Peanut product	Concentration (μg/g)
Raw peanuts	0.023–1.792
Boiled peanuts	1.779–7.092
Roasted peanuts	0.018–0.080
Peanut butter	0.148–0.504

Resveratrol is a phytosterol that is heat sensitive, so post-cultivation processes that involve heat tend to lower the resveratrol concentrations (Sobolev and Cole 1999)

4.1.6 Health Attributes

Peanut consumption in various forms (peanut butter, processed peanuts, and peanut oil) has shown positive effects toward health, mainly because of their ideal lipid content that consists of higher quantities of HDLs and lower quantities of LDLs. Consuming peanut along with their skins doubles the antioxidant effect (Ho et al. 2010). Similarly roasting them can also increase their effectivity. Another effective method is to boil the peanuts which increases their antioxidant potential by two- to fourfold (Yu et al. 2005). Research suggests that regular consumption peanut or peanut oil can reduce the risk of cardiovascular diseases (CVD), type II diabetes, insulin resistance, Alzheimer's disease, and colorectal cancer (Blomhoff et al. 2006). Bioactive compounds (such as flavonoids, phytosterols, and stilbenes) contained within peanuts impart antioxidant, cardioprotective, and anti-inflammatory properties to the peanuts. As a consequence, they reduce LDL oxidation, lower blood pressure, and reduce mortality rate by as much as 40% (Fraser et al. 1992). Details of the health benefits of peanut consumption are given below.

4.1.6.1 Role in Coronary Heart Disease (CHD)

Evidence that peanut consumption has a cardioprotective effect is implied through various epidemiological studies which largely indicate toward an antagonistic relationship between peanut consumption and risk of CHD (Hargrove et al. 2001; Nash and Nash 2008; Kelly and Sabaté 2006; Sabate and Ang 2009). Additionally some studies suggest that peanut consumption enhances the lipid profile which in turn lowers CHD risk (Nash and Nash 2008; Feldman 2002; Griel and Kris-Etherton 2006; Mukuddem-Petersen et al. 2005).

In order to investigate the benefit of peanut consumption in prevention of CHD, five large-scale surveys were conducted. The studies provide evidence that consumption of ground nuts have a cardioprotective effect on heart conditions such as coronary heart disease (CHD) and ischemia. These studies were Iowa Women's Health study, the Adventist Health study, the Cholesterol and Recurrent Events study, the Physician's Health study, and the Nurses' Health study (Ellsworth et al. 2001; Pfeffer et al. 1995; Bao et al. 2016). The Adventist Health study suggested that individuals who consumed groundnuts up to four times a week had a lower risk of fatal coronary heart disease and was crucial in the prevention of non-fatal myocardial infarctions (Fraser et al. 1992). Results of the Nurses' Health study showed that women who had maintained a steady consumption of more than 5 ounces of nuts per week had lower risk of developing CHD as compared to those that consumed lesser quantities or did not consume nuts (Bao et al. 2016). Similar results were observed in the Physicians' Health study that regular consumption of nuts was related to a lower risk of CHD and cardiac arrest (Li et al. 2009). Peanuts contain phytocompounds (such as poly-unsaturated fatty acids, phytosterols, tocopherols, resveratrol, and arginine) that reduce inflammation by downregulating the inflammatory marker proteins. Inflammation plays a key role in all stages of atherosclerosis; thus, the bioactive compounds (poly-unsaturated fatty acids antioxidants and L-arginine) in peanuts may prevent this by lowering inflammation. Nut consumption is inversely associated with the bloodstream concentrations of inflammatory molecules including C-reactive proteins, fibrinogen, and interleukin-6 (IL-6). Reduction in the inflammation factors is important in the prevention of CHD, the reason being that inflammation plays a key role in the development and progression of CHD. Another risk factor for CHD is endothelial dysfunction, which includes deformity of the endothelial lining of arteries, reduction in nitric acid production, and reduced cell adhesion. Poly-unsaturated fatty acids, vitamins, and antioxidants found in peanuts may reduce endothelial dysfunction by reducing hypercholesteremia (Kris-Etherton et al. 2008). Similarly, the use of peanut butter and peanut oil, as a replacement for regular oil, enhanced LDL oxidization as compared to a fat-rich diet (Hargrove et al. 2001).

4.1.6.2 Type 2 Diabetes

There has been recent interest in the role of peanuts with respect to type 2 diabetes. The Nurses' Health study indicated that regular consumption of peanuts reduces the risk of type 2 diabetes (Bao et al. 2016). Concrete evidence regarding the mechanisms through which peanut bioactive compound reduce the risk of type 2 diabetes still need to be investigated (Kris-Etherton et al. 2008). An overall 45% risk reduction for type 2 diabetes was observed in women consuming peanuts five times a week. However, no protective effect of peanut consumption against type 2 diabetes was indicated in men according to the Physicians' Health study (Li et al. 2009). Jiang et al. (2002) investigated the risk of type 2 diabetes in women and preventive role of peanuts in a prospective study. They concluded that peanut consumption (in the form of whole nut and peanut butter) inversely affected the relative risk of type II diabetes (Jiang et al. 2002). Barbour et al. investigated the role of regular peanut consumption for a period of 12 weeks on blood composition and diabetes risk markers such as BMI, waist circumference, blood glucose, lipids, and C-reactive protein. The results of the study indicated no significant difference in the participants' blood chemistry, insulin, and c-reactive protein levels. This may be due to the small duration of the trial in which the participants were given peanuts or possibly due to cohort type (Barbour et al. 2015).

4.1.6.3 Insulin Resistance and Glycemic Control

There is a lack in sufficient data regarding the underlying mechanisms that regulate insulin sensitivity and correlate with peanut consumption. Possibly regular consumption of peanuts gives the body a dose of specific high density fatty acids that are assimilated and incorporated in the cell membranes of skeletal muscles. The additional fatty acids in the membrane have the potential to alter insulin sensitivity which includes insulin receptor binding, altering the ion permeability of the cell membrane, and cell signaling (Kris-Etherton et al. 2008; Borkman et al. 1993).

4.1.6.4 Role in Cardiovascular Diseases

Several large-scale studies have been conducted which correlate peanut consumption with the risk of cardiovascular diseases. Supporting data from these studies further lead the investigation into the underlying mechanisms that confer peanuts with a cardioprotective effect. Peanut supplementation studies have a cross section that observes effects of peanut consumption on cardiovascular disease risk. Parameters for risk evaluation were lipid profile, presence of antioxidants, and inflammatory markers in the blood (Ellsworth et al. 2001; Duh and Yen 1997; Fraser et al. 1992; Pfeffer et al. 1995; Li et al. 2009). The following section will elaborate on the effect of peanut consumption on risk factors of cardiovascular disease.

Peanut Consumption and Lipid Profile

Lipid profile is a clear predictor of heart disease. Higher lipid content in the blood, especially cholesterol, and low-density lipoproteins (LDLs) are considered a greater risk as they have the potential to deposit in the blood vessels and initiate atherosclerosis. A large-scale survey consisting of 583 participants from 25 clinical trials from seven different countries was performed and was observed that peanut consumption had an inverse effect on the blood cholesterol levels. Participants consuming 67 g (per day) of nuts daily had a decrease in various types lipids in the bloodstream including cholesterol (5% decrease) and LDL cholesterol (7% decrease). Also, participants with a higher serum LDL profile had significant decline in LDL levels ($p < 0.001$) after peanut consumption (Sabaté et al. 2010). Flavonoids in peanuts are potent antioxidants which neutralize reactive oxygen species, prevent LDL oxidation and subsequent deposition of LDLs in the blood vessels (which can lead to atherosclerosis) and, thus, may be the bioactive compounds responsible for the cardioprotective properties of peanuts (Yen and Duh 1993).

Antioxidant Properties of Peanuts

Production of reactive oxygen and nitrogen species is a natural cellular process. However, highly reactive oxygen species such as singlet oxygen species, hydroperoxyl radical, superoxide anions, alkyl/peroxyl free radicals, and nitrogen superoxide radicals can cause more damage when produced in excess. Buildup of free radicals can lead to several conditions such as cancer, atherosclerosis, diabetes, and damage to cataracts (Knekt et al. 2002). Polyphenols found in peanuts have the ability to neutralize these free radicals such as singlet oxygen, superoxide anion, and lipid-peroxide radical by donating their free electrons. Peanut polyphenols can also prevent oxidative damage by chelating with ionized copper and iron which blocks the production of highly reactive hydroxyl species (Yang et al. 2005). Thus the antioxidants found in peanuts are effective against inflammation, oxidation damage, and removal of heavy metals from the body.

4.1.6.5 Anti-inflammatory Effects

Studies on the effect of peanut consumption on type 2 diabetes and cardiovascular diseases indicate a decrease in inflammatory molecules which in turn protect against these diseases (Yu et al. 2016; Barbour et al. 2015; Souza et al. 2015; Mazidi et al. 2016). A large-scale survey consisting of 5013 patients was conducted which correlated peanut consumption with the decrease in inflammatory biomarkers in the bloodstream (Jiang et al. 2005). The results indicated an inverse association with peanut consumption and inflammatory markers. Proinflammatory molecules such as CRP (C-reactive protein), fibrinogen, and IL-6 (interleukin-6) decreased in the bloodstream in individuals who had a regular peanut intake. Participants of the study included individuals with a high risk of cardiovascular disease. Their diet patterns were monitored and especially those who have a Mediterranean diet which included a regular intake of peanuts. Participants that adhered to the Mediterranean diet which included peanuts had significantly ($p < 0.003$) lower levels of IL-6, CRP, and ICAM (intracellular cell adhesion molecule-I) (Salas-Salvadó et al. 2008). Also, peanuts contain dietary fibers, vitamin E, phenols, and L-arginine which are implicated in anti-inflammatory pathways (Andoh et al. 1999).

4.1.6.6 Anticancer and Antitumor Potential of Peanuts

Nutritional components of peanuts including vitamins, minerals, and flavonoids are implicated in anticancer activity (González and Salas-Salvadó 2006), and phytosterols, abundant in peanuts, have known antioxidant properties and are studied as anticancer agents (Huang et al. 2010). Phytosterols isolated from peanut had an antitumor and antimetastatic effect on SCID (severe combined immunodeficiency) mice injected with human prostate cancer cells (PC-3 cell line). The mice were given a diet containing 2% phytosterols after inoculation with tumor cells, and tumor growth was monitored for 8 weeks. There was a 43% decrease in the number of tumors and 50% reduction in secondary metastasis as compared with control groups (Woyengo et al. 2009). Peanuts are a rich source of resveratrol (1920 ng) (Sanders et al. 2000), a compound that downregulates in vivo tumor angiogenesis and cell division in colon cancer, breast cancer, liver cancer, and pancreatic cancer (Fazel Nabavi et al. 2014; Chatterjee et al. 2011; Alfaras et al. 2010).

However, there are no studies published that include direct consumption of peanuts as anticancers, the reason being that daily recommended intake of peanuts (60–100 g) does not have the effective concentrations of resveratrol necessary for the anticancer effect (Shankar et al. 2011; Salado et al. 2011). However, peanut sprouts and roots offer a cheap source of the potent anticancer resveratrol (Xiong et al. 2014).

Stilbene compounds in peanuts have excellent antitumor properties such as inhibition of tumor cell growth, tumor progression, and metastasis. They are also potent antioxidants that reduce free radical formation which can be a cause of tumor formation. Stilbenes also have anti-mutagenic properties as they cause the production of quinine reductase enzyme that has the ability to reduce inflammation and detoxifying carcinogenic compounds. They are capable of partially inhibiting cyclooxygenase that in turn reduces prostaglandin formation which is responsible for tumor growth and activation of carcinogens (Cassidy et al. 2000). Nuclear factor kappa B (NFκB) is a transcription factor that has a known role in tumor development and progression. In several types of cancers, an altered activity of NFκB leads to DNA damage and altered cell cycles. Stilbenoids present in peanuts have a negative effect on NFκB-directed transcription in cell lines. Inhibition on NFκB activity usually results in cell apoptosis. However, the anticancer activity of stilbenes is not tumor specific; therefore, they are suggested as chemopreventive agent's due to their antioxidant ability (Sobolev et al. 2011).

4.1.6.7 Reducing the Risk for Alzheimer's Disease

High levels of niacin (13.52×10^{-3} g) and vitamin E (69.3×10^{-4} g) in peanuts, which both are effective against the development of Alzheimer's disease and slow down cognitive decline due to aging (Fata et al. 2014; Morris et al. 2004).Research suggests that niacin- and vitamin E-rich foods are have a protective role and are effective against cognitive decline (Morris et al. 2004). Resveratrol found in peanuts also exhibits protective properties against Alzheimer's disease by preventing nerve degeneration (Chen et al. 2005). Peanut intake can therefore slow down age-related neuronal degeneration by reducing oxidative stress in the neurons (Butterfield et al. 2002).

4.1.7 Adverse Effects and Individual Concerns

4.1.7.1 Allergies

Consumption of peanuts can lead to severe allergic reactions and, if left unattended, can even cause anaphylactic shock and ultimately death. The exact reason for such a strong immune response is unknown, but the allergy is mostly from the two types of proteins: albumins (water soluble) and globulins (saline soluble) present in the kernels (cotyledons) of the peanut (Barnett et al. 1983). Consumption of peanuts in allergic persons initiates an immune response mainly driven by immunoglobulin E (IgE) on mast cells that detects peanut allergens and releases massive quantities of histamine as a response. This in turn triggers constriction of the lung bronchioles leading to a bronchospasm. The result of such an excessive immune response is swelling of the face and face due to angioedema, vomiting, diarrhea, asthma, ectopic eczema, and anaphylactic shock (Galli et al. 2008).

Individuals allergic to peanuts cannot consume peanuts, inhale peanut dust, or even tolerate minor quantities in food. The prevalence of peanut allergy is on the rise with greater cases being reported each year (Grundy et al. 2002; Venter et al. 2010; Venter et al. 2016). Treatments to eliminate peanut allergies are underway including anti IgE therapy, Chinese medicine, intake of probiotics, engineered allergen immunotherapy, and bacterial adjuvants (Narisety et al. 2015; Wang et al. 2015).

4.1.7.2 Peanuts and Aflatoxins

Peanuts stored in moist dark environments can become infected with a fungus known as *Aspergillus flavus* that produces a potent aflatoxin, a known toxin and carcinogen. Other factors that contribute to toxin production are the soil temperature and surrounding weather during pod filling (Lavkor and Var 2017).

4.1.8 Food Applications

4.1.8.1 As Protein Sources (Protein Concentrates and Protein Isolates)

In the present time, there is a high demand for cheap protein to feed the masses in poverty-stricken regions, or there is a steady rise in individuals preferring the vegetarian diet (Gray 2015). In order to cope with the increasing demand for plant-based proteins, industries have turned to peanuts as they contain sufficient protein (26 g per 100 g peanut). Proteins are extracted from the peanut meal which is a by-product of peanut oil extraction and is rich in protein (50–55% of the husk). The peanut meal is used to formulate a high-quality protein concentrate (Table 4.10). High-quality protein concentrate obtained from the peanut flour has numerous applications in various food industries. "Alkali extraction" method is commonly used to isolate proteins from the meal, and then it is subjected to isoelectric precipitation (Adebowale et al. 2011). However, this method leaves a nutty and beany flavor in the protein concentrate. Recently a newer technique that uses membrane filtration to separate protein content from peanut flour has successfully reduced nutty/beany aroma to a marked extent and, thus, helped in producing a better quality of protein in terms of sensory characteristics and functionality (Jain et al. 2015).

Protein isolates contain 90% or more protein content than concentrates and meals. They are produced by solubilizing the proteins present in peanuts in an alkaline solution followed by precipitation with an acid. Peanut protein isolates are used to produce tasteless powder that have applications in breads and baked goods (Jain et al. 2015).

Table 4.10 Overview of different applications of safflower in food industries (Singh and Singh 1991)

Ingredient	Product	Principal constituent
Peanut flour (20%)	Chapatis	Wheat
Peanut flour (12.5%)	Bread	Wheat
Peanut flour (15%)	Noodle	Wheat
Peanut meal	Epa-ogi	Corn
Peanut meal (7%)	Extrudates	Corn
Peanut meal (20%)	Laddus	Millets
Peanut meal (10%)	Nshima	Corn
Peanut flour (30%)	Kisra	Sorghum
Peanut flour (15%)	"Toe"	Sorghum
Peanut flour (10%)	Chapatis	Wheat—chickpea
Peanut flour (20%)	Bread	Wheat—cowpea
Peanut flour (35%)	Cookies	Wheat—cowpea
Peanut grits (22%)	Snacks	Corn—oat
Peanut flour (10%)	Bread	Wheat—soybean
Peanut flour (10%)	Chapatis	Corn—chickpea
Peanut flour	Akara	Cowpea
Peanut grits (10%)	Gari	Cassava
Peanut (20%)	Tempeh	Soybean
Peanut (100%)	Oncorn	Peanut
Peanut	Snacks—chikki	Chickpea
Peanut flour (50%)	Wheat biscuits	Wheat biscuits
Protein isolate (22%)	Wheat biscuits	Wheat biscuits
Peanut paste	Wheat marzipan	Wheat marzipan
Peanut flour	Millets biscuits	Millets biscuits
Peanut flour (10%)	Wheat—cowpea cake	Wheat—cowpea cake
Peanut flour	Ragi—soybean biscuits	Ragi—soybean biscuits
Roasted peanuts	Sorghum—chickpea candles	Sorghum—chickpea candles
Peanut flour (100%)	Peanut muffin	Peanut muffin
Peanut milk	Corn muffin	Corn muffin
Peanut (8%)	Whey chocolate shake	Whey chocolate shake
Peanut concentrate	Carob milk cup confection	Carob milk cup confection
Protein isolate	Peanut ice cream	Peanut ice cream
Protein isolate	Buffalo milk ice cream	Buffalo milk ice cream
Partially defatted	Peanut beverage	Peanut beverage
Peanut meal	Peanut panned product	Peanut panned product
Peanut solids	Whey shaka	Whey shaka
Peanut meal (20%)	Beef sausage	Beef sausage
Roasted peanuts	Peanut crunch	PeanutcCrunch
Peanut flour (10%)	Sweet potato—soybean cake	Sweet potato—soybean cake

4.1.8.2 In Snacks and Bakery Products

Peanuts are used in numerous snacks and bakery products such as peanut butter crackers and sandwiches and in peanut biscuits, cereals, ice creams, and candies. Peanuts in combination with sugar, malt, yeast, lecithin, honey, and synthetic flavors have been used to make various bakery items (Zhao et al. 2012). Peanut flour can also be added to the bread to increase its nutritional value (Adeboye et al. 2018).

Peanut Flour and Its Use in Baked Goods

It is obtained by removing the skin from full fat peanuts and drying them till their moisture content is reduced to 2–4%. Dried peanuts are then ground to a fine powder and later on dried again until they have a moisture content of 2%. Peanut flours have multiple uses such as in breads, baked goods, cereals, and flakes for breakfast, snacks, drinks, soups, ice creams, spreads, meat patties, and frostings (Zhao et al. 2012; Lusas 1979).

Peanut Butter as an Alternate to Dairy Butter

The use of a combination of peanut butter and ghee (50:50) imparts an improved texture, flavor, and crispiness to biscuits when compared to using only dairy butter (Gajera et al. 2010).

4.1.8.3 In Chocolates and Confectionary

Peanuts are added to confections and chocolates to enhance their flavor. The sweet and savory combination is generally well liked and thus is an integral part of a lot of candies. Almost 60% of the chocolates and candies manufactured in America contain nuts. The use of peanuts in confections comes second only to chocolates (Lees 2012).

4.1.8.4 In Dairy Products

Protein isolates have been used to produce vegetable-based milk known as “Miltone” which has a long shelf life. Similarly, peanut isolate has been used to produce dairy free yoghurt and cheese (Lees 2012).

4.1.8.5 In Beverage Industry

Developing countries where large numbers of individuals do not have access to milk and other dairy products, peanut-derived beverages can be used to make protein-rich drinks. In one study a chocolate-flavored peanut protein-based drink was made from protein isolates (from peanut), butter, sugar, stabilizer, cocoa, and water. The taste, texture, and aroma were similar to commercially available milk products (Chompreeda et al. 1989).

4.1.8.6 In Meat and Meat Products

Peanuts are used in combination with meat for stew preparation such as in African cuisine. They can also be ground together as a base for soups such as in Philippine cuisine and used to thicken or add texture to spicy pepper sauces and meat such as in Latin cuisine. Crude peanut powder has been proposed as a meat stabilizer (for coating the meat); however, soy-based alternatives are more popular (Angelo and Mann 1973).

4.1.8.7 Use of Peanuts in Ready-to-Use Therapeutic Foods (RUTFs)

Ready-to-use therapeutic foods are used in famine or disaster-struck regions as a remedy for malnourishment. RUTFs are usually given to children that suffer from severe malnourishment. RUTF is helpful in rehabilitation of severely malnourished individuals so that they can be brought to a healthy weight. Essentially the RUTF is an oil-based emulsion, rich in calories, discourages bacterial growth, and is ready-to-eat without the need for cooking. Peanuts are a major component of RUTF since they can help with healthy weight gain. Additional components of RUTF include sugar, milk powder, vegetable oil, vitamins, and minerals. All the ingredients are ground up to a uniform paste, with particle size of less than 200 microns. The whole process is done without the addition of water. The major advantages of RUTF are the ease of large-scale production, cost-effective, and long shelf life (Manary 2006). Commercially available peanut-based RUTF is called Plumpy'nut, which is manufactured by a French company called Nutriset (Nackers et al. 2010). Similar to other RUTFs, the Plumpy'nut is used in emergency malnutrition cases and can be administered from home (virtually eliminates the need for hospitalization). The product is in the form of a peanut butter-like paste containing fats, carbohydrates, dietary fiber, proteins, vitamins, and minerals. The unique composition of this peanut-based RUTF causes healthy weight gain and also maintains it (Smith et al. 2013). The Plumpy'nut has been successfully given to malnourished children in real-life scenario in Niger (Nackers et al. 2010), where its administration led to a better recovery rate, healthier weight gain, and low rate of transfers to an intensive care unit. However a similar trial in Bangladesh (Ali et al. 2013) was not so successful, with the participants complaining of vomiting, diarrhea, and abdominal pain thus making it difficult to incorporate into the diet.

4.1.8.8 Peanut Oil

Peanut oil is produced through a cold press and subjected to further refinement. Peanut oil is highly aromatic and has a mild peanut flavor (Hathorn and Sanders 2012). It is mainly used in Chinese, American, and Southeast Asian foods. Unrefined peanut oil is used as a dressing or as a condiment, whereas refined peanut oil can be used to make fried goods in large scale since it is cost-effective (Wang 2016). Peanut oil was one of the first oils to be used to produce biodiesel (Gunstone 2011; Jazie et al. 2012).

4.1.8.9 Antioxidant Extracts

Antioxidant extracts from peanuts are a healthy alternative to synthetic preservatives that are added to cooking oil to prevent oxidation. Cooking oils when oxidized turn rancid and have an altered flavor: therefore they are unfit for consumption. On an industrial level, artificial preservatives are added to prevent oxidation of cooking oils. A healthy alternate to artificial antioxidants are natural ones extracted from various foods. Craft et al. through the process of high-performance liquid chromatography (HPLC) which yielded 80% v/v extracts of p-coumarin and p-coumarin derivatives (antioxidants) (Craft et al. 2010). Antioxidant extracts from peanuts offer a healthy alternate to synthetic preservatives. Peanut antioxidants, due to their thermal stability, are an ideal choice (Taghvaei and Jafari 2015).

4.1.9 Biomedical Applications

Protein extracts from peanuts were initially used to develop biofilms, which act as a support matrix for attachment of animal cells, e.g., fibroblasts. Peanut proteins are extracted from peanut meal and subjected to compression molding. This gave rise to thermoplastic protein films. These films have potential applications in biomedical devices or as attachment media for cells to grow (Reddy et al. 2013). However, they had a lower tensile strength, so the peanut protein biofilm mesh was cross-linked with biocompatible citric acid. This additional treatment solved the issue of tensile and wet strength; however it behaved poorly as a device for attachment to animal cells (such as fibroblasts). In conclusion, these peanut protein films did not support the attachment and growth of mouse fibroblast cells suggesting that peanut proteins were cytotoxic (Reddy et al. 2012). Further research is needed to curb the toxicity of peanut protein-based biofilms, as the material is quite cost-effective to produce. Another study developed and characterized oral disintegrating films based on gelatin and hydroxypropyl methylcellulose incorporated with peanut skin extract (PSE) to release phenolic compounds. The in vitro release profile showed that within 5 min 80% of phenolic compounds were released, and the films retained 60% of the total phenolics in the accelerated stability test. These films offer good alternative to deliver the active compounds of the peanut skin (Adeboye et al. 2018).

The modified peanut protein isolate-based films incorporating thymol showed increased total phenolic content, antioxidant capacity, and antimicrobial activity against Gram-positive *Staphylococcus aureus* and *Lactobacillus plantarum* than Gram-negative *Escherichia coli* and *Pseudomonas aeruginosa*. These edible, modified peanut protein films have the potential to preserve food products, active food packaging (Zhong et al. 2017).

4.1.10 Alternative Applications

There are various applications of peanut-derived products in the manufacturing industry. Peanut oil can be used to make varnishes, leather lotions, lubricants, paints/polishes, and nitroglycerin. Peanut oil is saponified to manufacture soap and also as a constituent of cosmetics such as in lotions and creams. Peanut shell is the major waste product after harvesting, which can be used in the manufacture of plastic, abrasives, wallboard, paper, and glue (Schnepf 2016).

4.1.11 Future Challenges

4.1.11.1 Excessive By-Product Waste

The biggest concern with peanut agriculture is the production of excessive by-products in the form of hulls and shells. With the gradual increase in production, China (the largest producer of peanuts) is expected to produce annual 5 million metric tons of peanuts per year. This also implies that the number of by-products will also increase and create a waste disposal issue. Therefore, this must be responsibly dealt with by either extracting useful components from the by-products or incinerating them (Zhao et al. 2012).

4.1.11.2 Global Warming

Increase in the frequency and intensity of droughts and heat waves due to climate change has greatly affected the peanut agriculture (Hatfield and Prueger 2015).

Grains like peanuts, wheat, oats, and barley require lower temperatures for grain filling. However, shorter winters and longer summers has greatly reduced the yield. Recent simulations of future climatic conditions revealed that peanut production will be almost down to zero with soaring daytime temperatures and low nighttime temperatures, even with the increased carbon dioxide in the atmosphere. This is because peanuts require a warm climate for optimum growth; however, extremely high temperature, heat waves, and drought are counterproductive for yield. Therefore, scientists are currently working on developing drought-resistant species of peanut plants that maintain a high yield even under hot and dry conditions (Kottapalli et al. 2009).

4.1.12 Conclusion

Arachis hypogaea, commonly known as the groundnut or peanut, is an annual herbaceous legume and is mostly grown in in tropical and temperate. Peanuts are a composite food consisting of a wide variety of nutrients, such as carbohydrates, proteins, lipids, ten types of vitamins, ten types of minerals, and a good dose of fiber. Bioactive compounds have also been isolated from peanuts that include flavonoids, phytosterols, amino acids, and stilbenes. Large-scale clinical studies have shown that regular peanut consumption has a positive effect on CHD, type 2 diabetes, Alzheimer's, and CVD. These bioactive compounds also have anti-inflammatory, antioxidant, anticancer, and antitumor activities. Potential health concerns regarding peanuts include allergies and contamination with aflatoxins. Peanuts are widely used in the food industry for the production of flour, protein concentrates and isolates, confectionaries, oils, and beverages. However, global warming is posing a great threat to peanut cultivation which can drastically reduce its yield over the coming years, and hence, necessary steps should be taken to control the situation.

References

Adebowale YA, Schwarzenbolz U, Henle T (2011) Protein isolates from Bambara groundnut (Voandzeia Subterranean L.): chemical characterization and functional properties. Int J Food Prop 14(4):758–775

Adeboye A, Fayemi O, Bamgbose A, Adewunmi A, Sobowale S (2018) Towards the development of peanut–wheat flour composite dough: influence of reduced-fat peanut flour on bread quality. J Food Process Pres 42(1):e13385

Agati G, Brunetti C, Di Ferdinando M, Ferrini F, Pollastri S, Tattini M (2013) Functional roles of flavonoids in photoprotection: new evidence, lessons from the past. Plant Physiol Biochem 72:35–45

Aires V, Delmas D, Le Bachelier C, Latruffe N, Schlemmer D, Benoist J-F, Djouadi F, Bastin J (2014) Stilbenes and resveratrol metabolites improve mitochondrial fatty acid oxidation defects in human fibroblasts. Orphanet J Rare Dis 9(1):79

Akhtar S, Khalid N, Ahmed I, Shahzad A, Suleria HAR (2014) Physicochemical characteristics, functional properties, and nutritional benefits of peanut oil: a review. Crit Rev Food Sci Nutr 54 (12):1562–1575

Alfaras I, Juan ME, Planas JM (2010) trans-Resveratrol reduces precancerous colonic lesions in dimethylhydrazine-treated rats. J Agric Food Chem 58(13):8104–8110

Ali E, Zachariah R, Dahmane A, Van den Boogaard W, Shams Z, Akter T, Alders P, Manzi M, Allaouna M, Draguez B (2013) Peanut-based ready-to-use therapeutic food: acceptability among malnourished children and community workers in Bangladesh. Public Health Action 3 (2):128–135

Andoh A, Bamba T, Sasaki M (1999) Physiological and anti-inflammatory roles of dietary fiber and butyrate in intestinal functions. J Parenter Enteral Nutr 23(5_suppl):S70–S73

Angelo A, Mann G (1973) Peanut culture uses. Am Peanut Res Ed Soc Stone Printing Co, Raonote, p 561

Arya SS, Salve AR, Chauhan S (2016) Peanuts as functional food: a review. J Food Sci Technol 53 (1):31–41

Atasie V, Akinhanmi T, Ojiodu C (2009) Proximate analysis and physico-chemical properties of groundnut (Arachis hypogaea L.). Pak J Nutr 8(2):194–197

Bao Y, Bertoia ML, Lenart EB, Stampfer MJ, Willett WC, Speizer FE, Chavarro JE (2016) Origin, methods, and evolution of the three Nurses' Health Studies. Am J Public Health 106 (9):1573–1581

Barbour JA, Howe PR, Buckley JD, Bryan J, Coates AM (2015) Effect of 12 weeks high oleic peanut consumption on cardio-metabolic risk factors and body composition. Nutrients 7 (9):7381–7398

Barnett D, Baldo BA, Howden ME (1983) Multiplicity of allergens in peanuts. J Allergy Clin Immunol 72(1):61–68

Baur JA, Sinclair DA (2006) Therapeutic potential of resveratrol: the in vivo evidence. Nat Rev Drug Discov 5(6):493–506

Blomhoff R, Carlsen MH, Andersen LF, Jacobs DR (2006) Health benefits of nuts: potential role of antioxidants. Br J Nutr 96(S2):S52–S60

Borkman M, Storlien LH, Pan DA, Jenkins AB, Chisholm DJ, Campbell LV (1993) The relation between insulin sensitivity and the fatty-acid composition of skeletal-muscle phospholipids. N Engl J Med 328(4):238–244

Bortolomeazzi R, Cordaro F, Pizzale L, Conte LS (2003) Presence of phytosterol oxides in crude vegetable oils and their fate during refining. J Agric Food Chem 51(8):2394–2401

Brunetti C, Di Ferdinando M, Fini A, Pollastri S, Tattini M (2013) Flavonoids as antioxidants and developmental regulators: relative significance in plants and humans. Int J Mol Sci 14 (2):3540–3555

Butterfield DA, Castegna A, Pocernich CB, Drake J, Scapagnini G, Calabrese V (2002) Nutritional approaches to combat oxidative stress in Alzheimer's disease. J Nutr Biochem 13(8):444–461

Byrd-Bredbenner C, Moe G, Beshgetoor D, Berning J, Kelley D (2014) Wardlaw's perspectives in nutrition: a functional approach. McGraw-Hill

Campos-Mondragón M, De La Barca AC, Durán-Prado A, Campos-Reyes L, Oliart-Ros R, Ortega-García J, Medina-Juárez L, Angulo O (2009) Nutritional composition of new peanut (Arachis hypogaea L.) cultivars. Grasas Aceites 60(2):161–167

Cassidy A, Hanley B, Lamuela-Raventos RM (2000) Isoflavones, lignans and stilbenes–origins, metabolism and potential importance to human health. J Sci Food Agric 80(7):1044–1062

Chatterjee M, Das S, Janarthan M, Ramachandran HK, Chatterjee M (2011) Role of 5-lipoxygenase in resveratrol mediated suppression of 7, 12-dimethylbenz (α) anthracene-induced mammary carcinogenesis in rats. Eur J Pharmacol 668(1–2):99–106

Chen J, Zhou Y, Mueller-Steiner S, Chen L-F, Kwon H, Yi S, Mucke L, Gan L (2005) SIRT1 protects against microglia-dependent amyloid-β toxicity through inhibiting NF-κB signaling. J Biol Chem 280(48):40,364–40,374

Chompreeda P, Haruthaithanasan V, Oupadissakoon C, Suknak K (1989) Development of a chocolate flavored peanut beverage. J Food Sci 54(5):1359–1360
Chung I-M, Park MR, Chun JC, Yun SJ (2003) Resveratrol accumulation and resveratrol synthase gene expression in response to abiotic stresses and hormones in peanut plants. Plant Sci 164 (1):103–109
Conde Nast UNND (version SR-21) (2014) Nutrition facts for peanuts, all types, raw, USDA Nutrient Data. Accessed 24 Oct 2017
Craft BD, Kosińska A, Amarowicz R, Pegg RB (2010) Antioxidant properties of extracts obtained from raw, dry-roasted, and oil-roasted US peanuts of commercial importance. Plant Foods Hum Nutr 65(3):311–318
Davis JN, Medbery C, Sharma S, Pablo J, Kimsey F, Perry D, Muacevic A, Mahadevan A (2015) Stereotactic body radiotherapy for centrally located early-stage non-small cell lung cancer or lung metastases from the RSSearch® patient registry. Radiat Oncol 10(1):113
Dorner J (2008) Management and prevention of mycotoxins in peanuts. Food Addit Contam 25 (2):203–208
Duh P-D, Yen G-C (1995) Changes in antioxidant activity and components of methanolic extracts of peanut hulls irradiated with ultraviolet light. Food Chem 54(2):127–131
Duh P-D, Yen G-C (1997) Antioxidant efficacy of methanolic extracts of peanut hulls in soybean and peanut oils. J Am Oil Chem Soc 74(6):745
Duh P-D, Yeh D-B, Yen G-C (1992) Extraction and identification of an antioxidative component from peanut hulls. J Am Oil Chem Soc 69(8):814–818
Ejigui J, Savoie L, Marin J, Desrosiers T (2005) Influence of traditional processing methods on the nutritional composition and antinutritional factors of red peanuts (Arachis hypogea) and small red kidney beans (Phaseolus vulgaris). J Biol Sci 5(5):597–605
Elkowicz K, Sosulski F (1982) Antinutritive factors in eleven legumes and their air-classified protein and starch fractions. J Food Sci 47(4):1301–1304
Ellsworth J, Kushi L, Folsom A (2001) Frequent nut intake and risk of death from coronary heart disease and all causes in postmenopausal women: the Iowa Women's Health Study. Nutr Metab Cardiovasc Dis 11(6):372–377
FAOSTAT (2016) Food and Agriculture Organization of the United Nations, Rome, Italy. Available from http://www.fao.org/faostat/en/#data
Fata GL, Weber P, Mohajeri MH (2014) Effects of vitamin E on cognitive performance during ageing and in Alzheimer's disease. Nutrients 6(12):5453–5472
Fazel Nabavi S, Li H, Daglia M, Mohammad Nabavi S (2014) Resveratrol and stroke: from chemistry to medicine. Curr Neurovasc Res 11(4):390–397
Feldman EB (2002) The scientific evidence for a beneficial health relationship between walnuts and coronary heart disease. J Nutr 132(5):1062S–1101S
Fenech M (2012) Folate (vitamin B9) and vitamin B12 and their function in the maintenance of nuclear and mitochondrial genome integrity. Mutat Res 733(1):21–33
Francisco MLD, Resurreccion A (2008) Functional components in peanuts. Crit Rev Food Sci Nutr 48(8):715–746
Fraser GE, Sabate J, Beeson WL, Strahan TM (1992) A possible protective effect of nut consumption on risk of coronary heart disease: the Adventist Health Study. Arch Intern Med 152 (7):1416–1424
Gajera HP, Kapopara MB, Patel VH (2010) Application of peanut butter to improve fatty acid composition of biscuits. J Food Sci Technol 47(3):285–289. https://doi.org/10.1007/s13197-010-0044-y
Galli SJ, Tsai M, Piliponsky AM (2008) The development of allergic inflammation. Nature 454 (7203):445–454
González CA, Salas-Salvadó J (2006) The potential of nuts in the prevention of cancer. Br J Nutr 96 (S2):S87–S94
Gray EK (2015) The Vegetarian Crusade: The Rise of an American Reform Movement, 1817–1921. J South Hist 81(2):451

Griel AE, Kris-Etherton PM (2006) Tree nuts and the lipid profile: a review of clinical studies. Br J Nutr 96(S2):S68–S78

Grundy J, Matthews S, Bateman B, Dean T, Arshad SH (2002) Rising prevalence of allergy to peanut in children: data from 2 sequential cohorts. J Allergy Clin Immunol 110(5):784–789

Gunstone F (2011) Vegetable oils in food technology: composition, properties and uses. Wiley, Chichester, West Sussex

Gylling H, Simonen P (2015) Phytosterols, phytostanols, and lipoprotein metabolism. Nutrients 7 (9):7965–7977

Hargrove RL, Etherton TD, Pearson TA, Harrison EH, Kris-Etherton PM (2001) Low fat and high monounsaturated fat diets decrease human low density lipoprotein oxidative susceptibility in vitro. J Nutr 131(6):1758–1763

Hatfield JL, Prueger JH (2015) Temperature extremes: Effect on plant growth and development. Weather Clim Extremes 10:4–10. https://doi.org/10.1016/j.wace.2015.08.001

Hathorn CS, Sanders TH (2012) Flavor and antioxidant capacity of peanut paste and peanut butter supplemented with peanut skins. J Food Sci 77(11):S407–S411. https://doi.org/10.1111/j.1750-3841.2012.02953.x

Havsteen BH (2002) The biochemistry and medical significance of the flavonoids. Pharmacol Ther 96(2):67–202

He S, Hekmat S (2014) Sensory evaluation of non-dairy probiotic beverages. J Food Res 4(1):186

He FJ, MacGregor GA (2008) Beneficial effects of potassium on human health. Physiol Plant 133 (4):725–735

Henriques BJ, Olsen RK, Bross P, Gomes CM (2010) Emerging roles for riboflavin in functional rescue of mitochondrial β-oxidation flavoenzymes. Curr Med Chem 17(32):3842–3854

Herzig K-H, Bardocz S, Grant G, Nustede R, Fölsch U, Pusztai A (1997) Red kidney bean lectin is a potent cholecystokinin releasing stimulus in the rat inducing pancreatic growth. Gut 41 (3):333–338

Ho C-T, Mussinan C, Shahidi F, Contis ET (2010) Recent advances in food and flavor chemistry: food flavors and encapsulation, health benefits, analytical methods, and molecular biology of functional foods. Royal Society of Chemistry, Cambridge, UK

Huang C-P, Au L-C, Chiou RY-Y, Chung P-C, Chen S-Y, Tang W-C, Chang C-L, Fang W-H, Lin S-B (2010) Arachidin-1, a peanut stilbenoid, induces programmed cell death in human leukemia HL-60 cells. J Agric Food Chem 58(23):12123–12129

Jacobson EL, Kim H, Kim M, Jacobson M (2012) Niacin: vitamin and antidyslipidemic drug. In: Stanger O (ed) Water soluble vitamins. Springer, Dordrecht, pp 37–47

Jahnen-Dechent W, Ketteler M (2012) Magnesium basics. Clin Kidney J 5(Suppl_1):i3–i14

Jain A, Prakash M, Radha C (2015) Extraction and evaluation of functional properties of groundnut protein concentrate. J Food Sci Technol 52(10):6655–6662

Jazie AA, Sinha A, Pramanik H (2012) Optimization of biodiesel production from peanut and rapeseed oils using response surface methodology. Int J Biomass Renew 1(2):9–18

Jiang R, Manson JE, Stampfer MJ, Liu S, Willett WC, Hu FB (2002) Nut and peanut butter consumption and risk of type 2 diabetes in women. JAMA 288(20):2554–2560

Jiang R, Jacobs DR Jr, Mayer-Davis E, Szklo M, Herrington D, Jenny NS, Kronmal R, Barr RG (2005) Nut and seed consumption and inflammatory markers in the multi-ethnic study of atherosclerosis. Am J Epidemiol 163(3):222–231

Kelly JH, Sabaté J (2006) Nuts and coronary heart disease: an epidemiological perspective. Br J Nutr 96(S2):S61–S67

Knekt P, Kumpulainen J, Järvinen R, Rissanen H, Heliövaara M, Reunanen A, Hakulinen T, Aromaa A (2002) Flavonoid intake and risk of chronic diseases. Am J Clin Nutr 76(3):560–568

Kochert G, Stalker HT, Gimenes M, Galgaro L, Lopes CR, Moore K (1996) RFLP and cytogenetic evidence on the origin and evolution of allotetraploid domesticated peanut, Arachis hypogaea (Leguminosae). Am J Bot 83:1282–1291

Kottapalli KR, Rakwal R, Shibato J, Burow G, Tissue D, Burke J, Puppala N, Burow M, Payton P (2009) Physiology and proteomics of the water-deficit stress response in three contrasting peanut genotypes. Plant Cell Environ 32(4):380–407

Kris-Etherton PM, Hu FB, Ros E, Sabaté J (2008) The role of tree nuts and peanuts in the prevention of coronary heart disease: multiple potential mechanisms. J Nutr 138(9):1746S–1751S

Ku K-L, Chang P-S, Cheng Y-C, Lien C-Y (2005) Production of stilbenoids from the callus of Arachis hypogaea: a novel source of the anticancer compound piceatannol. J Agric Food Chem 53(10):3877–3881

Lavkor I, Var I (2017) The control of aflatoxin contamination at harvest, drying, pre-storage and storage periods in peanut: the new approach. In: Abdulra'uf LB (ed) Aflatoxin-control, analysis, detection and health risks. InTech, Rijeka

Lees R (2012) Sugar confectionery and chocolate manufacture. Springer Science & Business Media, New York

Li TY, Brennan AM, Wedick NM, Mantzoros C, Rifai N, Hu FB (2009) Regular consumption of nuts is associated with a lower risk of cardiovascular disease in women with type 2 diabetes. J Nutr 139(7):1333–1338

Likhtenshtein G (2009) Stilbenes: applications in chemistry, life sciences and materials science. Wiley, Weinheim

Lou H, Yuan H, Yamazaki Y, Sasaki T, Oka S (2001) Alkaloids and flavonoids from peanut skins. Planta Med 67(04):345–349

Lusas E (1979) Food uses of peanut protein. J Am Oil Chem Soc 56(3):425–430

Manary MJ (2006) Local production and provision of ready-to-use therapeutic food (RUTF) spread for the treatment of severe childhood malnutrition. Food Nutr Bull 27(3_suppl3):S83–S89

Manzetti S, Zhang J, van der Spoel D (2014) Thiamin function, metabolism, uptake, and transport. Biochemistry 53(5):821–835

Mazidi M, Rezaie P, Ferns GA, Gao H-k (2016) Impact of different types of tree nut, peanut, and soy nut consumption on serum C-reactive protein (CRP): a systematic review and meta-analysis of randomized controlled clinical trials. Medicine 95(44):e5165

Mierziak J, Kostyn K, Kulma A (2014) Flavonoids as important molecules of plant interactions with the environment. Molecules 19(10):16240–16,265

Mora-Escobedo R, Hernández-Luna P, Joaquín-Torres IC, Ortiz-Moreno A, Robles-Ramírez MC (2015) Physicochemical properties and fatty acid profile of eight peanut varieties grown in Mexico. CyTA J Food 13(2):300–304

Moreau RA, Nyström L, Whitaker BD, Winkler-Moser JK, Baer DJ, Gebauer SK, Hicks KB (2018) Phytosterols and their derivatives: Structural diversity, distribution, metabolism, analysis, and health-promoting uses. Prog Lipid Res 70:35

Morris MC, Evans DA, Bienias JL, Scherr PA, Tangney CC, Hebert LE, Bennett D, Wilson RS, Aggarwal N (2004) Dietary niacin and the risk of incident Alzheimer's disease and of cognitive decline. J Neurol Neurosurg Psychiatry 75(8):1093–1099

Mukuddem-Petersen J, Oosthuizen W, Jerling JC (2005) A systematic review of the effects of nuts on blood lipid profiles in humans. J Nutr 135(9):2082–2089

Nackers F, Broillet F, Oumarou D, Djibo A, Gaboulaud V, Guerin PJ, Rusch B, Grais RF, Captier V (2010) Effectiveness of ready-to-use therapeutic food compared to a corn/soy-blend-based pre-mix for the treatment of childhood moderate acute malnutrition in Niger. J Trop Pediatr 56(6):407–413

Narisety SD, Frischmeyer-Guerrerio PA, Keet CA, Gorelik M, Schroeder J, Hamilton RG, Wood RA (2015) A randomized, double-blind, placebo-controlled pilot study of sublingual versus oral immunotherapy for the treatment of peanut allergy. J Allergy Clin Immunol 135(5):1275–1282

Nash SD, Nash DT (2008) Nuts as part of a healthy cardiovascular diet. Curr Atheroscler Rep 10 (6):529–535

Ostlund RE Jr (2004) Phytosterols and cholesterol metabolism. Curr Opin Lipidol 15(1):37–41

Pendse R, Rao AR, Venkataraman K (1973) 5, 7-Dihydroxychromone from Arachis hypogoea shells. Phytochemistry 12(8):2033–2034

Pfeffer MA, Sacks FM, Moyé LA, Brown L, Rouleau JL, Hartley LH, Rouleau J, Grimm R, Sestier F, Wickemeyer W (1995) Cholesterol and recurrent events: a secondary prevention trial for normolipidemic patients. Am J Cardiol 76(9):98C–106C

Racette SB, Lin X, Ma L, Ostlund RE Jr (2015) Natural dietary phytosterols. J AOAC Int 98 (3):679–684

Rackis J, Wolf W, Baker E (1986) Protease inhibitors in plant foods: content and inactivation. In: Friedman M (ed) Nutritional and toxicological significance of enzyme inhibitors in foods. Springer, New York, pp 299–347

Rebello C, Greenway F, Finley JW (2014) A review of the nutritional value of legumes and their effects on obesity and its related co-morbidities. Obes Rev 15(5):392–407

Reddy N, Jiang Q, Yang Y (2012) Preparation and properties of peanut protein films crosslinked with citric acid. Ind Crop Prod 39:26–30. https://doi.org/10.1016/j.indcrop.2012.02.004

Reddy N, Chen L, Yang Y (2013) Thermoplastic films from peanut proteins extracted from peanut meal. Ind Crop Prod 43:159–164. https://doi.org/10.1016/j.indcrop.2012.06.051

Rizvi S, Raza ST, Ahmed F, Ahmad A, Abbas S, Mahdi F (2014) The role of vitamin E in human health and some diseases. Sultan Qaboos Univ Med J 14(2):e157

Ros E, Hu FB (2013) Consumption of plant seeds and cardiovascular health: epidemiological and clinical trial evidence. Circulation 128(5):553–565

Sabate J, Ang Y (2009) Nuts and health outcomes: new epidemiologic evidence. Am J Clin Nutr 89 (5):1643S–1648S

Sabaté J, Oda K, Ros E (2010) Nut consumption and blood lipid levels: a pooled analysis of 25 intervention trials. Arch Intern Med 170(9):821–827

Salado C, Olaso E, Gallot N, Valcarcel M, Egilegor E, Mendoza L, Vidal-Vanaclocha F (2011) Resveratrol prevents inflammation-dependent hepatic melanoma metastasis by inhibiting the secretion and effects of interleukin-18. J Transl Med 9(1):59

Salas-Salvadó J, Garcia-Arellano A, Estruch R, Marquez-Sandoval F, Corella D, Fiol M, Gómez-Gracia E, Viñoles E, Arós F, Herrera C (2008) Components of the Mediterranean-type food pattern and serum inflammatory markers among patients at high risk for cardiovascular disease. Eur J Clin Nutr 62(5):651–659

Sanders TH, McMichael RW, Hendrix KW (2000) Occurrence of resveratrol in edible peanuts. J Agric Food Chem 48(4):1243–1246

Schnepf R (2016) US peanut program and issues. Congressional Research Service, CRS Report 44156

Settaluri V, Kandala C, Puppala N, Sundaram J (2012) Peanuts and their nutritional aspects—a review. Food Nutr Sci 3(12):1644

Shankar S, Nall D, Tang S-N, Meeker D, Passarini J, Sharma J, Srivastava RK (2011) Resveratrol inhibits pancreatic cancer stem cell characteristics in human and KrasG12D transgenic mice by inhibiting pluripotency maintaining factors and epithelial-mesenchymal transition. PLoS One 6 (1):e16530

Sharma KK, Bhatnagar-Mathur P (2006) Peanut (Arachis hypogaea L.). Methods Mol Biol 343:347–358

Shils ME, Shike M (2006) Modern nutrition in health and disease. Lippincott Williams & Wilkins, Philadelphia, PA

Shin E-C, Pegg RB, Phillips RD, Eitenmiller RR (2010) Commercial peanut (Arachis hypogaea L.) cultivars in the United States: phytosterol composition. J Agric Food Chem 58(16):9137–9146

Shu-bo W (2009) Opportunities facing peanut industry in China and strategies for its science and technology development. J Agric Sci Technol 1:002

Singh B, Singh H (1991) Peanut as a source of protein for human foods. Plant Foods Hum Nutr 41:165–177

Sitren H, Ahmed E, George D (1985) In vivo and in vitro assessment of antinutritional factors in peanut and soy. J Food Sci 50(2):418–423

Smith AF (2002) Peanuts: The illustrious history of the goober pea, vol 20. University of Illinois Press, Urbana

Smith MI, Yatsunenko T, Manary MJ, Trehan I, Mkakosya R, Cheng J, Kau AL, Rich SS, Concannon P, Mychaleckyj JC (2013) Gut microbiomes of Malawian twin pairs discordant for kwashiorkor. Science 339(6119):548–554

Sobolev VS, Cole RJ (1999) trans-Resveratrol content in commercial peanuts and peanut products. J Agric Food Chem 47(4):1435–1439

Sobolev VS, Khan SI, Tabanca N, Wedge DE, Manly SP, Cutler SJ, Coy MR, Becnel JJ, Neff SA, Gloer JB (2011) Biological activity of peanut (Arachis hypogaea) phytoalexins and selected natural and synthetic stilbenoids. J Agric Food Chem 59(5):1673

Souza RG, Gomes AC, Naves MM, Mota JF (2015) Nuts and legume seeds for cardiovascular risk reduction: scientific evidence and mechanisms of action. Nutr Rev 73(6):335–347

Stalker HT, Tallury SP, Seijo GR, Leal-Bertioli SC (2016) Biology, speciation, and utilization of peanut species. In: Stalker T, Wilson RF (eds) Peanuts. Elsevier, London, pp 27–66

Stansbury MF, Field ET, Guthrie JD (1950) The tannin and related pigments in the red skins (testa) of peanut kernels. J Am Oil Chem Soc 27(8):317–321

Strange RN, Rao PS (2017) Chemistry, biology, and role of groundnut phytoalexins in resistance to fungal attack. In: Daniel M (ed) Handbook of phytoalexin metabolism and action. Routledge, New York, pp 199–227

Taghvaei M, Jafari SM (2015) Application and stability of natural antioxidants in edible oils in order to substitute synthetic additives. J Food Sci Technol 52(3):1272–1282. https://doi.org/10.1007/s13197-013-1080-1

Tang K, Zhan J-C, Yang H-R, Huang W-D (2010) Changes of resveratrol and antioxidant enzymes during UV-induced plant defense response in peanut seedlings. J Plant Physiol 167(2):95–102

Tanwar B, Modgil R, Goyal A (2018) Antinutritional factors and hypocholesterolemic effect of wild apricot kernel (Prunus armeniaca L.) as affected by detoxification. Food Funct 9(4):2121–2135

Tanwar B, Modgil R, Goyal A (2019) Effect of detoxification on biological quality of wild apricot (Prunus armeniaca L.) kernel. J Sci Food Agric 99(2):517–528

Vellidis G, Ortiz B, Beasley J, Hill R, Henry H, Brannen H (2014) Reducing digging losses by using automated steering to plant and invert peanuts. Agronomy 4(3):337–348

Venter C, Hasan Arshad S, Grundy J, Pereira B, Bernie Clayton C, Voigt K, Higgins B, Dean T (2010) Time trends in the prevalence of peanut allergy: three cohorts of children from the same geographical location in the UK. Allergy 65(1):103–108

Venter C, Maslin K, Patil V, Kurukulaaratchy R, Grundy J, Glasbey G, Twiselton R, Dean T, Arshad SH (2016) The prevalence, natural history and time trends of peanut allergy over the first 10 years of life in two cohorts born in the same geographical location 12 years apart. Pediatr Allergy Immunol 27(8):804–811

Wang Q (2016) Peanuts: processing technology and product development. Academic Press, London

Wang J, Jones SM, Pongracic JA, Song Y, Yang N, Sicherer SH, Makhija MM, Robison RG, Moshier E, Godbold J (2015) Safety, clinical, and immunologic efficacy of a Chinese herbal medicine (Food Allergy Herbal Formula-2) for food allergy. J Allergy Clin Immunol 136(4):962–970.e961

Woyengo T, Ramprasath V, Jones P (2009) Anticancer effects of phytosterols. Eur J Clin Nutr 63(7):813–820

Xiong Q, Zhang Q, Zhang D, Shi Y, Jiang C, Shi X (2014) Preliminary separation and purification of resveratrol from extract of peanut (Arachis hypogaea) sprouts by macroporous adsorption resins. Food Chem 145:1–7

Yang J, Halim L, Liu R (2005) Antioxidant and antiproliferative activities of common nuts. In: Institute of Food Technologists Annual Meeting and Food Expo, New Orleans, LA, 035-005

Yen G-C, Duh P-D (1993) Antioxidative properties of methanolic extracts from peanut hulls. J Am Oil Chem Soc 70(4):383–386

Yu J, Ahmedna M, Goktepe I (2005) Effects of processing methods and extraction solvents on concentration and antioxidant activity of peanut skin phenolics. Food Chem 90(1):199–206
Yu J, Ahmedna M, Goktepe I, Dai J (2006) Peanut skin procyanidins: composition and antioxidant activities as affected by processing. J Food Compos Anal 19(4):364–371
Yu Z, Malik VS, Keum N, Hu FB, Giovannucci EL, Stampfer MJ, Willett WC, Fuchs CS, Bao Y (2016) Associations between nut consumption and inflammatory biomarkers. Am J Clin Nutr 104(3):722–728
Zeisel SH, Da Costa KA (2009) Choline: an essential nutrient for public health. Nutr Rev 67 (11):615–623
Zhao X, Chen J, Du F (2012) Potential use of peanut by-products in food processing: a review. J Food Sci Technol 49(5):521–529
Zhong T, Liang Y, Jiang S, Yang L, Shi Y, Guo S, Zhang C (2017) Physical, antioxidant and antimicrobial properties of modified peanut protein isolate based films incorporating thymol. RSC Adv 7(66):41610–41618

Chapter 5
Sunflower (*Helianthus annuus*) Seed

Sakshi Khurana and Ravinder Singh

Abstract Sunflower (*Helianthus annuus* L.) is an important agricultural crop grown for its seeds worldwide. Globally, sunflower is the fourth largest source of vegetable oil next to soybean, palm and rapeseed. Besides edible oil, it is grown for its fruits both for human and livestock consumption. Production of sunflower seed has increased over the years because of the increasing demand for its healthful oil. The health benefits of sunflower seeds and oil are attributed to its proteins, antioxidant (vitamin E), phytonutrients, minerals like selenium and magnesium and healthy lipid profile. Thus, sunflower seeds have multifaceted therapeutic benefits including attenuation of some of the widespread chronic diseases like cardiovascular (CVDs) and inflammatory diseases. In this chapter, the proximate composition of sunflower seed and oil, health benefits and food applications will be discussed in detail.

Keywords Sunflower oil · Cardiovascular diseases · Phytonutrients · Inflammatory diseases · Salad dressing

5.1 Origin and History

Sunflower (*Helianthus annuus* L.) belongs to genus *Helianthus* and is thought to be originated in North America. It has quite an extensive and diverse past as a commercial crop, though the time as well as location of its primary cultivation is doubtful. Earlier, English and French explorers found sunflower in widespread use by the American Indians, who acquainted it with their particular regions as well. Gradually, it found its way into the trade routes to Italy, Egypt, Afghanistan, India,

S. Khurana (✉)
Food Technology, Bhaskaracharya College of Applied Sciences, University of Delhi, New Delhi, India

R. Singh
Haldiram Snacks Pvt. Ltd., Noida, Uttar Pradesh, India

B. Tanwar, A. Goyal (eds.), *Oilseeds: Health Attributes and Food Applications*,
https://doi.org/10.1007/978-981-15-4194-0_5

China and Russia. Later on, sunflower emerged as a leading oilseed crop in Russia and established throughout the Europe (Berglund 2007). Sunflower was a usual harvest among American Indians all through North America. Certain facts propose that the plant was developed by American Indians in Arizona and New Mexico around 3000 BC. Sunflower is being utilized as food and/or an ingredient by the American Indians even before corn became well-known. Its seeds were used to beaten into flour to make cakes, mush or bread and sometimes eaten as a snack. There are evidences of crushing the seed for oil and further utilizing the oil in baking bread. Sunflower was additionally utilized as a therapeutic harvest and a source of dye. All through the current Western Europe, the plant became popular essentially as a decorative, though some therapeutic applications were developed. A milestone was believed to be achieved when in 1716 an English patent was conceded for pressing oil from sunflower seeds (Schneiter 1997).

Sunflower turned out to be extremely prevalent as a cultivated plant in the eighteenth century. By 1769, sunflower cultivation was driven majorly by oil production. By 1830, industrial manufacture of sunflower oil started. Through much of these years, sunflower was largely grown as a decorative plant. Until the early 1900s, it was grown for cattle feed as well. By the early nineteenth century, more than two million acres of land in Russia was under sunflower production. It was this time when two major types of sunflower were recognized: oilseed for oil production and non-oilseed for direct human consumption. The crop entered the US terrain by the end of the nineteenth century, wherein it was first used as ensilage for poultry on commercial scale. Around 1926, Missouri Sunflower Growers' Association began expressing oil from the seeds. The Canadian government began growing sunflower in 1930. The land under cultivation extended as a result of oil demand; consequently, the Canadian farmers fabricated a pilot pulverizing plant around 1946 and cultivation spread to various parts of the country. The Canadian government also certified the Russian variety of sunflower (Peredovik) that gave exceptional returns and high oil content (about 44%) compared to 28–30% oil content in other varieties like Arrowhead, Mingren, etc. (Robinson et al. 1967). The year 1966 brought about a revolutionary change in the US sunflower production as the acreage shifted from non-oilseed varieties to the commercially significant oilseed sunflower, thus increasing the area under sunflower cultivation in the USA. Hybridization of sunflower in the mid-1970s provided higher yield and higher oil percent as well as improved resistance against diseases like the downy mildew, white mold, rust, verticillium wilt and head rot (McMullen 1985). Another milestone was achieved in the early 1970s with the formation of the National Sunflower Association (NSA) in the USA.

The successful development of high-oil breeds as well as recent development of hybrids of sunflower by plant researchers was principally responsible for widespread global production of the crop. Need for high-quality oil has been the driving force for these advancements. Sunflower hybrids with valuable attributes such as insect and disease resistance have been produced that give higher oil content. By the early 1980s, sunflower's hybrid variety yielding high oleic acid was developed in the USA and commercialized by the end of the decade. Another variety, the mid-oleic sunflower, became commercial in the USA in 1998.

5.2 Production

Sunflower is native of North America, though it was commercialized in Russia. Sunflower oil is a preferred oil not only in Europe but also in Mexico and most of South America. According to Food and Agricultural Organization, total world production of sunflower seeds has increased from 26.49 in 2007 to 47.86 million tonnes in 2017 (FAOSTAT 2019). As per the report, Ukraine and Russia together produce 56.0% of the total world's production of sunflower seed (FAOSTAT 2019). India ranks 22nd in the production of sunflower with 0.211 million tonnes in 2017. The production of sunflower seed in 2017 by top ten producer countries and their percent share are shown in Figs. 5.1 and 5.2, respectively.

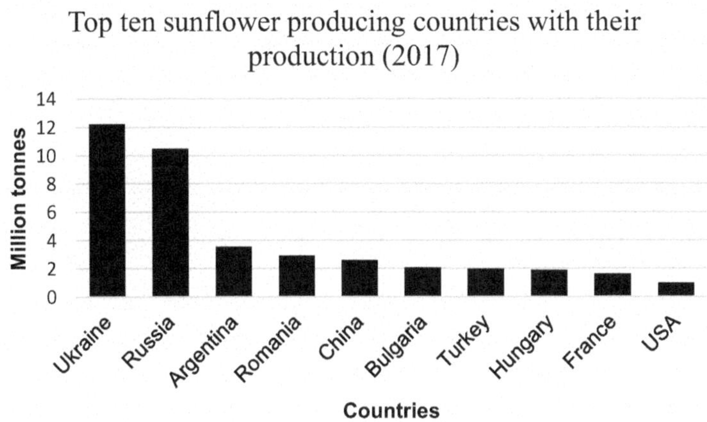

Fig. 5.1 Top ten sunflower-producing countries with their production in million tonnes (2017) (FAOSTAT 2019)

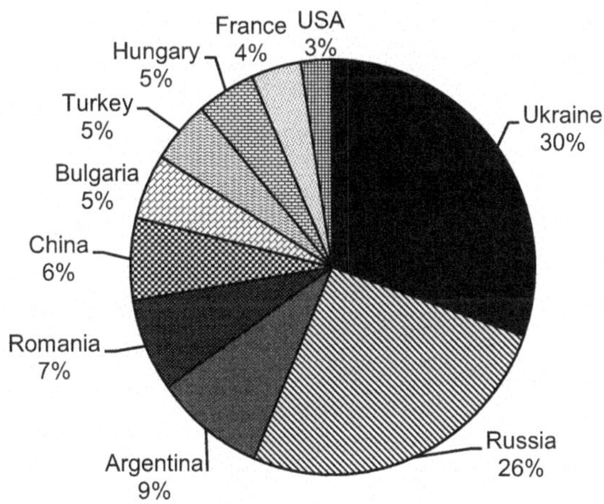

Fig. 5.2 Percent share in production of sunflower by top ten producing countries (2017) (FAOSTAT 2019)

5.3 Types of Sunflowers

Primarily, on the basis of oil content, sunflower can be divided into two types:

1. *Oilseed sunflower:* Which contain minimum 40% oil and are grown primarily for oil production. Examples: Peredovik, Armavirec, VNIIMK, Smena.
2. *Confection sunflower (non-oilseed):* These are used as snacks for human and as feed for birds. Examples: Mingren, Commander, Mennonite, Arrowhead.

On the other hand, the oilseed variety can also be divided into three types based on the major fatty acids present (Berglund 2007): (1) linoleic, (2) mid-oleic and (3) high oleic. Generally, seeds of oilseed variety have black coloured seeds and have adhered thin hull, difficult to remove from the kernels of oilseed. Their seeds contain approximately 38–50% oil and 20% protein. Some of these oilseeds are hulled to be used as bird's feed (Berglund 2007; OECD 2007). Non-oilseed sunflower or the confection sunflower generally has white stripes and larger seed size than the oilseed variety. It has a rather thick hull, which is loosely attached to the kernel, allowing complete separation of the hull. It is classified on the basis of seed size as follows (Schild et al. 1991):

(a) Largest seeds: Consumed as snack after roasting and salting within the shell.
(b) Mid-size seeds: Dehulled seeds (kernels) may be used as a snack and in bakery products or other foods for their crispness and/or for improving the nutrient levels.
(c) Small seeds: Particularly used as bird feed.

Sunflower fruit (seed) is approximately 10–15 mm in length (Fig. 5.3). The hull (pericarp) consists of pigmented and elongated cells. Next to the pericarp are multiple layers of sclerenchyma cells. The testa (seed coat) lies under this layer.

Fig. 5.3 Sunflower fruit (seed) (Armstrong 2007)

5.4 Chemical Composition

Sunflower seeds are rich source of nutrients providing significant amounts of vitamins and minerals, phytonutrients and essential fatty acids. The seed meal is a good source of protein, fibre, zinc, selenium, copper, iron, vitamin B_6 and folate. Sunflower seed is characterized by a high oil content (40–45%), moderate amount of proteins (14–18%) and good amount of hulls (25–30%) that predominantly consist of non-digestible crude fibre. The proximate composition of whole sunflower seed and deoiled meal (fully dehulled) is shown in Table 5.1.

5.4.1 Carbohydrates

The major carbohydrates present in sunflower seeds are dietary fibre. One particular characteristic of sunflower seeds carbohydrate is the very little starch content (0.42%). Dried sunflower seed kernels are rich in fibre with 8.6 g/per100 g (USDA 2019). This is similar to the fibre content of the oat bran cereals (8.6 g/100 g) and higher than an apple (2.8 g/100 g) (USDA 2018). Analysis of the molar sugar composition in the kernels (4.4–6.3%) revealed that sunflower contains high amounts of glucose (46%), followed by arabinose (16%), uronic acids (14%) and galactose (11%) (González-Pérez et al. 2002).

5.4.2 Proteins

The protein content of sunflower kernels ranges from 20 to 40%. Sammour et al. (1995) characterized sunflower seed proteins and reported 38.33% albumins, 39.04% globulins, 5.54% prolamins and 17.09% glutelins. An important parameter

Table 5.1 Proximate composition of whole sunflower seed and deoiled meal

Parameter	Whole seeds/ kernels[a]	Whole seeds/ kernels[b]	Deoiled meal (fully dehulled)[b]
Moisture (%)	4.73	6.5–6.7	11–12
Protein (%)	20.78	14–18	38–40
Fat (%)	51.46	40–45	0.5
Carbohydrates (%)	20.00	NR	NR
Crude fibres (%)	8.6[c]	13–20	14–17
Ash (%)	3.02	3.50	6

NR not reported
[a]USDA (2019)
[b]Le Clef and Kemper (2015)
[c]Dietary fibres

Table 5.2 Amino acid profile of sunflower oilseed protein

Amino acid	In whole oilseed[a] (g/100 g)	In seed protein[b] (g/100 g)
Tryptophan	0.348	1.57
Threonine	0.928	3.70
Isoleucine	1.139	4.40
Leucine	1.659	6.92
Lysine	0.937	3.87
Methionine	0.494	2.26
Cystine	0.451	0.31
Phenylalanine	1.169	5.32
Tyrosine	0.666	3.21
Valine	1.315	5.29
Arginine	2.403	10.12
Histidine	0.632	2.79
Alanine	1.117	4.59
Aspartic acid	2.446	10.02
Glutamic acid	5.579	22.45
Glycine	1.461	5.35
Proline	1.182	3.61
Serine	1.075	4.05

[a]USDA (2019)
[b]Robinson (1975)

in the evaluation of protein quality is the amino acid composition which is also helpful in feed formulation. Relative to soybean, sunflower proteins contain high quantities of sulphur containing amino acids (2.1% vs. 1.3% in soybean) (Piva 1992). However, the lack of lysine as in cereals can be made up by supplementation with pulses. The sunflower's proteins are not preferred as a protein supplement for human consumption probably due to two reasons. Firstly, the sunflower seed proteins get denatured during oil extraction process, and thus, the functionality of the altered (denatured) protein reduces. Secondly, the phenolic components, particularly chlorogenic acid (CGA), of seed bind with seed proteins and cause the discoloration of the final products (Wildermuth et al. 2016). The amino acid profile of sunflower oilseed and protein is given in Table 5.2.

5.4.3 *Lipids*

Regular sunflower whole seeds contain approximately 40–45% oil content, of which >98% comprises triacylglycerol fraction (in refined oil). It is important to note that regular sunflower oil is rich in linoleic acid (C18:2, omega-6) (48–74%), which is an essential fatty acid. Unlike soybean and canola oil, it contains low levels of saturated fatty acids (palmitic and stearic acid) and insignificant amount of α-linolenic acid

Table 5.3 Major fatty acids composition of sunflower oil from different type of seeds

Type of sunflower oil	Palmitic acid (C16:0)	Stearic acid (C18:0)	Oleic acid (C18:1)	Linoleic acid (C18:2)
Regular	6.3	4.6	26.7	61.1
Mid oleic	4.9	3.8	57.9	32.3
High oleic	3.8	4.1	82.1	8.7
High stearic	7.4	27.1	16.1	46.3
High stearic-high oleic (HSHO)	5.4	24.9	57.8	8.2
High palmitic	34.7	2.6	6.9	45.1
High palmitic-high oleic (HPHO)	31.7	2.0	50.5	2.7

Source: Salas et al. (2015)

(ALA, omega-3). The presence of very long-chain saturated fatty acids such as arachidic acid (C20:0) and behenic acid (C22:0) is also signature fatty acids of sunflower oil. However, Dorni et al. (2018) did not detect the presence of behenic acid in sunflower oil. With time, different varieties of sunflower seeds were developed through mutagenesis and/or breeding techniques. These new varieties were different to each other in just one or two fatty acids, which were developed to meet the different functional and/or nutritional properties for different food applications. The major fatty acids in different varieties of sunflower seeds are shown in Table 5.3.

Sunflower oil contains 0.5–1.5% minor components such as phospholipids, free fatty acids, tocopherols, pigments, alcohols, phytosterols, carotenoids, phenolic acids, etc. Sterols such as β-sitosterol, campesterol and stigmasterol are discussed in further section. Out of these minor components, tocopherols and phytosterols confer not only positive health effects but also provide oxidative stability to oil. The rest of the molecules are responsible for the oxidative rancidity and generation of off-flavour compounds during processing and storage. Being organic molecules, phospholipids are found in the oil fraction after seed extraction and could be divided into phosphatidylcholines, phosphatidylethanolamines, phosphatidylinositols and phosphatidic acids (Phillips 2000). Refined sunflower oil has low phospholipid content. The non-saponifiable components and their content in sunflower oil are shown in Table 5.4.

5.4.4 *Vitamins*

Sunflower kernels are one of the richest sources of B-complex vitamins with good amounts of vitamins B_1, B_2, B_3, B_5, B_6 and B_9 (Table 5.5). Vitamin E is present in exceptionally high amount and has been listed by several sources like the Micronutrient Information Center, Oregon State University (2015); Factsheet Food Sources of Vitamin E National Sunflower Association; Dietitians of Canada; and also in the scientific report of the US Dietary Guidelines as the 'best whole-food source of

Table 5.4 The content of non-saponifiable matter in sunflower oil

Component	In crude regular sunflower oil (mg/kg oil)[a]	In sunflower oil (mg/kg oil)
Unsaponifiable matter	NR	11,000–15,000[b]
Diacylglycerols	10,000–18,900	NR
Phospholipids	6000–12,000	1020–10,390[c]
Free fatty acids	3700–4500	NR
Phytosterols	2400–4600	2582.7–3662.7[b]
– Campesterol	NR	– 7.7–8.5%[d]
– Stigmasterol	NR	– 9.0–9.5%[d]
– β-sitosterol	NR	– 53.5–56.4%[d]
– Δ-7-Stigmasterol	NR	– 11.9–15.2%[d]
Hydrocarbons	1000	NR
Tocopherols	403–935	864.3–1458.3[b]
Waxes	200–3500	NR
Aliphatic alcohols	59–63	NR
Trace metals	45–95	NR
Carotenoids	6.5–15.3	NR
Phenolic compounds	4.8–16.4	3453.5 ± 17.3[e]
Chlorophyll	0–1	NR

NR not reported
[a]Velasco and Ruiz-Méndez (2015)
[b]Anastasi et al. (2010)
[c]Carelli et al. (1997)
[d]Indicates the percent of total phytosterols
[e]Weisz et al. (2009)

vitamin E'. Vitamin E is not only thought to help prevent Alzheimer's disease and dementia but also a natural antioxidant that may protect against heart diseases by scavenging the free radicals (the harmful molecules). Furthermore, owing to the presence of high amount of vitamin E (α-tocopherol), sunflower oil is naturally resistant to photo-oxidation. However, since low levels of γ-tocopherols are present in oil, it is not stable against autoxidation.

5.4.5 *Minerals*

Sunflower seeds are loaded with many essential minerals, viz. iron, zinc, manganese, copper, selenium and calcium (Anjum et al. 2010) (Table 5.6). These important minerals play a crucial role in hormone production, enzyme synthesis, red blood cell production, bone mineralization as well as regulation of metabolic, cardiac and skeletal muscle activities. Around 25% of the daily needs for selenium can be met by consuming 1 ounce of sunflower seeds. Selenium is a rare antioxidant, which acts synergistically with vitamin E and prevent cell damage that may cause cancer, heart disease or other health problems (Vogt et al. 2003).

Table 5.5 Vitamin content in sunflower oilseed

Vitamin	Quantity (per 100 g dry matter of whole seed)
L-ascorbic acid	1.40 mg
Thiamine	1.48 mg
Riboflavin	0.35 mg
Niacin	8.33 mg
Pantothenic acid	1.13 mg
Vitamin B_6	1.35 mg
Folate (total)	227.00 μg
Choline	55.10 mg
Vitamin B_{12}	0 μg
Vitamin A	50 IU
α-tocopherol (vitamin E)	35.17 mg
β-tocopherol	1.18 mg
γ-tocopherol	0.37 mg
δ-tocopherol	0.02 mg
Vitamin K (phylloquinone)	0 μg
Vitamin D	0 μg

Source: USDA (2019)
IU International Units

Table 5.6 Mineral content of sunflower oilseeds

Mineral	Whole seed (mg per 100 g)[a]	Whole seed (mg per 100 g)[b]
Calcium	78	100
Phosphorus	660	940
Magnesium	325	410
Potassium	645	740
Sodium	9	20
Copper	1.8	2.5
Iron	5.25	6.7
Manganese	1.95	2.6
Zinc	5	7.6
Selenium	53[c]	NR

NR not reported
[a]USDA (2019)
[b]Robinson (1975)
[c]In μg

5.5 Anti-nutritional Factors

Sunflower seeds have been reported not to contain considerable amount of anti-nutritional factors (OECD 2007; Le Clef and Kemper 2015). However, Ingale and Shrivastava (2011) reported very minute amount of some anti-nutritional factors such as cyanide content (4.026–4.175 mg HCN/100 g), tannins (0.623–0.651 g/100 g) and oxalates (0.098–0.113 g/100 g) with no trypsin inhibitor activity in sunflower seeds. Phytic acid and phytin are the storage forms of phosphorus in plants and may account for >70% of kernel phosphate. In another study, phytic acid (myoinositolhexaphosphate) content of sunflower kernel has been reported to be ~60 mg/100 g (Duke and Beckstrom 1999; CSA 1999). The primary biological effect of ingested phytate is chelation of divalent cations (e.g. Fe^{2+}, Ca^{2+}, Zn^{2+}), which decreases the bioavailability of essential minerals (Tanwar et al. 2018, 2019). However, binding of iron inhibits iron-catalysed hydroxyl radical formation (Zhou and Erdman 1995), and this mechanism has been cited as protective against colorectal cancer (Owen et al. 1996). Phytate may have other chemopreventive and therapeutic value against cancer by influencing mammalian cellular inositol phosphate metabolism (Shamsuddin 1995). Phytate has also been reported to reduce blood cholesterol and to serve as an antioxidant and preservative in foods (Zhou and Erdman 1995). Phytate is a rich dietary source of inositol, which is released by phytase during digestion (Holub 1982). Protease inhibitors bind to intestinal proteases (e.g. trypsin, chymotrypsin, pepsin) and thus interfere with protein digestion and decrease the bioavailability of proteins (Liener 1995). They also induce pancreatic hyperplasia when consumed in large quantities. However, few beneficial effects have also been attributed to protease inhibitors, including inhibition of tumour growth (Fjeld and Lawson 1999). Much information is not available on the antinutrients in sunflower seeds.

5.6 Phytonutrients and Phenolics

Numerous phytonutrients such as phenolic acids, phytosterols, phospholipids, triterpenes, isoflavones and lignans are present in sunflower kernel. They offer protection from heart diseases and by inhibit the risk of cancers (including colon, prostate and breast cancer) (Phillips 2000).

5.6.1 Phenolic Acids

The total phenolic content of sunflower seeds is reported within the range 1–4% on dry matter basis. The principal phenolic constituents are chlorogenic acid; smaller quantities of caffeic acid, cinnamic, coumaric, ferulic, sinapic and hydroxycinnamic acids; and trace amount of vanillic, syringic and hydroxybenzoic acids. 5-O-caffeoylquinic acid, a chlorogenic acid isomer, comprises 43–73% of phenolic compounds extracted from sunflower kernels (Pedrosa et al. 2000; González-Pérez and Vereijken 2007; Weisz et al. 2009). Karamać et al. (2012) have identified isomers of coumaroylquinic acid (probably 3-O-pcoumaroylquinicand 4-O-p-coumaroylquinic acids) and dicaffeoylquinic acids (probably 1,3-di-O-caffeoylquinic and1,4-di-O-caffeoylquinic acids) in addition to non-esterified

Table 5.7 Concentration of individual phenolic compounds in sunflower oilseed

Component	Amount (mg/100 g dry matter)[a]	Amount (mg/g fresh weight)[b]
Caffeic acid	26.7 ± 1.1	NR
Ferulic acid	12.4 ± 2.0	NR
Non-esterified phenolic acids	39.0 ± 2.3	NR
5-O-p-coumaroylquinic acid	11.3 ± 1.0	NR
5-O-feruloylquinic acid	11.3 ± 1.0	NR
Coumaric and ferulic acid derivatives	22.6 ± 1.4	NR
3-O-caffeoylquinic acid	439.9 ± 8.6	4.74 ± 0.76
4-O-caffeoylquinic acid	87.5 ± 4.1	0.83 ± 0.13
5-O-caffeoylqinic acid (chlorogenic acid)	2467.0 ± 13.9	11.82 ± 1.42
Caffeoylquinic acid	36.5 ± 2.2	NR
Monocaffeoylquinic acids	3030.9 ± 17.0	NR
3,4-Di-O-caffeoylquinic acid	28.8 ± 0.3	NR
3,5-Di-O-caffeoylquinic acid	211.2 ± 1.1	ND
4,5-Di-O-caffeoylquinic acid	120.9 ± 0.2	2.45 ± 0.27
Dicaffeoylquinic acids	360.9 ± 1.1	NR
Total amount	3453.5 ± 17.3	NR

NR not reported, *ND* not detected
[a]Weisz et al. (2009)
[b]Romani et al. (2017)

phenolic acids, isomers of caffeoylquinic acid, p-coumaroylquinic acid and dicaffeoylquinic acids in sunflower seeds. Phenolic acids also function as natural antioxidants, thus reducing the risk of several chronic diseases (Tanwar et al. 2018, 2019). The concentration of these compounds in sunflower meal has been represented in Table 5.7.

5.6.2 *Phytosterols and Triterpenes*

Phytosterol, a bioactive component resembling cholesterol, is found in ample amounts in sunflower seeds (Piironen et al. 2000). When compared with some selected nuts, legumes and seeds, sunflower seeds were next to wheat germ, sesame seeds and pistachio having substantially higher amount of phytosterols than walnuts, pecans, pistachio, almond, cashew, soybean, flaxseeds, peanuts and many more (Phillips et al. 2005; King and Young 1999).

The phytosterol content of sunflower oil/seed is shown in Table 5.4. Fernández-Cuesta et al. (2014) analysed the collection of 985 accessions and reported a large variation in the phytosterol content (1319–5119 mg kg^{-1}) of sunflower kernel. The authors also reported a wide variation in the concentrations of β-sitosterol

(448–755 g kg^{-1}), campesterol (32–197 g kg^{-1}), stigmasterol (41–128 g kg^{-1}) and 7-stigmastenol (8–274 g kg^{-1}). During absorption in the intestine owing to its similarity in structure with cholesterol, it competes with cholesterol and, thus, reduces the blood cholesterol levels (Lagarda et al. 2006). Research has demonstrated that even low levels of phytosterols (about 2 g per day) are helpful in lowering serum LDL-cholesterol absorption. Such low quantities of phytosterols can easily be obtained through diet loaded with plant foods (Racette 2017). Besides, they also offer protection against colon, breast and prostate cancer and improve immunity (Awad and Fink 2000). Sunflower oil has been reported to contain approximately 0.22% triterpenes (Fedeli et al. 1966). A comprehensive review published on the health effects of triterpenes suggested that triterpenes confers anti-microbial, anti-protozoal, anti-arthritic, anti-diabetic, anti-cancerous, anti-inflammatory and immunomodulatory effects (Siddique and Saleem 2011).

5.6.3 *Phospholipids*

Phospholipids, the essential components of the seed oil body, act as emulsifiers and stabilizers and, thus, improve the texture of multiphase food materials (Pasini et al. 2013). Furthermore, they can also act as antioxidants owing to their synergistic action in decomposing hydroperoxides and their metal scavenging activity (Koga and Terao 1994; Carelli et al. 1997). Sunflower seed is a good source of phospholipids, primarily phosphatidylcholine (38.5–230 mg/100 g) and phosphatidylethanolamine (61.5–123 mg/100 g), along with minor amounts of phosphatidylinositol and phosphatidic acid with a total concentration lower than 1.2% (Padley et al. 1994; CSA 1999; Duke and Beckstrom 1999). Lecithin, the major phospholipid, is hydrolysed in the intestine by phospholipase C to yield choline, a membrane phospholipid which provides choline reserve for neurotransmitter (acetylcholine) synthesis and is also important for normal liver function (Mihai and Steven 2013).

5.6.4 *Phytoestrogens*

Phytoestrogens fall mainly under two categories: isoflavones (e.g. genistein and daidzein) and lignans (e.g. secoisolariciresinol, matairesinol). Sunflower seeds have relatively low content of isoflavones. Kuhnle et al. (2008) observed phytoestrogens, isoflavones and lignans in sunflower oilseeds to the value of 111.0, 2.0 and 9.0 μg/100 g (wet basis), respectively. However, Tham et al. (1998) reported a very high amount of lignans (400 μg/100 g) in sunflower seeds. Phytoestrogens may have favourable estrogenic effects on the risk of cardiovascular disease, hypercholesterolemia, carcinogenic activity, osteoporosis and hormone-regulatory effects (Tham et al. 1998).

5.7 Health Attributes

As discussed previously, sunflower seed is packed with several phytonutrients and bioactive components; their documented health effects are discussed here:

5.7.1 *Reduction of Dyslipidaemia and Cardiovascular Diseases*

Dyslipidaemia is the elevation of plasma cholesterol, triglycerides or both or a low high-density lipoprotein (HDL) level that contributes to the development of atherosclerosis. It is an important independent adjustable risk factor for cardiovascular diseases (CVDs). CVDs are currently the primary cause of death in the world, with its morbidity and mortality growing consistently. Low physical activity, tobacco use, dyslipidaemia, hypertension and diabetes are some of the factors, which if taken care of at an early stage could considerably reduce the deaths caused by CVDs. Dyslipidaemia is one of the key factors responsible for the initiation and growth of atherosclerosis, and its association with risk of CVD is undoubted. Timely and efficient modification of plasma lipids can help inconsiderable reduction in the occurrence of CVDs. Oleic acid, present in significant amounts in nontraditional sunflower oil, is hypocholesterolemic and decreases LDL cholesterol. Research has revealed that diets rich in sunflower oil and/or rich in MUFA helped in considerable reduction in LDL, total cholesterol, triglyceride levels and risk of thrombosis, while significantly elevated the levels of HDL cholesterol (Binkoski et al. 2005; Junker et al. 2001; Pal 2011). On the other hand, low-fat, high-carbohydrate diet is likely to raise triglycerides and drop HDL cholesterol. NuSun™ sunflower oil, which holds the major US market share of sunflower oil, has been shown to significantly lower total and LDL cholesterol. It is thought that the fatty acid profile of NuSun™ renders the effect. NuSun™ has lower percentage of saturated fat in comparison with olive oil (9.6% against 14.3%). Besides, it has zero-gram *trans*-fat as per FDA (Food and Drug Administration) standards. *Trans*-fatty acid has been well studied to negatively affect blood lipid profile just like saturated fats and, therefore, amplify the risk of coronary heart diseases. It has also been demonstrated that diets in which at least 5–10% of energy is contributed by ω-6 or ω-3 PUFAs reduce the occurrence of CVDs and may reduce expected death from acute coronary syndrome. Besides, the presence of high amount of vitamin E in sunflower oil has a dual effect: (1) beneficial effect on serum lipids and (2) scavenging reactive oxygen free radicals, which have negative impact on atherosclerosis.

5.7.2 Prevention of Diabetes Mellitus

Hyperglycaemia produces reactive oxygen species (ROS), which then lays foundation for lipid peroxidation and membrane damage. The free radicals, in diabetic patients, lead to secondary complications like damage to the kidney, eye, blood vessels and nerves. Antioxidants may offer protection from the growth of diabetes by preventing the destruction of β-cells, thus inhibiting the peroxidation chain reaction. Sunflower seeds are naturally gifted with antioxidants, viz. vitamin E and C, phenolic acids and flavonoids, that can maintain β-cell function and avoid diabetes-induced ROS formation. Consuming sunflower seeds offers several critical micronutrients like folate and magnesium. Eating plenty of magnesium has been reported to lower risk of type 2 diabetes (Razzaque 2018; Sobczak et al. 2019). Research has suggested that consumption of sunflower seeds/oil resulted in a decrease in fasting blood glucose levels in diabetes mellitus II patients and animals (Madigan et al. 2000; Sebbagh et al. 2009; Pal 2011). Besides, sunflower seeds have been recommended as a domestic and natural cure to provide control over the blood sugar levels in patients with diabetes mellitus.

5.7.3 In Reducing the Risks of Cancer

Sunflower seeds are a rich source of phytochemicals like phenolic acids and lignans, which may help prevent heart disease and cancer (Al-Jumaily et al. 2013; Smith et al. 2016). The presence of high amounts of phytosterols like β-sitosterol in sunflower seeds, phenolic compounds and other bioactive components prevents the development of several types of tumour cells and reduces the size and scope of tumour metastases (Kapadia et al. 2002; Taha et al. 2011, 2012; Pal 2011). Selenium, another important component of sunflower seeds, gets attached to the active site of numerous proteins, for example, glutathione peroxidise (antioxidant enzyme), which has been observed to reduce the risk of prostate cancer (Nkengfack et al. 2019)

5.7.4 Role in Inflammatory Diseases and Immune Function

Sunflower seeds have shown their potential in preventing chronic inflammatory conditions like bronchial asthma, osteoarthritis and rheumatoid arthritis (Pal 2011; Odabasoglu et al. 2008; Macías et al. 2002). Substantial amount of zinc is also present in the seeds, which helps boost the immune system. Sunflower seed oil has been found to play an important role in the wound healing process, particularly in the inflammation phase (Lania et al. 2019; Maver et al. 2018). The oil stimulates re-epithelialization and the formation of collagen fibres in oral wound healing (Eichenfield et al. 2009). This effect has been attributed to the presence of high amounts of linoleic acid, phytosterols, and vitamin E in sunflower oil. Choline and the amino acid tryptophan in sunflower seeds are helpful in beating stress, anxiety and depression as well as memory enhancement (Erickson 2016).

5.8 Food Applications

The resourcefulness of sunflower oil is acknowledged by food manufacturers internationally. It is prized for its neutral taste, superior frying quality and basket full of health benefits. Quality of the oil, both for food and non-food applications, is mainly determined by its constituent fatty acids. However, sunflower seeds, when consumed in large quantities, have been reported to cause constipation, abdominal pain and rectal pain in healthy people (Lim et al. 2013; Manatakis et al. 2018). The general food applications of sunflower oil have been discussed below:

5.8.1 As Cooking Oil

Sunflower oil has traditionally been used as a cooking oil. It has low α-linolenic acid content (<1%), which makes it satisfactorily oxidative stable for frying. Oils carrying >3% linolenic acid are generally unstable in most food applications and develop an objectionable fishy and rancid flavour on heating and processing. Furthermore, the presence of high levels of γ- and δ-tocopherols, rather than α- and β-tocopherols, in sunflower oil makes it more stable to oxidation at high-temperature processing.

Frying is an admired cooking choice throughout the world. It quickly develops pleasing organoleptic attributes. Mid-oleic and high oleic sunflower oil has better oxidative stability than regular/conventional oil. As a result, applicability of sunflower oil at high temperatures as for frying has become more widespread. Besides, they produce fried foods with improved shelf life. The new sunflower oil (NuSun™) works exceptionally well in industrial processing and is being effectively utilized in the production of snack foods as well as in food service applications. Because of its neutral taste, allowing the particular flavours of foods to stand out, NuSun™ is liked by the customers. Besides, the presence of balanced amount of linoleic acid adds to the flavour of the products. As compared to other liquid vegetable oils, NuSun™ has a smoke point of 450 °F (232 °C) that is equivalent to the smoke point of the saturated palm oil, suggesting higher heat stability. Sunflower oil, as discussed previously, has been developed in various varieties such as high linoleic oil, mid-oleic and high oleic oil, etc. The food applications of these modified varieties are discussed here:

5.8.1.1 High Linoleic Oil

This variety of sunflower oil is rich in linoleic acid and is produced from the traditional/conventional type of sunflower, which has been grown for several years. Linoleic sunflower oil is used as salad oil in liquid form, utilized in the manufacturing of margarine and shortening. However, due to the presence of very high amount of linoleic acid in sunflower oil (68%), as well as low levels of γ-tocopherol, the oil is prone to oxidation during industrial processing, particularly frying. Hydrogenation of the oil can improve the stability just like any other PUFA rich oil, like soybean or canola. However, poor image of hydrogenated oils and limitations of non-hydrogenated oil in frying have resulted in production of only small volumes of this type of oil.

5.8.1.2 Mid Oleic Oil (NuSun®)

The development of this variety was made possible through traditional breeding techniques. This new mid-oleic sunflower oil surpasses other varieties for industrial use by offering more favourable health benefits, superior taste and better function. One of the very important attributes is that it is highly stable to frying temperatures in industrial processes and, thus, does not require hydrogenation and is free of *trans*-fats. Currently, it is the principal sunflower oil produced in USA and Canada and is rather the 'standard' sunflower oil in North America. NuSun oil has become a commodity in the USA. Because of its widespread cultivation, the oil is now available in large quantities and is cost competitive with other natively stable oils. Food manufacturers are ready to spend higher price for sunflower oil over soybean oil because of its healthy image, neutral taste and better performance. Commercial advantages of NuSun® oil are:

- Superior frying quality
- No hydrogenation required; free from *trans*-fat
- Low saturated fat content (<10%)
- Healthy fatty acid composition, high in monounsaturated fat
- Improved oxidative stability
- Shelf stable products
- Non transgenic; traditionally bred
- Excellent source of vitamin E

5.8.1.3 High Oleic Oil

Just like the mid-oleic variety, this variety was also developed by using traditional breeding techniques. It contains approximately 82% oleic acid with a few varieties producing oleic levels as high as 90%. This class of oil also has neutral taste profile. It is endowed with exceptional stability without being hydrogenated. Hence, customers get *trans*-fat-free oil. Besides its use in several types of frying, the oil finds various applications like in bakery; manufacturing of non-dairy creamers; spray coating oils for cereals, crackers and dried fruits; etc.

5.8.1.4 High Stearic/High Oleic Oil

This is the newest member of the sunflower oil family, developed by using traditional breeding techniques. Oil obtained from this variety is called Nutrisun™ with higher levels of saturated fatty acids (18% stearic acid). It is a perfect substitute for partially hydrogenated oils or tropical oils that have higher saturated fat level. Hence, it works as an excellent shortening agent in baked products and in the preparation of margarine, ice creams, chocolates and other products that demand for a high-melting (solid) oil.

5.8.2 In Bakery Industry

The use of a wide array of nuts and seeds in the bakery industry is well-known. However, sunflower seeds gained popularity of their use in baked products a little later. Pleasant flavour, nutty texture and healthy nutritional image of the seeds have attracted the manufacturers as well as consumers alike. Besides, small size of the kernels eliminates the need for chopping. However, one of the drawbacks of using them during baking is the darkening of colour because of the presence of chlorogenic acid in seeds. Chlorogenic acid causes an irreversible olive green/dark colouration, which may affect the overall acceptability of the final product. To use sunflower kernels in baked products, it is important to control the moisture, temperature and exposure to oxygen to prolong the shelf life of kernels. The kernels being high in PUFAs are susceptible to oxidation but at the same time are high in vitamin E which counteracts the oxidative reactions in processed product. Besides, baked products like bread and many others have a shorter shelf life than the sunflower kernels. So, shelf life of sunflower kernels is by and large not a concern in baked product.

The kernels can either be added in the formula on flour weight basis or sprinkled on the loaf after first brushing the baked loaf with ingredients such as honey. Such use of sunflower seeds can improve the nutritive value of baked foods, which are generally prepared from refined wheat flour that is low in fibre and protein, has poor balance of essential amino acids and lacks some essential vitamins and minerals. The use of high oleic sunflower oil (TriSun® extra) in the production of plastic fats and its consequent use in baked products is one of the very important applications of sunflower oil. It can also be used in combination with lard and other hydrogenated or partially hydrogenated vegetable oils. A 60:40 interesterified mixture of lard/TriSun has been reported to resemble soft-type margarine (Seriburi and Akoh 1998).

5.9 Conclusion

Sunflower is one of the most common oilseeds grown worldwide. The seeds are rich in oil, protein, phytonutrients, phytoestrogens and many more bioactive components, which have been reported to reduce the risk of several non-communicable diseases such as cardiovascular diseases, obesity, cancer, hypercholesterol, diabetes, arthritis and chronic inflammation. When consumed at recommended levels, sunflower oil, which is rich in omega-6 fatty acids, maintains omega-6/omega-3 ratio in human body and confers abovementioned health benefits. Different varieties of sunflower oil are developed through breeding techniques and are used for various food applications such as cooking oil, salad dressing, shortening and margarines.

References

Al-Jumaily RMK, Al-Shamma NMJ, Al-Halbosiy MMF, Al-Shamma LMJ (2013) Anticancer activity of sunflower (Helianthus annuns L.) seeds oil against cell lines. Iraqi J Sci 54 (4):1003–1009

Anastasi U, Santonoceto C, Giuffrè AM, Sortino O, Gresta F, Abbate V (2010) Yield performance and grain lipid composition of standard and oleic sunflower as affected by water supply. Field Crop Res 119(1):145–153

Anjum FM, Nadeem M, Khan MA, Hussain S (2010) Nutritional and therapeutic potential of sunflower seeds: a review. Br Food J 114(4):544–552

Armstrong WP (2007). https://www2.palomar.edu/users/warmstrong/fruitid5.htm. Accessed 03 Feb 2019

Awad AB, Fink CS (2000) Phytosterols as anticancer dietary components: evidence and mechanism of action. J Nutr 130:2127–2130

Berglund DR (2007) Sunflower production. https://www.ag.ndsu.edu/extensionentomology/recent-publications-main/publications/A-1331-sunflower-production-field-guide. Accessed 10 Sep 2019

Binkoski AE, Kris-Etherton PM, Wilson TA, Mountain ML, Nicolosi RJ (2005) Balance of unsaturated fatty acids is important to a cholesterol-lowering diet: comparison of mid-oleic sunflower oil and olive oil on cardiovascular disease risk factors. J Am Diet Assoc 105 (7):1080–1086

Carelli AA, Brevedan MI, Crapiste GH (1997) Quantitative determination of phospholipids in sunflower oil. J Am Oil Chem Soc 74(5):511–514

CSA (1999) Cambridge Scientific Abstracts, Internet database service (1999). Cambridge Scientific Abstracts (CSA), Bethesda, MD

Dorni C, Sharma P, Saikia G, Longvah T (2018) Fatty acid profile of edible oils and fats consumed in India. Food Chem 238:9–15

Duke J, Beckstrom S (1999) Dr. Duke's phytochemical and ethnobotanical databases. US Department of Agriculture-Agriculture Research Service

Eichenfield LF, McCollum A, Msika P (2009) The benefits of sunflower oleodistillate (SOD) in pediatric dermatology. Pediatr Dermatol 26(6):669–675

Erickson ML (2016) Inventor. System and method for increasing learning through eating protein, exercising, and singing and preventing bullying. United States patent application US 14/914,242

FAOSTAT (2019). http://www.fao.org/faostat/en/#data/QC/visualize. Accessed 20 Mar 2019

Fedeli E, Lanzani A, Capella P, Jacini G (1966) Triterpene alcohols and sterols of vegetable oils. J Am Oil Chem Soc 43(4):254–256

Fernández-Cuesta A, Jan CC, Fernández-Martínez JM, Velasco L (2014) Variability for seed phytosterols in sunflower germplasm. Crop Sci 54(1):190–197

Fjeld CR, Lawson RH (1999) Food, phytonutrients and health: rationale for the conference and workshop. Nutr Rev 57:S3–S6

González-Pérez S, Vereijken JM (2007 Sep) Sunflower proteins: overview of their physicochemical, structural and functional properties. J Sci Food Agric 87(12):2173–2191

González-Pérez S, Merck KB, Vereijken JM, Van Koningsveld GA, Gruppen H, Voragen AGJ (2002) Isolation and characterization of Undenatured Chlorogenic acid-free sunflower (Helianthus annuus) proteins. J Agric Food Chem 50:1713–1719

Holub BJ (1982) The nutritional significance, metabolism and function of myo-inositol and phosphatydilinositol in health and disease. Adv Nutr Res 4:107–141

Ingale SA, Shrivastava SK (2011) Chemical, nutritional and anti-nutritional study of new varieties of oil seeds from sunflower, safflower and groundnut. Int J Biotechnol Appl 3(4):118–129

Junker R, Kratz M, Neufeld M, Erren M, Nofer JR, Schulte H, Nowak-Göttl U, Assmann G, Wahrburg U (2001) Effects of diets containing olive oil, sunflower oil, or rapeseed oil on the hemostatic system. Thromb Haemost 85(02):280–286

Kapadia GJ, Azuine MA, Tokuda H, Takasaki M, Mukainaka T, Konoshima T, Nishino H (2002) Chemopreventive effect of resveratrol, sesamol, sesame oil and sunflower oil in the Epstein–Barr virus early antigen activation assay and the mouse skin two-stage carcinogenesis. Pharmacol Res 45(6):499–505

Karamać M, Kosińska A, Estrella I, Hernández T, Duenas M (2012 Aug 1) Antioxidant activity of phenolic compounds identified in sunflower seeds. Eur Food Res Technol 235(2):221–230

King A, Young G (1999) Characteristics and occurrence of phenolic phytochemicals. J Amer Diet Assoc 99:213–218

Koga T, Terao J (1994 Jun) Antioxidant activity of a novel phosphatidyl derivative of vitamin E in lard and its model system. J Agric Food Chem 42(6):1291–1294

Kuhnle GG, Dell'Aquila C, Aspinall SM, Runswick SA, Mulligan AA, Bingham SA (2008) Phytoestrogen content of beverages, nuts, seeds, and oils. J Agric Food Chem 56 (16):7311–7315

Lagarda MJ, García-Llatas G, Farré R (2006) Analysis of phytosterols in foods. J Pharm Biomed Anal 41(5):1486–1496

Lania BG, Morari J, de Almeida AR, da Silva MN, Vieira-Damiani G, de Almeida LK, César CL, Velloso LA, Maia NB, Cintra ML, Velho PE (2019) Topical essential fatty acid oil on wounds: local and systemic effects. PLoS One 14(1):e0210059

Le Clef E, Kemper T (2015) Sunflower seed preparation and oil extraction. In: Martínez-Force E, Dunford NT, Salas JJ (eds) Sunflower: chemistry, production, processing, and utilization. AOCS Press, Urbana, IL, pp 187–226

Liener IE (1995) Possible adverse effects of soybean anticarcinogens. J Nutr 125:744S–750S

Lim RG, Schultz SL, Rawlings AL, Bechtold ML (2013) Unusual cause of constipation: sunflower seeds bezoar. Endoscopy 45(S02):E46

Macías FA, Varela RM, Torres A, Galindo JL, Molinillo JM (2002) Allelochemicals from sunflowers: chemistry, bioactivity and applications. In: Chemical ecology of plants: allelopathy in aquatic and terrestrial ecosystems. Birkhäuser, Basel, pp 73–87

Madigan C, Ryan M, Owens D, Collins P, Tomkin GH (2000) Dietary unsaturated fatty acids in type 2 diabetes: higher levels of postprandial lipoprotein on a linoleic acid-rich sunflower oil diet compared with an oleic acid-rich olive oil diet. Diabetes Care 23(10):1472–1477

Manatakis D, Sioula M, Passas I, Zerbinis H, Dervenis C (2018) Rectal seed bezoar due to sunflower seed: a case report and review of the literature. Pan Afr Med J 31:157

Maver T, Maver U, Pivec T, Kurečič M, Persin Z, Stana KK (2018) Active substances for acceleration of wound healing. In: Maver T, Maver U, Pivec T, Kurečič M, Persin Z, Stana

Kleinschek K (eds) Bioactive polysaccharide materials for modern wound healing, SpringerBriefs in Molecular Science. Springer, Cham

McMullen MP (1985) Sunflower production and pest management extension bulletin 25 (rev.). North Dakota State University, pp 18

Mihai GM, Steven HZ (2013) Choline's role in maintaining liver function: new evidence for epigenetic mechanisms. Curr Opin Clin Nutr Metab Care 16(3):339–345

Nkengfack G, Englert H, Haddadi M (2019) Selenium and immunity. In: Mahmoudi M, Rezaei N (eds) Nutrition and immunity. Springer, Cham, Switzerland, pp 159–165

Odabasoglu F, Halici Z, Cakir A, Halici M, Aygun H, Suleyman H, Cadirci E, Atalay F (2008) Beneficial effects of vegetable oils (corn, olive and sunflower oils) and α-tocopherol on anti-inflammatory and gastrointestinal profiles of indomethacin in rats. Eur J Pharmacol 591 (1–3):300–306

OECD (2007) Series on the Safety of Novel Foods and Feeds No. 16, Consensus document on compositional considerations for new varieties of the sunflower: key food and feed nutrients, anti-nutrients and toxicants, OECD Environment, Health and Safety Publications, Environment Directorate, Organisation for Economic Co-operation and Development, Paris 2007

Owen RW, Weisgerber UM, Spiegelhalder B, Bartsch H (1996) Faecal phytic acid and its relation to other putative markers of risk of colorectal cancer. Gut 38:591–597

Padley FB, Gunstone FD, Harwood JL (1994) Occurrence and characteristics of oils and fats. In: Gunstone FD, Harwood JL, Padley FB (eds) The lipid handbook. Chapman and Hall, London, pp 53–146

Pal D (2011) Sunflower (Helianthus annuus L.) seeds in health and nutrition. In: Preedy VR, Watson RR, Patel VB (eds) Nuts and seeds in health and disease prevention. Academic Press, London, UK, pp 1097–1105

Pasini F, Riciputi Y, Verardo V, Caboni MF (2013) Phospholipids in cereals, nuts and some selected oilseeds. Recent Res Dev Lipids 9:139–201

Pedrosa MM, Muzquiz M, Garcia-Vallejo C, Burbano C, Cuadrado C, Ayet G et al (2000) Determination of caffeic and chlorogenic acids and their derivatives in different sunflower seeds. J Sci Food Agric 80:459–464

Phillips KM (2000) Phytochemicals in sunflower kernel. Literature review. Final Report. https://www.sunflowernsa.com/.../phytochemicalssfkernel_april2000_k. phillips.pdf

Phillips KM, Ruggio DM, Ashraf-Khorassani M (2005) Phytosterol composition of nuts and seeds commonly consumed in the United States. J Agri Food Chem 53:9436–9445

Piironen V, Lindsay DG, Miettinen TA, Toivo J, Lampi AM (2000) Plant sterols: biosynthesis, biological function and their importance to human nutrition. J Sci Food Agric 80(7):939–966

Piva G (1992) Nutritional value of sunflower seed and sunflower meal for livestock animal. Instituto de ScienzedellaNutrizione. Univ. Catt. Sacro Cuore, Piacenza, Italy. T1992Feed01

Racette SB (2017) Phytonutrients, micronutrient information center. Linus Pauling Institute. Copyright 2005–2019. Oregon State University

Razzaque M (2018) Magnesium: are we consuming enough? Nutrients 10(12):1863. https://doi.org/10.3390/nu10121863

Robinson RG (1975) Amino acid and elemental composition of sunflower and pumpkin seeds. Agron J 67(4):541–544

Robinson RG, Johnson FK, Soine OC (1967) The sunflower crop in Minnesota. Agriculture extension service, University of Minnesota. Extension Bulletin 299, p 7

Romani A, Pinelli P, Moschini V, Heimler D (2017) Seeds and oil polyphenol content of sunflower (Helianthus annuus L.) grown with different agricultural management. Adv Hortic Sci 31(2):85

Salas JJ, Bootello MA, Garcés R (2015) Food uses of sunflower oils. In: Martínez-Force E, Dunford NT, Salas JJ (eds) Sunflower: chemistry, production, processing, and utilization. AOCS Press, Urbana, IL, pp 441–464

Sammour RH, MN EI-S, Abo-Shady AM, Abasary AM (1995) Characterization of sunflower seed proteins by electrophoretic techniques. Arab Gulf J Scient Res 13(3):591–604

Schild J, Baltensperger P, Lyon P, Hein G, Kerr E (1991) G91–1026 Sunflower production in Nebraska. Historical Materials from University of Nebraska-Lincoln Extension. 772:1–7. https://digitalcommons.unl.edu/extensionhist/772
Schneiter AA (ed) (1997) Sunflower technology and production, Agronomy Monograph 35. The American Society of Agronomy, Madison, WI, pp 1–19
Sebbagh N, Cruciani-Guglielmacci C, Ouali F, Berthault MF, Rouch C, Sari DC, Magnan C (2009) Comparative effects of Citrullus colocynthis, sunflower and olive oil-enriched diet in streptozotocin-induced diabetes in rats. Diabetes Metab 35(3):178–184
Seriburi V, Akoh CC (1998) Enzymatic interesterification of lard and high-oleic sunflower oil with Candida antarctica lipase to produce plastic fats. J Am Oil Chem Soc 75(10):1339–1345
Shamsuddin AM (1995) Inositol phosphates have novel anticancer function. J Nutr 125:725S–732S
Siddique HR, Saleem M (2011) Beneficial health effects of lupeol triterpene: a review of preclinical studies. Life Sci 88(7–8):285–293
Smith LF, Patterson J, Walker LT, Verghese M (2016) Chemopreventive potential of sunflower seeds in a human colon cancer cell line. Int J Cancer Res 12(1):40–50
Sobczak AI, Stefanowicz F, Pitt SJ, Ajjan RA, Stewart AJ (2019) Total plasma magnesium, zinc, copper and selenium concentrations in type-I and type-II diabetes. Biometals 32(1):123–138
Taha FS, Mohamed GF, Mohamed SH, Mohamed SS, Kamil MM (2011) Optimization of the extraction of total phenolic compounds from sunflower meal and evaluation of the bioactivities of chosen extracts. Am J Food Technol 6(12):1002–1020
Taha FS, Wagdy SM, Hassanein MM, Hamed SF (2012) Evaluation of the biological activity of sunflower hull extracts. Grasas Aceites 63(2):184–192
Tanwar B, Modgil R, Goyal A (2018) Antinutritional factors and hypocholesterolemic effect of wild apricot kernel (Prunus armeniaca L.) as affected by detoxification. Food Funct 9(4):2121–2135
Tanwar B, Modgil R, Goyal A (2019) Effect of detoxification on biological quality of wild apricot (Prunus armeniaca L.) kernel. J Sci Food Agric 99(2):517–528
Tham DM, Gardner CD, Haskell WL (1998) Clinical review 97: potential health benefits of dietary phytoestrogens: a review of the clinical, epidemiological and mechanistic evidence. J Clin EndocrinolMetab 83:2223–2235
USDA (2018) Nutrient database for standard reference, release. Full report (all nutrients): 45209868, APPLES, UPC: 033383086927, Basic report: 08352, Cereals ready-to-eat, QUAKER, MOTHER'S Toasted Oat Bran cereal
USDA (2019). https://fdc.nal.usda.gov/fdc-app.html#/food-details/170562/nutrients. Accessed 20 Aug 2019
Velasco L, Ruiz-Méndez MV (2015) Sunflower oil minor constituents. In: Martínez-Force E, Dunford NT, Salas JJ (eds) Sunflower: chemistry, production, processing, and utilization. AOCS Press, Urbana, IL, pp 297–329
Vogt TM, Ziegler RG, Graubard BI, Swanson CA, Greenberg RS, Schoenberg JB (2003) Serum selenium and risk of prostate cancer in U.S. blacks and whites. Int J Cancer 103(5):664–670
Weisz GM, Kammerer DR, Carle R (2009 Jul 15) Identification and quantification of phenolic compounds from sunflower (Helianthus annuus L.) kernels and shells by HPLC-DAD/ESI-MSn. Food Chem 115(2):758–765
Wildermuth SR, Young EE, Were LM (2016 Sep) Chlorogenic acid oxidation and its reaction with sunflower proteins to form green-colored complexes. Compr Rev Food Sci Food Saf 15 (5):829–843
Zhou JR, Erdman JW (1995) Phytic acid in health and disease. Crit Rev Food Sci Nutr 35:495–508

Chapter 6
Palm/Palm Kernel (*Elaeis guineensis*)

Shalini Sehgal and Vasudha Sharma

Abstract Palm (*Elaeis guineensis*) is one of the most important oilseed crops in the world. Crude palm and palm kernel oil are rich in phytonutrients such as sterols, vitamin E, squalene, phospholipids, etc. It is the richest source of tocopherols and carotenoids among all the commonly used vegetable oils. Palm oil is produced from fleshy pulp (mesocarp), which contains approximately 29.32% moisture (on wet basis), 2.03% protein, 68.09% lipids and 1.11% ash content (on dry basis). On the other hand, palm kernel oil is extracted from the inner kernel and contains about half of the oil content than that of palm fruit. Palm and palm kernel oils are rich in saturated fatty acids (40–92%), free from trans fats, and have high amount of bioactive components. Palm protein has been reported to have higher amino acids score (except for lysine and tryptophan) than that of FAO/WHO reference protein. In this chapter, the proximate composition of palm and palm kernel, minor oil components, health effects and food applications are discussed in detail.

Keywords Palm (*Elaeis guineensis*) oil · Palm kernel oil · Phytonutrients · Tocopherols · Squalene · Cardiovascular diseases · Cooking oil

6.1 Introduction

Palm (*Elaeis guineensis*) belongs to the genus *Elaeis* and is one of the most important oilseed crop. The palm oil produced by the palm fruit is the second most consumed vegetable oil after soybean oil (Saad et al. 2007; Siddique et al. 2010) and is one of the most important and versatile raw material for both food and

S. Sehgal (✉)
Department of Food Technology, Bhaskaracharya College of Applied Sciences, University of Delhi, Dwarka, New Delhi, India

V. Sharma
Department of Food Technology, Jamia Hamdard (Deemed to be University), New Delhi, India

B. Tanwar, A. Goyal (eds.), *Oilseeds: Health Attributes and Food Applications*, https://doi.org/10.1007/978-981-15-4194-0_6

Fig. 6.1 Palm (fleshy fruit) and palm kernel (seed)

non-food industries. Palm oil is produced from fleshy mesocarp which is used for edible purposes, whereas palm kernel oil is produced from innermost part of the fruit, i.e. kernel, which finds application in oleochemical industry (Sambanthamurthi et al. 2000) (Fig. 6.1). The kernel constitutes about 45–48% (by weight) of the nut. Palm oils have emerged as the fastest-growing oil within the span of four decades. Commercially, palm oil is majorly grown in Africa, South America, Southeast Asia, South Pacific and other tropical areas. The palm has been confined to West and Central Africa and exists in a wild, semi-wild and cultivated state. Palm cultivation expanded from Africa to the Southeast Asian regions and strengthened the entry of palm oil into the world of oils and fats trade in the end of the eighteenth century (Shahidi 2005). Malaysia and Indonesia are the two largest producers of palm oil and together account for roughly 85% of the world palm oil production (Abdullah and Sulaiman 2013). Practically, with the co-generation system all palm oil mills generate their own heat and power (Abdullah and Sulaiman 2013; Noor et al. 1999).

6.2 Origin and History

The African oil palm, native of West Africa, was first illustrated in 1763 by Nicholas Jacquin, hence its name, *Elaeis guineensis Jacq*. It is said that *E. guineensis* originated in West Africa in the fifteenth century; and Portuguese introduced it in to Brazil and other tropical countries. In the nineteenth century, its propagation surged when the Dutch brought seeds from West Africa to Indonesia in 1848. The oil palm was initially planted as an ornamental in West Malaysia, and the first commercial planting was done only in 1917. *E. oleifera* originated in South America and became popular due to its higher oleic and linoleic acid content in the mesocarp oil and lower content of palmitic and other saturated acids (Sambanthamurthi et al. 2000). The oil palm tree is an important source of vegetable oil and the most important cash crops in the region (Akinyeye et al. 2011).

6.3 Production

In terms of land use, the oil palm tree is the most efficient oil crop, and the fruits are harvested monthly, when the oil palm trees are 3–4 years old. It has the highest yield per hectare of land with lower amount of fertilizers as compared to other oil crops. More than 80% of the total world production of oil palm fruit is produced by Indonesia and Malaysia. Percent share of top five 'oil palm fruit'-producing countries is shown in Fig. 6.2. According to a recent report, total world production of palm oil was estimated approximately 58.8 MMT in 2018 (Sawe 2018). Indonesia is the largest producer of palm oil with the total production of 36 MMT. The production of palm oil (in MMT) by the world and top five producing countries is shown in Fig. 6.3. India has been reported to produce approximately 0.2 MMT of palm oil and 2000 metric tonnes of palm kernel oil in 2018 (USDA 2019).

6.4 Proximate Composition

Proximate composition of palm fruit and palm kernel vary greatly depending on the species, soil composition and other agroclimatic factors. Palm fruit has been reported to contain approximately 29.32% moisture (on wet basis), 2.03% protein, 68.09% lipids and 1.11% ash content (on dry basis) (Li et al. 2012). Majority of the researchers have reported the proximate composition of palm fruit and palm kernel on dry basis (Bora et al. 2003; Akpanabiatu et al. (2001); Atasie and Akinhanmi (2009), which has been shown in Table 6.1. Palm fruit contains approximately

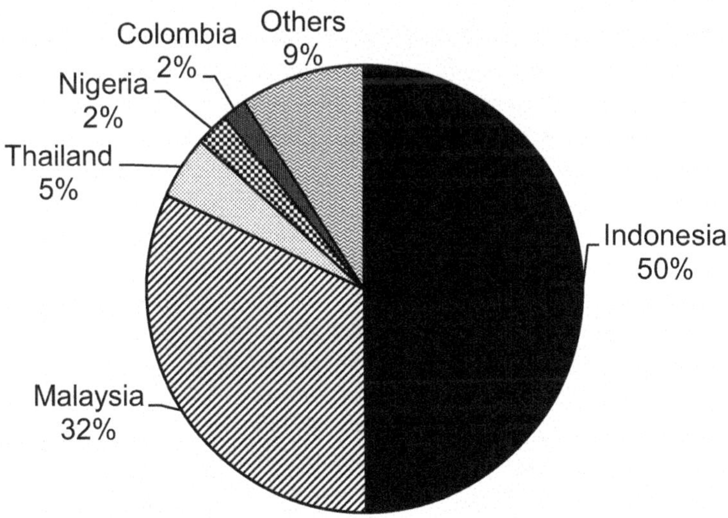

Fig. 6.2 Percent share of top five oil palm fruit-producing countries in 2017 (FAOSTAT 2019)

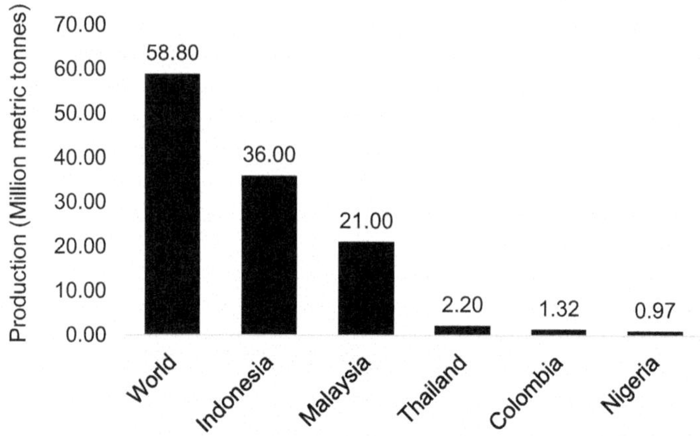

Fig. 6.3 Palm oil production (MMT) by world and top five producing countries (Sawe 2018)

Table 6.1 Proximate composition of palm (fruit pulp), palm kernel and defatted palm kernel (meal)

Constituents (%)	Palm (pulp) (dwb)	Palm kernel (dwb)	Defatted palm kernel
Lipids	68.0–73.2	32.6–49.4	–
Crude protein	2.0–3.4	7.5–10.9	14.0–19.2
Carbohydrate	13.3	33.4–35.1	NR
Crude fibre	4.3	11.0–15.6	14.7–17.9
Ash	1.1–1.9	1.5–1.9	3.0–4.3

Sources: Bora et al. (2003), Akpanabiatu et al. (2001), Li et al. (2012), Atasie and Akinhanmi (2009). *dwb* Dry weight basis

double of the total lipids present in kernel (Bora et al. 2003). Akpanabiatu et al. (2001) evaluated the nutrient composition of Nigerian palm kernel from the dura and tenera varieties of the oil palm and reported approximately 6–14% moisture content, 41–49% lipids and 7.5–8.1% crude proteins content on dry basis.

6.4.1 Lipids

Palm oil is a yellowish orange colour with the characteristic palm fruit flavour. Palm oil is obtained from fleshy pulp (mesocarp) of the palm fruit, while the kernel is the source of kernel oil. Oil content varies from 68.0% to 73.2% and 32.6% to 49.4% in palm fruit and palm kernel oil, respectively. Bora et al. (2003) characterized the Brazilian oil palm fruits and observed approximately 24 and 18 fatty acids in palm and palm kernel oil, respectively. The fatty acid composition of both types of oils is presented in Table 6.2. Palm kernel oil has been reported to contain a very high amount of saturated fatty acids (>80% of total fatty acids), even higher than palm oil

Table 6.2 Major fatty acid profile of palm and palm kernel oil

Fatty acid (% of total fatty acids)	Palm oil	Palm kernel oil
Caproic (C6:0)	Traces	Traces-0.10
Caprylic (C8:0)	Traces	3.00–3.09
Capric (C10:0)	Traces	2.20–3.37
Lauric (C12:0)	Traces-0.08	46.40–53.20
Myristic (C14:0)	0.53–1.15	16.40–19.30
Palmitic (C16:0)	36.90–39.93	9.40–10.35
Stearic (C18:0)	4.04–4.68	2.30–2.34
Oleic (C18:1)	35.99–45.29	5.50–18.00
Linoleic (C18:2)	10.69–14.53	0.61–2.70
α-Linolenic C18:3	ND-0.27	ND
SFA	42.79	92.62
MUFA	45.73	6.71
PUFA	10.96	0.61

Sources: Bora et al. (2003), Akpanabiatu et al. (2001), Li et al. (2012). *SFA* Saturated fatty acids, *MUFA* Monounsaturated fatty acids, *PUFA* Polyunsaturated fatty acids, *ND* Not detected

(>40%) (Bora et al. 2003; Gibon 2012). Although both types of oils are extracted from the same plant, the fatty acid profile of both oils is substantially different. The most abundant fatty acids in palm oil are the oleic acid (C18:1, 35.99–45.29%) and palmitic acid (C16:0, 36.90–39.93%). On the other hand, palm kernel oil is reported to contain maximum amount of lauric acid (46.40–53.20%) and thus belongs to the class of lauric fats (CODEX 1999). Due to the difference in dominating fatty acids, the physical properties of both the oils are also different. Palm kernel oil is substantially harder than that of palm oil at lower temperatures. Unlike soybean, canola, flaxseed and chiaseed oil, palm and/or palm kernel oil contains no or insignificant amount of α-linolenic acid (omega-3) (Table 6.2). Moreover, another essential fatty acid, i.e. linoleic acid, is also not present in a very high amount in palm kernel oil unlike other regular vegetable oils.

6.4.2 Proteins

The data on protein composition suggested that both palm and palm kernel protein have high content of glutamic acid, aspartic acid and arginine (Kok et al. 2011). Similar observations have also been reported by Santoso et al. (1996). Kok et al. (2011) evaluated the nutrient composition of oil palm kernel and reported glutamic acid, arginine and aspartic acid in the range of 30.4–40.2, 20.3–26.9 and 13.6–17.2 mg/g (dry weight), respectively. When the content of essential amino acids of palm kernel protein was compared with those of the standard protein as per FAO/WHO (1991), it was found that the contents of both lysine and tryptophan were lower in palm kernel protein than that of standard protein. Thus, the amino acid score

Table 6.3 Amino acids composition of palm and palm kernel protein with standard protein

Amino acid (g/100 g protein)	Palm (pulp) protein	Palm kernel protein	Standard protein (FAO/WHO)[a]
Essential amino acids			
Isoleucine	3.93	2.94–4.82	2.8
Leucine	7.08	6.21–9.10	6.6
Lysine	2.84	2.99–4.40	5.8
Methionine	0.37	0.90–3.04	2.5[b]
Phenylalanine	3.73	3.12–5.89	6.3[c]
Threonine	3.74	3.74–4.38	3.4
Tryptophan	NR	0.60–0.66	1.1
Valine	6.14	5.11–7.52	3.5
Non-essential amino acids			
Alanine	8.36	5.80–5.82	–
Arginine	8.15	13.87–16.10	–
Aspartic acid	10.19	8.50–11.60	–
Cystine	2.87	1.60–2.40	–
Glutamic acid	13.64	9.47–27.2	–
Glycine	10.54	6.64–9.41	–
Histidine	1.49	1.53–2.32	–
Proline	4.70	3.61–4.72	–
Serine	7.11	5.06–6.16	–
Tyrosine	2.67	1.55–3.31	–

Sources: Bora et al. (2003), Sreedhara and Kurup (1998), Abdollahi et al. (2015)
NR Not reported
[a]FAO/WHO (1991)
[b]Indicates Methionine + Cystine
[c]Phenylalanine + Tyrosine

of these two amino acids, i.e. lysine (~49%) and tryptophan (59.8%), is lower than 100% (Kok et al. 2011). In other words, both palm and palm kernel proteins are limiting in lysine and tryptophan amino acids. Similar observation has also been reported by Bora et al. (2003) for Brazilian palm and kernel's protein. It is worthwhile to mention that remaining amino acids such as histidine, isoleucine, leucine, threonine and valine in palm and/or palm kernel protein have amino acid score of >100%, i.e. 111–130%, 122.6–125.3%, 108.6–110.0%, 103.0–106.2% and 147.0–150.9%, respectively (Kok et al. 2011). It implies that palm and palm kernel proteins should be supplemented with legumes or other proteins which are rich in lysine and tryptophan, for the maximal benefits (Table 6.3).

Table 6.4 Mineral content (mg/100 g dry matter) of palm and palm kernel meal (defatted)

Element (mg/100 g dry matter)	Palm meal	Palm kernel meal
Calcium	203.48	107.0–390.0
Magnesium	85.27	280.0–554.0
Potassium	326.40	570.0–660.0
Sodium	21.94	0.7–120.0
Total phosphorus	30.00	350.0–470.0
Iron	3.19	13.0–41.0
Copper	1.30	2.0–2.7
Manganese	0.14	26.0–61.0
Zinc	0.55	2.7–4.3

Sources: Abdollahi et al. (2015), Akpanabiatu et al. (2001), Li et al. (2012)

6.4.3 Ash and Minerals

Palm and palm kernel are reported to contain ash content in the range of 1.1–1.9% and 1.5–1.9%, respectively (Bora et al. 2003; Akpanabiatu et al. 2001). However, Sreedhara and Kurup (1998) reported slightly higher ash content in palm kernel (2.1%) (on dry basis). Kok et al. (2011) studied the nutrient composition in kernel of tenera and clonal materials of oil palm and reported 1.93–2.64% ash content (on dry basis). In general, palm fruits are rich in K, Ca and Mg, whereas palm kernels have been reported to be rich in K, Ca, Mg, Mn and P (Table 6.4). The concentration of different minerals present in palm and palm kernel meal is presented in Table 6.4. It is worthy to note that palm kernels have substantially higher content of all of the minerals as compared to their respective palm fruit. Akpanabiatu et al. (2001) evaluated the proximate composition of Nigerian palm kernel from the dura and tenera varieties and reported higher mineral values in later variety. However, lead (Pb) was absent in both of the varieties.

6.4.4 Minor Components and Phytonutrients

The minor constituents which make up approximately 1% of total palm oil include carotenoids, tocopherols, squalene, polyphenols, coenzyme Q, sterols and phospholipids. Crude palm oil is one of the richest known sources of bioactive carotenoids (500–1000 ppm), of which α- and β-carotene account for ~35% and 56%, respectively. Palm oil also contains appreciable amounts of vitamin E (150–1500 ppm)—both the tocopherols and tocotrienols—and is one of the richest natural sources of the latter. The contents of several minor components such as sterols, vitamin E, carotenoids, phospholipids, etc. in palm and palm kernel oil are presented in Table 6.5.

Table 6.5 Concentration of minor components of crude palm and palm kernel oil

Constituent	Crude palm oil	Palm kernel oil
Total sterols (mg/kg oil)	300–700	700–1400
– Cholesterol[a]	2.6–6.7	0.6–3.7
– Campesterol[a]	18.7–27.5	8.4–12.7
– Stigmasterol[a]	8.5–13.9	12.0–16.6
– β-Sitosterol[a]	50.2–62.1	62.6–73.1
– Δ-5-Avenasterol[a]	ND-2.8	1.4–9.0
Total tocopherols + tocotrienols (mg/kg)	150–1500	ND-260
– α-Tocopherol (mg/kg)	4.0–193.0	ND-44.0
– β-Tocopherol (mg/kg)	ND-234.0	ND-248.0
– γ-Tocopherol (mg/kg)	ND-526.0	ND-257.0
– δ-Tocopherol (mg/kg)	ND-123.0	ND
Carotenoids (ppm)	500–1000	4.3–11.8
Squalene (ppm)	280–800	–
Phospholipids (ppm)	20–130	3000–3900
Coenzyme Q10/ubiquinones (ppm)	10–80	–
Polyphenols (ppm)	40–70	–

Source: CODEX (1999), Aparicio et al. (2018), Goh et al. (1985), Gibon (2012), Amri (2011)
ND Not detected
[a]% of total sterols

6.5 Health Effects

Palm oil is one of the most widely used vegetable oil in the world, and its nutritional and health properties are well documented in the literature. Palm oil has a high demand and strong appeal globally as a cooking medium, as it is free from *trans* fats which are reported to be the leading cause of clogging of arteries. It is more oxidative stable and has better handling properties. Also, food processors prefer it over other vegetable oils as it has distinct quality and requires little or no hydrogenation and extends the shelf life of different food products (Mukherjee and Mitra 2009). Palm oil is one of the largest sources of tocotrienol. It contains about 10% linoleic acid, which is a PUFA (ω-6 fatty acid) and one of the two essential fatty acids. Animal studies as well as human studies have reported several health benefits of the phytonutrients present in crude palm oil such as reduced risk of arterial thrombosis and atherosclerosis, cholesterol lowering effects, anti-ageing, anti-inflammatory and anticarcinogenic effects, inhibition of platelet aggregation and reduced blood pressure (Loganathan et al. 2008, 2010; Boon et al. 2013; Boateng et al. 2016; Obahiagbon 2012). However, palm oil in the oxidized state is reported to have adverse health effects (Mukherjee and Mitra 2009).

Red palm oil has abundant carotenoids (500–700 mg/L) giving it the characteristic colour in crude oil. The carotenoids, vitamin E, ascorbic acid, enzymes and proteins collectively convert highly reactive radicals and free fatty peroxy radicals to less active species, thereby protecting against oxidative damage to cells. Red palm oil is rich in β-carotene, a precursor of vitamin A important in the visual process, regulation of cell division, enhancement of immune response, lung development, genetic regulation and differentiation of cellular epithelium (Kritchevsky 2000; Oguntibeju et al. 2009; Scrimshaw 2000). Crude palm oil is one of the richest sources of vitamin E (tocopherols and tocotrienols) (Table 6.5). Vitamin E acts as potential antioxidants and protects cellular membranes from free radical catalysed lipid peroxidation. Palm oil is the only vegetable oil which contains appreciable quantities of tocotrienols. Tocotrienols have been reported to be natural inhibitors of cholesterol synthesis, promote anti-thrombotic state by reducing platelet aggregation and modulating prostanoid synthesis and prevent some forms of cancer (Edem 2002; Morales-Gonzalez 2018; Sailo et al. 2018; Peh et al. 2016).

6.5.1 Reduced Risks of Arterial Thrombosis and Atherosclerosis

Globally cardiovascular diseases have been reported to be one of the leading causes of death. Blood plasma cholesterol levels are affected by the degree of saturation, chain length of dietary fatty acids and their effect on the synthesis, processing and catabolism of plasma lipoproteins. Saturated fatty acids and cholesterol when consumed in high amounts in the diet are found to be associated with a high incidence of blockage of arteries and injuries involving both atherosclerosis and thrombosis. Also, the quality and quantity of fat present in the diet is a major aspect involved in the regulation of blood pressure and incidence of hypertension (Edem 2002).

Few reports testing palm oil in animals for atherosclerosis claim that palm oil does not promote atherosclerosis. In the Netherlands, palm oil was tested in a rabbit model; it was fed in the diet for over 1.5 years and was observed to induce the least atherosclerosis as compared to fish oil, olive oil, linseed oil and sunflower oil (Klurfeld et al. 1990). Another research studied the effects of palm oil and compared them with that of cottonseed oil, coconut oil and an American fat blend made from tallow, lard, butterfat, salad oils, peanut oil, corn oil and shortenings in a rabbit model. At the end of 14 months of feeding, it was observed that rabbits fed with coconut oil were observed to have the highest aortic lesions. The effects of palm oil were considerably lower than coconut oil and similar to those of edible oils. This confirmed the fact that consumption of palm oil at 32% of fat energy did not increase chances of atherosclerosis (Chong and Ng 1991).

6.5.2 Inhibition of Cholesterol Biosynthesis

Palm oil and its fractions are rich in tocopherols, particularly γ-tocotrienol. These are natural antioxidants and possess free radical scavenging properties, protecting the biological cell against oxidative and carcinogenic stress in addition to inhibiting platelet aggregation and inhibition of endogenous cholesterol biosynthesis (Morales-Gonzalez 2018; Sailo et al. 2018; Peh et al. 2016). Elson and Qureshi (1995) studied the impact of palm oil on cardiovascular disease and cancer and found that monoterpenes, sesquiterpenes, carotenoids and tocotrienols downregulate 3-hydroxy-3-methylglutaryl coenzyme A reductase activity, thereby decreasing cholesterol biosynthesis followed by decreased LDL cholesterol. Palm oil is observed to have a lipid lowering effect possibly because of a high content of tocotrienols present in palm oil (780–1080 μg/g). Qureshi et al. (1991) conducted a double-blind, crossover, 8-week study to compare effects of tocotrienol-enriched fraction of palm oil (200 mg per day) with those of 300 mg corn oil per day on serum lipids of hypercholesterolemic human subjects. A significant decrease in the serum total cholesterol (−15%), low density lipoprotein (−8%), thromboxane (−25%) and platelet aggregation (−16%) was found when the tocotrienol-rich fraction was administered for 4 weeks to hypercholesterolemic subjects. Serum cholesterol concentrations of seven hypercholesterolemic subjects decreased 31% during the 4-week period.

6.5.3 Prevention of Platelet Aggregation

An important risk factor for cardiovascular disease is the tendency for a thrombus (clot) to be formed in the blood wall vessel. Kooyenga et al. (1997) studied the antioxidant properties of α- and γ-tocopherol-enriched fractions of palm oil with carotid atherosclerosis. Serum lipids, fatty acid peroxides, platelet aggregation and carotid artery stenosis were measured over a 24-month period in 50 patients with cerebrovascular disease. The studies revealed apparent carotid atherosclerotic regression in 8 and progression in 2 of the 25 antioxidant patients, while none of the control group exhibited regression, and 10 of 25 showed progression ($P < 0.01$). Serum α-tocopherol doubled, while tocotrienols were undetectable throughout the study. Several studies report that saturated fats such as coconut oil and beef fat increase the blood clotting tendency, whereas unsaturated fats such as fish oils decrease platelet aggregation. Surprisingly, palm oil is found to have effects similar to polyunsaturated oils, i.e. it reduced platelet aggregation and decreased blood clotting. Thromboxane (TxA_2) is a powerful platelet aggregating hormone and also vasoconstrictive substance that promotes clotting; on the other hand, prostacyclin (PGI_2) is a hormone which acts as a platelet aggregation inhibitor. The balance between these local hormones results in linked with the arterial thrombotic tendency. Research reports have revealed that palm oil diet in animals increases the production of anti-clotting hormone prostacyclin and decreases the formation of thromboxane (Chong and Ng 1991). Qureshi et al. (2011) conducted a study to determine the extent to which tocotrienols inhibit platelet aggregation and reduce coronary thrombosis. The study determined the comparative effects of α-tocopherol and α-tocotrienol or tocotrienol-rich fraction (TRF) on in vivo platelet thrombosis and ex vivo platelet aggregation after intravenous injection in anaesthetized dogs by using Folts cyclic flow model. It was observed that collagen-induced platelet aggregation in platelet-rich plasma was decreased markedly after treatment with tocotrienol (59%; $P < 0.001$) and TRF (92%; $P < 0.001$). α-Tocopherol produced only a 22% decrease in platelet aggregation.

6.5.4 Reduction in Blood Pressure

It has been observed that saturated fats caused an increase in blood pressure, while consumption of mono- and poly-unsaturated fatty acids causes a decrease in blood pressure (Hall 2009; Salonen et al. 1988). However, studies on the effect of palm oil diets on mean arterial pressure of rats reveals that diets containing 15% (w/w) red palm oil fed for 18 weeks did not increase the blood pressure (Edem 2002). Cheng et al. (2017) investigated the effects of tocotrienol-rich fractions (TRF) of palm oil in high-fat diet-treated rats. Male, post-weaning Sprague Dawley rats were provided high-fat (60% kcal) diet for 8 weeks followed by a TRF (60 mg/kg) treatment for another 4 weeks. Physical, metabolic and histological changes were compared to those on control and high-fat diets, respectively. TRF reversed stolic and diastolic hypertension, hypercholesterolaemia, hepatic steatosis, impaired antioxidant defence and myeloperoxidase hyperactivity triggered by the high-fat diet. Palm oil is observed to have a protective effect on the endothelium of blood vessels. Bayorh et al. (2005) investigated the cardiovascular effects of natural vitamin-rich palm oil using Dahl salt-sensitive hypertension model. Male rats were fed a high-salt (8% NaCl) or low-salt (0.3% NaCl) diet with or without palm oil for 4 weeks. The study revealed that palm oil significantly reduced the progression of salt-induced hypertension and mortality, via mechanisms involving modulation of endothelial function and reduction in oxidative stress (Bayorh et al. 2005).

6.5.5 Palm Oil and Coronary Heart Diseases

Approximately 90% of palm oil produced is consumed by humans, and therefore, nutritional properties of palm oil and its fractions need to be demonstrated and understood properly. A large number of studies conducted in animals and humans demonstrate the nutritional adequacy of palm oil and its products in dietary trials. For several years, palm oil was considered to be a source of saturated fat, and assumption was made that like all other saturated fats, palm oil also raises the blood cholesterol levels. However, several human and animal feeding experiments report that no increase in the levels of blood cholesterol and LDL cholesterol was observed with palm oil; moreover, it was found to decrease these levels as compared to saturated fats from plant and animal sources. Lucci et al. (2016) examined the effect of palm oil supplementation on human plasma lipids related to cardiovascular disease risk factors. One hundred sixty participants were randomized and assigned to one of the two treatments 25 mL hybrid palm oil (HPO) or 25 mL extra virgin olive oil (EVOO) daily for 3 months. Fasting venous samples were obtained at baseline and after 1, 2 and 3 months for measurement of plasma lipids (TC, LDL-C, HDL-C and TAGs). Overall reduction in total cholesterol (7.4%, $p < 0.001$) and in LDL-C (15.6%, $p < 0.001$) was observed. HPO had similar effects on palm oil as EVOO in terms of measurements of plasma lipids.

In a study, conducted in a group of 40 Dutch male volunteers, it was observed that the maximal replacement of fats in the diet with palm oil had no significant effect on the blood cholesterol levels. The cholesterol levels were found to be 190 mg/dL for Dutch fat blend and 191 mg/dL for palm oil diet. Moreover, a significant increase in the beneficial HDL cholesterol and significant reduction in LDL triglycerides was observed in the subjects fed with palm oil diet (Hornstra and Sundram 1989). Another study conducted in Malaysia, compared the effects of diet containing coconut oil, corn oil and palm oil in three groups of adult volunteers. It was found that palm oil followed by coconut oil consumption reduced serum cholesterol levels by 36 mg/dL; corn oil feeding followed by coconut oil consumption reduced serum cholesterol levels by 68 mg/dL, and whereas coconut oil diet throughout did not reduce the serum cholesterol levels, the final cholesterol levels for palm oil, corn oil and coconut consumption were at 155, 122 and 190 mg/dL, respectively (Ng et al. 1991). In another study, 110 student volunteers aged between 11 and 17 years were fed with a palm olein diet for 5 weeks, and the plasma cholesterol levels were found averaging around 149 mg/dL. This was followed by a wash out period of 6 weeks and then fed a soybean oil diet for 5 weeks where the plasma cholesterol levels were found averaging around 153 mg/dL, indicating that plasma cholesterol levels were comparable in palm oil and soybean oil groups (Chong and Ng 1991).

6.6 Adverse Effects and Individual Concerns

Several benefits of palm oil such as reduced platelet aggregation, risk of atherosclerosis, coronary heart diseases, blood pressure and similar effects have been studied and discussed so far. However, it is also important to note that considerable amount of palm oil is used in the oxidized state which possesses potential dangers to the physiological and biochemical functions of the body. Thermal processing of oil in combination with other additives increases the chances of deteriorative changes in the oil and may result in an oxidized state and formation of toxic new chemical species. Many researchers have shown that oxidized palm oil induces an adverse plasma lipid profile, free fatty acids, phospholipids and cerebrosides (Osim et al.

1996). And it induces reproductive toxicity and organotoxicity particularly of the kidneys, lungs, liver and heart in oxidized state (Owu et al. 1998). Available evidence suggests that at least part of the oxidized oil impact on health reflects generation of toxicants due to oxidation. The reduction of the dietary level of oxidized oil and/or the level of oxidation may reduce the health risk associated with consumption of oxidized fats (Ebong et al. 1999).

6.7 Food Applications

Palm oil holds several properties that make it a favourable food choice in commercial food applications. Over 95% of palm oil consists of mixtures of triglycerides, with a high-solid glyceride content, which gives it the required consistency without hydrogenation. It has longer shelf stability as it is resistant to oxidation, due to the presence of natural antioxidants: tocopherols and tocotrienols. The content of high melting point triglycerides along with its relatively low solid content at 10 °C facilitates preparation of foods with a wide plastic range appropriate for high temperatures. Its tendency to crystallize in small β' crystals makes it suitable for margarine and cakes. It has a wide plastic range due to its slow melting properties. A major use of palm oil is in manufacture of different types of margarine: cake, pastry and tub margarines (Cottrell 1991).

6.7.1 Cooking Oil

Natural antioxidants tocopherols and tocotrienols confer a good oxidative stability to palm oil making it a preferable choice. It is inexpensive to use and good for producing fried foods with a longer shelf stability and good flavour. Foods produced with palm oil compare well with those produced with groundnut oil and deteriorate less rapidly as compared to soybean oil and sunflower oil due to its oxidative stability. However, on repeated frying, it results in formation of a brown colour due to its phenolic components, and this discoloration is unfavourable. A double fractionated palm olein which has a low cloud point has also been manufactured which is suited for use as a salad oil (Cottrell 1991).

6.7.2 Bakery Industry

Palm oil blends, hydrogenated palm oils and palm stearin are widely used as shortenings in bakery industry due to their smooth consistency and spread ability. Margarine blends from palm stearin are popularly used in the bakery industry. It is used as a dough improver in bakery industry (Cottrell 1991).

6.7.3 Chocolate and Confectionary

Palm mid fraction is used as the main component (50–70%) of cocoa butter equivalents along with sal, shea or illipe. It is used a cocoa butter extender and should demonstrate properties similar to cocoa butter. Few products based on palm olein have also been used as cocoa butter extenders. It is used as a crystallization starter in after glycerolysis in food emulsifier preparations. Blended palm oil is also used in non-dairy ice creams, chocolate coatings, etc. (Cottrell 1991).

6.8 Alternative Applications

Palm oil is known to be a multipurpose vegetable oil with applications varying from food to biodiesel. Cost of palm oil being cheaper than most other oils makes it a more cost-effective candidate for biofuel production compared to other vegetable oils. Therefore, palm oil has a potential of becoming a source of renewable energy and biodiesel (Tan et al. 2009).

6.9 Future Challenges

Palm oil has long been used as a source of oil for food and other uses. However, in the long term, it has resulted in percolation of effluents from palm mills into water bodies and our environment, thereby disrupting the ecosystem. Therefore, the need of the hour is production of palm oil through sustainable plantations and developing of value chains in processing palm oil which completely utilize the by-products left after oil extraction, possibly through biofuel product, biomass production or biohydrogen production.

References

Abdollahi MR, Hosking B, Ravindran V (2015) Nutrient analysis, metabolisable energy and ileal amino acid digestibility of palm kernel meal for broilers. Anim Feed Sci Technol 206:119–125

Abdullah N, Sulaiman F (2013) The oil palm wastes in Malaysia. In: Matovic MD (ed) Biomass now—sustainable growth and use. In Tech, Novi Sad. https://doi.org/10.5772/55302. https://www.intechopen.com/books/biomass-now-sustainable-growth-and-use/the-oil-palm-wastes-in-malaysia

Akinyeye RO, Adeyeye EI, Fasakina O, Agboolaa A (2011) Physico-chemical properties and anti-nutritional factors of palm fruit products (Elaeis guineensis jacq.). J Environ Agri Food Chem 10:2190–2198

Akpanabiatu MI, Ekpa OD, Mauro A, Rizzo R (2001) Nutrient composition of Nigerian palm kernel from the dura and tenera varieties of the oil palm (Elaeis guineensis). Food Chem 72 (2):173–177

Amri IN (2011) The lauric (coconut and palm kernel) oils. In: Gunstone FD (ed) Vegetable oils in food technology composition, properties and uses, 2nd edn. Wiley, Chichester, UK, pp 169–197. ISBN: 978-1-4443-3268-1

Aparicio R, González DLG, Aparicio-Ruiz R (2018) Vegetable oils. In: Morin J-F, Lee M (eds) Handbook on food authenticity issues and related analytical techniques, ERA Science Ltd., UK, pp 1–23. https://doi.org/10.32741/fihb.19.vegetableoil

Atasie VN, Akinhanmi TF (2009) Extraction, compositional studies and physico-chemical characteristics of palm kernel oil. Pak J Nutr 8(6):1680–5194

Bayorh MA, Abukhalaf IK, Ganafa AA (2005) Effect of palm oil on blood pressure, endothelial function and oxidative stress. Asia Pac J Nutr 14(4):325

Boateng L, Ansong R, Owusu W, Steiner-Asiedu M (2016) Coconut oil and palm oil's role in nutrition, health and national development: a review. Ghana Med J 50(3):189–196

Boon CM, Ng MH, Choo YM, Mok SL (2013) Super, red palm and palm oleins improve the blood pressure, heart size, aortic media thickness and lipid profile in spontaneously hypertensive rats. PLoS One 8(2):e55908. https://doi.org/10.1371/journal.pone.0055908

Bora PS, Rocha RV, Narain N, Moreira-Monteiro AC, Moreira RA (2003) Characterization of principal nutritional components of Brazilian oil palm (Elaeis guineensis) fruits. Bioresour Technol 87(1):1–5

Cheng HS, Ton SH, Tan JBL, Abdul Kadir K (2017) The ameliorative effects of a Tocotrienol-rich fraction on the AGE-RAGE Axis and hypertension in high-fat-diet-fed rats with metabolic syndrome. Nutrients 9(9):984

Chong YH, Ng TK (1991) Effects of palm oil on cardiovascular risk. Med J Malaysia 46(1):41–50

CODEX (1999) Codex standards for fats and oils from vegetable sources. http://www.fao.org/3/y2774e/y2774e04.htm#fn3

Cottrell RC (1991) Introduction: nutritional aspects of palm oil. Am J Clin Nutr 58:989S–1009S

Ebong PE, Owu DU, Isong EU (1999) Influence of palm oil (Elaeis guineensis) on health. Plant Foods Hum Nutr 53(3):209–222

Edem DO (2002) Palm oil: biochemical, physiological, nutritional, hematological and toxicological aspects: a review. Plant Foods Hum Nutr 57(3–4):319–341

Elson CE, Qureshi AA (1995) Coupling the cholesterol-and tumor-suppressive actions of palm oil to the impact of its minor constituents on 3-hydroxy-3-methylglutaryl coenzyme A reductase activity. Prostaglandins Leukot Essent Fat Acids 52(2–3):205–208

FAO/WHO (1991) Protein quality evaluation. Food and Agriculture, Rome, Italy. https://apps.who.int/iris/bitstream/handle/10665/38133/9251030979_eng.pdf?sequence=1&isAllowed=y

FAOSTAT (2019). http://www.fao.org/faostat/en/#data/QC. Accessed 10 Oct 2019

Gibbon V (2012) Palm oil and palm kernel oil refining and fractionation technology. In: Lai OM, Tan CP, Akoh CC (eds) Palm oil: Production, Processing, Characterization, AOCS Press; Urbana, IL, 2012; pp 329–375

Goh SH, Choo YM, Ong SH (1985) Minor constituents of palm oil. J Am Oil Chem Soc 62:237–240

Hall WL (2009) Dietary saturated and unsaturated fats as determinants of blood pressure and vascular function. Nutr Res Rev 22(1):18–38

Hornstra G, Sundram K (1989) The effect of dietary palm oil on' cardiovascular risk in man. Abstracts 1989 PORIM International Development Conference, 5–9 September 1989, Kuala Lumpur N3

Klurfeld D, Davidson LM, Lopez-Guisa JM (1990) Palm and other edible oils: atherosclerosis study in rabbits. FASEB 4:23–26

Kok S, Ong-Abdullah M, Ee GC, Namasivayam P (2011) Comparison of nutrient composition in kernel of tenera and clonal materials of oil palm (Elaeis guineensis Jacq.). Food Chem 129 (4):1343–1347

Kooyenga DK, Geller M, Watkins TR, Gapor A, Diakoumakis E, Bierenbaum ML (1997) Palm oil antioxidant effects in patients with hyperlipidaemia and carotid stenosis-2 year experience. Asia Pac J Clin Nutr 6(1):72–75
Kritchevsky D (2000) Impact of red palm oil on human nutrition and health. Food Nutr Bull 21 (2):182–188
Li R, Xia Q, Tang M, Zhao S, Chen W, Lei X, Bai X (2012) Chemical composition of Chinese palm fruit and chemical properties of the oil extracts. Afr J Biotechnol 11(39):9377–9382
Loganathan R, Selvaduray KR, Radhakrishnan A, Nesaretnam K (2008) Palm oil: rich in health promoting phytonutrients. Palm Oil Dev 50:16–26. http://palmoilis.mpob.gov.my/publications/POD/pod50-radhika.pdf
Loganathan R, Selvaduray KR, Nesaretnam K, Radhakrishnan AK (2010) Health promoting effects of phytonutrients found in palm oil. Malaysian J Nutr 16:309–322
Lucci P, Borrero M, Ruiz A, Pacetti D, Frega NG, Diez O, Angel M (2016) Palm oil and cardiovascular disease: a randomized trial of the effects of hybrid palm oil supplementation on human plasma lipid patterns. Food Funct 7(1):347–354
Morales-Gonzalez JA (ed) (2018) Vitamin E in health and disease. BoD–Books on Demand. IntechOpen, UK. ISBN: 978-1-78984-246-3
Mukherjee S, Mitra A (2009) Health effects of palm oil. J Hum Ecol 26(3):197–203
Ng TKW, Hassan K, Lim JH, Lye MS, Ishak R (1991) Non-hypercholesterolemic effects of a palm oil-diet in Malaysian volunteers. Am J Clin Nutr 53:1015S–1020S
Noor MA, Ani FN, Kannan KS (1999) Renewable energy scenario in Malaysia. International seminar on renewable energy for poverty alleviation at IEB, Dhaka, Bangladesh
Obahiagbon FI (2012) A review: aspects of the African oil palm (Elaeis guineensis jacq.) and the implications of its bioactives in human health. Am J Biochem Mol Biol 2(3):106–119
Oguntibeju OO, Esterhuyse AJ, Truter EJ (2009) Red palm oil: nutritional, physiological and therapeutic roles in improving human wellbeing and quality of life. Br J Biomed Sci 66 (4):216–222
Osim EE, Owu DU, Etta KM (1996) Arterial pressure and lipid profile in rats following chronic ingestion of palm oil diets. Afr J Med Med Sci 25:335–340
Owu DU, Osim EE, Ebong PE (1998) Serum liver enzymes profile of Wistar rats following chronic consumption of fresh or oxidized palm oil diets. Acta Trop 69(1):65–73
Peh HY, Tan WD, Liao W, Wong WF (2016 Jun 1) Vitamin E therapy beyond cancer: tocopherol versus tocotrienol. Pharmacol Ther 162:152–169
Qureshi AA, Qureshi N, Wright JJK, Shen Z, Kramer G, Gapor A et al (1991) Lowering of serum cholesterol in hypercholesterolemic humans by tocotrienols (palmvitee). Am J Clin Nutr 53 (4):1021S–1026S
Qureshi AA, Karpen CW, Qureshi N, Papasian CJ, Morrison DC, Folts JD (2011) Tocotrienols-induced inhibition of platelet thrombus formation and platelet aggregation in stenosed canine coronary arteries. Lipids Health Dis 10(1):58
Saad B, Ling CW, Jab MS, Lim BP, Ali ASM, Wai WT, Saleh MI (2007) Determination of free fatty acids in palm oil samples using non- aqueous flow injection titrimetric method. Food Chem 102:1407–1414
Sailo BL, Banik K, Padmavathi G, Javadi M, Bordoloi D, Kunnumakkara AB (2018 Apr 1) Tocotrienols: the promising analogues of vitamin E for cancer therapeutics. Pharmacol Res 130:259–272
Salonen JT, Salonen R, Ihanainen M, Parviainen M, Seppänen R, Kantola M, Seppänen K, Rauramaa R (1988 Nov 1) Blood pressure, dietary fats, and antioxidants. Am J Clin Nutr 48 (5):1226–1232
Sambanthamurthi R, Sundram K, Tan YA (2000) Chemistry and biochemistry of palm oil. Prog Lipid Res 39(6):507–558
Santoso U, Kubo K, Ota T, Tadokoro T, Maekawa A (1996) Nutrient composition of kopyor coconuts (Cocos nucifera L.). Food Chem 57(2):299–304

Sawe BE (2018) Top palm oil producing countries in the world. WorldAtlas, https://www.worldatlas.com/articles/top-palm-oil-producing-countries-in-the-world.html

Scrimshaw NS (2000) Nutritional potential of red palm oil for combating vitamin a deficiency. Food Nutr Bull 21(2):195–201

Shahidi F (2005) Bailey's industrial oil and fat products, 6th edn. Wiley-Interscience, New York, 3616 pages

Siddique BM, Ahmad A, Ibrahim MH, Hena S, Rafatullah M, Omar AKM (2010) Physico-chemical properties of blends of palm olein with other vegetable oils. Grasa y Aceites 61:423–429

Sreedhara N, Kurup PA (1998 Aug) Effect of hydrochloric acid treatment of palm kernel for dehulling on the nutritional quality of kernel protein. J Sci Food Agric 77(4):435–440

Tan KT, Lee KT, Mohamed AR, Bhatia S (2009) Palm oil: addressing issues and towards sustainable development. Renew Sust Energ Rev 13(2):420–427

USDA (2019). https://www.indexmundi.com/agriculture/?country=in&commodity=palm-kernel-oil&graph=production

Chapter 7
Coconut (*Cocos nucifera*)

Pankaj T. Parmar, Ashish Kumar Singh, and Sanket G. Borad

Abstract This chapter is about the history, components, their health attributes, and food applications of coconut. The exact origin of coconut is not known, but it was believed to be more than 80 million years old. Asia, especially south India is leading the coconut production in the world. The fruit is made up of exocarp, mesocarp, endocarp, testa, and at the center coconut water surrounded by the kernel. The coconut water is a rich source of minerals and used as energy drink, especially for athletes. The kernel part is an abundant source of vitamins, phytochemicals, protein, carbohydrate, and fat – rich in saturated fat, medium-chain triglycerides, MUFA, and PUFA. The presence of large amounts and variety of phytochemicals especially lauric acid and α-tocopherol makes coconut a potential fruit for the treatment of a wide range of ailments such as inflammation, cholera, oral carries, diabetes, skin infection, hypertension, cancer, CVD, obesity, and many more. Coconut comes in the top five food allergens in India. It's food applications include bakery products, confectionery products, chocolate making, and frozen desserts manufacture, while non-food applications include soap making, pharmaceutical products, cosmetics products, and biodiesel manufacture. Coconut has a broad scope in food and non-food applications, and it has to be studied, tried, and implemented for the wellness of mankind.

Keywords *Cocos nucifera* · Health attributes · Phytonutrients · Saturated fat · Testa · Virgin coconut oil

---

P. T. Parmar (✉)
Verghese Kurien Centre of Excellence, Institute of Rural Management Anand (IRMA), Anand, Gujarat, India

A. K. Singh
Dairy Technology Division, ICAR-National Dairy Research Institute, Karnal, Haryana, India

S. G. Borad
Havmor Ice cream Pvt. Ltd., Ahmedabad, Gujarat, India

B. Tanwar, A. Goyal (eds.), *Oilseeds: Health Attributes and Food Applications*,
https://doi.org/10.1007/978-981-15-4194-0_7

7.1 Origin and History

The origin of coconut is believed to be as old as 80 million years which could be due to the landmass of the Southern Hemisphere, Gondwana. The displacement of earth's tectonic plates moved the land surfaces which comprise Australia, North and northeast New Zealand, Madagascar, Africa, Southern America, and South Asian countries especially India, Bangladesh, and Pakistan. Humans started using this extraordinary fruit half a million years ago even though the exact time of first use is still unknown. About 4000 years ago, colonizing mariners began moving eastwards from the coast of Southeast Asia to the islands of the South Pacific, extending 10,000 km from New Guinea to Tahiti and also to southern India and Sri Lanka. More than 2000 years ago, the traders carried coconut from islands on their journey towards home islands which comprises Indonesia in the present era. Coconut's voyage towards the Indian Ocean to Madagascar and then to East Africa is assumed to be started by the people of Borneo.

If we talk about recent times, in 1498, Portuguese mariners, starting with Vasco da Gama, took the coconut from India and East Africa to the tropical eastern Atlantic. Westwards, coconuts were food and drink sources on slave trading ships traveling to Cuba and other islands. Eastwards, they were spread from the Cape Verde Islands to the coast of West Africa, from Senegal to Angola. Before these interesting events, West African people were using local forest palm oil as a food ingredient. Its easy planting and management got it dispersed in Caribbean islands. Before coconuts were discovered by European mariners, they were already growing on coasts of Pacific and nearby islands. With the completion of the planting of coconut palm in the Caribbean in the sixteenth century, its journey towards the globe was completed. This makes it the most widespread and widely used palm in the world. Modern coconut palm shows diversity in its size, shape, and appearance. The coconut fruit color ranges from dark to light brown to light dark green, orange, and yellow. Humans selected this ancestral fruit for pleasure and profit. It has been selected due to its attractive color and thinner husk and for increased nut diameter.

7.2 Production

Coconuts grow into a palm under suitable conditions and start fruiting after 3 years. On the ancient palm, the fruits are usually in a bunch and the bunch may contain up to 12 nuts at a time. The dried kernel of coconut is called *copra* which is used for the production of coconut oil. Asia, Central and South America, and the Pacific islands are the leaders in coconut production as they share 97.3% of the global coconut production. India is the leader in coconut production with 24.4 million coconut units per year followed by Indonesia and the Philippines. In 2013, the global coconut production was estimated to be 73.81 million coconuts or 2,896,709 MT of *copra* equivalent (Chan 2016).

7.3 Chemical Composition

The coconut fruit is composed of five parts (Fig. 7.1). The refreshing and highly nutritious coconut water is in the center of the coconut surrounded by five parts, i.e., kernel (coconut meat), testa, endocarp, mesocarp, and exocarp (inside out). The fruit is usually ovoid in shape, and the bunch of coconuts grows monthly. A single unit of coconut can grow up to 2 kg. The main products of coconut are coconut water and coconut kernel.

7.3.1 Coconut Water

The proximate composition of coconut water of the Malayan Tall coconuts is presented in Table 7.1.

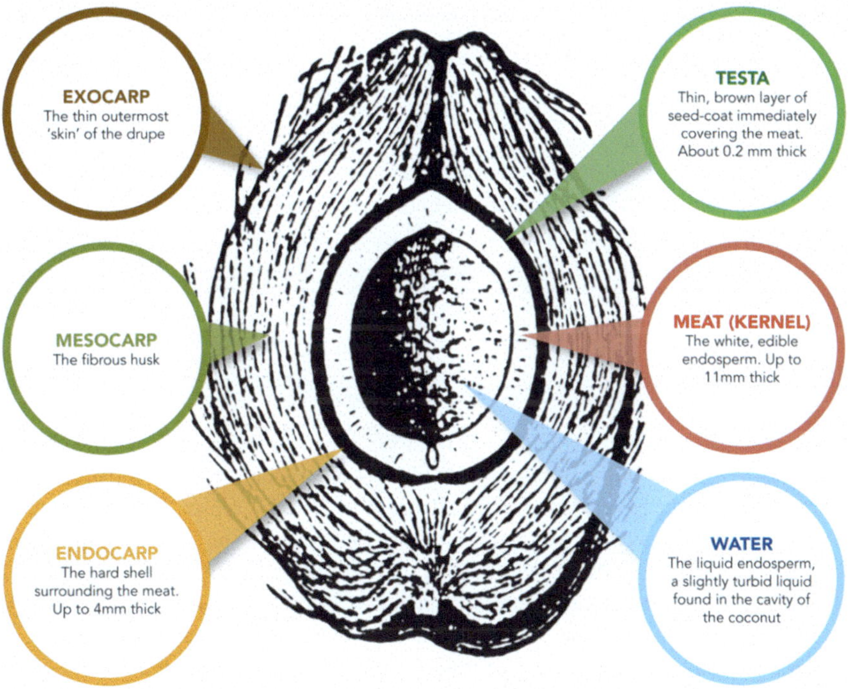

Fig. 7.1 Different parts of the coconut (Chan 2016)

Table 7.1 Proximate composition of coconut water

	Maturity stage (months)		
Physico-chemical properties	5–6 (Immature)	8–9 (Mature)	≥12 (Overly mature)
Volume of water (mL)	684 ± 27.0	518 ± 14.20	332 ± 19.90
Total soluble solids (TSS) (°brix)	5.60 ± 0.14	6.15 ± 0.21	4.85 ± 0.17
Titratable acidity (% malic acid)	0.09 ± 0.004	0.08 ± 0.01	0.06 ± 0.00
pH	4.78 ± 0.13	5.34 ± 0.12	5.71 ± 0.10
Turbidity	0.03 ± 0.013	0.34 ± 0.11	4.05 ± 0.32
Sugar content			
Fructose (mg/mL)	39.04 ± 0.82	32.52 ± 0.23	21.48 ± 0.21
Glucose (mg/mL)	35.43 ± 0.51	29.96 ± 0.24	19.06 ± 0.19
Sucrose (mg/mL)	0.85 ± 0.01	6.36 ± 0.06	14.37 ± 0.25
Minerals			
Potassium (mg/100 mL)	220.94 ± 0.32	274.32 ± 0.14	35.11 ± 0.13
Sodium (mg/100 mL)	7.61 ± 0.04	5.60 ± 0.02	36.51 ± 0.02
Magnesium (mg/100 mL)	22.03 ± 0.07	20.87 ± 0.02	31.65 ± 0.04
Calcium (mg/100 mL)	8.75 ± 0.05	15.19 ± 0.03	23.98 ± 0.05
Iron (mg/L)	0.29 ± 0.08	0.308 ± 0.01	0.32 ± 0.05
Protein (mg/mL)	0.04 ± 0.01	0.042 ± 0.00	0.22 ± 0.02
Total phenolic content (mg GAE/L)	54.00 ± 3.14	42.59 ± 0.83	25.70 ± 1.76

Source: Tan et al. (2014)

7.3.1.1 Total Soluble Solids

Total soluble solids (TSS) indicate the richness of a liquid substance. It is found to be the highest in mature coconuts. TSS content of coconut water increases in the early months of maturing followed by a gradual decrease (Jackson et al. 2004).

7.3.1.2 Carbohydrates

Principal sugars present in coconut are fructose, glucose, and sucrose which impart sweetness to the coconut and its products like coconut water, coconut milk, coconut cream, and coconut beverages (Campbell-Falck et al. 2000). Fructose is highest in coconut water followed by glucose and sucrose (Table 7.1). The sucrose content can reach up to as high as 90% of the total sugar content of coconut water (Solangi and Iqbal 2011). It is evident from Table 7.1 that sucrose content of coconut water within the fruit increases with time which could be due to the formation of sucrose, a non-reducing sugar at an expense of monosaccharides, glucose, and fructose.

7.3.1.3 Proteins

Coconut water is not a rich source of protein (Table 7.1), but it should not be neglected. The presence of free amino acids such as lysine, tryptophan, glutamic acid, alanine, glycine, and aspartic acid along with reducing sugars may trigger Maillard reaction during its thermal processing for preparation of ready-to-drink coconut water. Maillard reaction in coconut water can lead to browning or discoloration of coconut water and subsequent impaired sensory properties (Jayalekshmy and Mathew 1990).

7.3.1.4 Vitamins

Vitamins are very essential for a better healthy life. They are complex compounds and are not synthesized in the human body. Coconut water is a rich source of water-soluble vitamins particularly ascorbic acid (vitamin C), thiamin, riboflavin, niacin, pantothenic acid, pyridoxine, and folic acid.

7.3.1.5 Minerals

Generally, coconut water is rich in potassium and low in sodium content. The presence of magnesium, calcium, and iron though in a lesser amount than potassium and sodium makes it a better replacement of sports drinks as it helps in fulfilling the loss of electrolytes lost during exercise (Saat et al. 2002). More evident from Table 7.1 is that coconut water of matured fruit has a more balanced mineral content than immature one which makes the former suitable replacement of rehydration and sports drinks.

7.3.1.6 Total Phenolic Content (TPC)

Aging causes changes in the total phenolic content of coconut water as well (Table 7.1). As coconut fruit ages, its TPC content decreases which is the measure of antioxidant properties. The reduction of TPC content coconut water reduces its free radical scavenging ability (Mantena et al. 2003).

7.3.1.7 Titratable Acidity and pH

The titratable acidity of coconut water is expressed as percentage malic acid as it is the major organic acid present in coconut water. Titratable acidity of coconut water decreases with time (Jackson et al. 2004; Terdwongworakul et al. 2009). Similarly, as the maturity time of coconut progresses, pH also increases (Table 7.1).

7.3.2 Coconut Kernel

Coconut kernel also called coconut meat is the endosperm of the fruit covered by a brown-colored layer called testa. Generally, the dried coconut kernel is used for the extraction of coconut oil. The virgin coconut oil is the one that is extracted from dried kernel or *copra* and doesn't go any chemical refining process. The percentage oil in the kernel is the measure of the richness of coconut kernel. The weight of different parts of coconut at various stages of maturity is presented in Table 7.2.

The proximate composition of copra and coconut kernel is presented in Table 7.3 which includes percentage fat, protein, carbohydrates, ash, crude fiber, and moisture content.

It is evident from Table 7.3 that moisture content is highest in CW among *copra* parts, while in wet parts like WCW, WCWK, and WCT, it is highest in WCWK. The rest are discussed hereunder.

7.3.2.1 Carbohydrates

The main constituent of coconut kernel and testa has always been carbohydrate, and it is clearly evident from Table 7.3. The carbohydrate content of *copra* parts is almost double as the amount present in the wet kernel.

Table 7.2 Weight of different parts of coconut at various stages of maturity

	Coconut maturity (months)			
Parts (gm)	7	9	12	15
Husk	1190.0	740.0	518.5	269.0
Shell	140.0	189.1	156.6	134.3
Kernel/meat	20.3	180.5	244.5	160.5
Water	425.0	255.0	165.0	35.0
Total	1775.3	1365.0	1084.6	598.7

Source: Banzon and Velasco (1982)

Table 7.3 Proximate composition of copra and coconut kernel

	Content (%)					
Sample	Moisture	Fat	Protein	Carbohydrates	Crude fiber	Ash
CW	4.3	59.8	10.2	24.3	7.0	1.4
CWK	3.8	63.6	8.1	22.4	6.6	2.1
CT	4.0	59.0	9.3	26.3	11.6	1.4
WCW	42.2	37.0	7.5	12.3	14.3	1.0
WCWK	43.5	38.8	6.2	10.6	11.7	0.9
WCT	32.9	34.7	7.1	24.6	17.2	0.7

Source: Appaiah et al. (2014)

CW Copra whole, *CWK* copra white kernel, *CT* copra testa, *WCW* wet coconut whole, *WCWK* wet coconut white kernel, *WCT* wet coconut testa

7.3.2.2 Proteins

Coconut kernel is rich in protein too, though not as high as carbohydrate and fat content, but still a considerably high. The protein content of the dried kernel is ranging from 8.1 to 10.2, while the wet kernel is 6.2 to 7.5.

7.3.2.3 Lipids

The fat content of the *copra* part is in the range of 59.0–63.6 while that of the wet kernel is 34.7–38.8. It is highest in CWK among all six parts. The fatty acid composition of oils extracted from copra and coconut kernel is presented in Table 7.4.

Caprylic and capric acid content of oils extracted from CT and WCT are lesser than the other samples (Table 7.4). Lauric acid is the major fatty acid present in coconut which can be easily observed from Table 7.4 as it is present in the highest amount than other fatty acids. Here oils extracted from CT and WCT contain lesser amount of lauric acid than CW, CWK, WCW, and WCWK. However, myristic, palmitic, oleic, and linoleic acid content of oils extracted from CT and WCT is higher than other parts of coconut kernel and testa. The stearic acid content of oils extracted from *copra* is in the range of 1.1–1.9, while that of wet coconut parts is 1.2–2.3. Coconut is the richest in SFA content of oils extracted from coconut kernel and testa as it is ranging from 82.5 to 94.6 for copra and 71.6 to 92.7 for oils extracted from wet coconut parts. MUFA and PUFA content of oils extracted from coconut follow similar trends as of myristic, palmitic, oleic, and linoleic acid since

Table 7.4 Fatty acid composition of oils extracted from copra and coconut kernel

Fatty acids (%)	CW	CWK	CT	WCW	WCWK	WCT
Caprylic acid ($C_{8:0}$)	9.6	6.7	3.9	8.1	5.6	1.6
Capric acid ($C_{10:0}$)	6.4	6.2	3.8	7.8	5.8	2.2
Lauric acid ($C_{12:0}$)	51.5	52.6	40.9	50.5	52.8	32.4
Myristic acid ($C_{14:0}$)	19.1	18.9	20.9	16.1	19.2	20.2
Palmitic acid ($C_{16:0}$)	6.9	7.4	11.3	6.8	7.4	14.1
Stearic acid ($C_{18:0}$)	1.1	1.9	1.6	2.3	1.9	1.2
Oleic acid ($C_{18:1}$)	4.3	4.8	12.2	5.6	5.5	17.8
Linoleic acid ($C_{18:2}$)	1.1	1.6	5.3	1.8	1.0	10.6
SFA	94.6	93.7	82.5	92.6	92.7	71.6
MUFA	4.3	4.8	12.2	5.6	5.5	17.8
PUFA	1.1	1.6	5.3	1.8	1.0	10.6
MCFA	67.5	65.5	48.3	66.3	64.2	36.2

Source: Appaiah et al. (2014)
SFA Saturated fatty acid, *MUFA* monounsaturated fatty acid, *PUFA* polyunsaturated fatty acid, *MCFA* medium-chain fatty acid, *CW* Copra whole, *CWK* copra white kernel, *CT* copra testa, *WCW* wet coconut whole, *WCWK* wet coconut white kernel, *WCT* wet coconut testa

CT and WCT contain a higher amount of MUFA and PUFA than that of CW, CWK, WCW, and WCWK. Oils extracted from copra and wet coconut testa are poor in MCFA content than other parts. The composition of oils extracted from the whole coconut kernel is in accordance with the results obtained by Bhatnagar et al. (2009).

7.3.2.4 Crude Fiber

The crude fiber content of *copra* is lesser than the wet kernel which is evident from Table 7.3. It was in the range of 6.6–11.6% for *copra,* while that of the wet kernel, it was in the range of 11.7–17.2. Coconut testa is richer in crude fiber than dried kernel and whole wet coconut which is the reason behind high fiber values of CT and WCT which is 11.6% and 17.2%, respectively. A healthy coconut kernel should have a good amount of crude fiber which is around 13.13% (Rao et al. 1998).

7.3.2.5 Proteins

Coconut kernel contains 5–10% protein. It is a mixture of essential, non-essential, and conditionally essential amino acids. The amino acid composition of fresh coconut kernel is presented in Table 7.5. According to USDA, the amino acids which cannot be synthesized are called essential amino acids, and therefore they should be supplemented in the human diet. Histidine, isoleucine, leucine, lysine, methionine, phenylalanine, threonine, tryptophan, and valine are the examples of essential amino acids. Those amino acids which can be synthesized from their precursors, usually essential amino acids, are called non-essential amino acids. Alanine, aspartic acid, asparagine, glutamic acid, and serine are examples of non-essential amino acids. Those amino acids that can be synthesized under regular conditions, in the human body when the body feels relaxed are called the conditional

Table 7.5 Amino acid composition of fresh coconut kernel

Vitamin	Content (g/100 g fresh kernel)	Vitamin	Content (g/100 g fresh kernel)
Essential amino acids		Non-essential amino acids	
Tryptophan	0.039	Alanine	0.170
Threonine	0.121	Aspartic acid	0.325
Isoleucine	0.131	Glutamic acid	0.761
Leucine	0.247	Serine	0.172
Lysine	0.147	*Conditional essential amino acids*	
Methionine	0.062	Arginine	0.546
Phenylalanine	0.169	Cystine	0.066
Valine	0.202	Glycine	0.158
Histidine	0.077	Proline	0.138
		Tyrosine	0.103

Source: USDA National Nutrient Database (2019)

essential amino acids, e.g., arginine, cysteine, glutamine, glycine, proline, and tyrosine.

7.3.2.6 Vitamins

Matured fruit kernel is a source of both water and oil-soluble vitamins which are presented in Table 7.6. Among water-soluble vitamins present in matured coconut kernel, ascorbic acid is the major one which is present in highest proportion (3.3 mg/ 100 g fresh kernel) followed by niacin (0.540), pantothenic acid (0.300), thiamine (0.066), pyridoxine (0.054), folate (0.026), and riboflavin (0.020). There are only two oil-soluble vitamins present in coconut oil extracted from matured coconut kernel, i.e., α-tocopherol (0.24 mg/100 g fresh kernel) and phylloquinone (0.20 μg/100 g fresh kernel).

7.3.2.7 Minerals

The mineral composition of copra and coconut kernel is presented in Table 7.7. Coconut kernel is a rich source of potassium followed by sodium, calcium, iron, and zinc. CW, CWK, CT, WCW, WCWK, and WCT have 120.3, 124.1, 120.3, 122.1, 123.8, and 107.8 mg of potassium per 100 g, respectively. Sodium, calcium, iron, and zinc are in the range of 15.5–29.8, 14.0–18.1, 1.5–7.9, and 1.6–3.0 mg/100 g. Similar observations were reported by Solangi et al. (2011).

Table 7.6 Vitamin content of matured coconut kernel

Vitamin	Unit	Content (Per 100 g fresh kernel)
Water-soluble vitamins		
Thiamin (B_1)	mg	0.066
Riboflavin (B_2)	mg	0.020
Niacin (B_3)	mg	0.540
Pantothenic acid (B_5)	mg	0.300
Pyridoxine (B_6)	mg	0.054
Folate	mg	0.026
Ascorbic acid (C)	mg	3.3
Oil-soluble vitamins		
α-Tocopherol (E)	mg	0.24
Phylloquinone (K)	μg	0.20

Source: USDA National Nutrient Database (2019)

Table 7.7 Mineral composition of copra and coconut kernel

Samples	Minerals (mg/100 g)				
	K	Na	Ca	Fe	Zn
CW	120.3	15.5	14.6	7.9	2.9
CWK	124.1	21.0	18.1	3.2	1.6
CT	120.3	22.4	17.0	6.2	3.0
WCW	122.1	21.6	18.1	7.9	2.2
WCWK	123.8	20.3	14.0	1.5	1.8
WCT	107.8	29.8	16.7	1.9	1.6

Source: Appaiah et al. (2014)
K Potassium, *Na* sodium, *Ca* calcium, *Fe* iron, *Zn* zinc

7.4 Phytonutrients

Coconut has an abundant amount of phytonutrients present in it. The ethanolic extract of coconut mesocarp shown the presence of a range of phytonutrients such as tannins, phenols, flavonoids, triterpenes, steroids, leucoanthocyanidins, and alkaloids (Matos 1997), while butanol extract revealed the presence of saponins, condensed tannins, and triterpenes (Costa et al. 2010). A different part of the coconut tree, as well as fruit, contains a good amount of phytochemicals such as coconut oil which is a rich source of lauric acid (Tangwatcharin and Khopaibool 2012) and α-tocopherol (Arlee et al. 2013). Liquid albumen of coconut fruit has an adequate amount of ascorbic acid (Yong et al. 2009) and L-arginine (Yong et al. 2009; Salil and Rajamohan 2001). Coconut leaves (epicuticular wax) are rich in lupeol-methylether, skimmiwalin, and iso- skimmiwalin (Erosa et al. 2002). The roots of the palm contain saponins (Pal et al. 2011), while coconut fiber contains catechins (Matos 1997; Freitas et al. 2011). Coconut fiber root inflorescence is rich in flavonoids (Matos 1997; Pal et al. 2011; Renjith et al. 2013), and coconut fiber inflorescence contains tannins (Freitas et al. 2011; Renjith et al. 2013).

7.5 Health Attributes

Various researchers have exploited the health benefits of the tropical palm over the years. It has got many health-promoting applications. The traditional uses of coconut in the treatment of various diseases are presented in Table 7.8.

Numerous studies have been conducted to identify the active ingredients present in coconut and their health attributes. Some of the important pharmacological activities of active molecules present in coconut are explained hereunder.

Table 7.8 Traditional uses of coconut in treatment of various diseases

Coconut part	Preparation	Use	Country	Reference
Shell fiber	Tea	Diarrhea	Brazil	Esquenazi et al. (2002)
		Amenorrhea	Haiti	Weniger et al. (1986)
		Venereal diseases treatment	Trinidad	Wong (1976)
	Extract	Antipyretic, kidney inflammation	Guatemala	Caceres et al. (1987)
		Diuretics, gonorrhea treatment	Peru	Ramirez et al. (1988)
		Urogenital inflammation caused by *Trichomonas vaginalis*	Mexico	Calzada et al. (2007)
		Amenorrhea, dysmenorrhea	Trinidad	Wong (1976)
		Diabetes treatment	Jamaica	Morrison (1994), Mitchell and Ahmad (2006)
		Asthma treatment	Haiti, Peru	Hope et al. (1993), Ramirez et al. (1988)
	Cream	Abscesses, dermatitis treatment, and injuries	Guatemala	Caceres et al. (1987)
		Burns	Haiti	Weniger et al. (1986)
Root	Tea	Diarrhea and stomach pains	Papua New Guinea	Holdsworth and Wamoi (1982), Holdsworth (1992)
	Extract	Antipyretic, diarrhea treatment	Indonesia	Brondegaard (1973)
Coconut pulp (solid albumen)	Oil	Preventing hair loss, wound healing	Fiji, Indonesia	Singh (1986), Sachs et al. (2002)
	Milk	Diarrhea treatment	Ghana	Yartey et al. (1993)
		Oral contraceptive	Indonesia	Hirschhorn (1983)
	Pulp	Aphrodisiac	Mozambique	Amico (1977)
		Relief to rashes caused by HIV-AIDS infections	Kenya	Nagata et al. (2011)
	A decoction of the pulp	Treatment of fever and malaria	Malaysia	Al-Adhroey et al. (2011)
Coconut water inflorescence	Water	Treatment of renal diseases	Fiji	Singh (1986)
	Tea	Treatment of changes in the menstrual cycle	India	Bhandary et al. (1995)

Source: Lima et al. (2015)

7.5.1 Anti-atherosclerotic Effect

Atherosclerosis, a phenomenon generally induced by *Chlamydia pneumonia*, plays a major role in initiating an inflammatory process that oxidizes lipoproteins with the induction of cytokines and the production of proteolytic enzymes (Enig 2004). Esquenazi et al. (2002) observed the inhibitory effect of aqueous extract of coconut husk fiber against acyclovir-resistant herpes simplex virus (HSV), and the responsible components would be catechins, epicatechins, and tannins. Almost all the HSVs are believed to be destroyed by monoglycerides (MG) and fatty acids ranging from C_6 to C_{14} which constitutes around 80% of fatty acids and MGs present in coconut oil.

7.5.2 Antibacterial, Antifungal, and Antiviral Effect

Tender coconut water (TCW) is preferred in the treatment of cholera because of its saline and albumen content (Effiong et al. 2010). MCFA found in coconut especially lauric acid and its derivatives are effective in killing bacteria consisting of lipid membranes as it breaks down their lipid membrane. They are also effective against bacteria that cause gastric ulcers, food poisoning, dental caries, and infection in the urinary tract.

The comparison of aqueous, ethanolic, and dry distilled extracts of coconut endocarp with gentamicin and ciprofloxacin for their antimicrobial activity showed a strong antimicrobial effect against *S. aureus*, *M. luteus*, *B. subtilis*, and *P. aeruginosa* (Nagata et al. 2011). The dry distilled extract was effective against *R. oligosporus* at 10, 50, 100, and 300 μg/disc (Singla et al. 2011). Verma et al. (2012) studied the antimicrobial activity of coconut mesocarp powder against *E. coli* and *S. typhi* by extracting them in ethanol, chloroform, diethyl ether, benzene, acetone, and formaldehyde. They noticed the highest antimicrobial activity against *E. coli* with benzene while diethyl ether for *S. typhi*. The identified components present in coconut mesocarp responsible for antimicrobial activity against these two pathogens were β-sitosterol, cycloartenol, alcohol palmitoyl, and tocopherol.

Esquenazi et al. (2002) conducted trials for the antimicrobial study using crude extract and TLC fractions (I-V) of coconut mesocarp fiber, and TLC fractions II–V were analyzed against *S. aureus*. Antifungal activity was tested against *Cryptococcus neoformans* or *Fonsecaea pedrosoi* and *Candida albicans* by checking their growth inhibition. Furthermore, antiviral activity was checked with crude extract and only TLC fraction-II. It was observed that all three effects were exhibited by catechins and condensed tannins coconut itself and thereby in extracts especially fraction II. The MCSFA present in coconut oil is very effective against viruses especially cytomegalovirus, Epstein-Barr virus, and visna virus, and it is achieved by the destruction of their cell wall membranes and thereby interfering in viral maturation (Arora et al. 2011).

7.5.3 Anticaries Effect

Traditionally, a decoction prepared from coconut tree roots is being used for mouthwash and gargling to reduce the risk of dental caries. Since the use of the aqueous extract of coconut husk fiber does not prompt any ocular or dermic reactions, it could be a potential substitute for drugs used to treat oral diseases (Alviano et al. 2008). Mouth sores can be treated by coconut flour as coconut is rich in lauric acid, and it is already being used for oral treatments (Taheri et al. 2010). Coconut contains sucrose monolaurate which is a potential anticaries substance since it reduces glycolysis and oxidation of sucrose caused by *Streptococcus mutans*. Therefore, it can be used to treat dental plaque. Soap prepared using coconut was used along with sodium hypochlorite solution (0.05%) in the treatment of stomatitis and denture biofilm (Barnabé et al. 2004).

7.5.4 Antidiabetic Effect

Diabetes mellitus became a prominent disease in the past few decades due to a drastic change in food habits as well as reduced physical activity. Diabetes is characterized by deviations in carbohydrate, protein, and lipid metabolism due to complete or relative insufficiency of insulin secretion from pancreatic β cells and/or defects in insulin action (Unger and Foster 1998). The protein present in coconut kernel exhibits antidiabetic activity, and that can be backed by three mechanisms exhibited as a result of action of coconut kernel protein, viz., increase in glycogen and insulin level and reduction in serum glucose level, efficient degradation of carbohydrate by β-D-galactosidase, and inhibitory action of arginine on regeneration of β-pancreatic cells resulting in increased insulin secretion and subsequent increase in glycogen level in blood serum (Salil et al. 2011).

Preetha et al. (2014) studied the effects of mature coconut water on alloxan-induced diabetes in rats and observed reduced blood glucose level and lipid level in blood and tissues and reduction inactivity of 3-hydroxy-3-methylglutarylcoenzyme A (HMG CoA). In a study to assess antidiabetic effects of coconut inflorescence on streptozotocin-induced diabetes in Sprague-Dawley rats, Renjith et al. (2013) observed that a dose of methanolic extract of coconut inflorescence at 200 mg/kg body weight was effective in reducing streptozotocin-induced diabetes in rats which could be due to arginine, leucine, isoleucine, polyphenols, and dietary fibers present in coconut itself.

7.5.5 Antidermatophytic

Coconut oil is being used since centuries as moisturizer in different parts of the world. In a double-blind study conducted by Agero and Verallo-Rowell (2004) for the moisturizing effect of extra virgin coconut oil on the skin surface for mild to more xerosis and observed a significant improvement in skin hydration and increased lipid level of the skin surface. Monolaurin is a derivative of lauric acid which is a major fatty acid present in coconut kernel and has been investigated for its in vitro antibacterial activity against skin infection-causing organisms by Carpo et al. (2007), and they observed no resistance of gram-positive and gram-negative bacteria isolated from superficial skin infections. Another study of VCO and monolaurin on atopic dermatitis in adults showed broad-spectrum activity against *S. aureus*, fungi, and viruses which can be useful in the treatment of atopic dermatitis colonization in humans (Verallo-Rowell et al. 2008).

7.5.6 Antihypertensive Effect

Hypertension is defined as systolic blood pressure above 140 mm of Hg or diastolic blood pressure above 90 mm of Hg and is a major growing health problem across the globe (Go 2013; Go et al. 2013), Weber et al. (2014), Mancia et al. (2013)). Phenolic compounds present in plants especially flavonoids are known for their effectiveness against hypertension (Perez-Vizcaino et al. 2009). Bankar et al. (2011) studied the antihypertensive activity of an ethanolic extract of coconut endocarp was studied in the deoxycorticosterone acetate salt-induced rats and observed a significant reduction in blood pressure of model rats which could be attributed to the activation of nitric oxide/guanylate cyclase pathway as well as presence of flavonoids in the ethanolic extract prepared from coconut endocarp.

7.5.7 Anti-inflammatory Effect

A decoction of coconut husk fiber is widely used traditionally in Brazil for the treatment of arthritis and diarrhea (Esquenazi et al. 2002). Silva et al. (2013) conducted a study on anti-inflammatory effect of coconut husk fiber extract on carrageenan, histamine, and serotonin-induced rat paw edema and reported that crude extract significantly reduced rat paw edema induced by histamine (150 mg/kg) and serotonin (100 and 150 mg/kg). Fraction 1 (molecular weight lesser than 1 kDa, 1, 10, and 50 mg/kg) also had significant inhibitory effect on histamine and serotonin-induced edema which may be due to inhibitory effect of husk fiber extract through either formation or liberation of inflammatory mediators such as serotonin, histamine, and bradykinin or by direct blocking of receptors.

7.5.8 Antineoplastic Effect

Koschek et al. (2007) studied the anti-neoplastic activity of coconut husk fiber on human leukemia cells K562 and Lucena 1, by the 3-[4,5-dimethylthiazol-2-yl]-2,5-diphenyltetrazolium bromide (MTT) assay. The cells were treated against MTT at concentrations of 0, 5, 50, or 500 μg/mL for 48 h. The results showed antitumoral activity against both leukemia cells K562 and Lucena 1. The variety used in the study is grown extensively in Brazil, and the husk is usually thrown away as waste which can be a potential economical source to prepare antineoplastic drugs as well as solves environmental problems.

7.5.9 Antioxidant Effect

Free radicals are continuously produced in the human body and are causing tissue damage in many diseases (Levy et al. 1998). The free radicals break phospholipids present in cell membranes and initiate lipid peroxidation (Ferrari et al. 1992), and it could lead to atherosclerosis (Eder and Kirchgessner 1997). Apart from this, they can degrade enzymes, DNA that can lead to cancer (Parthasarathy et al. 1998). L-arginine and ascorbic acid present in TCW hasveantioxidant activity (Loki and Rajamohan 2003). VCO showed efficient antioxidant activity in Sprague-Dawley rats (Nevin and Rajamohan 2006).

The total phenolic content of virgin coconut oil is almost seven times that of commercial coconut oil as the oil refining process destroys many of the biologically active components (Seneviratne and Dissanayake 2008) which was evident from the 1,1-diphenyl-2-picrylhydrazyl (DPPH) test too (Marina et al. 2009). DPPH radical scavenging, nitric oxide radical scavenging, and alkaline dimethyl sulfoxide (DMSO) analysis of *C. nucifera* endocarp extracts showed strong antioxidant activity.

Coconut water has a good antioxidant activity which was evaluated by administering it in a dose of 6 mL/100 g of body weight to female rats intoxicated with carbon tetrachloride, and results revealed that lipid peroxidation decreased and antioxidant enzymes improved their antioxidant power (Loki and Rajamohan 2003). In general, different parts of coconut contain flavonoids and phenolic compounds which showed potential antioxidant activity.

7.5.10 Antiparasitic Effect

Costa et al. (2010) studied the antihelminthic activity of aqueous and butanol extract of green coconut bark on mouse intestinal nematodes. In this study naturally infected mice were distributed into six groups as follows: group I (1000 mg/kg) aqueous extract; group II (2000 mg/kg) aqueous extract; group III (500 mg/kg) of butanol extract; group IV (1000 mg/kg) of butanol extract; group V (0.56 mg/kg) febendazole; and group VI (3% dimethylsulfoxide). Among all the extracts, only butanol extract at 500 and 1000 mg/kg showed an efficiency of 62.72% and 98.36%, respectively ($P < 0.05$). They also studied the ovicidal and larvicidal activity of the aqueous and butanolic extract of the coconut husk against *Haemonchus contortus*. In egg hatching and larval development tests, 2.5 mg/mL aqueous and 10 mg/mL butanolic extract showed 100% ovicidal activity, while larvicidal effects were 81.30% and 99.80% at 65 and 80 mg/mL, respectively. These results suggest that different parts of coconut can be used to treat gastrointestinal nematodes, and more studies are needed to evaluate their potential in humans.

7.5.11 Antithrombotic Effect

Virgin coconut oil has better antithrombotic activity than copra oil (Nevin and Rajamohan 2008). As compared to high MUFA and PUFA diet, a high SFA diet helps in lowering the postprandial tissue plasminogen activator antigen concentration and that affects the lipoprotein A concentration and fibrinolytic system as well. Coconut oil is a rich source of SFA, if included in the diet, it can have a good antithrombotic effect as dietary SFA has more lipoprotein A lowering capacity than percentage saturated fat energy (Muller et al. 2003).

7.5.12 Cardioprotective Effect

As discussed earlier in this chapter, coconut has an abundant amount of MCFA (Table 7.4) which is directly absorbed in the intestine and goes to the liver for energy generation and, therefore, does not take part in cholesterol biosynthesis and transport (Enig 2004). Coconut water is a rich source of potassium which possesses cardioprotective effect in myocardial infarction. Nevin and Rajamohan (2004) reported a significant increase in HDL cholesterol and a significant decrease in LDL cholesterol levels of Male Sprague-Dawley rats when fed with VCO diet. Polyphenols present in VCO prevents oxidation of LDL cholesterol in vitro.

Cardioprotective activity of coconut water was evaluated in rats by inducing myocardial infarction using isoproterenol. The study revealed that feeding of tender coconut water to experimental animals reduced the myocardial infarction and mitochondrial lipid peroxidation (Rajamohan and Anurag 2003). It can reduce VLDL, LDL, TGs, and total cholesterol levels in blood serum which is evident from the findings of Sandhya and Rajamohan (2006) who reported a reduction of these components in rats when they were fed with coconut water at the rate of 4 mL/100 g body weight. Feeding of dietary coconut sprout of West Coast Tall variety to the rats having an isoproterenol-induced myocardial infarction (Chikku and

Rajamohan 2012) showed decreased levels of CK-MB and troponin-T (a cardiac marker) in serum when rats were fed with the sprouts at 50, 100, or 200 mg/100 g body weight for 45 days. Furthermore, it reduced oxidative stress in the heart and improved antioxidant status which could be due to the presence of polyphenols, vitamins, and alkaloids in coconut. These results suggest that coconut can be used to treat dyslipidemia in humans too.

7.5.13 Hepatoprotective Effect

Hepatotoxicity is a covalent bonding of toxic intermediates and hepatocytes which causes centrilobular hepatic necrosis or lipid peroxidation as well as thiol group oxidation. A study on the effect of TCW on CCl_4 induced liver injury in female rats showed no necrosis or any fatty infiltration (Loki and Rajamohan 2003). This is the sole report that states the hepatoprotective effect of coconut water which generates the need for further study to discover the role of the coconut in the same.

7.5.14 Hypolipidemic Effect

VCO has the ability to reduce lipid oxidation (Nevin and Rajamohan 2008). L-arginine is a prominent amino acid responsible to reduce the risk of hyperlipidemia (Mini and Rajamohan 2004). Polyphenols present in coconut helps in maintaining the optimum level of lipid in the blood serum and tissues (Nevin and Rajamohan 2004). It entraps reactive oxygen species in human interstitial fluid and plasma in arterial walls and inhibits the oxidation of LDL thereby reduces the absorption of cholesterol in intestines as well as reverses the transport of cholesterol in humans body (Von Eckardstein et al. 2001). This mechanism of polyphenols helps in reducing the risk of hyperlipidemia.

7.5.15 In Reducing the Risks of Abdominal Obesity

Obesity is a prevalent disease usually caused by multiple endogenous and environmental factors. Among all the factors, excessive caloric intake is the prominent one (Wisse et al. 2007). Obesity is the root cause of other chronic diseases such as CVD, diabetes, hypertension, and even cancer (Poirier et al. 2006). To manage or reduce obesity, one should do modifications in the regular diet by reducing the calorie intake, exercise regularly, and maintain a healthy lifestyle. A report published a couple of years back showed that coconut water reduced adipose tissue mass by boosting metabolism accomplished by a decrease in leptin levels in animals fed high-fat and high-fructose diets (Imaga et al. 2016).

In context to the above report, Mohamad et al. (2017) studied the anti-obesity and anti-inflammatory effects of coconut water vinegar on high-fat-diet (HFD)-induced obese mice. C57/BL mice were fed with an HFD diet continuously for 33 weeks to persuade obesity. Obese mice were fed with coconut water vinegar at a rate of 0.08 and 2 mL/kg body weight from week 24 to 33. The oral ingestion reduced the body weight, serum lipid profile, and fat-pad weight of obese mice significantly ($p < 0.05$). This would be achieved by alterations in the gut microbiota through an increase in the populations of the *Bacteroides* and *Akkermansia* genera by the coconut water vinegar that may have helped to overcome the obesity and inflammation caused by the HFD.

With a fact that consumption of MCTGs in overweight adults reduces obesity through increased thermogenesis, LaBarrie and St-Onge (2017) conducted a double-blind, cross-over study on 15 children of 13–18 years of age by feeding them two test meals containing 20 g fat which was corn oil or coconut oil-enriched baking fat. They observed a significant effect of fat type on leptin ($P = 0.027$), peptide YY ($P = 0.0085$), and triglycerides ($P = 0.020$). The concentration of leptin and triglyceride was lowered, while peptide YY concentrations were increased in comparison with corn oil consumption.

From the two studies cited above, it can be concluded that coconut oil and coconut water can be used as a body weight management tool, but some more research is needed to demonstrate it.

7.5.16 Renal Protective Effect

Urolithiasis is a common occurrence affecting up to 10%–15% of the globe at some point during their lifetime (Long and Park 2007). Due to improved living standards, race, ethnicity, and geographical location increased the incidence of kidney stones in the world (Stamatelou et al. 2003). Many medicinal plants have been used to treat urinary stones since ancient times. Coconut water serves the same. To evaluate its renal protective ability, Gandhi et al. (2013) studied the prophylactic effect of coconut water against nephrolithiasis-induced kidney stones in Wistar rats. They were divided into three groups: control (standard diet), induced nephrolithiasis rats fed with 0.75% ethylene glycol in drinking water, and in third group coconut water instead of ethylene glycol for 7 weeks. The urine sample analysis showed a tremendous reduction in calcium oxalate crystals in coconut water-fed rats. It significantly ($p < 0.05$) reduced urea and creatinine levels in group 3. Based on the results of the study, it can be suggested that coconut water has renal protective activity, and coconut water can be investigated for its potential in the treatment of kidney stones in humans.

7.6 Adverse Effects and Individual Concerns

Coconut can be a food allergen, and in India, it comes in the top five food allergens (Teuber and Peterson 1999).

7.7 Food Applications

Coconut (Cocos nucifera) milk is being used by confectionaries, bakeries, biscuits, and ice cream industries worldwide to enhance the flavor and taste of various products (Persley 1992). Coconut juice was found to be rich in calcium (800 mg), while the protein and fat contents were 50 and 65 g, respectively. The energy content was 61.0 kcal and the total available carbohydrate was 300 g. The milk was reported to be high in minerals and vitamin content (Nieuwentus and Nieuwelink 2002), while total saturated fat was 10% of the total energy (Thai Food Composition 2004). Percentage energy distribution from protein, total fat, and carbohydrate were 10:30:60.

The potential of coconut milk to treat the protein and calorie malnutrition in Africa, Belewu and Belewu (2007) evaluated the potential of tiger nut, coconut and soybean in inclusion to the confectionery products. They extracted milk from all three sources and evaluated them for chemical composition. They observed that coconut milk has 7.87% crude protein, 24.10% fat, 9.40% calcium, and 2.14% phosphorous present. The total energy was 332.20 kcal per 100 g coconut milk. Coconut milk was found to be rich in lauric and capric acid. The chemical composition of coconut milk revealed that it is a calorie-dense milk alternate containing a good amount of dietary crude protein, fat, and minerals like calcium and phosphorous, and it can be included in human diet as well as livestock diets.

Out of the total applications of coconut oil, 61% is used in food applications (Gunstone and Harwood 2007), while the major portion from the remaining 39% is in the oleochemical industry. The coconut oil is largely used in frying, production of margarine, filled milk, frozen desserts, etc.

7.7.1 Frying

In tropical countries especially in India and the Philippines, the most common use of coconut oil is frying as it is a local product and has a local preference too (Rossel 2001). It is mainly suitable for shallow frying domestically. Furthermore, it is unsuitable for industrial frying since it is rich in lauric acid and other fatty acids with less than 14 carbon atoms. The reason for its unsuitability for industrial frying is the liberation of free fatty acids upon hydrolysis of glycerol esters while used for moisture-rich foods. Kochhar (2001) reported that use of lauric rich oils in frying at the temperature range of 180–200 °C results in unsatisfactory frying.

7.7.2 Margarine

Margarine means an emulsion of edible oils and fats with water (FSSR 2011). As per FSSR (2011), there are types of margarine such as table margarine, bakery, and industrial margarine, fat spreads, and blended spreads. In the past, partially hydrogenated fats were the principle ingredient used in the production of margarine, but it is not much in use now due to major helath concerns associated with trans fatty acids. Lauric acid has a low melting point which helps in higher spreadability compared to butter. Moreover, Lauric oils contain a higher amount of MCT which are readily metabolized, absorbed, and cleared from the blood than long-chain fatty acids (Deckere and Verschuren 2000).

Medium-chain triacylglycerols (MCTs) are obtained by the distillation of coconut fatty acids. They are low molecular weight triglycerides hence readily absorbed and used immediately as an energy source in the body and therefore do not get stored in the adipose tissues and thereby helps in reducing the risk of obesity. They have higher oxidative stability as they contain saturated fatty acids which makes them more suitable to be used in preventing sticking of bakery products on the pan surface (Idris and Samsudin 2005). They are being used for decades to avoid the unpleasant taste of spices and mixes by spraying onto them.

7.7.3 Filled Milk

Filled milk is milk in which the natural fat has been replaced by another fat source. In countries where fresh milk availability is very low, filled milk is a substitute which is more economical than fresh milk import. Since milk has a very unique flavor, the primary criteria for selection of vegetable oil for the preparation of filled milk would be less flavor. Coconut oil is largely used in filled milk due to its oxidative stability, bland taste, low melting point, and good mouthfeel. The oxidative stability of hydrogenated coconut oils is very high, therefore they are widely used for the preparation of filled milk.

7.7.4 Frozen Dessert

Frozen dessert or frozen confection means the product obtained by freezing a pasteurized mix prepared with edible vegetable oils or fats, having a melting point of not more than 37 °C or vegetable protein products, or both. It may also contain milk fat and other milk solids with the addition of nutritive sweeteners and other permitted non-dairy ingredients. The said product may contain incorporated air and may be frozen hard or frozen to a soft consistency (FSSR 2011). Coconut oil, natural or modified, have high solid fat content at 0 °C, low melting point, and bland taste, hence preferred as a fat source in frozen desserts.

7.8 Alternative Applications

One of the major non-edible applications of coconut oil is in the soap industries; one important chemical derivative of coconut oil is methyl esters of coconut fatty acids, which are produced by treating coconut oil with methyl alcohol. These methyl esters constitute an important raw material for the chemical industries as they are more stable and are easier to separate by fractional distillation. Coconut oil has many other industrial uses in pharmaceuticals, cosmetics, plastics, rubber substitutes, synthetic resins, etc. Coconut oil has also been found useful for mixing with diesel. This mixture in the proportion as 30:70 has given excellent road performance of diesel vehicles. Methyl esters of coconut oil fatty acids are also being used as lubricants and biodiesel in the aviation industry.

7.9 Conclusion

Coconut has been an intrinsic part of the diet of Asia, Africa, Central America, and the South Pacific Islands since ancient times. Apart from its health attributes, it has been used in various food and non-food preparations. In the past few years, coconut has been criticized for increasing the rate of heart diseases as it is rich in saturated fatty acids. The tide has turned, and now its health benefits especially VCO's health benefits have been exploited. Moreover, further research is needed to claim its health benefits. A different part of coconut should be studied for application in other foods too which can be useful in inventing new functional products that can be future scope for the miraculous fruit.

References

Agero AL, Verallo-Rowell VM (2004) A randomized double-blind controlled trial comparing extra virgin coconut oil with mineral oil as a moisturizer for mild to moderate xerosis. Dermatitis 15:109–116. https://doi.org/10.1111/j.0105-1873.2004.00309ew.x

Al-Adhroey AH, Nor ZM, Al-Mekhlafi HM, Amran AA, Mahmud R (2011) Evaluation of the use of Cocos nucifera as antimalarial remedy in Malaysian folk medicine. J Ethnopharmacol 134:988–991. https://doi.org/10.1016/j.jep.2011.01.026

Alviano WS, Alviano DS, Diniz CG, Antoniolli AR, Alviano CS, Farias LM, Carvalho MAR, Souza MMG, Bolognese AM (2008) In vitro antioxidant potential of medicinal plant extracts and their activities against oral bacteria based on Brazilian folk medicine. Arch Oral Biol 53:545–552. https://doi.org/10.1016/j.archoralbio.2007.12.001

Amico A (1977) Medicinal plants of Southern Zambesia. Fitoterapia 48:101–139

Appaiah P, Sunil L, Prasanth Kumar PK, Gopalakrishna AP (2014) Composition of coconut testa, coconut kernel and its oil. J Am Oil Chem Soc 91:917. https://doi.org/10.1007/s11746-014-2447-9

Arlee R, Suanphairoch S, Pakdeechanuan P (2013) Differences in chemical components and antioxidant-related substances in virgin coconut oil from coconut hybrids and their parents. Int Food Res J 20:2103–2109

Arora R, Chawla R, Marwah R, Arora P, Sharma RK, Kaushik V, Goel R, Kaur A, Silambarasan M, Tripathi RP, Bhardwaj JR (2011) Potential of complementary and alternative medicine in preventive management of novel H1N1 flu (Swine flu) pandemic: thwarting potential disasters in the bud. Evid Based Complement Alternat Med 2011:586506. https://doi.org/10.1155/2011/586506

Bankar GR, Nayak PG, Bansal P, Paul P, Pai KSR, Singla RK, Bhat VG (2011) Vasorelaxant and antihypertensive effect of Cocos nucifera Linn. endocarp on isolated rat thoracic aorta and DOCA salt-induced hypertensive rats. J Ethnopharmacol 134(1):50–54. https://doi.org/10.1016/j.jep.2010.11.047

Banzon JA, Velasco JR (1982) Coconut: production and utilization. Philippine Coconut Res. and Dev. Foundation, Manila, pp 325–244

Barnabé W, de Mendonça Neto T, Pimenta FC, Pegoraro LF, Scolaro JM (2004) Efficacy of sodium hypochlorite and coconut soap used as disinfecting agents in the reduction of denture stomatitis, Streptococcus mutans and Candida albicans. J Oral Rehabil 31(5):453–459. https://doi.org/10.1111/j.1365-2842.2004.01254.x

Belewu MA, Belewu KY (2007) Comparative physico-chemical evaluation of tiger-nut, soybean and coconut milk sources. Int. J Agri Biol 9(5):785–787

Bhandary MJ, Chandrashekar KR, Kaveriappa KM (1995) Medical ethnobotany of the Siddis of Uttara Kannada district, Karnataka, India. J Ethnopharmacol 47:149–158. https://doi.org/10.1016/0378-8741(95)01274-H

Bhatnagar AS, Prasanth Kumar PK, Hemavathi J, Gopala Krishna AG (2009) Fatty acid composition, oxidative stability, and radical scavenging activity of vegetable oil blends with coconut oil. J Am Oil Chem Soc 86:991–999. https://doi.org/10.1007/s11746-009-1435-y

Brondegaard VJ (1973) Contraceptive plant drugs. Planta Med 23:167–172. https://doi.org/10.1055/s-0028-1099428

Caceres A, Giron LM, Alvarado SR, Torres MF (1987) Screening of antimicrobial activity of plants popularly used in Guatemala for the treatment of dermatomucosal diseases. J Ethnopharmacol 20:223–237. https://doi.org/10.1016/0378-8741(87)90050-X

Calzada F, Yepez-Mulia L, Tapia-Contreras A (2007) Effect of Mexican medicinal plant used to treat trichomoniasis on Trichomonas vaginalis trophozoites. J Ethnopharmacol 113:248–251. https://doi.org/10.1016/j.jep.2007.06.001

Campbell-Falck D, Thomas T, Falck TM, Tutuo N, Clem K (2000) The intravenous use of coconut water. Am J Emerg Med 18:108–111. https://doi.org/10.1016/S0735-6757(00)90062-7

Carpo BG, Verallo-Rowell VM, Kabara JJ (2007) Novel antibacterial activity of monolaurin compared with conventional antibiotics against organisms from skin infections: an in vitro study. Drugs Dermatol 6(10):991–998

Chan SY (2016) Coconut handbook. Tetra Pak South East Asia Pte Limited, Coconut Knowledge Centre, Singapore

Chikku A, Rajamohan T (2012) Dietary coconut sprout beneficially modulates cardiac damage induced by isoproterenol in rats. Bangladesh J Pharmacol 7:258–265

Costa CT, Bevilaqua CM, Morais SM, Camurca-Vasconce-los AL, Maciel MV, Braga RR, Oliveira LMB (2010) Anthelmintic activity of Cocos nucifera L. on intestinal nematodes of mice. Res Vet Sci 88:101–103. https://doi.org/10.1016/j.rvsc.2009.05.008

Deckere EAM, Verschuren PM (2000) Functional fats and spreads. In: Gibson GR, Williams CM (eds) Functional foods: concept to product. Woodhead Publishing Limited, Cambridge, pp 234–257. https://doi.org/10.1533/9781855736436.3.233

Eder K, Kirchgessner M (1997) Concentrations of lipids in plasma and lipoproteins and oxidative susceptibility of low density lipoproteins in zinc deficient rats fed linseed oil or olive oil. J Nutr Biochem 8:461–468. https://doi.org/10.1016/S0955-2863(97)00057-0

Effiong GS, Ebong PE, Eyong EU, Uwah AJ, Ekong UE (2010) Amelioration of chloramphenicol induced toxicity in rats by coconut water. J Appl Sc Res 6(4):331–335

Enig MG (2004) Coconut: in support of good health in the 21st Century. http://coconutoil.com/coconut_oil_21st_century/. Accessed 27 June 2018

Erosa FE, Gamboa-León MR, Lecher JG, Arroyo-Serralta GA, Zizumbo-Villareal D, Oropeza-Salín CO, Peña-Rodríguez LM (2002) Major components from the epicuticular wax of Cocos nucifera. Rev Soci Quımica México 46:247–250

Esquenazi D, Wigg MD, Miranda MM, Rodrigues HM, Tostes JB, Rozental S, da Silva AJ, Alviano CS (2002) Antimicrobial and antiviral activities of polyphenolics from Cocos nucifera Linn. (Palmae) husk fiber extract. Res Microbiol 153(10):647–652. https://doi.org/10.1016/S0923-2508(02)01377-3

Ferrari R, Ceconi C, Curello S, Cargnoni A, De Guili F, Visioli O (1992) Occurrence of oxidative stress during myocardial reperfusion. Mol Cell Biochem 111:61–69. https://doi.org/10.1007/BF00229575

Freitas JCC, Nunes-Pinheiro DCS, Pessoa AWP, Silva LCR, Girão VCC, Lopes-Neto BE, Agostinho MS, Abreu CRA (2011) Effect of ethyl acetate extract from husk fiber water of Cocos nucifera in Leishmania braziliensis infected hamsters. Rev Bras Farmacogn 21:1006–1011. https://doi.org/10.1590/S0102-695X2011005000138

FSSR (2011) Food safety and standards (food products standards and food additives) Regulations. Version-IX, p 2, 56

Gandhi M, Aggarwal M, Puri S, Singla SK (2013) Prophylactic effect of coconut water (Cocos nucifera L.) on ethylene glycol induced nephrocalcinosis in male wistar rat. Int Braz J Urol 39:108–117

Go AS (2013) Heart disease and stroke statistics-2013 update: a report from the American Heart Association. Circulation 127:E841–E841

Go AS, Mozaffarian D, Roger VL, Benjamin EJ, Berry JD et al (2013) Heart disease and stroke statistics-2013 update a report from the American heart association. Circulation 127:E6–E245

Gunstone FD, Harwood JL (2007) Occurrence and characterization of oils and fats. In: Gunstone FD, Harwood JL, Dijkstra AJ (eds) Lipid handbook, 3rd edn. CRC Press, Boca Raton, FL, pp 37–141. https://doi.org/10.1201/9781420009675

Hirschhorn HH (1983) botanical remedies of the former Dutch East Indies (Indonesia). Part i: eumycetes, pteridophyta, gymnospermae, angiospermae (monocotyledones only). J Ethnopharmacol 7:123–156. https://doi.org/10.1016/0378-8741(83)90016-8

Holdsworth D (1992) Medicinal plants of the Gazelle peninsula, New Britain Island, Papua New Guinea, part I. Int J Pharmacog 30:185–190. https://doi.org/10.3109/13880209209053992

Holdsworth D, Wamoi B (1982) Medicinal plants of the Admiralty Islands, Papua New Guinea. Part I. Int J Crude Drug Res 20:169–181

Hope BE, Massey DG, Fournier-Massey G (1993) Hawaiian material medica for asthma. Hawaii Med J 52:160–166

Idris NA, Samsudin S (2005) Edible uses of palm kernel oil and its products. In: Basiron Y, Darus A, Ngan MA, Weng CK (eds) Palm kernel products: characteristics and applications. Malaysian Palm Oil Board, Kuala Lumpur, pp 70–94

Imaga NA, Obafemi A, Onadeko O, Okafor UA (2016) Leptin levels attenuated by coconut water in high-fat and high-fructose diet fed animal models. Faseb J 30:1

Jackson JC, Gordon A, Wizzard G, McCook K, Rolle R (2004) Changes in chemical composition of coconut (Cocos nucifera) water during maturation of the fruit. J Sci Food Agric 84:1049–1052. https://doi.org/10.1002/jsfa.1783
Jayalekshmy A, Mathew AG (1990) Changes in the carbohydrates and proteins of coconut during roasting. Food Chem 37:123–134. https://doi.org/10.1016/0308-8146(90)90086-J
Kochhar SP (2001) The composition of frying oils. In: Rossell JB (ed) Frying: improving quality. Woodhead Publishing, Cambridge, pp 87–114
Koschek PR, Alviano DS, Alviano CS, Gattass CR (2007) The husk fiber of Cocos nucifera L. (Palmae) is a source of anti-neoplastic activity. Braz J Med Biol Res 40(10):1339–1343. https://doi.org/10.1590/S0100-879X2006005000153
LaBarrie J, St-Onge MP (2017) A coconut oil-rich meal does not enhance thermogenesis compared to corn oil in a randomized trial in obese adolescents. Insights Nutr Metab 1(1):30–36
Levy Y, Bartha P, Ami Ben-Amotz J, Brook G, Dankner G, Lin S, Hammerman H (1998) Plasma antioxidants and lipid peroxidation in acute myocardial infraction and thrombolysis. J Am Coll Nutr 17:337–341. https://doi.org/10.1080/07315724.1998.10718772
Lima EBC, Sousa CNS, Meneses LN, Ximenes NC, Santos Júnior MA, Vasconcelos GS, Lima NBC, Patrocínio MCA, Macedo D, Vasconcelos SMM (2015) Cocos nucifera (L.) (Arecaceae): a phytochemical and pharmacological review. Braz J Med Biol Res 48(11):953–964. https://doi.org/10.1590/1414-431X20154773
Loki AL, Rajamohan T (2003) Hepatoprotective and antioxidant effect of tender coconut water on carbon tetrachloride induced liver injury in rats. Indian J Biochem Biophys 40:354–357
Long LO, Park S (2007) Update on nephrolithiasis management. Minerva Urol Nefrol 59 (3):317–325
Mancia G, Fagard R, Narkiewicz K, Redon J, Zanchetti A, Böhm M, Christiaens T, Cifkova R, Backer GD, Dominiczak A, Galderisi M, Grobbee DE, Jaarsma T, Kirchhof P, Kjeldsen SE, Laurent S, Manolis AJ, Nilsson PM, Ruilope LM, Schmieder RE, Sirnes PA, Sleight P, Viigimaa M, Waeber B, Zannad F (2013) 2013 ESH/ESC Guidelines for the management of arterial hypertension. Blood Pressure 22(4):193–278. https://doi.org/10.3109/08037051.2013.812549
Mantena SK, Jagadish BSR, Siripurapu KB, Unnikrishnan MK (2003) In vitro evaluation of antioxidant properties of Cocos nucifera Linn. water. Nahrung 47:126–131. https://doi.org/10.1002/food.200390023
Marina AM, Che Man YB, Nazimah SAH, Amin I (2009) Chemical properties of virgin coconut oil. J Am Oil Chem Soc 86:301–307. https://doi.org/10.1007/s11746-009-1351-1
Matos FJA (1997) Introdução à fitoquımica experimental, 2nd edn. Imprensa Universitária, Fortaleza
Mini S, Rajamohan T (2004) Influence of coconut kernel protein on lipid metabolism in alcohol fed rats. Indian J Exp Biol 42(1):53–57
Mitchell SA, Ahmad MH (2006) A review of medicinal plant research at the University of the West Indies, Jamaica, 1948–2001. West Indian Med J 55(4):243–269
Mohamad NE, Yeap SK, Ky H, Ho WY, Boo SY, Chua J, Beh BK, Sharifuddin SA, Long K, Alitheen NB (2017) Dietary coconut water vinegar for improvement of obesity-associated inflammation in high-fat-diet-treated mice. Food Nutr Res 61(1):1368322. https://doi.org/10.1080/16546628.2017.1368322
Morrison EYSA (1994) Local remedies, Yeh or nay. West Indian Med J 43(Suppl 2):9
Muller H, Lindman AS, Blomfeldt A, Seljeflot I, Pedersen JI (2003) A diet rich in coconut oil reduces diurnal postprandial variations in circulating tissue plasminogen activator antigen and fasting lipoprotein (a) compared with a diet rich in unsaturated fat in women. J Nutr 133 (11):3422–3427. https://doi.org/10.1093/jn/133.11.3422
Nagata JM, Jew AR, Kimeu JM, Salmen CR, Bukusi EA, Cohen CR (2011) Medical pluralism on Mfangano Island: use of medicinal plants among persons living with HIV/AIDS in Suba District, Kenya. J Ethnopharmacol 135:501–509. https://doi.org/10.1016/j.jep.2011.03.051

Nevin KG, Rajamohan T (2004) Beneficial effects of virgin coconut oil on lipid parameters and in vitro LDL oxidation. Clin Biochem 37(9):830–835. https://doi.org/10.1016/j.clinbiochem.2004.04.010

Nevin KG, Rajamohan T (2006) Virgin coconut oil supplemented diet increases the antioxidant status in rats. Food Chem 99(2):260–266. https://doi.org/10.1016/j.foodchem.2005.06.056

Nevin KG, Rajamohan T (2008) Influence of virgin coconut oil on blood coagulation factors, lipid levels and LDL oxidation in cholesterol fed Sprague-Dawley rats. Eur e-J Clin Nutr Metab 3(1): e1–e8. https://doi.org/10.1016/j.eclnm.2007.09.003

Nieuwentus R, Nieuwelink J (2002) Agrodok series no. 10. CTA, The Netherlands, pp 28–30

Pal D, Sarkar A, Gain S, Jana S, Mandal S (2011) CNS depressant activities of roots of Coccos nucifera in mice. Acta Pol Pharm 68:249–254

Parthasarathy S, Santanam N, Auge N (1998) Antioxidants and low density lipoprotein oxidation. In: Papas AM (ed) Antioxidant status, diet, nutrition and health, 1st edn. CRC Press, Boca Raton, pp 189–210

Perez-Vizcaino F, Duarte J, Jimenez R, Santos-Buelga C, Osuna A (2009) Antihypertensive effects of the flavonoid quercetin. Pharmacol Rep 61:67–75

Persley GJ (1992) Replanting the Tree of Life. Towards an international agenda for coconut palm research. Redwood Press Ltd., Melksham

Poirier P, Giles TD, Bray GA, Hong Y, Stern JS, Pi-Sunyer FX, Eckel RH, American Heart Association, Obesity Committee of the Council on Nutrition, Physical Activity, and Metabolism (2006) Obesity and cardiovascular disease: pathophysiology, evaluation, and effect of weight loss: an update of the 1997 American Heart Association Scientific Statement on Obesity and Heart Disease from the Obesity Committee of the Council on Nutrition, Physical Activity, and Metabolism. Circulation 113:898–918. https://doi.org/10.1161/CIRCULATIONAHA.106.171016

Preetha PP, Devi VG, Rajamohan T (2014) Antihyperlipidemic effects of mature coconut water and its role in regulating lipid metabolism in alloxan-induced experimental diabetes. Comp Clin Pathol 23(5):1331–1337. https://doi.org/10.1007/s00580-013-1784-7

Rajamohan T, Anurag P (2003) Cardioprotective effect of tender coconut water in experimental myocardial infarction. Plant Food Hum Nutr 58:1–12. https://doi.org/10.1023/B:QUAL.0000040363.64356.05

Ramirez VR, Mostacero LJ, Garcia AE, Mejia CF, Pelaez PF, Medina CD et al (1988) Vegetales empleados em medicina tradicional Norperuana. Banco Agrario del Peru & NACL Univ Trujillo, Trujillo, Peru, p 54

Rao CV, Newmark HL, Reddy BS (1998) Chemo-preventive effect of Squalene on colon cancer. Carcinogenesis 19:287–290. https://doi.org/10.1093/carcin/19.2.287

Renjith RS, Chikku AM, Rajamohan T (2013) Cytoprotective, antihyperglycemic and phytochemical properties of Cocos nucifera (L.) inflorescence. Asian Pac J Trop Med 6:804–810. https://doi.org/10.1016/S1995-7645(13)60142-X

Rossel JB (2001) Factors affecting the quality of frying oils and fats. In: Rossell JB (ed) Frying: improving quality. Woodhead Publishing Limited, Cambridge, pp 115–164

Saat M, Singh R, Sirisinghe RG, Nawawi M (2002) Rehydration after exercise with fresh young coconut water, carbohydrate-electrolyte beverage and plain water. J Physiol Anthropol Appl Human Sci 21(2):93–104

Sachs M, von Eichel J, Asskali F (2002) Wound management with coconut oil in Indonesian folk medicine. Chirurg 73(4):387–392. https://doi.org/10.1007/s00104-001-0382-4

Salil G, Rajamohan T (2001) Hypolipidemic and antiperoxidative effect of coconut protein in hypercholesterolemic rats. Indian J Exp Biol 39(10):1028–1034

Salil G, Nevin KG, Rajamohan T (2011) Arginine rich coconut kernel protein modulates diabetes in alloxan treated rats. Chem Biol Interact 189(1-2):107–111. https://doi.org/10.1016/j.cbi.2010.10.015

Sandhya VG, Rajamohan T (2006) Beneficial effects of coconut water feeding on lipid metabolism in cholesterol-fed rats. J Med Food 9(3):400–407. https://doi.org/10.1089/jmf.2006.9.400

Seneviratne KN, Dissanayake DMS (2008) Variation of phenolic content in coconut oil extracted by two conventional methods. Inter J Food Sci Technol 43:597–602. https://doi.org/10.1111/j.1365-2621.2006.01493.x

Silva RR, Silva O e, Fontes HR, Alviano CS, Fernandes PD, Alviano DS (2013) Anti-inflammatory, antioxidant, and antimicrobial activities of Cocos nucifera var. typica. BMC Complement Altern Med 13:107. https://doi.org/10.1186/1472-6882-13-107

Singh YN (1986) Traditional medicine in Fiji: some herbal folk cures used by Fiji Indians. J Ethnopharmacol 15(1):57–88. https://doi.org/10.1016/0378-8741(86)90104-2

Singla RK, Jaiswal N, Bhat GV, Jagani H (2011) Antioxidant and antimicrobial activities of Cocos nucifera Linn. (Arecaceae) endocarp extracts. Indo Global J Pharm Sci 1:354–361

Solangi AH, Iqbal MZ (2011) Chemical composition of meat (kernel) and nut water of major coconut (Cocos nucifera) cultivars at coastal area of Pakistan. Pak J Bot 43(1):357–363

Stamatelou KK, Francis ME, Jones CA, Nyberg LM, Curhan GC (2003) Time trends in reported prevalence of kidney stones in the United States: 1976–1994. Kidney Int 63:1817–1823. https://doi.org/10.1046/j.1523-1755.2003.00917.x

Taheri JB, Espineli FW, Lu H, Asayesh M, Bakshi M, Nakhostin MR (2010) Antimicrobial effect of coconut flour on oral microflora: an in vitro study. Res J Biol Scs 5(6):456–459. https://doi.org/10.3923/rjbsci.2010.456.459

Tan TC, Cheng LH, Bhat R, Rusul G, Easa AM (2014) Composition, physicochemical properties and thermal inactivation kinetics of polyphenol oxidase and peroxidase from coconut (Cocos nucifera) water obtained from immature, mature and overly-mature coconut. Food Chem 142:121–128. https://doi.org/10.1016/j.foodchem.2013.07.040

Tangwatcharin P, Khopaibool P (2012) Activity of virgin coconut oil, lauric acid or monolaurin in combination with lactic acid against Staphylococcus aureus. Southeast Asian J Trop Med Public Health 43(4):969–985

Terdwongworakul A, Chaiyapong S, Jarimopas B, Meeklangsaen W (2009) Physical properties of fresh young Thai coconut for maturity sorting. Biosyst Eng 103:208–216. https://doi.org/10.1016/j.biosystemseng.2009.03.006

Teuber SS, Peterson WR (1999) Systemic allergic reaction to coconut (Cocos nucifera) in 2 subjects with hypersensitivity to tree nut and demonstration of cross-reactivity to legumin-like seed storage proteins: new coconut and walnut food allergens. J Allergy Clin Immunol 103 (6):1180–1185. https://doi.org/10.1016/s0091-6749(99)70196-x

Thai Food Composition (2004) Institute of Nutrition Mahidol University (INMU), Thailand

Unger RH, Foster DW (1998) Diabetes mellitus. In: Wilson JD, Foster DW, Kronenberg HM, Larsen PR (eds) Williams text book of endocrinology, 9th edn. Saunders, Philadelphia, pp 973–1059

USDA National Nutrient Database (2019) Nuts, coconut meat, raw. SR legacy, 170169. https://fdc.nal.usda.gov/fdc-app.html#/food-details/170169/nutrients

Verallo-Rowell VM, Dillague KM, Syah-Tjundawan BS (2008) Novel antibacterial and emollient effects of coconut and virgin olive oils in adult atopic dermatitis. Dermatitis 19(6):308–315

Verma V, Bhardwaj A, Rathi S, Raja RB (2012) Potential antimicrobial agent from Cocos nucifera mesocarp extract. ISCA J Biol Sci 1:48–54

Von Eckardstein A, Nofer JR, Assmann G (2001) High density lipoproteins and arteriosclerosis: role of cholesterol efflux and reverse cholesterol transport. Arterioscler Thromb Vasc Biol 21 (1):13–27. https://doi.org/10.1161/01.atv.21.1.13

Weber MA, Schiffrin EL, White WB, Mann S, Lindholm LH, Kenerson JG, Flack JM, Carter BL, Materson BJ, Ram CV, Cohen DL, Cadet JC, Jean-Charles RR, Taler S, Kountz D, Townsend RR, Chalmers J, Ramirez AJ, Bakris GL, Wang J, Schutte AE, Bisognano JD, Touyz RM, Sica D, Harrap SB (2014) Clinical practice guidelines for the management of hypertension in the community: a statement by the American Society of Hypertension and the International Society of Hypertension. J Clin Hypertens (Greenwich) 16(1):14–26. https://doi.org/10.1111/jch.12237

Weniger B, Rouzier M, Daguilh R, Henrys D, Henrys JH, Anton R (1986) Traditional medicine in the Central Plateau of Haiti. 2. Ethnopharmacologic inventory. J Ethnopharmacol 17:13–30. https://doi.org/10.1016/0378-8741(86)90070-x

Wisse BE, Kim F, Schwartz MW (2007) Physiology. An integrative view of obesity. Science 318 (5852):928–929. https://doi.org/10.1126/science.1148032

Wong W (1976) Some folk medicinal plants from Trinidad. Econ Bot 30:103–142. https://doi.org/10.1007/BF02862958

Yartey J, Harisson EK, Brakohiapa LA, Nkrumah FK (1993) Carbohydrate and electrolyte content of some homeavailable fluids used for oral rehydration in Ghana. J Trop Pediatr 39(4):234–237. https://doi.org/10.1093/tropej/39.4.234

Yong JW, Ge L, Ng YF, Tan SN (2009) The chemical composition and biological properties of coconut (Cocos nucifera L.) water. Molecules 14(12):5144–5164. https://doi.org/10.3390/molecules14125144

Chapter 8
Mustard (*Brassica nigra*) Seed

H. K. S. De Zoysa and Viduranga Y. Waisundara

Abstract *Brassica nigra* plays an important role in global agriculture, horticulture, health and wellness aspects due to its culinary and medicinal values. *B. nigra* plant is also grown to obtain oil for industrial purposes as well as a nutritionally valued seed meal. The seed primarily contains oligosaccharides belonging to the raffinose family; amino acids; fatty acids such as palmitic, stearic, oleic, linoleic, linolenic, eicosenoic and erucic acids; vitamins; minerals (mostly iron); anti-nutritional factors (in particular, enzyme inhibitors); glucosinolates; and a wide range of phenolic compounds. *B. nigra* seeds have demonstrated to impart antidiabetic, anticonvulsant, antithrombotic, antibacterial, antifungal and antioxidant activities, as well as immunomodulatory and inflammatory effects. It is also recognized to provide protection against factors leading to gastrointestinal cancer. Given the presence of antioxidants in *B. nigra* seeds, it may be hypothesized as being useful for cardiac disorders as well. When it comes to safety aspects, *B. nigra* seeds contain storage proteins of the 2S albumin class and, hence, have been the cause food allergies which were mostly reported in Europe. During the last two decades, rapid developments have taken place in the agricultural breeding of *B. nigra* owing to the advancements in plant biology and biotechnology applications. These developments facilitated the adoption of *B. nigra* plant models which are of economic importance and commercial value.

Keywords Black mustard seed · Brassicaceae · Condiments · Food · Glucosinolates · Oilseeds

H. K. S. De Zoysa
Department of Bioprocess Technology, Faculty of Technology, Rajarata University of Sri Lanka, Mihintale, Sri Lanka

V. Y. Waisundara (✉)
Australian College of Business & Technology—Kandy Campus, Kandy, Sri Lanka

B. Tanwar, A. Goyal (eds.), *Oilseeds: Health Attributes and Food Applications*,
https://doi.org/10.1007/978-981-15-4194-0_8

8.1 Origin and History of Usage

Brassica nigra or black mustard seeds belong to the family Brassicaceae and order Brassicales. The family Brassicaceae itself has 338 genera and 3709 species (Monsalve et al. 2001). The word 'mustard' was derived from the European practice of mixing the sweet 'must' of old wine with crushed seeds of *B. nigra* to form a hot paste ('hot must' or 'mustum ardens'), hence the modern term 'mustard' (Hemingway 1976). Plants belonging to the genus *Brassica* including *B. nigra* are primarily grown as condiments (Alam et al. 2011; Augustine et al. 2014; Warwick et al. 2006; Monsalve et al. 2001). These plants can be easily distinguished from other flowering plants as they have cruciferous flowers (Augustine et al. 2014). The *B. nigra* plant grows up to 2 m and requires no vernalization for flower induction. The flowers are yellow in colour, and the seeds occur in shades ranging from dark to brown with a diameter of 1–2 mm (Fig. 8.1) (Rakow 2004; Van Dam et al. 2004; Veromann et al. 2012). The origin of *Brassica* species has been proven by evidence of their cultivation during ancient times in the Middle East, India, China, Rome and Greece, where *B. nigra* seed was used as a medicine as well as a spice (Augustine et al. 2014; Warwick et al. 2006; Amare et al. 2015). *B. nigra* grows in the form of a wild weed in cultivated areas of the Mediterranean region (both plains and hilly areas) and is even found on roadsides and fields in many countries such as Tangiers and Morocco (Rakow 2004; Tawaha and Turk 2003). *B. nigra* is also grown under semi-cultivated conditions in Rhodes, Crete, Sicily, Turkey and Ethiopia (Rakow 2004). This plant is believed to have evolved separately from the other two diploid *Brassica* species, *B. rapa* and *B. oleracea*, while cytological, isozyme, nuclear and chloroplast DNA restriction site and sequence data have shown *B. nigra* to possess a closer genetic relationship with *Sinapis arvensis* (Warwick 2011).

Fig. 8.1 Representative image of *B. nigra* seeds

8.2 Production of *Brassica* Species and *B. nigra*

The recent increase in the agricultural production of *B. nigra* has been mostly due to its usage as a medicine and home remedy (Amare et al. 2015). Additionally, the *B. nigra* plant is also widely grown to obtain oil for industrial purposes as well as a nutritionally valued seed meal (Hirani et al. 2012; Mourato et al. 2015). *Brassica* oilseed species are the third most cultivated oilseed crops globally, although *B. nigra* is grown on a very small scale in contrast with its other species (*B. campestris*, *B. napus*, *B. juncea* and *B. carinata*) (Ashraf and McNeilly 2004). All of its plant parts including its terminal buds, stem, inflorescences, leaves, roots and seeds are edible. The higher nutritional and functional values of *B. nigra* oil are some of the key areas which are considered for crop improvement (Augustine et al. 2014). The seeds produce their maximum yield under normal soil and environmental conditions, although their growth, seed yield and oil production may reduce remarkably by environmental stresses such as drought, waterlogging, salinity, low or high temperature, nutrient deficiency or excess (Ashraf and McNeilly 2004).

In Europe, the overall production of oils from seeds belonging to the family Brassicaceae has been estimated to be approximately 70 million tons/annum with an average consumption per person of approximately 3–5 kg per annum (Björkman et al. 2011). Canada was the highest exporter of *B. nigra* during 2004–2008 (Alvarez and Boye 2012). *B. nigra* seed oil has a variety of food applications (which have been elaborated in Table 8.2). The seed extracts of *B. nigra* are also used in the industry as antimicrobial agents and for industrial biofilms (Jahangir et al. 2009). In addition, inorganic phosphorous was also recorded in the *B. nigra* seeds (Chanda and Chakrabarti 1996), and the *B. nigra* leaf was recorded to contain other types of minerals such as Fe (241.20 mg/100 g), Zn (5.50 mg/100 g), Cu (1.90 mg/100 g), Mn (102. 73 mg/100 g), Na (62.06 mg/100 g), K (73.83 mg/100 g), Ca (67.13) and Mg (15.96 mg/100 g) (Saha et al. 2015). All of these minerals in *B. nigra* are very important for building strong bones, teeth, skin, hair, nerve function and muscles and maintaining good blood circulation for metabolic processes.

8.3 Chemical Composition

The chemical composition of various parts of the *B. nigra* plant in terms of carbohydrates, proteins (in crude form and amino acids) and lipids is shown in Table 8.1. Studies which focus only on the composition of the seed are scarce, while most studies have been carried out on the leaf of the plant, and thus, the information provided in the table has been arranged accordingly.

Table 8.1 Chemical composition and other constituents which are present in various parts of the *B. nigra* plant (AII and BII indicates protein fractions)

Chemical composition/ constituents		Part of the plant	Amount present		Reference
Carbohydrates	–	Seed	0.74 w/w of seed weight		MacKenzie and Blakely (1972)
	Sucrose		74.7 mol/g and 2.6 w/w of seed weight		Andersen et al. (2005)
	Raffinose		9.0 mol/g		
	Stachyose		19.4 mol/g		
	Verbascose		ND		
	Ajugose		ND		
	–	Leaf	5.56 g/100 g		Saha et al. (2015)
Proteins (amino acids)	Lysine	Seed	0.27 mol/g AII protein	0.57 mol/g BII protein	MacKenzie and Blakely (1972)
	Histidine		0.15 mol/g AII protein	0.47 mol/g BII protein	
	Ammonia		2.06 mol/g AII protein	1.42 mol/g BII protein	
	Arginine		0.43 mol/g AII protein	0.40 mol/g BII protein	
	Aspartic acid		0.88 mol/g AII protein	0.18 mol/g BII protein	
	Threonine		0.38 mol/g AII protein	0.26 mol/g BII protein	
	Serine		0.52 mol/g AII protein	0.39 mol/g BII protein	
	Glutamic acid		1.22 mol/g AII protein	1.44 mol/g BII protein	
	Proline		0.78 mol/g AII protein	1.64 mol/g BII protein	
	Glycine		1.01 mol/g AII protein	0.56 mol/g BII protein	
	Alanine		0.63 mol/g AII protein	0.52 mol/g BII protein	
	Valine		0.54 mol/g AII protein	0.55 mol/g BII protein	
	Isoleucine		0.43 mol/g AII protein	0.38 mol/g BII protein	
	Leucine		0.79 mol/g AII protein	0.79 mol/g BII protein	
	Tyrosine		0.16 mol/g AII protein	0.06 mol/g BII protein	
	Phenylalanine		0.33 mol/g AII protein	0.23 mol/g BII protein	
Crude protein	–	Leaf	4.03 g/100 g		Saha et al. (2015)
Lipids	–	Seed	36.5 w/w of seed weight		MacKenzie and Blakely (1972)
	–	Leaf	4.19 g/100 g		Saha et al. (2015)

8.4 Anti-nutritional Factors

Anti-nutritional factors are constituents, which are naturally generated in food elements during the metabolism of particular species. These anti-nutritional factors are generated in food by various mechanisms such as inactivation of some nutrients, diminution of the digestive process or metabolic utilization of feed and, thus, reduce the availability of nutrients of particular food types causing growth inhibition (Andersen et al. 2005; Thangaraj 2016). *B. nigra* seeds consist of anti-nutritional factors primarily in the form of enzymes inhibitors. Protease inhibitors, phytic acid, cyanogen, lectins, phenolic compounds, phytates, oxalic acid, tannin and R-galactosides are also considered as anti-nutritional factors, and most of them are found in *B. nigra* seeds (Andersen et al. 2005; Thangaraj 2016; Saha et al. 2015). The anti-nutritional effects caused by these components result in goitres, growth retardation and liver damage in humans, and one of the reasons for poor egg production in animals as well (Avato and Argentieri 2015).

8.5 Phytonutrients/Glucosinolates

Several studies have been conducted for investigating the Brassicaceae family for phytonutrients and have revealed the presence of glucosinolates, tocopherols and vitamin E forms (mostly α-tocopherol) in *B. nigra* seeds (Avato and Argentieri 2015). Glucosinolates contain a cyano and a sulphate group, which are derived from amino acid biosynthesis and are water-soluble. In addition, glucosinolates are considered as secondary metabolites of the Brassicaceae family and defend plants against pests, pathogenic invaders and diseases (Augustine and Bisht 2015; Jahangir et al. 2009; Avato and Argentieri 2015; Björkman et al. 2011; Mazumder et al. 2016). Glucoraphanin – which is a widely investigated glucosinolate due to several health benefits – is present in *B. nigra* seeds albeit in minute amounts as compared with other *Brassica* species (Augustine and Bisht 2015).

Phenolic/polyphenolic compounds (sinapine), flavonoids (anthocyanins (cyanidin-3-sophoroside-5-glucoside), myricetin, fisetin, morin, quercetin, kaempferol, isorhamnetin), melatonin (N-acetyl-5-methoxytryptamine) and lignans are known for their potential benefits on human health such as antiviral, antiallergic, antiplatelet, anti-inflammatory, antitumor and antioxidant activities (Iriti and Faoro 2006; Kapusta-Duch et al. 2012; Lako et al. 2007). These compounds are able to prevent oxidative stress, induce detoxification enzymes, stimulate the immune system, decrease the risk of cancers, inhibit malignant transformation and carcinogenic mutations and reduce proliferation of cancer cells (Kapusta-Duch et al. 2012). The presence of these phytochemicals is especially common for seeds present in the plants belonging to the Brassicaceae family, including *B. nigra* (Alam et al. 2011; Avato and Argentieri 2015). In addition, *B. nigra* was also shown to contain glucosides such as sinigrin, alkaloids such as nortropane, flavonoids such as isorhamnetin and tannins (Alam et al. 2011; Obi et al. 2009).

Sinigrin (2-propenyl-glucosinolate) is the major glucosinolate found in *B. nigra*; due to its volatility, it has the ability to provide a pungent aroma as well as taste and flavour to the extracted oil (Amare et al. 2015; Avato and Argentieri 2015; Talapatra and Talapatra 2015; Borş et al. 2015; Hirani et al. 2012; Mazumder et al. 2016). In addition, sinigrin has the ability to produce isothiocyanates by the metabolic activation, which contributes to antitumour activities (Jie et al. 2014). *B. nigra* has other types of glucosinolates as well, namely, glucoraphanin, sinalbin (4-hydroxybenzyl), 4-hydroxyglucobrassicin, glucobrassicin, 4-methoxyglucobrassicin and neoglucobrassicin. The production of glucosinolates depends on the availability of light in the environment in which the plants are grown. It has been observed that *B. nigra* grown under dark conditions produce more glucosinolates than under light conditions (Borş et al. 2015).

The use of mustard and rapeseed meal utilization as a food for domestic animals has increased during past few decades. Different types of heating and extraction methods, including toasting and microwave processing, are used in the preparation of detoxified meals. It has to be highlighted that the main purposes of these processing methods are to remove the glucosinolates or their hydrolysis products and combined products such as myrosinase. However, it leads to losses in proteins, partial removal of glucosinolates and their degradation products and reduced functional or nutritional properties of the detoxified products as well as high processing costs. Other methods have also been developed to detoxify mustard seeds, such as the extraction of glucosinolates with water, using aqueous bases and aqueous alcohols (Pecháček et al. 2000).

8.6 Health Benefits of *B. nigra* Seed and Oil

The overall medicinal properties of primarily *B. nigra* seed and oil are summarized and presented as a schematic as shown in Fig. 8.2. The properties are discussed individually along with evidence in Sects. 8.6.1 to 8.6.10.

8.6.1 Antidiabetic Effects and Preventing Hepatic and Renal Damage

B. nigra seeds have been revealed to possess antidiabetic effects. This ability has been proven in streptozotocin (STZ)-induced diabetic animals, where the rats were treated with the aqueous extract of *B. nigra* seeds leading to reduced serum glucose levels. The extracts also helped in maintaining reduced glycosylated haemoglobin and serum lipids (Anand et al. 2007). In a follow-up study by Anand et al. (2009)

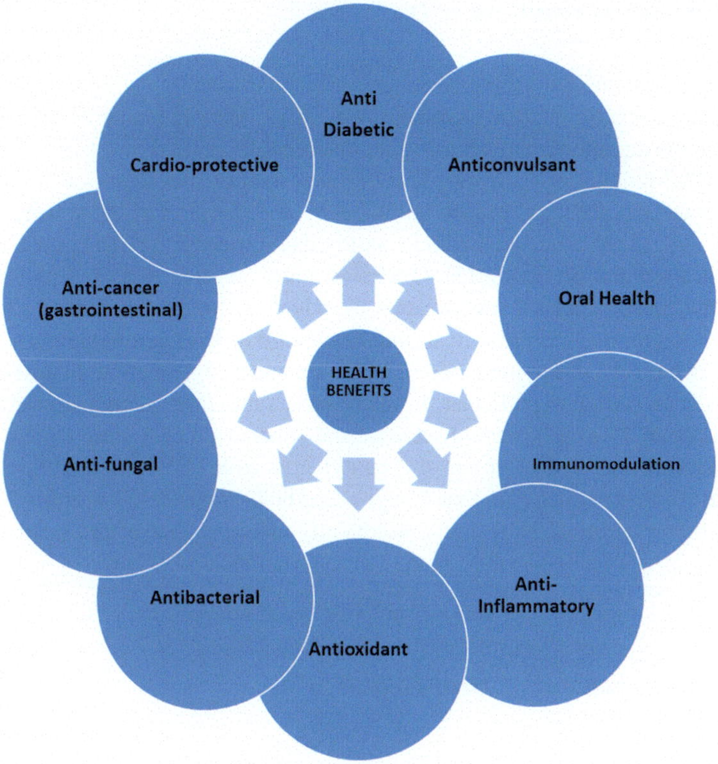

Fig. 8.2 Schematic depiction of the health benefits and medicinal properties of *B. nigra* seeds and oil

which was meant to better understand the mechanism of antidiabetic action of the aqueous extract of *B. nigra* seeds, the effect of oral administration of the extract for 2 months on glycolytic and gluconeogenic enzymes was studied in the liver and kidney of STZ-induced diabetic rats. It was observed in this study that the activities of the gluconeogenic enzymes were higher and the activities of the glycolytic enzymes had decreased in both the liver and kidney tissues of the rats. Anand et al. (2009) had attributed these effects to the release of insulin from the pancreas and the change of glucose-metabolizing enzyme activities to normal levels. Rajamurugan et al. (2012a, b) evaluated hepatic and nephroprotective effects of the *B. nigra* leaf extract against d-GalN-induced liver and nephrotoxicity in Wistar rats. The methanol extract of *B. nigra* leaves exhibited a protective effect, which was associated with the presence of antioxidant compounds. *B. nigra* has also been purported to stimulate bile production in the liver which is a beneficial characteristic towards mitigating hepatotoxicity resulting from increased glucose levels (Srinivasan 2005).

8.6.2 Anticonvulsant Activity

Anticonvulsants are a highly diverse group of pharmacological agents which are used to treat epileptic seizures (Kiasalari et al. 2012; Manohar et al. 2009), which is one of the most predominant neurological disorders found in 0.5–1% of the world's population (Manohar et al. 2009). The *B. nigra* seed extract (hydro-alcoholic) has demonstrated the ability to reduce the intensity and duration of the seizure, which has once again been associated with its antioxidant properties (Kiasalari et al. 2012; Abdollahi Fard and Shojaii 2013). Oxidative stress and free radical production are two of the most important mechanisms by which neurological disorders such as epileptic seizures occur. Free radicals induce membrane lipid peroxidation and tissue injury which results in cell membrane destruction and its dysfunction. In the study by Kiasalari et al. (2012) in particular, it was observed that *Brassica nigra seed extract at a dosage of* 150 mg/kg could increase the superoxide dismutase (SOD) levels compared with the control group of 2-pentylentetrazole (PTZ)-kindled mice. This result has led to the conclusion that *B. nigra* seed helps in the deletion of free radicals through preservation of SOD and consequently affecting the seizure intensity and duration.

8.6.3 Management of Oral Health

B. nigra oilseed has the ability to stimulate and enhance the blood circulation through restoration of health of the gingiva and reducing the occurrence of its inflammation (Markose et al. 2016). This is an important property when it comes to oral health, where Markose et al. (2016) have suggested that massaging with *B. nigra* oil is able to stimulate blood circulation and restore the health of gingiva. It may be implied from this particular study that incorporation of *B. nigra* seed oil into dental floss or toothpaste might be an effective means of maintaining oral health.

8.6.4 Immunomodulation Properties

Immunomodulatory compounds have the ability to stimulate or suppress the immune system and may help the body to fight against cancer, infection and other similar diseases. Melatonin (N-acetyl-5-methoxytryptamine) has immunomodulatory and cytoprotective properties and has been found in the *B. nigra* dry plant seeds ranging from 120 to 150 mg/g (Manchester et al. 2000; Iriti and Faoro 2006; Posmyk and Janas 2009). This compound has the ability to act on the either physiological or pathological conditions including immunity and cancer, as well as sleeping, vascular tone, calcium homeostasis, bone turnover and sexual development. The presence of melatonin is much higher in *B. nigra* seeds in comparison with other edible seeds such as sunflower seeds, fenugreek, coriander and fennel that have less than 100 mg/g (Manchester et al. 2000; Iriti and Faoro 2006). Melatonin is a known immunomodulator, and its presence in *B. nigra* seeds further ascertains its role as an agent with the ability to enhance the immune system (Manchester et al. 2000).

8.6.5 Anti-inflammatory Effects

Proteins and enzymes play an essential role in inflammation and other functions of the immune system. Proteolytic enzymes such as bromelain, papain, trypsin and chymotrypsin are important regulators and modulators of the inflammatory response in the body (Alam et al. 2011). Inflammatory diseases such as rheumatism can be reduced by treatment with *B. nigra* plants while simultaneously reducing congestion among internal organs (Alam et al. 2011; Obi et al. 2009). The whole plant – including seeds – has been traditionally used for neuralgia, spasms, alopecia, epilepsy, snakebite and toothache (Alam et al. 2011; Markose et al. 2016). Moreover, sinigrin in *B. nigra* has the ability to act against atherosclerosis, a recognized chronic inflammatory disease (Mazumder et al. 2016).

Studies have revealed that the anti-inflammatory activity of *B. nigra* is initiated by protease inhibition. Flavonoids (gallic acid, followed by quercetin, ferulic acid, caffeic acid and rutin) in *B. nigra* have been demonstrated to inhibit trypsin as well as act as an anti-inflammatory agent (Alam et al. 2011). The anti-inflammatory activity of *B. nigra* also depends on the presence of sinigrin which is able to prevent the phosphorylation of mitogen-activated protein kinase (MAPK) expression as well as NOD-like receptors (NLRP-3) and p65, thereby lowering the production of pro-inflammatory mediators (Mazumder et al. 2016). In addition, *B. nigra* extract has been able to demonstrate a membrane stabilization effect, while many sesquiterpenes found in the *B. nigra* seed extract are known to possess anti-inflammatory properties. The presence of terpenoids is another reason for its demonstrated anti-inflammatory activity, and this compound was revealed to be present in both seeds and leaf of *B. nigra* (Rajamurugan et al. 2012b).

8.6.6 Antioxidant Activity

Antioxidants prevent deleterious physiological effects imparted by free radicals by acting as electron donors, thereby neutralizing the radicals as well as quenching singlet and triplet Oxygen (Alam et al. 2011; Radhakrishnan et al. 2014; Diengdoh 2017). Glucosinolate breakdown products present in *B. nigra* seeds such as allyl isothiocyanate, goitrin and thiocyanate ion act as naturally occurring antioxidants. Also, the breakdown products enhance the synthesis of glutathione, the most abundant intracellular antioxidant found in *B. nigra* (Jahangir et al. 2009). Some of the phenolic compounds with known antioxidant properties reported to be found in *B. nigra* seed are catechin, epicatechin, gallic acid, caffeic acid, ferulic acid, myricetin, quercetin and rutin (Lee et al. 2015; Rajamurugan et al. 2012a).

8.6.7 Antibacterial Activity

In developing countries, infectious diseases are one of the major health issues for which traditional medicinal plants are known to be of paramount help in the past as well as in the present. *B. nigra* plant has received much attention in this aspect due to its antibacterial activity. Several studies have revealed the positive response of *B. nigra* plant seed extracts in terms of antibacterial activity (Amare et al. 2015; Ng et al. 2013; Markose et al. 2016). According to these studies, seed extracts reacted positively by inhibiting the growth of *Staphylococcus aureus*, *Bacillus cereus*, *Enterococcus faecalis*, *Pseudomonas aeruginosa*, *Shigella sonnei*, *E. coli* and *Salmonella* spp. (Amare et al. 2015; Obi et al. 2009; Bhatia and Sharma 2012). The major component found in the essential oil of *B. nigra* is allyl isothiocyanate, a non-phenolic sulphur compound known for its ability to inactivate the extracellular enzymes of bacteria by breaking disulphide bonds (Bhatia and Sharma 2012). Moreover, the flour of *B. nigra* seeds is considered as an antiseptic, as it acts against the growth of many disease-causing microorganisms such as *E. coli* and *Streptococcus* spp. (Alam et al. 2011). Glucosinolates have also been purported to be responsible for the antibacterial activity demonstrated by *B. nigra* seeds, especially *Shigella sonnei* and *Salmonella* spp. (Mazumder et al. 2016). In addition, *B. nigra* also contains sinapine, sinapic acid and sinigrin which have the ability to impart antimicrobial activities (Obi et al. 2009).

8.6.8 Antifungal Activity

The antifungal activity of *B. nigra* seeds has been owed to the presence of allyl isothiocyanate (Mejia-Garibay et al. 2015). *B. nigra* seeds, as well as many other types of seeds belonging to the Brassicaceae family, produces a significant amount of antifungal compounds such as sinigrin and sinapin (Mazumder et al. 2016). In addition, *B. nigra* oilseed possesses the ability to act as antifungal compounds especially against the growth of *Aspergillus niger*, *Aspergillus ochraceus* and *Penicillium citrinum* (Mejia-Garibay et al. 2015).

8.6.9 Protection Against Gastrointestinal Cancer

Sulphoraphane present in *B. nigra* seeds (10–20 μg/g) has healing properties and is best known for its treatment against breast, cervical, prostrate, colon and stomach cancers and rectum carcinomas (Augustine and Bisht 2015; Avato and Argentieri 2015). Studies also revealed that a diet rich with sulphoraphane acts as a defence against *Helicobacter pylori*, causes stomach ulcers as well as provides protection against other physiological disorders such as cystic fibrosis, ageing, rhinitis, arthritis, asthma and other lung disorders. In addition, *B. nigra* is useful in the management of abdominal swelling, intestinal worms and wounds (Manohar et al. 2009). Sinigrin present in *B. nigra* has the ability to act as an anti-proliferative agent by inhibiting the proliferation of liver tumour cells and apoptosis without causing any toxicity to the liver (Jie et al. 2014; Mazumder et al. 2016). Plants in the *Brassica* genus, including *B. nigra*, are also known to contain cyanides, thiocyanates and phenols, which help inhibit the mutagenicity of benzo[a]pyrene (B[a]P) and 3-methylcholanthrene in vivo and, therefore, are considered as carcinogenic metabolites (Polasa et al. 1994). Thus, the consumption of mustard powder and oil at very low level is recommended to be effective with the ability to inhibit carcinogenesis (Polasa et al. 1994).

8.6.10 Protection Against Cardiovascular Disease

The presence of antioxidants in *B. nigra* is known to help decrease the risk of cardiovascular disease in humans (Cartea et al. 2011; Diengdoh 2017). It is hypothesized that the presence of kaempferol imparts a protective effect for the prevention of coronary heart disease. Despite the presence of antioxidants in *B. nigra* seeds, excessive consumption in view of this particular health benefit may also cause deranging of blood, primarily due to the presence of minor amounts of cyanides and thiocyanates (Manohar et al. 2009).

8.7 Adverse Effects Resulting from Excessive Consumption of *B. nigra*

B. nigra seeds contain storage proteins belonging to the 2S albumin class. These proteins have been associated with the numerous allergic symptoms which were reported in Europe due to overconsumption of *B. nigra* seed oils. Some of the symptoms include acute giant urticaria with oedema of the glottis, vesicular hand eczema, recurrent cases of acute severe generalized urticaria and angioneurotic oedema, severe anaphylactic symptoms, contact dermatitis symptoms (vesicular dermatitis), contact urticaria, bronchial asthma and allergic rhinitis, anaphylactic shock, gastric pain, rhinitis, atopic dermatitis, contact dermatitis, angioedema, urticaria or anaphylactic shock, irritant contact dermatitis, acute generalized urticaria, facial and throat swelling, chest tightness, anaphylactic reaction symptoms, atopic dermatitis with respiratory symptoms, respiratory symptoms, atopic dermatitis, chronic hand eczema, oral pruritus, wheezing with shortness of breath, oedema and pruritus in the lips, oral mucosa and the pharynx, facial urticarial, atopic dermatitis, laryngeal oedema, angioedema, asthma and anaphylaxis and oral allergy syndrome (Monsalve et al. 2001).

8.8 Food Applications

B. nigra seeds and its oil have been used to produce a wide variety of commercial products – mostly food, some which are summarized in Table 8.2. The oilseed of *B. nigra* is primarily used in food products as a source of edible protein in the form of protein concentrates and protein isolates (Jahangir et al. 2009; Biswas et al. 2007). Other specific applications of *B. nigra* in the food industry are explained in detail from Sects. 8.8.1 to 8.8.4.

Table 8.2 Commercial products made by using *B. nigra* seeds and its oil

Product/Food	References
Babies and toddlers commercial foods	Monsalve et al. (2001)
Barbecue sauce	
Curry sauce	
Cumberland sauce	
Condiment sauce	Amare et al. (2015)
Flours for flavouring fried fish or meat	Monsalve et al. (2001)
Fruit juice	Amare et al. (2015)
Ketchup/tomato sauce	Monsalve et al. (2001)
Marinades	
Mayonnaise	
Meat (processed/sausages)	
Mustard paste (homemade, combining the powder with	
Wine or vinegar and oil)	
Mustard powder (additive to foods)	
Piccalilli/mustard pickle	Monsalve et al. (2001), Sõukand et al. (2015)
Salad oil	Gokhale et al. (2004), Monsalve et al. (2001)
Salad dressing	Monsalve et al. (2001)
Table mustard	Amare et al. (2015)
Vinaigrettes	Monsalve et al. (2001)
Industrial applications	
Industrial biofilms	Jahangir et al. (2009)
Lubricants	Monsalve et al. (2001)
Soap	Small (2009)

8.8.1 Culinary Preparations

Brassica oilseed species are important as a cooking source (Ashraf and McNeilly 2004). *B. nigra* oil, in particular, contains sinigrin, which is highly volatile and provides a pungent aroma and a unique taste and flavour to food products. This flavour itself stimulates gastric mucosa and increases pancreatic secretion, which helps in the digestion of food (Talapatra and Talapatra 2015).

8.8.2 Preservative Effects

B. nigra acts as a bio-preservative by eliminating food spoilage through antioxidant effects (Radhakrishnan et al. 2014). It also has the ability to act as an antibacterial agent in this aspect due to high content of sulphur (60 mg/kg) (Bhatia and Sharma 2012; Mazumder et al. 2016). In particular, due to the presence of sinigrin, it is able to act against pathogenic microorganisms by causing enzymatic inhibition and membrane damage. Also, allyl isothiocyanate, which is one of the degradation products of sinigrin, is related to the inhibition of deoxyribonucleic acid (DNA) synthesis of microbes. In addition, the antifungal activities of the *B. nigra* oilseed help in the inhibition of fungal attacks such as *Aspergillus niger*, *Aspergillus ochraceus* and *Penicillium citrinum* in food (Mazumder et al. 2016). The nonaqueous solvent extracts of *B. nigra* oilseeds contain a high amount of carotenes and carotenoids which help control food-borne pathogens such as *Staphylococcus aureus*, *Bacillus cereus* and *Clostridium perfringens* and prevent lipid oxidation (Radhakrishnan et al. 2014). Thus, *B. nigra* seed is commonly incorporated to fish, meat and vegetables as a bio-preservative.

8.8.3 Beverage Industry

Hardaliye, a famous lactic acid-containing fermented beverage commonly consumed in Turkey, is made from grapes and contains *B. nigra* seeds, sour cherry leaves and benzoic acid as flavouring agents (Coskun 2017; Gucer et al. 2009). In addition to being a flavouring agent, *B. nigra* also reduces the yeast fermentation due to the presence of etheric oils, which are able to block the formation of alcohol by inhibiting the growth of yeast. However, due to the presence of allyl isothiocyanate, it also imparts a bitter taste. Tocopherols (mainly α and β) in *B. nigra* seeds are also purported to increase the shelf life of hardaliye (Coskun 2017).

8.9 Other Applications of *B. nigra*

Various parts of the *B. nigra* plant, including its seed and oil, have been used for a wide variety of environmental applications, mostly for remedial purposes, some of which are described in Sects. 8.9.1–8.9.3. These applications demonstrate the versatility of the plant and its products for more than simply medicinal and condimental purposes.

8.9.1 *Phytoremediation*

In general, the Brassicaceae family—including *B. nigra* have the ability to accumulate heavy metals Pb, Cu, Zn, Cd, Cr, Fe, Mn and Ni in its roots (Angelova and Ivanov 2009; Mourato et al. 2015). Apart from the roots, other vegetative and reproductive organs of the plant are also able to accumulate these heavy metals at a significant level. Due to this characteristic, *B. nigra* is able to grow on contaminated soil and, therefore, would be a noteworthy candidate in its application for phytoremediation. The presence of cysteine (3–5 mg/g), a key component of phytochelatins, has the ability to detoxify metals and, thus, identified as the primary reason behind this beneficial characteristic of the plant (Angelova and Ivanov 2009; Purakayastha et al. 2008; Bharagava et al. 2008).

8.9.2 *Weed Control and Interaction with Herbivores*

The allelopathic activity of the *B. nigra* plant and its seed oil owes to the presence glucosinolates (Bialy et al. 1990; Harvey et al. 2003; Siemens et al. 2002; Tawaha and Turk 2003). *B. nigra* seeds also contain allyl isothiocyanate, which is a glucosinolate breakdown product known to be inhibitory to wheat, barnyard grass and lettuce but also able to depress seedling growth and to a smaller effect, wheat germination (Bialy et al. 1990; Siemens et al. 2002). Therefore, *B. nigra* seed oil could be used as a weed-controlling agent in agriculture due to its allelopathic properties. The young leaves of *B. nigra* are primarily known to possess allelochemicals in abundance (Merritt 1996; Tawaha and Turk 2003; Van Dam et al. 2005).

8.9.3 *Bio-fumigation*

As a bio-fumigation agent, *B. nigra* can be used to control pests including soilborne species in the agriculture and horticulture sector. The *B. nigra* plant contains many biologically active phytonutrients such as glucosinolates and their breakdown products (isothiocyanates) which have the potential to be used for bio-fumigation purposes (Mazumder et al. 2016). The presence of isothiocyanates in *B. nigra* is from glucosinolates hydrolysed by thioglucosidase (myrosinase) and a β-D-thioglucosidase which results in an aglycone, leading to breakdown products such as isothiocyanates, thiocyanates, nitriles, epithionitriles and oxazolidine-2-thione. Volatile compounds such as allyl cyanide and allyl nitrile and other products with similar chemical structures originating from the breakdown of glucosinolates are most suitable for use against soilborne pathogens such as *Aspergillus* spp. (Mazumder et al. 2016).

8.9.4 Disease Resistance

Most of the *Brassica* species including *B. nigra* are considered to possess naturally available disease-resistant traits which are essential for crop improvement. *B. nigra* is resistant to several common plant diseases such as white rust (*Albugo candida*), black leaf spot (*Alternaria* spp.), blackleg (*Leptosphaeria maculans*) and black rot (*Xanthomonas campestris*) (Warwick 2011).

8.10 Future Developments and Challenges

During the last two decades, novel developments in plant biology and biotechnology applications have been incorporated to expedite the agricultural breeding of *B. nigra*. These advancements facilitate the adoption of plant models which are economically important and help in enriching genetic, genomic, transcriptomic and metabolomic resources. Therefore, many sequencing projects on *B. nigra* have been recently initiated, to understand its evolution, gene functions and genomic structure. Usage of novel molecular biotechnological techniques and molecular cloning has led to the increase of tolerance factors of *B. nigra* plants, primarily against abiotic factors, as well as improving nutritional value and developing heterosis for yield enhancement. Biotechnological tools which have been introduced open new opportunities to the manipulation of genetic factors which lead to increased potential to produce hybrid seeds. One such developed technique is the barnase-barstar system, which has been widely studied for the development of male sterile (barnase) and fertility restorer (barstar) lines in crop plants including *B. nigra* (Augustine et al. 2014).

The reverse genetic approach for the improvement of *Brassica* crops, including *B. nigra*, has become possible during the past few decades, and this approach deals with the phenotypic effects of mutations in a gene. In addition, application of this technology is still in its initial stages and requires a high level of standardization (Augustine et al. 2014). Moreover, the use of introgression of genetic/genomic information from basic diploids and allopolyploid species into one type of plants to form a secondary gene pool is hypothesized to improve both the yield and resistance to disease in *B. nigra* cultivars. The advantage in this aspect when it comes to *B. nigra* is that the *Brassica* species can be crossbred with other members, thereby leading to better crops and superior agricultural produce (Pradhan and Pental 2011).

8.11 Conclusions

B. nigra remains an important crop when it comes to the oilseed industry. While its medicinal properties have been made evident through several in vitro and in vivo studies, through ascertaining the existence of bioactive compounds such as sinigrin and sinapin, its food and agricultural applications have been widely recognized in the recent past. However, when it comes to the functional properties of the plant, its seed and the oil, further investigations need to be targeted at isolation and identification of bioactive compounds of interest, following which clinical trials need to be conducted on their efficacy and dosages. Further research needs to be targeted towards the production of disease-resistant *B. nigra* crops with a higher produce as well. In terms of safety and consumption, awareness needs to be raised on possible toxic and allergenic effects of the seed and the oil. This aspect is relatively less known and, thus, warrants more attention in order to safeguard consumers from any adverse effects extending from consumption of the seed and the oil.

Conflicts of Interest The authors have no conflicts of interest to report, financial or otherwise.

References

Abdollahi Fard M, Shojaii A (2013) Efficacy of Iranian traditional medicine in the treatment of epilepsy. BioMed Res Int 2013:1–8. https://doi.org/10.1155/2013/692751

Alam MB, Hossain MS, Haque ME (2011) Antioxidant and anti-inflammatory activities of the leaf extract of *Brassica nigra*. Int J Pharm Sci Res 2(2):303–310

Alvarez PA, Boye JI (2012) Food production and processing considerations of allergenic food ingredients: a review. J Allergy 2012:1–14. https://doi.org/10.1155/2012/746125

Amare A, Hadush A, Aregawi H, Kide N (2015) Antibacterial activity of oil extracts of black mustard (*Brassica nigra*) seeds against bacteria isolated from fresh juice in selected areas of Axum town. Int J Integr Sci Innov Technol (IJIIT) 4(4):15–18

Anand P, Murali K, Tandon V, Chandra R, Murthy P (2007) Preliminary studies on antihyperglycemic effect of aqueous extract of *Brassica nigra* (L.) Koch in streptozotocin induced diabetic rats. Indian J Exp Biol 45(8):696–701

Anand P, Murali YK, Tandon V, Murthy PS, Chandra R (2009) Insulinotropic effect of aqueous extract of *Brassica nigra* improves glucose homeostasis in streptozotocin induced diabetic rats. Exp Clin Endocrinol Diabetes 117(06):251–256. https://doi.org/10.1055/s-2008-1080917

Andersen KE, Bjergegaard C, Møller P, Sørensen JC, Sørensen H (2005) Compositional variations for α-galactosides in different species of leguminosae, brassicaceae, and barley: a chemotaxonomic study based on chemometrics and high-performance capillary electrophoresis. J Agric Food Chem 53(14):5809–5817. https://doi.org/10.1021/jf040471v

Angelova V, Ivanov K (2009) Bio-accumulation and distribution of heavy metals in black mustard (*Brassica nigra* Koch). Environ Monit Assess 153(1):449–459. https://doi.org/10.1007/s10661-008-0370-y

Ashraf M, McNeilly T (2004) Salinity tolerance in Brassica oilseeds. Crit Rev Plant Sci 23 (2):157–174. https://doi.org/10.1080/07352680490433286

Augustine R, Bisht NC (2015) Biofortification of oilseed *Brassica juncea* with the anti-cancer compound glucoraphanin by suppressing GSL-ALK gene family. Sci Rep 5:2–12. https://doi.org/10.1038/srep18005

Augustine R, Arya GC, Nambiar DM, Kumar R, Bisht NC (2014) Translational genomics in Brassica crops: challenges, progress, and future prospects. Plant Biotechnol Rep 8(2):65–81. https://doi.org/10.1007/s11816-013-0298-8

Avato P, Argentieri M (2015) Brassicaceae: a rich source of health improving phytochemicals. Phytochem Rev 14(6):1019–1033. https://doi.org/10.1007/s11101-015-9414-4

Bharagava RN, Chandra R, Rai V (2008) Phytoextraction of trace elements and physiological changes in Indian mustard plants (*Brassica nigra* L.) grown in post methanated distillery effluent (PMDE) irrigated soil. Bioresour Technol 99(17):8316–8324. https://doi.org/10.1016/j.biortech.2008.03.002

Bhatia M, Sharma A (2012) *Brassica nigra* and *Cuminum cyminum*: inhibitors of food borne pathogens. Int J Appl Biol Pharm Technol 3(3):114–120

Bialy Z, Oleszek W, Lewis J, Fenwick G (1990) Allelopathic potential of glucosinolates (mustard oil glycosides) and their degradation products against wheat. Plant Soil 129(2):277–281. https://doi.org/10.1007/BF00032423

Björkman M, Klingen I, Birch AN, Bones AM, Bruce TJ, Johansen TJ, Meadow R, Mølmann J, Seljåsen R, Smart LE (2011) Phytochemicals of Brassicaceae in plant protection and human health–influences of climate, environment and agronomic practice. Phytochemistry 72 (7):538–556. https://doi.org/10.1016/j.phytochem.2011.01.014

Borş MD, Socaci S, Vicas SI, Anamaria C, Semeniuc AVPC, Nagy M, Tofană M (2015) Effect of light conditions on sulfur-containing phytochemicals from radish and mustard sprouts. J Agroaliment Process Technol 21(3):272–278

Cartea ME, Francisco M, Soengas P, Velasco P (2011) Phenolic compounds in *Brassica* vegetables. Molecules 16:251–280. https://doi.org/10.3390/molecules16010251

Chanda S, Chakrabarti S (1996) Plant origin liquid waste: a resource for single cell protein production by yeast. Bioresour Technol 57(1):51–54. https://doi.org/10.1016/0960-8524(96)00053-3

Coskun F (2017) A traditional Turkish fermented non-alcoholic grape-based beverage, “Hardaliye”. Beverages 3(1):1–11. https://doi.org/10.3390/beverages3010002

Diengdoh D (2017) Evaluation of phytochemicals and antioxidant activity in underutilised wild edible plants of Meghalaya state, India. Int J ChemTech Res 10(5):674–680

Gokhale S, Joshi P, Tatiya A, Bakliwal S, Fursule R (2004) Some neglected spices in India. Nat Prod Radiance 3(5):1–4

Gucer Y, Aydogdu H, Durgun T (2009) A traditional Thracian beverage: ‘hardaliye’. Trakia J Sci 7 (2):208–210

Harvey JA, Van Dam NM, Gols R (2003) Interactions over four trophic levels: food plant quality affects development of a hyperparasitoid as mediated through a herbivore and its primary parasitoid. J Anim Ecol 72(3):520–531. https://doi.org/10.1046/j.1365-2656.2003.00722.x

Hemingway JS (1976) Mustards: *Brassica* spp. and *Sinapis alba* (Cruciferae). In: Simmonds NW (ed) Evolution of crop plants. Longman, London, pp 56–59

Hirani AH, Li G, Zelmer CD, McVetty PB, Asif M, Goyal A (2012) Molecular genetics of glucosinolate biosynthesis in Brassicas: genetic manipulation and application aspects. In: Goyal A (ed) Crop plant. InTech, pp 189–216

Iriti M, Faoro F (2006) Grape phytochemicals: a bouquet of old and new nutraceuticals for human health. Med Hypotheses 67(4):833–838. https://doi.org/10.1016/j.mehy.2006.03.049

Jahangir M, Kim HK, Choi YH, Verpoorte R (2009) Health-affecting compounds in Brassicaceae. Compr Rev Food Sci Food Saf 8(2):31–43. https://doi.org/10.1111/j.1541-4337.2008.00065.x

Jie M, Cheung WM, Yu V, Zhou Y, Tong PH, Ho JW (2014) Anti-proliferative activities of sinigrin on carcinogen-induced hepatotoxicity in rats. PLoS One 9(10):1–9. https://doi.org/10.1371/journal.pone.0110145

Kapusta-Duch J, Kopec A, Piatkowska E, Borczak B, Leszczynska T (2012) The beneficial effects of Brassica vegetables on human health. Rocz Panstw Zakl Hig 63(4):389–395

Kiasalari Z, Khalili M, Roghani M, Sadeghian A (2012) Antiepileptic and antioxidant effect of *Brassica nigra* on pentylenetetrazol-induced kindling in mice. Iran J Pharm Res 11 (4):1209–1217

Lako J, Trenerry VC, Wahlqvist M, Wattanapenpaiboon N, Sotheeswaran S, Premier R (2007) Phytochemical flavonols, carotenoids and the antioxidant properties of a wide selection of Fijian fruit, vegetables and other readily available foods. Food Chem 101(4):1727–1741. https://doi.org/10.1016/j.foodchem.2006.01.031

Lee YH, Choo C, Waisundara VY (2015) Determination of the total antioxidant capacity and quantification of phenolic compounds of different solvent extracts of black mustard seeds (*Brassica nigra*). Int J Food Prop 18(11):2500–2507. https://doi.org/10.1080/10942912.2014.986331

MacKenzie SL, Blakely JA (1972) Purification and characterization of seed globulins from Brassica juncea, *B. nigra*, and *B. hirta*. Can J Bot 50(9):1825–1834. https://doi.org/10.1139/b72-229

Manchester LC, Tan DX, Reiter RJ, Park W, Monis K, Wenbo Q (2000) High levels of melatonin in the seeds of edible plants: possible function in germ tissue protection. Life Sci 76:3023–3029

Manohar P, Pushpan R, Rohini S (2009) Mustard and its uses in Ayurveda. Indian J Tradit Knowl 8 (3):400–404

Markose A, Krishnan R, Ramesh M (2016) Management of oral health through ayurvedic methods. J Dent Appl 3(2):319–321

Mazumder A, Dwivedi A, Du Plessis J (2016) Sinigrin and its therapeutic benefits. Molecules 21 (4):416. https://doi.org/10.3390/molecules21040416

Mejia-Garibay B, Palou E, Lopez-Malo A (2015) Composition, diffusion, and antifungal activity of black mustard (*Brassica nigra*) essential oil when applied by direct addition or vapor phase contact. J Food Prot 78(4):843–848. https://doi.org/10.4315/0362-028X.JFP-14-485

Merritt SZ (1996) Within-plant variation in concentrations of amino acids, sugar, and sinigrin in phloem sap of black mustard, *Brassica nigra* (L.) Koch (Cruciferae). J Chem Ecol 22 (6):1133–1145. https://doi.org/10.1007/BF02027950

Monsalve R, Villalba M, Rodríguez R (2001) Allergy to mustard seeds: the importance of 2S albumins as food allergens. Int Symp Food Allergens 3(2):57–69

Mourato MP, Moreira IN, Leitão I, Pinto FR, Sales JR, Martins LL (2015) Effect of heavy metals in plants of the genus Brassica. Int J Mol Sci 16(8):17975–17998. https://doi.org/10.3390/ijms160817975

Ng TB, Ng CCW, Wong JH (2013) Health benefits of *Brassica* species. In: Lang M (ed) Brassicaceae: characterization, functional genomics and health benefits. Nova Science Publishers, New York, pp 1–18

Obi R, Nwanebu F, Ndubuisi U, Orji N (2009) Antibacterial qualities and phytochemical screening of the oils of Cucurbita pepo and *Brassica nigra*. J Med Plants Res 3(5):429–432

Pecháček R, Velíšek J, Davídek J (2000) Decomposition of sinigrin by methanol/ammonia/water treatment in model systems and mustard (*Brassica nigra* L.) seed meal. Eur Food Res Technol 210(3):196–201. https://doi.org/10.1007/PL00005511

Polasa K, Kumar P, Krishnaswamy K (1994) Effect of *Brassica nigra* on benzo [a] pyrene mutagenicity. Food Chem Toxicol 32(8):777–781. https://doi.org/10.1016/S0278-6915(09)80011-2

Posmyk MM, Janas KM (2009) Melatonin in plants. Acta Physiol Plant 31(1):1–11. https://doi.org/10.1007/s11738-008-0213-z

Pradhan AK, Pental D (2011) Genetics of *Brassica juncea*. In: Schmidt R, Bancroft I (eds) Genetics and genomics of the Brassicaceae, vol 9. Springer, New York, NY, pp 323–345. https://doi.org/10.1007/978-1-4419-7118-0_11

Purakayastha T, Viswanath T, Bhadraray S, Chhonkar P, Adhikari P, Suribabu K (2008) Phytoextraction of zinc, copper, nickel and lead from a contaminated soil by different species of Brassica. Int J Phytoremediat 10(1):61–72. https://doi.org/10.1080/15226510701827077

Radhakrishnan K, Babuskin S, Babu PAS, Fayidh MA, Sabina K, Archana G, Sivarajan M, Sukumar M (2014) Bio protection and preservation of raw beef meat using pungent aromatic plant substances. J Sci Food Agric 94(12):2456–2463. https://doi.org/10.1002/jsfa.6580

Rajamurugan R, Selvaganabathy N, Kumaravel S, Ramamurthy C, Sujatha V, Thirunavukkarasu C (2012a) Polyphenol contents and antioxidant activity of *Brassica nigra* (L.) Koch. Leaf extract. Nat Prod Res 26(23):2208–2210. https://doi.org/10.1080/14786419.2011.637215

Rajamurugan R, Suyavaran A, Selvaganabathy N, Ramamurthy C, Reddy GP, Sujatha V, Thirunavukkarasu C (2012b) *Brassica nigra* plays a remedy role in hepatic and renal damage. Pharm Biol 50(12):1488–1497. https://doi.org/10.3109/13880209.2012.685129

Rakow G (2004) Species origin and economic importance of *Brassica*. In: Pua EC, Douglas CJ (eds) Biotechnology in agriculture and forestry, vol 54. Springer, Berlin, Heidelberg, pp 3–11. https://doi.org/10.1007/978-3-662-06164-0_1

Saha J, Biswal A, Deka S (2015) Chemical composition of some underutilized green leafy vegetables of Sonitpur district of Assam, India. Int Food Res J 22(4):1466–1473

Siemens DH, Garner SH, Mitchell-Olds T, Callaway RM (2002) Cost of defense in the context of plant competition: *Brassica rapa* may grow and defend. Ecology 83(2):505–517. https://doi.org/10.1890/0012-9658(2002)083[0505:CODITC]2.0.CO;2

Small E (2009) Top 100 food plants: the world's most important culinary crops. NRC Research Press, Ottawa, Ontario

Sõukand R, Pieroni A, Biró M, Dénes A, Dogan Y, Hajdari A, Kalle R, Reade B, Mustafa B, Nedelcheva A (2015) An ethnobotanical perspective on traditional fermented plant foods and beverages in Eastern Europe. J Ethnopharmacol 170:284–296. https://doi.org/10.1016/j.jep.2015.05.018

Srinivasan K (2005) Plant foods in the management of diabetes mellitus: spices as beneficial antidiabetic food adjuncts. Int J Food Sci Nutr 56(6):399–414. https://doi.org/10.1080/09637480500512872

Talapatra SK, Talapatra B (2015) Organic phytonutrients, vitamins, and antioxidants. In: Talapatra SK, Talapatra B (eds) Chemistry of plant natural products. Springer, Berlin, Heidelberg, pp 1011–1037. https://doi.org/10.1007/978-3-642-45410-3_34

Tawaha A, Turk M (2003) Allelopathic effects of black mustard (*Brassica nigra*) on germination and growth of wild barley (*Hordeum spontaneum*). J Agron Crop Sci 189(5):298–303. https://doi.org/10.1046/j.1439-037X.2003.00047.x

Thangaraj P (2016) Anti-nutritional factors. In: Rainsford KD (ed) Pharmacological assays of plant-based natural products, vol 71. Springer, Cham, pp 43–47. https://doi.org/10.1007/978-3-319-26811-8_7

Van Dam NM, Witjes L, Svatoš A (2004) Interactions between aboveground and belowground induction of glucosinolates in two wild *Brassica* species. New Phytol 161(3):801–810. https://doi.org/10.1111/j.1469-8137.2004.00984.x

Van Dam NM, Raaijmakers CE, Van Der Putten WH (2005) Root herbivory reduces growth and survival of the shoot feeding specialist Pieris rapae on *Brassica nigra*. Entomol Exp Appl 115 (1):161–170. https://doi.org/10.1111/j.1570-7458.2005.00241.x

Veromann E, Metspalu L, Williams I, Hiiesaar K, Mand M, Kaasik R, Kovacs G, Jogar K, Svilponis E, Kivimagi I (2012) Relative attractiveness of *Brassica napus*, *Brassica nigra*, *Eruca sativa* and *Raphanus sativus* for pollen beetle (*Meligethes aeneus*) and their potential for use in trap cropping. Arthropod Plant Interact 6(3):385–394. https://doi.org/10.1007/s11829-012-9191-6

Warwick SI (2011) Brassicaceae in agriculture. In: Schmidt R, Bancroft I (eds) Genetics and genomics of the Brassicaceae, vol 9. Springer, New York, NY, pp 33–65

Warwick S, Francis A, Al-Shehbaz I (2006) Brassicaceae: species checklist and database on CD-rom. Plant Syst Evol 259(2–4):249–258. https://doi.org/10.1007/s00606-006-0422-0

Chapter 9
Olive (*Olea europaea*)

Nicola Caporaso and Dimitrios Boskou

Abstract Olive oil is the product obtained from the olive fruit. Several grades can be obtained, extra virgin olive oil being the top category. The balanced fatty acid composition of olive oil, particularly its high content of oleic acid, makes this product nutritionally important. In addition, virgin olive oil contains bioactive compounds such as unsaponifiable constituents and phenolic compounds. Polar phenols and volatile compounds are two of the most interesting classes of compounds in olive oil. The scientifically proven beneficial health effects related to VOO consumption and the unique sensory properties of this product make it versatile for several culinary uses. This chapter reviews olive oil composition, focusing on the flavour-active compounds and their changes due to the variety, environmental conditions, processing and technological factors. The culinary uses of olive oils are then presented, with emphasis on traditional uses, possible food pairing and innovative applications including substitution of other fats and molecular gastronomy.

Keywords Olive oil · EVOO · Olive oil culinary use · Olive oil flavour · Natural phenolic compounds · Olive and cardiovascular diseases

N. Caporaso (✉)
Department of Agricultural and Food Science, University of Naples Federico II, Naples, Italy

Department of Food Sciences, School of Biosciences, University of Nottingham, Nottingham, UK

D. Boskou
Aristotle University of Thessaloniki, Hellas, Thessaloniki, Greece
e-mail: boskou@chem.auth.gr

B. Tanwar, A. Goyal (eds.), *Oilseeds: Health Attributes and Food Applications*,
https://doi.org/10.1007/978-981-15-4194-0_9

9.1 Origin and History

Olive is perhaps the most ancient plant known to be cultivated for its oil. The origin of the olive plants go back into the Greek mythology. The first evidence of cultivation is dated back to 7000 years ago in the Mediterranean Basin. 3000 years BC, this plant was already cultivated and exploited commercially in Crete. Considering the dramatic influence of olive and its products in the life and history of people living in the Mediterranean Basin, as well as its symbolic connotations, economical significance and influence on gastronomy, books on this plant have been written centuries or millennia ago. References to olive are found in Ancient Greek literature, e.g. Homer poems, as well as in the Ancient Roman writers, e.g. Pliny the Elder and Vitruvius who tried to "scientifically" report information on this plant, its cultivation and uses.

A chapter providing information on the origin of the olive plant has been authored by Polymerou-Kamilakis in the book *Olive Oil: Chemistry and Technology*, edited by Boskou (2006). It contains details about its history and myths related to the origin of olive growing, as well as the multiple traditional uses of products obtained from the olive plant. The oil extracted from the olive fruits was used in many applications, including heating and lighting with torches, as well as in cosmetics, spreading on the body or for haircare. Its symbolic meaning made olive oil a precious element in rituals and worship ceremonies.

Olive oil has been used since millennia for human nutrition, and extensive trade has been described since the Roman Empire, e.g. when the region of Andalucía in Spain produced olive oil that was exported to other regions including Rome.

A possible reason for which olive oil is probably one of the most ancient vegetable oils used for human consumption is the relative ease of oil extraction. The simplest mean to obtain olive oil was to crush the olive fruits and naturally separate the lipid fraction from the water and solid fraction by natural decantation. During the Roman age, a millstone crusher was used, and the crushed meal was separated by presses. A great innovation was obtained in the seventeenth century with the use of hydraulic press, while obvious advantages were given by the electricity in the modern era. One of the latest innovations in the olive oil technology was the introduction of centrifugal decantation systems during the 1960s, allowing continuous separation of the oil, water and solid fractions. The following sections of this chapter describe its composition, health benefits and food applications, as well as details on the systems currently available for olive oil production.

9.2 Production

Olive oil has a limited production with respect to other vegetable oils, but it has a paramount importance in terms of economic significance for the producing countries and interest for consumers worldwide. This is due to its health properties and unique sensory characteristics. The major olive oil production is localised in the Mediterranean area. The European Union (EU) is the first producer with almost 2.5 million tons in 2013–2014. From the non-European countries, Tunisia, Syria, Turkey and Morocco are the other important producing countries, whereas a great year-by-year variability is observed because of the olive plant physiology. The EU accounts for 80–90% of the world olive production. Spain, Italy and Greece specifically produce 45%, 25% and 20% of the world olive oil production, respectively (IOC 2015a).

Olive oil is contained mostly as vacuoles inside the mesocarp cells of the fruit, and very little amounts are found in the seed. The extraction herein reported is referred to as "virgin olive oil" (VOO), whose name comes from the fact that no chemical solvent or other chemical means is allowed for its extraction. Thus, by definition, VOO is obtained by only physical treatment of the olive fruits. The first processing step implies crushing of the olives, using a metallic "hammer"-based system (or other crushers) or a more traditional mill stone, to give the so-called olive paste. This is a mixture of solids, water contained in the drupe and the oil. The solids are removed by either pressing the olive paste using traditional hydraulic press systems or by more modern horizontal centrifuges that separate the different fractions according to their density. Two main centrifuge systems exist, i.e. "three-phase" or "two-phase", which are improperly named after the number of products obtained, i.e. water, oil and solids in the first case and oil and wet slurry named *alperujo* in the second case. A second centrifugation is often carried out to remove residual water in the oil. The final product is a turbid opalescent liquid that is finally clarified by deposition in the tanks or filtration. Unfiltered—veiled—virgin olive oil is generally richer in phenolic compounds, and it is popular among some consumers and chefs who judge the opalescent appearance as an indicator of higher wholesomeness and flavour. However, the unfiltered oil can undergo fermentative processes with consequent appearance of some sensory defects. Presently, large-scale production of cloudy oil is prevented because it has a limited stability due to sedimentation of the solid residues and water, with consequent quality deterioration and appearance of sensory defects (Tsimidou et al. 2005; Zullo and Ciafardini 2017).

A crucial step between olive crushing and the solid-liquid separation is "malaxation". This process is specific for virgin olive oil production and consists of a slow mixing of the olive paste at relatively high temperature (possibly not exceeding 28 °C to avoid excessive oxidation). This process is aimed to promote coalescence of fat droplets so that a higher extraction yield is obtained in the subsequent centrifugation. It has also a crucial importance in terms of phenolic and flavour compound formation, as most of the lipid oxidation reactions take place during malaxation, with formation of aroma compounds and dramatic change of phenolic compound composition. The influence of each of these steps on EVOO flavour is reported in the following paragraphs.

Detailed description of the different equipment available for olive oil extraction is reported in a book recently published by Peri (2014), which covers all aspects related to processing condition and olive mill equipment. The reader can also refer to Clodoveo et al. (2015) for information about the production steps in relation to minor bioactive constituents and new strategies to develop continuous plants.

In addition to VOO, olive pomace oil can be obtained by treatment of the residue or olive pomace, once VOO has been mechanically extracted. Olive pomace oil is obtained by solvent extraction and has to undergo refinement.

The following products can be obtained from the lipid fraction of the olive fruit: olive pomace oil, refined olive oil, olive oil, lampante olive oil, virgin olive oil (VOO), extra virgin olive oil (EVOO). Additionally, the International Olive Council (IOC) recognises another category named ordinary olive oil, which ranks between "virgin olive oil" and "lampante olive oil". Virgin olive oil is obtained by solvent extraction of the residue contained in the solids after the extraction of virgin olive oils. "Olive oil" is defined as the mixture of virgin olive oil with a non-well-defined amount of refined olive oil. The latter cannot be sold to the final consumer as it is, but it is used commercially for the preparation of admixtures with virgin olive oils. Similarly, consumers do not find "lampante olive oil" in the retail market. Lampante olive oil is obtained by mechanical extraction only but has such low-quality indices that cannot be consumed directly. VOO and EVOO are the top products in the family of oils obtained from the olive fruit. They are exclusively obtained by means of physical extraction of the oil fraction from the olive fruits. EVOO has more restrictive limits related to its physical properties and chemical composition and particularly to its sensory characteristics. Poor physical and chemical indices may be due to low-quality olive fruits, poor pre-extraction handling, incorrect extraction conditions or poor storage and packaging of the oil.

9.3 Chemical Composition

Extensive literature reports are available on olive oil composition(Aparicio & Harwood 2013); therefore, this paragraph will just give a brief summary and focus on the most prominent compounds. For a comprehensive overview on the topic, the reader can refer to Lanza (2012) and Boskou et al. (2006).

Olive oil—contrary to other vegetable fats which usually derive from seeds—is obtained from a fruit. The olive fruit is a drupe, which can be classified into two parts, namely, flesh and seed.

The major olive components are water (40–70%) and fat (6–25%), mainly present in the mesocarp. The fruit also contains simple sugars (2–5%), represented by glucose, fructose and mannose (Montaño et al. 2010). Other components are cellulose (approximately 6%), protein (1–2%) and ash (1–2%). Organic acids, nitrogen compounds, flavonoids and anthocyanins are also present in olive fruits. Particularly important for olive composition is the phenolic fraction, mainly constituted by oleuropein, dimethyloleuropein and verbascoside. These compounds are present in concentrations between 0.5% and 2.5% of the fresh weight. The oil fraction is present as oil droplet in the pulp (16.5–23.5% fresh weight), while small amounts are also found in the seed (1–1.5%).

While the olive fruit can be consumed after fermentation and other treatment aimed to reduce its strong bitterness caused by phenolic compounds, the main use of this product worldwide is for olive oil extraction.

The majority of the oil is contained in the olive flesh, and approximately 5% is contained in the seed. The composition of olive oil is dramatically different from the olive drupe. Triacylglycerols represent the majority of compounds in olive oil. Other constituents are “free” fatty acids (not bound to a glyceride molecule) and mono- and diacylglycerols present at low concentration, together with hydrocarbons, sterols, aliphatic alcohols, tocopherols and pigments (Boskou et al. 2006). Most of the so-called minor constituents of olive oils are often regarded as particularly important in terms of health properties and sensory effects of virgin olive oils (VOOs).

Tocopherols are antioxidant compounds found in appreciable amounts in olive oil, especially α-tocopherol, found at 150–250 mg kg^{-1}. Similar amounts have been reported in EVOOs for phytosterols, with β-sitosterol representing 90–95% the total sterolic fraction (Peri 2014). The analysis of sterols can be useful for authenticity purposes as a tool to differentiate oils from different sources, and in fact strict limits are applied also regarding the content of specific phytosterols in olive oils.

Oleic acid is the major fatty acid found in olive oil lipid fraction, and it is approximately 55.0–83.0% of the total lipid content (Boskou et al. 2006; Beltrán et al. 2004), with variations depending on the olive variety and agronomical factors. Palmitic (7.5–20.0%), palmitoleic (0.3–3.5%), stearic (0.5–5.0%), linoleic (3.5–21.0%) and linolenic acids are also present in appreciable amounts, while myristic, heptadecanoic and eicosanoic acids are found in trace amounts (Boskou et al. 2006).

Fatty acid composition of olive oils is known to vary within certain limits depending on several factors, including the olive variety, climate, altitude and other agronomic conditions, as well as the ripening degree of the olive fruits at harvest (Table 9.1). Established limits for fatty acid composition are included in the International Olive Council (IOC 2013) and the Codex Alimentarius (2003) norms as well as the European regulations (1991, 2002, 2013).

Tocopherols are found in relatively abundant concentrations in olive oils, with α-tocopherol being the most abundant one. The range of tocopherols can vary from approximately 50 mg/kg to above 350 mg/kg (Boskou et al. 2006; Caporaso et al. 2015a, b).

Sterols have an interest from a chemical point of view because they can be used as genuineness indicators, due to their dependence upon genetic factors, e.g. detection of sunflower oil added to olive oil. Limits of total sterols are set between 1000 and 2000 mg/kg. High levels are an indicator of admixture with lampante oils, while a lower content reveals the presence of refined olive oils (a process forbidden for virgin olive oils) (Table 9.2).

Table Olives

In addition to olive oil production, table olives are another product obtained from the olive tree. They can be consumed as they are, after a fermentation process aimed to lower their strong bitterness and make them edible (Montaño et al. 2010). Table olive processing is based on either using a brine treatment or alkali treatment

Table 9.1 Fatty acid and sterol composition of olive oil depending on the category (modified from EC Reg. 1348/2013)

	Minor fatty acid composition (% of total fat)						Sterols (% of total sterols)						
Category	Myristic (%)	Linolenic (%)	Arachidic (%)	Eicosenoic (%)	Behenic (%)	Lignoceric (%)	Cholesterol (%)	Brassicasterol (%)	Campesterol (%)	Stigmasterol (%)	β-Sitosterol (%)	Δ-7-Stigmastenol (%)	Total sterols (mg/kg)
1. Extra virgin olive oil	≤0.03	≤1.00	≤0.60	≤0.40	≤0.20	≤0.20	≤0.5	≤0.1	≤4.0	<Camp.	≥93.0	≤0.5	≥1000
2. Virgin olive oil	≤0.03	≤1.00	≤0.60	≤0.40	≤0.20	≤0.20	≤0.5	≤0.1	≤4.0	<Camp.	≥93.0	≤0.5	≥1000
3. Lampante olive oil	≤0.03	≤1.00	≤0.60	≤0.40	≤0.20	≤0.20	≤0.5	≤0.1	≤4.0	–	≥93.0	≤0.5	≥1000
4. Refined olive oil	≤0.03	≤1.00	≤0.60	≤0.40	≤0.20	≤0.20	≤0.5	≤0.1	≤4.0	<Camp.	≥93.0	≤0.5	≥1000
5. "Olive oil"	≤0.03	≤1.00	≤0.60	≤0.40	≤0.20	≤0.20	≤0.5	≤0.1	≤4.0	<Camp.	≥93.0	≤0.5	≥1000
6. Crude olive-pomace oil	≤0.03	≤1.00	≤0.60	≤0.40	≤0.30	≤0.20	≤0.5	≤0.2	≤4.0	–	≥93.0	≤0.5	≥2500
7. Refined olive-pomace oil	≤0.03	≤1.00	≤0.60	≤0.40	≤0.30	≤0.20	≤0.5	≤0.2	≤4.0	<Camp.	≥93.0	≤0.5	≥1800
8. Olive-pomace oil	≤0.03	≤1.00	≤0.60	≤0.40	≤0.30	≤0.20	≤0.5	≤0.2	≤4.0	<Camp.	≥93.0	≤0.5	≥1600

Table 9.2 Quality parameters for olive oil depending on its category (modified from EC Reg. 1348/2013)

Category	Acidity (%)	Peroxide index	K_{232}	K_{270}	ΔK	Sensory evaluation of defects	Sensory evaluation of fruity
Extra virgin olive oil	≤ 0.8	≤ 20	≤ 2.50	≤ 0.22	≤ 0.01	Md = 0	Mf > 0
Virgin olive oil	≤ 2.0	≤ 20	≤ 2.60	≤ 0.25	≤ 0.01	Md $\leq$ 3.5	Mf > 0
Lampante olive oil	> 2.0	–	–	–	–	Md > 3.5	–
Refined olive oil	≤ 0.2	≤ 5	–	≤ 1.10	≤ 0.16	–	–
"Olive oil"	≤ 1.0	≤ 15	–	≤ 0.90	≤ 0.15	–	–
Crude olive-pomace oil	–	–	–	–	–	–	–
Refined olive-pomace oil	≤ 0.3	≤ 5	–	≤ 2.00	≤ 0.20	–	–
Olive-pomace oil	≤ 1.0	≤ 15	–	≤ 1.70	≤ 0.18	–	–

Acidity is calculated as % oleic acid. Peroxide index is expressed as mEq O_2/kg. K_{232}, K_{270}, and ΔK are obtained by spectrophotometric measurement. The values of sensory scores indicate the median of each attribute

(e.g. NaOH); however, many variants exist depending on the location, traditional practices and the desired characteristics of the final product (Piperno et al. 2004). The composition of table olives can vary depending on the processing, as shown in Table 9.3.

Table olives can show a wide range of colours, from green to turning colour or completely black. They can undergo fermentation, but it is not necessary when methods involving only the use of alkaline treatments are applied. Sevillan of Spanish-style olives are green olives debittered with NaOH, followed by several washing treatments to remove the sodium hydroxide. The olives are eventually stored in brine to allow fermentation. Castelvetrano-style olives are produced by adding NaOH and salt almost at the same time, by leaving the alkaline brine for 8–10 days. Compared to the Spanish style, Castelvetrano-style olives are softened more quickly and have a deeper green colour. Other traditional treatments include the use of lime and ash, which have been historically used as a way to alkalinise the olives in order to debitter the product. The "natural turning colour olives" are produced using sodium chloride only, to stimulate the microflora that starts the fermentation process, typically under anaerobic conditions. Itrana-style green olives are produced by using an initial immersion in water (without salt) for 10–30 days, after which salt is added and fermentation takes place up to 6 months. Other typical table olives are produced by manually crushing the fruit after harvesting and washing the olives several times by leaving them in water. Another production method includes the use of very well ripe olives (black colour), which are packed

Table 9.3 Chemical composition of table olives depending on the production process (from Lanza 2012)

Nutrients/ 100 g e.p.	Sevillan green olives *Intosso*	Sevillan green olives *Bella di Cerignola*	Castelvetrano green olives *Nocellana B.*	Ferrandina black olives *Majafica*	Natural black olives *Taggiasca*	Natural black olives *Peranzana*	Natural black olives *Itrana*	Natural green olives *Itrana*	Natural black olives *Cellina N.*
Energy (kcal)	190	164	204	455	226	247	235	193	223
Proteins (g)	1.0	1.2	1.0	2.2	1.5	1.7	1.4	1.5	1.3
Carbohydrates (g)	2.8	2.5	3.6	n.d.	8.9	5.8	6.5	5.0	7.2
Sugars (g)	tr	<0.6	0.4	4.4	tr	0.6	0.3	0.6	1.7
Fats (g)	17.5	15.5	19.8	46.9	19.9	23.2	21.7	17.7	19.9
SFA (g)	2.7	2.1	3.9	6.3	3.7	4.1	2.7	2.8	4.4
MUFA (g)	13.6	12.5	13.9	36.7	15.2	17.0	17.7	14.0	14.5
PUFA (g)	1.2	0.9	2.0	4.0	0.9	2.1	1.3	0.9	1.0
Fibre (g)	2.6	4.8	3.8	3.4	2.6	4.1	4.0	3.6	4.8
Sodium (g)	1.3	1.1	0.9	0.9	1.8	1.4	1.5	1.2	1.5
Calcium (mg)	33.6	34.9	n.d.	168.1	92.7	83.1	28.9	21.9	58.7
Polyphenols (mg)	168	104	24	263	206	334	211	109	299

in salt and spices. Additionally, other typical Italian olives are used as stuffed olives (Ascolana-style), from a variety that gives big fruits with thick and crunchy mesocarp (Lanza 2012).

Despite the fact that the olive drupe is technically a fruit, it has very low carbohydrate content and can be considered as sugar-free when it is in the form of edible table olives, due to the loss of sugars during the fermentation process. Table olives are also a "source of fibre" according to the European legislation, the minimum limit being 3 g fibre per 100 g product. In terms of mineral content, table olives contain a good amount of calcium, of about 100 mg per 100 g. It is important to note that salt is always added for table olive preparation so the sodium content is high, usually above 1.5 g/100 g. Table olives are rich in antioxidants and vitamins, including vitamins B, A and E. The content of vitamin C is low (<1 mg/kg), but some producers add it as a preservative, thus higher contents might be found. The amount of phenolic compounds in table olives is higher than the one in virgin olive oils, with concentrations ranging from 100 to 350 mg per 100 g of product. Organic acids such as oxalic, succinic, malic, citric and lactic are found in rather low content (4–10 g/kg, expressed as citric acid). The energy value of table olives has been reported to be 200–250 kcal per 100 g, with a few exceptions including *Majatica* olives (above 350 kcal) and *Bella di Cerignola* (160 kcal) (Lanza 2012).

9.3.1 Olive Oil Minor Constituents

The study of volatile and non-volatile minor compounds in olive oils has been extensively used to evaluate their quality or to describe peculiarities of some olive varieties (Boskou 2008). For example, some differences in the concentration of phenolic compounds, sterols and the linoleic/oleic acid ratio have been reported among the Spanish cultivar Arbequina compared to the Italian cultivar Coratina and the Greek cultivar Koroneiki (Aparicio et al. 1997).

Hundreds of volatile compounds have been identified in virgin olive oils, but similar to other food products, only a limited number of compounds exert an odour impact. The number of odour-active compounds at the usual concentrations found in VOOs is approximately 60 (Angerosa et al. 1996), while the number of "potent odorants" is more limited (Guth and Grosch 1991).

Extensive research carried out over the past 20 years, mostly by Italian, Spanish and Greek research groups, allowed to get an insight into the complex VOO volatile composition and to understand the formation pathways of aroma compounds. Some volatile compounds are still present in the fruit, as they originate from the lipid or amino acid metabolism (Conde et al. 2008; Kalua et al. 2007). However, the majority of the olive aroma derives from enzymatic oxidation of the fatty acids linoleic and linolenic (Angerosa 2002).

VOO volatile compounds are mostly C5 and C6 molecules, the majority of which arises from the so-called lipoxygenase pathway. This name was given because lipoxygenase is one of the first enzymes to act, by catalysing the deoxygenation of polyunsaturated fatty acids when the olive drupe is crushed, and it is released from the cell membrane, so it is exposed to the lipids in the vacuole, after the action of acyl-hydrolase. Volatile compounds such as hexanal, 1-hexanol, hexyl-acetate, cis-/trans-3-hexenal, cis-/trans-3-hexen-1-ol and cis-/trans-2-hexenyl-acetate are biosynthesised in this reaction. Compounds such as hexanal, (E)-2-hexenal, (Z)-3-hexenal, hexan-1-ol, (Z)-3-hexen-1-ol, hexyl acetate and (Z)-3-hexenyl acetate are the most abundant volatiles in VOOs (Angerosa et al. 1999; Vichi et al. 2003).

Among volatile compounds, those related to the sensory note described as "green" are C6 aliphatic compounds, hexanal, cis-3-hexenal, trans-2-hexenal, hexanol, cis-3-hexenol, trans-2-hexen-1-ol and the corresponding esters (Gomes da Silva et al. 2011).

The following volatile compounds account for approximately 80% of the total:

- C6 aldehydes (hexanal, *cis*-3-hexanal, *trans*-2-hexanal)
- Alcohols (hexanol, *cis*-3-hexenol, *trans*-2-hexenol)
- Esters (hexyl acetate, *cis*-3-hexenyl acetate (Sánchez and Harwood 2002; Conde et al. 2008; Angerosa 2002)

C6 volatile compounds are the major components of VOO volatiles and mainly contribute to its green odour (fruity and/or green olives) (Guth and Grosch 1991; Morales et al. 1997). The essential role played by hexanal and *trans*-2-hexenal on the green attributes was confirmed by Angerosa et al. (2000), who also reported the correlation between fruity, leaf, almond, banana, freshly cut grass, walnut husk, wild flowers, tomato, bitter, pungent and sweet attributes and the concentrations of compounds deriving from the LOX pathway and the total amount of the secoiridoids.

Some correlations between the level of total phenolic compounds and aroma compounds with sensory descriptors were also found in VOO. In particular, higher concentration of 1-penten-3-one and phenolic compounds causes the increase of leaf odour, and an increase in the phenolic concentration causes the increase of walnut husk odour (Angerosa et al. 2000).

9.3.1.1 Phenolic Compounds

Olive oil is rich in bioactive compounds, which are commonly classified as non-polar, found in the unsaponifiable fraction (squalene, tocopherols, sterols and triterpenic compounds), and polar compounds. The second group comprises phenolic compounds, which are often improperly referred to as "polyphenols". Other authors refer to them as "biophenols" to emphasise their natural occurrence, or generally phenolic compounds (Visioli et al. 2002).

The characteristic bitter taste of fresh olive fruits is due to the high content of phenolic compounds. VOOs also possess a considerable charge of phenolic compounds resulting in a more or less pronounced bitter-pungent character. The bitter sensation of olive oils has been associated with secoiridoid derivatives of hydroxytyrosol, 3,4-dihydroxyphenolethanol (3,4-DHPEA) (Gutiérrez-Rosales et al. 2003). On the other hand, the pungent note in olive oil has been mainly attributed to the dialdehydic form of decarboxymethylelenolic acid linked to tyrosol, oleoocanthal, the dialdehydic form of decarboxymethyl ligstroside aglycone (p-HPEA-EDA) (Andrewes et al. 2003). These authors reported that the deacetoxy-ligstrosideaglycone has strong burning pungent sensation, while oleuropein aglycone has slightly weaker pungent notes, and tyrosol shows astringent but not pungent mouth feelings. On the contrary, the bitterness has been attributed to the aldehydic form of oleuropein aglycone (Mateos et al. 2004).

Several types of phenolic compounds are found in the olive fruits, including anthocyanins (cyanidin glucosides); flavonols (mainly quercetin-3-rutinoside); flavones (luteolin and apigenin glucosides); phenolic acids (hydroxybenzoic, hydroxycinnamic, others); phenolic alcohols (tyrosol, hydroxytyrosol, 3,4-dihydroxyphenylglycol); secoiridoids (oleuropein, demethyloleuropein, ligstroside, nuzhenide); verbascoside, a hydroxycinnamic acid derivative; and lignans and oleoside-11-methylester (Boskou 2006, 2015). The secoiridoid compounds found in the olive drupe, ligstroside or oleuropein glucoside, undergo degradation, in particular spontaneous hydrolysis, and therefore the phenolic compounds found in the resulting olive oils show a different profile. The phenolic fraction of olive oil is constituted by several dozens of molecules, commonly classified into phenolic acids, phenolic alcohols, secoiridoids, flavonoids and lignans. A wide range of total phenolic compounds has been described for EVOOs, from approximately 100 mg kg^{-1} up to almost 1000 mg kg^{-1} (Kotsiou and Tasioula-Margari 2016; Caporaso et al. 2015a, b). A scheme reporting the structure of several phenolic compounds found in olive oil is shown in Fig. 9.1.

Oleuropein is easily degraded into simpler phenolic compounds during olive oil extraction and storage of the product. In particular, hydrolysis due to the enzyme β-glucosidase takes place at the molecular bond between the sugar β-glucopyranose and the remaining molecule, giving the oleuropein aglycone. Similarly, an aglycone originates from ligstroside, another compound with bitter-pungent character. A second hydrolysis leads to the formation of hydroxytyrosol, which has been reported to exert strong antioxidant capacity. In VOOs, the main phenolic compounds are the dialdehydic forms of elenolic acid linked to tyrosol (p-HPEA-EDA) and hydroxytyrosol (3,4-DHPEA-EDA), and ligstroside aglycons, with more limited concentrations of phenolic acids. Tyrosol, hydroxytyrosol and their secoiridoid derivatives are the most abundant compounds, reaching up to 90% of the total phenolic content.

Generally, olive oil phenolic compounds with the hydroxylic groups in *ortho* position show higher antioxidant capacity; thus, compounds such as 3,4-dihydroxyphenylethanol-elenolic acid and 3,4-dihydroxyphenylethanol-elenolic acid dialdehyde are the molecules with the highest expected antioxidant activity in virgin olive oils.

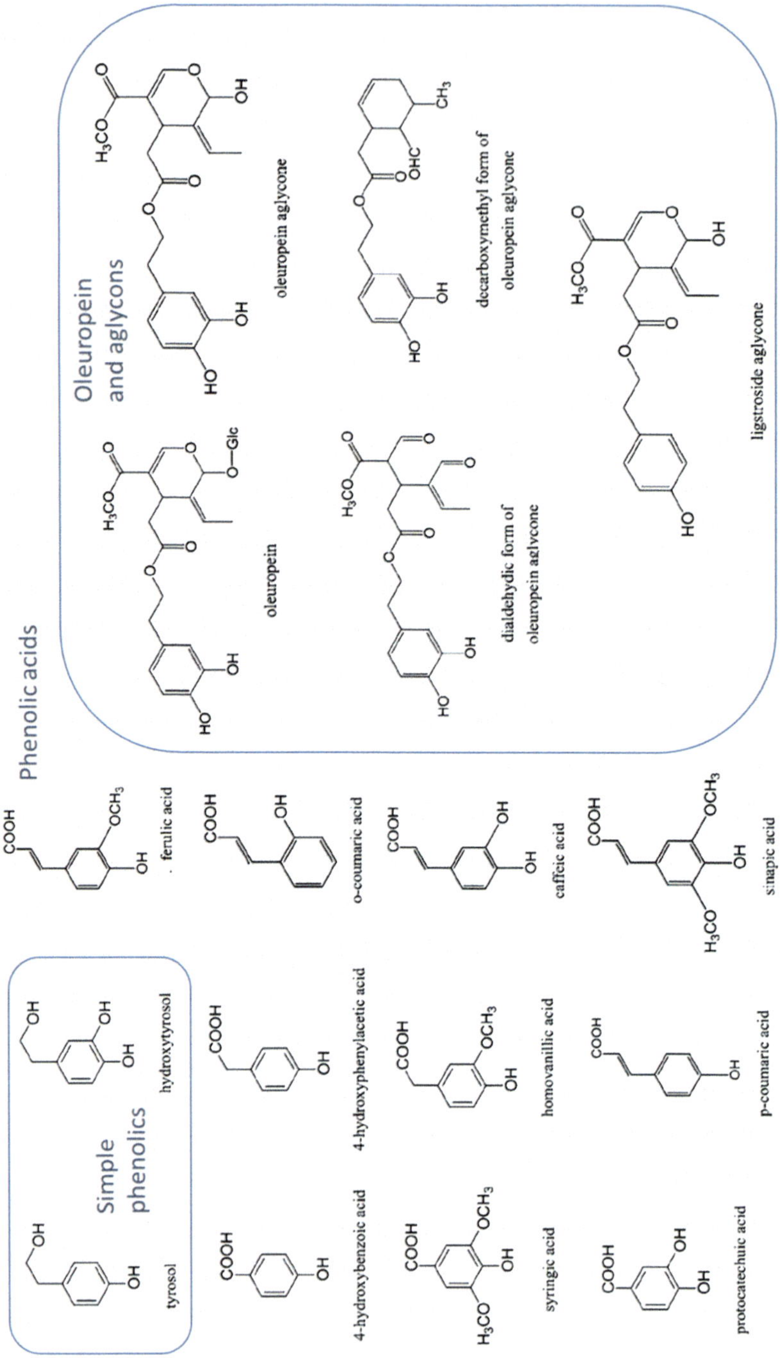

Fig. 9.1 Structure of some phenolic compounds found in virgin olive oil (modified from Boskou et al. 2006)

9.3.2 Compositional Changes due to Botanical and Agronomical Factors

9.3.2.1 Variety

Many factors can influence VOO composition, including the agronomical practices and extraction technology. The olive variety, or "cultivar", the climatic conditions and geographical location of the olive orchard and olive ripening degree influence the final composition of the oil and therefore its flavour (Angerosa et al. 1999; Malheiro et al. 2015; Kalogeropoulos and Kaliora 2015; García-González et al. 2010; Issaoui et al. 2009).

Olive cultivar is one of the most crucial factors responsible for olive oil flavour variability (Tura et al. 2008). Virgin olive oil extracted in the same conditions can show dramatic differences in its composition and sensory characteristics when obtained from different varieties. Volatile compounds C6, especially trans-2-hexenal, have been regarded as good indicators for differentiation of single-variety olive oils. The olive variety also influences the concentration of phenolic compounds; the major influence has been attributed to the olive ripening degree and environmental conditions (Škevin et al. 2003). At the same ripening degree, cv. Coratina and Moraiolo were reported to show higher phenolic concentration than other Italian varieties such as *Leccino*, *Frantoio*, *Carolea* and *Ogliarola Barese* (Gucci et al. 2007). Dhifi et al. (2004) reported the concentration of aroma compounds in VOO extracted from four Tunisian olive cultivars *Chetoui*, *Chemlali*, *Chemchali* and *Oueslati*. A wide range of differences was described depending on the variety. Other researchers compared Tunisian cultivars with Italian ones, reporting that cv. *Chetoui* is characterised by a low content of C6 aldehydes, responsible for the sensory notes "green" and "fruity" (Baccouri et al. 2008). According to Tura et al. (2008), the major volatile compounds affected by the olive cultivars are the following: ethanol, 2-methyl propanol, pentanol, cis-2-penten-1-ol, cis-3-hexenol and octanol. These compounds have been associated with the following sensory notes, respectively: floral, banana, apple, walnut, hay, butter, sweet and fruity.

9.3.2.2 Ripening Degree

The oil accumulation in the olive drupe starts about 5 weeks after anthesis, and its major increase is between 60 and 120 days after flowering (Conde et al. 2008). During the olive maturation phase, the fruit undergoes several modifications: the exocarp colour changes from green to violet-black. A decrease in fruit consistency, phenolic compound concentration and pigmentation takes place, with an increase of free acidity.

The oil quality is strictly linked to the fruit physiological conditions, as the progress of the ripening phase causes higher production of free fatty acids with a parallel decrease of the bitter note linked to the phenolic compounds (Garcia, Seller and Perez-Camino 1996).

Olive maturity degree also influences the generation of aroma compounds. The maximum concentration is observed when the colour changing phase starts in the fruit.

It is known that at higher ripening stages, the aromatic note described as "fruity-grassy" is significantly lower than in oils obtained from greener olives. The sensory notes of bitter-pungent and other positive aromatic notes decrease with increasing ripening degree. The increase in olive maturity leads to the increase in 1-penten-3-ol content. The compounds hexanal, *trans*-3-hexenol, *cis*-3-hexen-1-ol and *cis*-2-hexenol were reported to have high selectivity for unripe olives. The intermediate maturation phase was characterised using hexyl acetate, while overripe drupes can be discriminated using *trans*-3-hexenal, *cis*-2-hexenal, *trans*-3-hexenol and *cis*-3-hexenol (Aparicio and Morales 1998).

Some triterpenic alcohols were proposed as markers of fruit maturity. Sakouhi et al. (2009) demonstrated that triterpenic alcohols and 4-monomethyl sterol concentrations in the olive fruit increase from 18% to 30% during ripening of the cv. Picholine.

A negative correlation ($R^2 = -0.88$) between the drupe ripening degree and phenolic content in VOO was also reported by Rotondi et al. (2004). Gomez-Rico et al. (2006) studied the combined effect of maturity and irrigation on the phenolic and volatile compounds of VOOs obtained from cv. *Cornicabra*, showing a significant loss of the most complex phenolic compounds, also linked to the irrigation conditions in the field.

During drupe maturation, an increase of the triterpenic alcohol 24-methylene-cycloartenol has been reported. In addition, sterols undergo significant changes, e.g. increasing of Δ5-avenasterol and decreasing of β-sitosterol (Luna 2002).

Tura et al. (2008) reported the importance of the maturity degree on the content of phenolic and volatile compounds, but fruit storage can also result in a change of the flavour profile, due to the lowering of esters and aldehydes responsible for the positive notes (Kalua et al. 2007).

Longer storage of the olive fruits and increased ripening degree lead to a decrease of the content of volatile compounds, particularly those responsible for the sensory notes of fruity, almond, green (*trans*-2-hexenal, hexanal, *cis*-3-esenolo) and sweet (1-penten-3-one, 3-pentanone), with a contemporary increase of some alcohols (3-methyl-1-butanol, 2-methyl-1-propanol).

The olive fly (*Bactrocera oleae*) is the most important biotic stress factor affecting the drupe quality. VOO produced from fruits attacked by the olive fly has higher free acidity and peroxide number and different fatty acid composition and concentration of phenolic compounds (Gomez-Caravaca et al. 2008). Significant differences from the sensory panel of these oils have also been reported (Angerosa et al. 1992). The sensory profile of such VOOs has lower scores when the olives are harvested at full maturity stage, characterised by an increase of negative sensory notes such as fusty, mouldy and winey-vinegary. Olive fly attack causes a significant decrease of the attribute "pungent-bitter" (Tamendjari et al. 2009).

9.3.2.3 Environmental Conditions

Olive is a xerophyte plant, and it needs a mild climate, as it is particularly susceptible to fast temperature drops below −8 °C. Its full potential is described to be between 22 and 32 °C. At higher latitudes, an increase in oleic acid content and a higher ratio of unsaturated to saturated fatty acids are typically expected. The environmental conditions can decrease the activity of certain hydrolytic enzymes which act on phenols (Patumi et al. 1999).

Several studies have been conducted on the effect of soil salinity and irrigation water on the production and VOO fatty acid profile, but so far scarce and contradictory results have been published in relation to the effects on organoleptic characteristics. Ahmed et al. (2009) reported an increase in the concentration of phenolic compounds such as tyrosol, hydroxytyrosol and vanillic acid, and a lowering in the ratio of unsaturated to saturated fatty acids, with an increase in water salinity.

The influence of the organic agronomical practice on the VOO composition and flavour characteristics has been poorly investigated. Ninfali et al. (2008) reported higher concentrations of phenolic compounds from organic than non-organic VOOs in the first year of the experiment, while this difference was not significant over the second and third year. Organic VOOs have shown a stronger note of "hay" and "artichoke". Non-organic oils presented more markedly floral notes of "fresh grass" and "fruity". However, the results are not consistent in all the years and are strongly influenced by climate change over the years (Ninfali et al. 2008).

Water deficit represents one of the main environmental limiting factors on the olive production potential, caused by the decrease of the fruit endogenous esterases. Rainfall has prominent influence on EVOO volatiles. Compounds such as hexanal and isobutyl acetate were negatively correlated to the rainfall. The amount of water applied to the orchard can modify the fatty acid ratio; a higher concentration of monounsaturated and polyunsaturated fatty acids is obtained at higher water volumes applied (Salas and Sanchez 1997). A clear inverse correlation was also reported between the amount of irrigation water applied and the intensity of the EVOO bitter note. Gomez-Rico et al. (2006) indicated that phenolic compounds are strongly affected by the water availability during fruit development. The authors used three water management systems: no added water, fully water addition and controlled deficit irrigation according to the method proposed by FAO. It was shown that polyphenol content decreases at increased water supplied to the orchard, and secoiridoid derivatives of hydroxytyrosol decreased more than those of tyrosol. However, increase of volatile compounds was obtained under irrigation compared to rain-fed conditions.

Greater water availability seems also to affect favourably the concentration of volatile compounds, especially trans-2-hexenal, cis-3-hexen-1-ol and hexanal. Some other volatile compounds were not reported to be subject to variations; these are 1-penten-3-one and 1-penten-3-ol, which are linked more to seasonal factors and ripening degree than to the irrigation practices (Gomez-Rico et al. 2006). In subsequent studies, the same authors (Gomez-Rico et al. 2009) verified that the hexanal concentration also shows an inverse correlation with increasing water availability.

Opposite conclusions were reported by Stefanoudaki et al. (2009). Lower values of 1-penten-3-ol and 1-penten-3-one were found as a result of the irrigation, for cv. Leccino. A decrease of total volatile compounds was reported in irrigated olive oils. The C5 compounds undergo a drastic decrease, but, in contrast to other authors, no increases were reported in this study for C6 compounds.

The irrigation strategy has been reported to be more effective than the total amount of rainfall during the year, as the concentration of phenolic compounds depends on the water stress (Gomez-Rico et al. 2006; Stefanoudaki et al. 2009). Also fatty acid composition undergoes modification, with an increase in the ratio of monounsaturated to polyunsaturated fatty acids (Tognetti et al. 2007). The sensory evaluation of VOOs from irrigated orchards seems to show lower intensity of the attributes fruity, bitter and pungent (Stefanoudaki et al. 2009).

Montedoro et al. (1978) reported the possibility to discriminate VOOs from different Italian regions through the analysis of volatile compounds. VOOs from Italy were found to be richer in C6 aldehydes and lower in fruity esters than Moroccan oils (Kalua et al. 2007).

9.3.3 Compositional Changes due to Processing

9.3.3.1 Harvesting Systems and Crushing

The olive harvesting system has a dramatic influence on the final flavour of VOO. Oils obtained from olives picked from the soil are qualitatively poorer as shown by the chemical parameters (acidity, peroxides, UV) and the organoleptic analysis. The concentration of *trans*-2-hexenal and *cis*-3-hexenyl acetate ("green notes") in the oil decreases, while hydrocarbons, acetic acid and other carbonyl compounds are found at higher concentrations when there is a longer contact of the olives with the soil (Angerosa et al. 2004).

Mechanical harvesting of the olives has been reported to exert little influence on the chemical composition of VOOs compared to manual harvest. Olive harvesting needs to be performed in a way that damage of the fruit, spontaneous fermentation and mould growth are avoided.

The majority of VOO aroma compounds are produced during and after the crushing phase, when the enzymatic oxidation of linoleic and linolenic acid takes place (Baccouri et al. 2008). The type of mill influences the final oil quality, by changing the quali-quantitative composition of some minor constituents, with consequent influence on the organoleptic properties and product stability.

There are two major types or olive crushers: granite stone mill and metallic crusher. This latter can be divided into two types, i.e. disc and hammer crushers. The hammer crushers have stronger effects and cause a strong emulsification and an increase of the paste temperature, due to friction phenomena. They have the disadvantage of lower yield. It should be also noted that the stone mill is a discontinuous system, while the metallic crusher systems are continuous and therefore their performance in terms of processed olive is much higher.

VOO obtained with stone crusher is more aromatic and harmonic than oils obtained using the other two types of crushers. The disc crusher, however, leads to the formation of oil richer in phenolic compounds, more bitter and more stable to lipid oxidation over storage (Angerosa and Di Giacinto 1995).

Stone mill can lead to higher concentration of volatile compounds than metallic crushers, e.g. trans-2-hexenal, hexanal and cis-3-hexenol. For cv. Coratina and Oliarola, Servili et al. (2003) reported that disc crusher caused higher concentrations of C6 aldehydes and some esters such as hexyl acetate, 3-hexenyl acetate and cis-4-hexenyl acetate than hammer crusher.

Olive stoning and other challenges for the development of new crushing systems to reduce undesirable oxidation reactions and minimise enzymatic reactions have been reviewed by Clodoveo et al. (2015) and Veneziani et al. (2016).

9.3.3.2 Malaxation and Extraction

Malaxation is a fundamental stage in VOO extraction, as it influences its yield and composition and sensory profile of the product. It involves the continuous slow mixing of the olive paste. Malaxation time and temperature are parameters that are usually controlled by the industry during the VOO extraction process (Kalua et al. 2007; Clodoveo et al. 2015; Veneziani et al. 2016).

Malaxation time has been reported to be positively correlated to the total content of volatile compounds, but negatively correlated with the concentration of total phenolics (Ranalli et al. 2001). Longer malaxation times promote the accumulation of alcohols and C5 compounds, particularly hexanal. The temperature increase speeds up the activity of oxidative enzymes such as polyphenol oxidase, lipoxygenase and peroxidase (Angerosa 2002).

There is a loss of volatile compounds when high malaxation temperatures are applied, attributed to the enzyme's deactivation, especially of hydroperoxide lyase (Sánchez and Harwood 2002). This causes a considerable decrease of C6 compounds, cis-3-hexenol and C5 metabolites, as well as an increase in hexanol and *trans*-2-hexen-1-ol and a loss of the bitter-pungent notes (Angerosa et al. 2000).

Angerosa et al. (2000) studied the combined effect of malaxation temperature and time on the VOO aroma from cv. Coratina and Frantoio. They reported that malaxation negatively affects the total amount of secoiridoids. Some phenolic compounds, such as tyrosol, seem to undergo minor changes as affected by the malaxation time, but they are only affected by the temperature. The opposite trend was reported for the phenolic compound 3,4-dihydroxyphenylethanol-elenolic acid dialdehyde (3,4-DHPEA-DEDA), for which the maximum concentration was obtained at shorter malaxation time. The loss of phenolic compounds during malaxation is due to the enzymatic oxidation of the secoiridoid derivatives, catalysed by polyphenol oxidase and peroxidase (Servili et al. 2008).

The hydrolysis of complex phenols (oleuropein, dimethyl oleuropein, ligstroside, verbascoside, rutin and luteolin-7-glucoside) is fostered by the endogenous glucosidase. The non-hydrolysed oleuropein has low partition index toward the resulting aglycones, and simple phenols dissolve appreciably in the oil phase, while the non-hydrolysed phenol glycosides are more hydrophilic (Rodis et al. 2002).

The following compounds were proposed as markers for malaxation temperature: 1-penten-3-ol, cis-3-hexenal and octane, useful to differentiate malaxation temperatures of 15, 30 and 35 °C, respectively (Kalua et al. 2006).

The oxygen concentration in the paste headspace during malaxation can be manipulated to achieve significant modification of the VOO aroma, in relation to the desired characteristics and industrial needs (Servili et al. 2008).

The separation of the liquid phase and solid particles from the olive paste is usually performed using two major systems, pressure and centrifugation. The subsequent step involves the separation of the oil from the oil-water mixture, which is done by centrifugation.

When separation systems based on pressure are applied, VOOs tend to be fruitier and with a higher concentration of volatile alcohols, but possible fermentation and/or degradation phenomena can take place, with the consequent appearance of sensory defects (Angerosa 2002).

Differences in phenolic content in the oils were also observed between the conventional centrifugal ("three-phase") and the "two-phase". The latter does not involve the addition of water (except in very small amounts), which results in a lower loss of hydrophilic phenolic compounds, particularly ortho-diphenols.

The use of two-phase centrifuge, compared to the three-phase one, allows the production of VOOs with higher concentrations of *trans*-2-hexenal and greater total aromatic content, but with lower concentration of pigments, aliphatic and triterpene alcohols, sterols and waxes (Aparicio and Luna 2002).

The use of the three-phase centrifuge causes decrease in the content of C6 aldehydes, hexanol and *trans*-2-hexenol compared with the pressure extraction, probably because of the addition of hot water (Angerosa et al. 2004). Incorrect or improper management of the pressure extraction system may lead to olive paste fermentation and the subsequent formation of off-flavours.

9.3.3.3 Olive Oil Filtration and Storage

VOO filtration may have important effects on the product sensory properties and shelf life. Unfiltered oils are in an emulsion-dispersion form and contain a certain amount of water (from 2 to 4 g/kg).

It has been recently reported that the industrial-scale filtration of highly bitter-pungent extra VOO has an influence on the release of key aroma compounds, and therefore filtration should be regarded as one of the possible parameters that can potentially affect VOO phenolic composition. A significantly lower content of hydroxytyrosol and 3,4-DHPEA-EA was found after filtration and storage of the Italian cv. Ravece. In terms of VOO volatile compounds, 2-methylbutanal, 6-methyl-5-hepten-2-one and heptanol increased after filtration, while t,t-2,4-hexadienal, t-2-hexen-1-ol and c-2-penten-1-ol significantly decreased. These results indicate lower "green grass", "leaves" and "pungent" notes, with stronger "ripe fruit" and "malty" odour in the filtered product (Sacchi et al. 2015). However, the relatively little loss due to filtration might be compensated by the better stability of the filtered oil compared to the cloudy one; thus, filtration is encouraged.

VOO quality, similar to other vegetable oils, is affected by its storage conditions. Oxygen and light exposure, high temperatures and presence of traces of metal ions can accelerate the lipid oxidation and then shorten the shelf life of the product.

Among the saturated aldehydes, nonanal and hexanal suffer a sharp increase as oxidation rate increases. Some authors suggested the relationship hexanal/nonanal as an indicator of the oxidation status. Others reported that trans-2-heptenal is associated with the perception of rancid defect. *Trans*-2-hexenal, the most abundant volatile compound in VOOs, undergoes similar changes to those of phenolic compounds.

The type of container where the oil is stored was also reported to influence the lipid oxidation status and the chemical composition of VOOs under normal retail storage conditions (Savarese et al. 2013). Therefore, this parameter should be carefully evaluated to avoid quality loss over storage. Storage and packaging have been reviewed by Piscopo and Poiana (2012).

9.4 Health Attributes

Scientific evidence for the health beneficial effects of olive oil and in particular of EVOO have been broadly reported in the literature, especially in terms of its antimicrobial, antioxidant and anti-inflammatory activity, as well as potential chemopreventive and protective activity on gastric or intestinal cells (Fini et al. 2007; Sangiovanni et al. 2012; Vitaglione et al. 2015; Preedy & Watson 2010). Most of the work published is related to the protection against cardiovascular diseases, attributed to both the fatty acid composition (particularly the high level of oleic acid) and the presence of beneficial minor compounds such as phenolic compounds. The present paragraph briefly presents some general evidence without describing specific studies in detail. It has been pointed out that regular consumption of moderate amount of olive oil with high phenolic content is associated with a reduced risk or heart diseases (Covas et al. 2006).

Olive oil health effects are attributed to some major and minor constituents, and the reader can refer to Boskou (2006, 2015)for a comprehensive review of the recent finding related to the health effects of olive oil bioactive constituents.

9.4.1 *Effects on Oxidative Damage*

A large body of research has been published in relation to the antioxidant properties of olive oil constituents, particularly minor compounds such as phenolic compounds. Olive oil, in particular EVOO, is rich in vitamin E, which acts as an antioxidant toward free radicals.

Phenolic compounds are able to scavenge radical species by acting as chain-breaking antioxidants. Consequently, this suppresses lipid peroxidation by recycling other antioxidants, such as α-tocopherol (Perona and Botham 2013). Phenolic compounds act as relatively strong antioxidants, and their effect is due to their action on low-density lipoprotein cholesterol, so that atherosclerotic plaques have a more limited formation. Consumption of olive oil has also been associated with the reduction of the metabolic syndrome, in a context of appropriate "diet", which is not limited to just eating habits but extends to the whole idea of "Mediterranean diet". Protective effects of virgin olive oil against some type of cancers have been also reported. The anti-inflammatory activity of VOO bioactive compounds has been demonstrated by many authors. Its mechanism has been partly elucidated and has been studied also in relation to epigenetics. Effects are often attributed to the ensemble of oleic acid, polar phenolic compounds and unsaponifiable compounds such as α-tocopherol and β-sitosterol (Cardoso et al. 2014).

Several health effects, either confirmed or putative, have been attributed to VOO biophenols: scavenging and reduction of reactive species in human cells; reduction of oxidative stress; induction of endogenous antioxidant enzymes; effect on inflammation by diminishing of leukosides leukotriene B4; inhibition of pro-inflammatory enzymes; antihypertensive and potent antioxidant activities; platelet aggregation and endothelial function; anti-atherogenic activity; effects on the central nervous system effects; antimicrobial and chemotherapeutic effects (antibacterial, antifungal, antiviral properties); gastrointestinal effects, i.e. chemoprevention of peptic ulcers and gastric cancers, ulcerative colitis and Crohn's disease; anticancer and chemopreventive effects through induction of apoptosis, inhibition of proliferation of tumour cells; inhibition of oxidative DNA damage in human leukocytes; and other effects including immunomodulatory, endocrine and anti-ageing activity and respiratory effect (Boskou 2006, 2015).

9.4.2 *In Cardiovascular Diseases*

From studies published in 1954, Keys et al. (1954) suggested a possible link between the dietary habits, in particular the use of monounsaturated fatty acids, with lower cholesterol levels and lower risk of mortality due to cardiovascular diseases. It was then demonstrated, from epidemiological studies started in Southern Italy and then involving several countries, that the lowering of consumption of saturated fatty acids with mono- and polyunsaturated fatty acids, in particular olive oil, leads to low incidence of vascular diseases. Besides the effects on cholesterol, the consumption of olive oil was also shown to prevent thrombosis and platelet aggregation, while polyunsaturated fatty acids showed no effects (Sirtori et al. 1986). The consumption of olive oil has been shown to lower low-density lipoprotein (LDL) but not high-density lipoprotein (LDL) (Mattson and Grundy 1985). Consumption of olive oil has been reported to lower blood pressure, which has been overall related to its high content of oleic acid and minor constituents (Rasmussen et al. 1993; Pérez-Jiménez et al. 2007; Pérez-López et al. 2009). However, the mechanism of action has not been fully understood (Perona and Botham 2013).

9.4.3 Effects on Cellular Function

Consumption of olive oil has been also shown to improve the treatment of diabetes mellitus (Bonanome et al. 1991), as it prevents insulin resistance (Tierney and Roche 2007). Several studies have also shown a positive effect on the metabolic syndrome; however, it is difficult to differentiate between the Mediterranean diet and olive oil by itself (Perona and Botham 2013).

Investigations in humans, in animals and in vitro reported anti-inflammatory properties of olive oil. This effect has been mostly linked to oleic acid, which is able to inhibit insulin production (Vassiliou et al. 2009). This is probably due to the quenching of intracellular reactive oxygen species (ROS) (Perona et al. 2006).

Olive oil has been also associated with the prevention of cognitive loss and dementia, probably linked to its antioxidant effects (Pitt et al. 2009; Féart et al. 2010).

9.4.4 Anti-carcinogenic Activity

Several studies have reported on the effects of olive oil on the protective effects toward certain type of cancer (Trichopoulou and Vasilopoulou 2000; Trichopoulou et al. 2003; Menendez et al. 2009). It has been suggested that olive oil could lead to prevention of colorectal, breast, pancreas and endometrial cancers (Trichopoulou and Vasilopoulou 2000). Bosetti et al. (2002) reported an inverse relationship between olive oil consumption and breast and ovarian cancer risk, not observed with other types of fats.

9.4.5 Health Claims

A positive effect of olive oil in lowering the risk of coronary heart disease has been accepted by the Food and Drug Administration (FDA, https://www/fda/gov/media/

118199/download), when consuming approximately 23 g of olive oil daily. The health claim related to oleic acid should be formulated as follows: "Replacing saturated fats in the diet with unsaturated fats contributes to the maintenance of normal blood cholesterol levels" (EC Reg. No 1924/2006).

Scientifically proved effects of olive oil consumption have been reported widely, and the European Food Safety Authority (EFSA) issued an opinion that lead to the Commission Regulation (EU) 432/2012 (EU Regulation 2012). VOOs can potentially be labelled by also adding a health claim referred in relation to the beneficial effects of olive oil polyphenols: "Olive oil polyphenols contribute to the protection of blood lipids from oxidative stress. The claim may be used only for olive oil, containing at least 5 mg of hydroxytyrosol and its derivatives (e.g. oleuropein complex and tyrosol) per 20 g of olive oil". Roselli et al. (2017) gave a detailed overview on the health claims applicable to olive oil, by stressing the "unicity" of this product in contrast with other vegetable oils.

A consistent body of literature based on "randomized, crossover, controlled human studies in healthy subjects and in patients showed that the protective effects of VOO phenol compounds are clearly detected in humans under oxidative stress conditions" (Vitaglione et al. 2015). A dose-dependent effect has been shown in terms of protective effects of olive oil phenolic compounds in low-density lipoprotein (LDL) peroxidation.

It has been pointed out that the majority of EVOOs on the market usually has a phenolic content below 200 mg/kg, except a few high-quality PDO products (Caporaso et al. 2015b); therefore, efforts are needed for producers and the industry to increase the phenolic content of EVOOs. It is also important that the consumer will correctly evaluate VOOs and appreciate the benefits and uses of a product with bitter-pungent notes. It should be also highlighted that harmonisation in terms of official methods for olive oil phenolic compound analysis is still ongoing. However, the majority of literature uses colorimetric methods, such as the Folin-Ciocalteu technique, to quantify the total amount of phenolic compounds or HPLC-based methods which allow the quantification of individual phenolic compounds after hydrolysis.

9.5 Olive Oil Quality and Authenticity

The "basic" quality parameters for olive oils are those related to the free fatty acids, peroxide value and absorbance in the ultraviolet region, in addition to the sensory assessment. The reader can refer to the EU Commission Regulation (EC) 2568/91 with the recent modifications (EU Commission Regulation 2015, 1830).

Olive oil quality is tested from a physico-chemical point of view, using the following indices:

- Peroxide value
- Acidity (free fatty acids)

- K_{232}
- K_{270}
- ΔK

In addition, sensory analysis (panel test) is compulsory, as described in previous sections.

In addition to these parameters, one can investigate the content of chlorophylls, fatty acid composition to assess the level of oleic acid, triacylglycerol composition and composition and level of phenolic compounds, sterols and other minor constituents.

For authenticity and safety purposes, parameters that can be tested include contaminants, the level of α-tocopherol (as a possible indicator of deodorisation), volatiles and specific phenolic compounds, monoacylglycerols and diacylglycerols (the ratio of 1,2- to 1,3-diacylglycerols is an indicator of VOO freshness), pigments, etc.

Several methods are also used to verify the genuineness of olive oils: these include fatty acid composition; presence of trans-fatty acids (indicators of thermally processed oils); fatty acids in the 2-position of the triacylglycerol (to identify oils prepared by esterification); equivalent chain number (ECN); sterol composition (to verify adulteration of olive oil with other vegetable oils, as certain sterols are related to the plant of origin); content of erythrodiol, uvaol and waxes (to exclude oils obtained by solvent treatment); content of aliphatic alcohols (indicators of olive pomace oil added to olive oil); and level of stigmastadienes (indicators of thermally treated oils or products that undergo bleaching process) (Boskou 2006, 2015).

Other analytical methods can also be used, some non-official methods, for authenticity purposes. All these methods have been reviewed and explained in details by Angerosa et al. (2006) and Lercker and Rodriguez-Estrada (2000).

It has been highlighted that the current issues related to olive oil authenticity are mostly related to the fraudulent marketing of deodorised virgin olive oils treated with mild conditions that are not easily detectable by instrumental methods, the addition of VOOs obtained from a second mechanical extraction after the first pass, the incorrect use of geographical origin declaration and the misclassification of VOO (declassed for sensory defects of physico-chemical characteristics) that are sold as EVOOs.

In addition to these issues, the olive variety identification is also of interest for producers, consumers and control bodies. Some Protected Designation of Origin (PDO) products from well-defined origins or of mono-variety EVOOs might be perceived as oils with higher quality. Mono-variety EVOOs have been studied in the literature, but from a practical point of view, there are two issues related to the application of this knowledge. Olive plants require pollination, while the use of many varieties in the field is often recommended: thus, obtaining a "pure" mono-variety EVOO is difficult especially for certain cultivars. In addition, blending is one of the most common industrial practice, used to obtain consistent product and for economic reasons. Discrimination based on DNA methods or analysis of volatile compounds is much more difficult for blends of several varieties and from different origins (Raieta et al. 2015).

Authentication is a term used to confirm that a claimed statement is true. Therefore, it might be applied to olive oil, to verify whether it is "olive oil" or not, as stated in the regulations, or whether it is "virgin olive oil" or "extra virgin olive oil". It can be also applied to geographical indications, to mono-variety EVOOs or to specific products such as PDO EVOOs.

The issue of geographical origin traceability is of particular interest due to fact that products of the same category, e.g. EVOOs, obtained from different locations are sold at notably different prices. For these reasons, an analytical means to detect the geographical origin would be of great importance for the industry, control bodies and consumers.

Some studies have focused on the use of volatile compounds and cluster analysis to find markers that could discriminate VOOs from different areas. Luna et al. (2006) analysed 39 samples and reported successful discrimination between Italian and Spanish VOOs based on the following compounds: 2-methylbutyl acetate, 2-methyl-4-pentenal and hexanol. For the differentiation between Italian and Greek oils, pentane-3-one, cis-3-hexenal and a hydrocarbon were reported. Regression analysis allowed to classify oils derived from macro-areas of origin (Spain, Italy and Greece) by the following volatile compounds: hexyl acetate, ethyl benzene, 2-methyl-4-pentenal, ethyl furan, trans-2-pentenal, 1,2,3-trimethylbenzene, 3-methyl butanol and a hydrocarbon (Luna et al. 2006).

Olive oil traceability is a complex issue, due to the interference of many factors such as variety, year-to-year variations, agronomic practices and technological conditions applied for the extraction that can lead to compositional variations. Traceability and authentication require good knowledge of the composition and characteristics of the samples, their production process and theoretically their full history. To get a comprehensive picture and be able to build and apply a classification model to discriminate EVOOs from different geographical regions, the experimental design should include all the possible expected variability from several years, several growing conditions and other factors that can potentially affect olive oil composition. Many cases, in the literature, describe olive oil classification models based on a limited number of samples, or they are based on commercial samples whose origin is not clear, and therefore limited conclusions can be drawn. New methods, however, have been developed over the last few years. For example, the research field of DNA-based authentication of olive oil seems to be a new very active area (Bazakos et al. 2016).

Major or minor compounds can be used to characterise samples, as well as several chemometrics strategies. To reach solid classification models, correct sampling, correct and consistent analytical methods and appropriate statistical treatment of the data are required.

Several olive oil adulterations are still difficult to detect, despite the advanced analytical methods available nowadays, and although methods for authenticity of VOOs have been proposed, there are issues related to the geographical origin or olive cultivar.

9.6 Sensory Assessment and Volatile Compounds

The International Olive Council (IOC) is the intergovernmental organisation that brings together olive oil and table olive producing and consuming stakeholders. Its main mission is the harmonisation of rules and quality control among the olive oil sector, also proposing new methods and limits to assure high quality for the olive products and consumer protection. Its member states represent approximately 97% of the olive oil producers worldwide. The IOC regulation regarding the classification, physico-chemical analyses and sensory assessment of olive oils is compulsorily applied in the European Union, the world's first olive oil producer, and in many other countries.

Different from other food products, VOOs require compulsory sensory analysis. This is done by approved sensory panels of at least eight expert assessors, trained to describe and evaluate virgin olive oils based on a defined protocol and a quantitative method in which assessors classify the presence and intensity of defects and positive attributes (fruity, bitter, pungent). The training requirements, assessment protocol and data analysis to be applied on the results are strictly regulated (IOC 2015b, c; Reg. EU 1991 and its amendments). The visual appearance of olive oil is not a quality parameter for market or trade, but—similar to other products—is important for many consumers, as it gives a "first impression" about a food. However, this impression is misleading as olive oil's visual appearance has little relation with quality. Similarly, fresh VOOs often have a green colour due to chlorophylls, which degrade over storage to produce pheophytin and then chlorophyllides, with yellow-brown colour. The yellow colour of olive oils is due to carotenoids, which undergo similar transformation during storage, but some carotenoids are more stable than others.

The official method for VOO sensory assessment completely excludes any colour evaluation as a quality indicator, and it emphasises the importance of flavour, i.e. the combination of aroma and taste. The vocabulary for VOO evaluation is rather limited, when the purposes is to assess its commercial category. This is mostly related to the presence/absence of defects, which can fall into five categories, and three positive attributes, which are "fruity", "pungent", and "bitter". Few consumers are used to vegetable oils that naturally possess bitter and pungent notes. Even in producing countries, sometimes consumers tend to prefer "light" olive oils, i.e. those with limited bitter-pungent notes and relatively low intensity of fruitless. However, the current opinion is that consumers should be educated to positively evaluate more pungent-bitter VOOs, as this is related to higher quality.

Sensory defects of VOOs have different origins. When olive fruits are not processed within a limited period of time, often 12–24 h would be preferable—and especially when the olives are amassed and aeration is not guaranteed—fermentation takes place during storage. This can lead to the generation of a series of volatile compounds related to the aerobic or anaerobic fermentation. The first phenomenon is observed mainly when the olives are stored for prolonged time before processing, while the second one mainly occurs during the storage of the obtained VOO which still contains residues of the vegetable material (olive pulp, etc.) and emulsified water. For practical reasons, the official sensory method clusters the two descriptors "fusty" and "muddy" into one category. The defect named "musty-humid-earthy" comes from olives with high development of fungi and yeasts, or those collected from the ground in contact with earth or mud, or even when the washing phase is not carried out correctly (dirty washing water with excessive soil residues). Musty defect occurs when mould naturally present on the olive skin develops, especially under humid conditions. In certain cases, alcoholic fermentation can take place, resulting in the appearance of the "winery-vinegary" defect, which can be easily recognised due to the similarity with vinegar pungent note. It is correlated with high levels of acetic acid, ethyl acetate and ethanol, due to yeast or *Acetobacter* development. Rancid is a typical defect of olive storage, due to lipid oxidation. This is common to all vegetable oils or generally fats, and its intensity is related to the fatty acid composition and presence of antioxidants as well as storage conditions (light, oxygen exposure and high temperature). A defect named "frostbitten olives", or "wet wood", is due to the presence of frozen olives; it is encountered in areas where low overnight temperatures before harvest damage the olive fruits due to water crystallisation. Its sensory impact resembles the odour of wet wood, sometimes similar to chocolate aroma, in the case of intense damage. Other negative attributes can be added to the sensory evaluation form during the panel test. In the case of negative attributed listed as "others", a minimum number of assessors detecting a specific defect are required to judge the oil as defective.

According to the IOC norm, widely accepted in producing and importing countries, VOO sensory assessment is carried out by a panel of at least eight trained judges, who are required to pass specific training and assessment. Training of future panellists includes tests to evaluate their ability to detect basic flavours and adequately describe olive oil sensory defects. Accreditation of sensory testing laboratories is critical to guarantee that the panel act properly and give robust and valid results. The official method describes the settings of the laboratory, the light conditions, room and sample temperatures and the specifics for the blue glass designed for olive oil tasting, as well as the heater conditions (the sample must be tested at a temperature of 28 ± 2 °C) and the sheet to record the results. The positive and negative attributes are scores on a 10 cm unstructured scale. In case of sensory defects that do not belong to one of the five classes previously mentioned, the assessors can describe it, using an agreed vocabulary. For the positive attribute of "fruity", the assessor is asked to specify whether it is "green" or "ripe" fruity. The importance of using a panel instead of evaluation of single experts comes from the fact that subjectivity in the assessment should be removed as much as possible. For this purpose, appropriate statistical methods are used, e.g. application of the median instead of the mean value for each attribute.

In addition to virgin olive oils of various origins, there is a niche market for EVOOs with typical characteristics. These oils belong to the Protected Designation of Origin (PDO) labels, and they have to comply with additional EU regulations. In terms of sensory assessment, the disciplinary of the product can include compulsory presence of intensity limits for some attributes, e.g. almond, artichoke, fig leaf, etc.

However, sensory notes other than the "fruity", "pungent" and "bitter" that are assessed during the panel test for olive oil classification are not official, and they are not required for assessing the market category of the olive oil. In particular cases, the presence of peculiar sensory notes at a certain level of intensity is required for some extra virgin olive oils with Product Designation of Origin (PDO) labels. A PDO product has to provide distinctive characteristics from a sensory point of view in addition to providing evidence related to the history of the product (tradition) and the link between the people of a particular area and the product. In this case, the regulations for PDO products make it compulsory to perform sensory analysis to assess the presence and intensity of particular sensory descriptors, e.g. "tomato", "tomato leaves", "artichoke", "rosemary", "almond", etc.

Several hundreds of volatile compounds have been reported in olive oils, especially in VOOs. Aldehydes and C6 alcohols have been related to positive sensory attributes, and to the note of "sweetness", which is just the aroma recalling sweet attributes but does not refer to an actual taste sensation. The odour threshold of volatile compounds is dramatically different depending on the chemical structure and physico-chemical characteristics of each molecule. For example, cis-3-hexenal, which gives the typical odour of freshly cut grass and despite its low concentration, greatly contributes to the aroma as it has a very low odour threshold. On the contrary, trans-2-hexenal has much higher perception threshold, and it contributes minimally to the final VOO aroma, although it is the most abundant volatile compound in the oil (Kalua et al. 2007).

Trans-2-hexenal and *trans*-2-hexenol have been associated with VOOs obtained from olives in good conditions. The following compounds contribute mostly to the "green note": *cis*-3-hexenal, *cis*-3-hexenol (grass and banana) and *cis*-3-hexenyl acetate (fruity and green leaves). Important in defining the complex flavour are also *trans*-2-hexenal, hexanol and *trans*-3-hexenol. Hexyl acetate contributes to perceptions of fruity and sweet, while hexanal is responsible for the "green" and apple note.

Cis-3-hexenol has been associated with the bitter note of VOOs, as well as the attributes of "apple", "tomato", "vegetable bitter", "grass" and "fruity olive oil", along with some C5 alcohols (*cis*-2-penten-1-ol and 1-penten-3-ol) and *trans*-2-hexenal (Caporale et al. 2004). Other compounds such as toluene, octane, octene and 3-methyl butanol arise from different routes from the LOX pathway and are not related to positive sensory attributes of VOO. The "fruity" note of VOO was found to be positively correlated with the cis-3-hexenol and negatively correlated with 3-pentanone. The "ripe fruit" sensory note is correlated with 3-pentanone, while "leaf" is positively correlated with hexyl acetate, 1-penten-3-ol and *cis*-2-penten-1-ol and negatively correlated with hexanol, associated with the "grass" note. The note of "almond" positively correlates with the hexyl acetate, 1-penten-3-ol and *cis*-2-penten-1-ol and negatively with the hexanol; *cis*-3-hexenol correlates positively with the scent of "tomato" and "bitter vegetable"; 3-pentanone correlates negatively with the attributes of "bitter", "spicy", "tomato" and "vegetable bitter". The attributes of bitter and spicy oil are due to the presence of phenolic compounds. The bitter taste is attributed to aglycones of the dialdehydic form of the decarboxymethyl oleuropein and other forms of the oleuropein aglycone; the "pungent" note has been attributed to the aglycone from the dialdehydic form of the decarboxymethyl ligstroside (Servili et al. 2009).

The sensory defect of "winey-vinegary" is associated with fermentative processes especially from *Lactobacillus*, when olives are left on the ground for prolonged times. The consequent production of acetic acid has been indicated as a marker of olives collected from the ground. A high concentration of acetic acid and octane was correlated to the defect of "reheating", a consequence of the activity of *Enterobacteriaceae* of the genus *Aerobacter* and *Escherichia*. The genera *Pseudomonas*, *Clostridium* and *Serratia* appear if the olives are left in bags for a long time after harvesting. The activity of these microorganisms results in the presence of some volatile compounds at high concentrations. The defect of "mould-humidity" has been correlated to the presence of 3-methyl-1-butanol.

Other compounds that would be important in defining the overall aroma oil have been reported to occur in low concentrations in the olive paste: hexyl acetate (fruity odour) could be a marker of good quality oils, but it does not contribute to the aroma of the paste due to its high perception threshold (~1.04 mg/kg) and low concentration in the paste (maximum 1.00 mg/kg) (Garcia-Gonzales et al. 2007).

9.7 Food Applications

The use of olive oil is solely found as a condiment, which means that different from other products such as wine or cheese, it is never consumed by itself. The versatility of its use as a condiment or ingredient makes it difficult to choose the most appropriate type of olive oil for different recipes or dishes.

Olive oil can be used as a "raw" ingredient or for cooking. It can be included into baked goods and used for stir-frying or deep-frying. In the traditional area where olives are grown, i.e. around the Mediterranean Basin, olive oil is the main fat used in a wide range of food preparations. EVOO, the top category within the olive oil family, is often used as raw (without the application of high temperatures) to avoid degradation of phenolic compounds and loss of aroma during cooking; however, there is no harm in using it even for frying (Saghir et al. 2005). In this case, attention should be paid not to excessively heat the oil, as the presence of a small amount of residual water and solid particles (especially in unfiltered oil) slightly lower the so-called smoking point (Galeone et al. 2006). In terms of stability during prolonged heating, olive oil generally shows good stability due to its high content of oleic acid and the presence of natural antioxidants.

Olive oil can be used in salads or mixed into dressings; it can be applied to meat of fish for marinades or can be add at the end of a dish preparation to impart extra flavour and appearance. Due to its health benefits, it can also replace butter, margarine or other animal fats. Similarly, it can be used in sauces instead of other vegetable oils, giving a nutritionally balanced product with a unique flavour.

Table 9.4 Culinary uses of olive oils

Food applications of olive oil	Raw/cooked	Pairing or use (product, dish)
Dressing	Raw	Salads
Sauces	Raw	Dried herbs/spices
	Raw	Mayonnaise
	Cooked	Tomato sauce, etc.
Frying medium	Cooked (pan-frying or deep-frying)	Potential any food (mostly medium-low moisture)
Marinades	Raw, then cooked	Meat or fish
Bakery products	Cooked	Cakes, etc.
Animal fat replacement	–	Sausages, cakes, *taralli*, etc.

A scheme of possible uses of olive oil discussed in this section is reported in Table 9.4.

Food pairing is an interesting aspect for consumers, restaurants and catering industry, also becoming a topic for scientific studies in the area of food science and technology. Olive oil pairing, similar to other food pairing, does not follow particular "rules", but it is mostly related to the consumer or chef's creativity. As the creation of a dish is somehow more an art than a science, it is often difficult to define consistent and appropriate experimental designs or approaches to study food pairing from a holistic perspective. However, when trying to classify food pairing, one can use olive oil to enhance particular notes that are in the other ingredients, or work by "contrast", e.g. using a bitter-pungent EVOO on a dish with marked sweet notes.

9.7.1 Olive Oil Food Pairing

In a recent research paper, Genovese et al. (2015) emphasised that "generally, there are two distinct ways to pair VOO with food: complementary and contrasting approaches. A complementary flavour is obtained when two similar ingredients are blended. This results in the enhancement of primary flavours, while a contrasting approach consists in tasting each ingredient separately. Pairings of VOO with salads, vegetables, pesto, tomato sauces, etc. are examples of the first case. The latter example could be the case of pairing VOO and fresh mozzarella cheese in salads, such as the "insalata Caprese". In fact, the unique flavour of a "strong" VOO (bitter, pungent and fruity) contrasts with the delicate texture and taste of a fresh cheese like mozzarella, which shows sweet, acid and milk notes. Frequently, cheese is used as an ingredient in many recipes with VOO to obtain a sweetening effect of bitter-pungent notes in VOOs with an intense flavour. It is not surprising that renowned chefs pair VOO with many different foods, in order to obtain new sensations and create new culinary experiences. When VOO is combined with dairy products, an oil-in-water emulsion or dispersion is obtained. The presence of different phases and the concentration of volatile compounds affect their partitioning. In fact, lipophilic aroma compounds tend to move in the oil phase and their concentration considerably decreases in the continuous phase. To understand the complexity of interactions among volatile and non-volatile compounds in food ingredients during their interaction and pairing, the authors investigated the presence of whey protein and olive oil phenolic compounds, to assess the release of VOO aroma compounds (Genovese et al. 2015).

Traditional VOO food pairing is based on the use of products with marked herb notes, while oils with apple notes are suitable for salads, and peppery notes are desired for grilled meats (Boskou 2006, 2015). Herbs, as dry ground products or fresh (e.g. mint, basil), are often added to give both flavour and complexity, but also longer "shelf life" to the dish, by enhancing its antioxidant properties due to the presence of phenolic compounds. They can be added directly during the preparation of the dish, as cooked or raw ingredients, or can be used in olive oil flavouring. A niche market for olive oils is represented by flavoured olive oil "condiments", e.g. using dried oregano, rosemary, chilli pepper (Antoun and Tsimidou 1997; Caporaso et al. 2013; Gambacorta et al. 2007) and even fresh lemons directly added during milling (Sacchi et al. 2017). From a production point of view, adequate research is required to understand the stability of flavoured oils. A flavoured oil does not always show stronger antioxidant activity than unflavoured oils, due to the pro-oxidant effects of certain compounds released from the spice or herbs, e.g. chlorophyll.

From a consumer point of view, it is important to test several combinations that satisfy the palate, as personal preferences should guide this choice rather than a general "rule". Consumer education on different grades of "olive oils" is necessary in order to get the best from each category of olive oil in terms of cooking use, health effect and expected sensory impact. This can in turn lead to improvement in the industry and farmer benefits, which is necessary for the olive oil sector due to the current limited profitability for the sector, especially for the olive growers.

Harmony is a concept frequently used in culinary science and technology, and it is defined as the pleasant effect obtained when combining different ingredients of a whole (Spinelli 2014). Harmonic pairing is an important parameter in the "art" of cooking, which is reflected in the choice of the correct ingredients. Several theories have been proposed regarding the food pairing concept, e.g. the pairing of ingredients that share similar volatile compounds (for review see Caporaso and Formisano (2016)).

Traynor et al. (2013) evaluated new food pairing combinations, from an experimental work. The creation of certain flavour equilibria in the volatile profiles of food pairings is important for a positive hedonic response. Extra virgin olive oil was paired with banana, as well as with other foods (basmati rice and bacon), by applying the "flavour network" theory for food pairing, based on common volatile compounds. The authors evaluated both the volatile composition of the food ingredients and the sensory scores (hedonistic) obtained by a sensory panel. Results suggested that the success of food pairings is not just based on a mere volatile compound sharing, as previous theoretical works suggested.

Interactions between volatile compounds and the non-volatile matrix occur during oven-cooking, which uses very high temperatures, much higher than restaurant or home ovens. For example, a key ingredient in good quality traditional pizza is olive oil, which has been shown to interact with the other ingredient during the short cooking time (~180 s) required in the wood-fired ovens. Strong differences in the volatile and phenolic profiles of pizzas cooked using olive oil or other vegetable oils have been reported (Caporaso et al. 2015a). This demonstrates the importance of the interaction between volatile compounds and the food matrix during cooking.

Olive oil can also be used as a mixture (dispersion) or emulsions to formulate products such as sauces or mayonnaise to be used in salads or cold dishes. A body of research includes the study of the stability (physical and oxidative stability) and shelf life of products formulated with olive oil. For example, Paraskevopoulou et al. (2005) reported on the stabilisation methods for emulsions formulated with olive oil and lemon juice using polysaccharides, while in a subsequent work, the authors reported on the oxidative stability (Paraskevopoulou et al. 2007). Caporaso et al. (2016a) used EVOO to formulate low-fat emulsions stabilised by whey protein isolate and xanthan gum. In a publication that followed, a statistical model was proposed to describe the physical and chemical stability of the emulsion over storage (Caporaso et al. 2016b).

9.7.2 Olive Oil in Cooking and Frying

Physical and chemical phenomena have been described when olive oil is used in cooking, especially VOOs. The polar minor constituents of olive oil, especially the most reactive phenolic compounds, undergo hydrolysis reactions, but they also exert their antioxidant properties in such a complex matrix. During cooking, high temperatures are applied: phenolic compounds are hydrophilic substances and tend to rapidly migrate toward the water phase (Al-Saghir et al. 2004; Andrikopoulos et al. 2002). Complex phenolic compounds are degraded into simpler molecular structures, and their antioxidant activity might increase in certain conditions, for example, when hydroxytyrosol is released, as this compound is often reported to exert the strongest antioxidant activity among VOO phenolics.

Frying takes place at high temperatures and implies rapid dehydration of the food being cooked. Pan-frying differs from deep frying, as the former is often carried out at the beginning of many traditional Mediterranean recipes, to give extra flavour to the main ingredient. A small amount of olive oil is heated in the pan together with flavouring vegetable (or animal) ingredients, such as garlic, onion, spices, carrots, leaks, etc., or chopped meats. This is often used as the "basis" of the recipe, while further ingredients are added. It can be also used as the sauce for pasta, as many pasta-based dishes are prepared using an initial pan-frying step. Although an empirical and traditional method that probably arises from observation and creativity of our predecessors, it has in fact a scientific basis. The oil acts both as a lubricant and as a "solubiliser" for flavour molecules and potentially bioactive molecules (especially lipophilic compounds) and gives flavour to the dish by itself. This flavour comes from the initial characteristics of fruity, bitter and pungent notes of the VOO, as well as the additional aroma arising from oxidation and degradation during cooking.

Kalogeropoulos et al. (2007) studied the migration of EVOO phenolic compounds in finfish, during pan-frying. The authors showed a loss of antioxidant compounds—α-tocopherol, polar phenolic compounds and terpenic acids—in the oil phase, with a contemporary migration of these compounds into the fish. A similar experiment with vegetables, typical of the Mediterranean cuisine, showed similar results. In particular, certain loss of EVOO antioxidants was found in the frying oil, i.e. up to 70% for phenolic compounds, 60% for α-tocopherol and 80% for hydroxyl pentacyclic triterpene acids (Kalogeropoulos et al. 2007).

The use of VOOs as a frying medium may be suggested because of its high stability due to its fatty acid composition and the presence of natural antioxidants. However, an excessive number of frying operations has to be avoided when health effects are expected from the presence of natural antioxidants. Hydroxytyrosol and its derivatives are degraded with various rates depending on the number of heating operations and the temperature.

Perhaps one of the most known pairing of olive oil is with tomato or tomato sauce, typically to prepare sauces for pasta (e.g. ragù). Sacchi et al. (2014) reported a study of tomato sauce and EVOO cooked for a long time (up to 8 h). This very long time and low temperatures were chosen to simulate the typical Southern Italian cooking time for the "Sunday dish" based on tomato sauce, meat and olive oil, which is then added as the condiment to pasta. A protective effect of EVOO on the tomato sauce was obtained in terms of antioxidant capacity, due to the migration of bioactive compounds among the two ingredients used in the model system, including carotenoids and some flavonoids.

The addition of olive oil to vegetables during cooking has been shown to exert a positive effect on the final properties of the dish in terms of nutritional value. Hornero-Méndez and Mínguez-Mosquera (2007) studied the effects of carotenes' bio-accessibility when EVOO was added to cooking carrots. Despite the lowering of carotenoid content caused by the cooking process, a positive effect on carotene bio-accessibility was found.

It is suggested to use EVOO directly raw on the dishes; this should be generally preferred to preserve the unique flavour and to avoid loss of phenolic compounds. A typical use of olive oil can be the replacement of other fats in dressings and sauces, an easy application at the industrial level or home preparation.

Olive oil can be also conveniently used for marinating muscle foods, by preparing a mixture of VOO, herbs and acid liquids such as vinegar, wine or lemon juice. The raw meat or fish is immersed in abundant amount of this dispersion, which gives extra flavour. The acidic pH also tends to give a tender product, but the presence of EVOO has been shown to also give further benefits in terms of health properties. Monti et al. (2001) used EVOO to marinate meat and analysed the formation of heterocyclic amines, which are possible human carcinogens formed during meat cooking. They demonstrated a protecting effect of EVOO phenolic compounds. EVOO phenolic compounds have been also shown to exert a protective effect against acrylamide formation, which was attributed to the presence of ortho-diphenols (Napolitano et al. 2008).

In addition to its health effects, EVOO gives a strong contribution to the overall flavour of the product. Some molecular mechanisms that can potentially explain the interaction of olive oil compounds with other ingredients have been highlighted. For example, it was demonstrated that milk proteins can covalently interact with some volatile compounds such as hexanal and *trans*-2-hexenal. These compounds are found at relatively abundant concentrations in EVOOs. Low levels of hexanal are described to give positive odour notes similar to freshly cut grass, and its presence characterises good quality EVOOs. However, the same compound arises from oxidation of unsaturated fatty acids and hydroperoxide degradation; rancid oils have high levels of hexanal.

9.7.3 Other Uses as Replacement of Fats

Many strategies have been applied to produce healthier lipid formulations in meat products, in particular by substitution of animal fat with vegetable oils. The aim is to lower the amount of saturated fatty acids and cholesterol and to increase the content of unsaturated fatty acids, particularly oleic acid. Bulk olive oil or olive oil-in-water (O/W) emulsions were used to substitute animal fat. The addition of other health-promoting constituents has also been tried. López-López et al. (2010) reported the use of Wakame seaweed added to beef patties in addition to O/W emulsions to replace pork fat, to obtain low-fat and low-salt beef patties. The authors showed that the partial or total replacement of pork backfat by olive O/W emulsion resulted in a similar cooking weight loss and gave a softening effect. This effect of water and fat binding properties could be due to the stabilising effect of the oil in the emulsion.

A similar study was reported by Jiménez-Colmenero et al. (2010), who evaluated the influence of olive oil used as a pork backfat replacement in frankfurters. Products with olive oil emulsions stabilised with different types of protein systems showed harder texture characteristics, with slightly lower juiciness but generally similar acceptability compared to the control.

The addition of olive oil-in-water emulsions also modified the total fat content of the meat, which was significantly lower: this means that a healthier meat could be formulated. This latter aspect should be further investigated by nutritionists.

Not only meat but also other types of foods and recipes have been investigated by scientists to modify their lipid profile by adding olive oil or extra virgin olive oil (EVOO). The latter has been used as a fat replacement in several food products. Matsakidou et al. (2010) reported on the replacement of margarine with EVOO in Madeira cakes. The use of EVOO increased batter density and cake volume; the weight loss during cooking was reduced. As EVOO contains a wide variety of volatile compounds that are responsible for the unique flavour of this fat, its addition in the cake greatly modifies its aromatic profile. Due to the difference of margarine and EVOO in terms of chemical composition and physico-chemical state and properties (density, melting point, etc.), differences in the batter are expected: batter viscosity, surface tension and the size of the air bubbles inside the cake are modified, resulting in lower density and higher volume. Regarding volatile composition, some important changes were observed. Certain alcohols, such as (Z)-2-pentenol, (Z)-3-hexenol, (E)-2-hexenol, (Z)-2-hexenol, phenylethanol and nerolidol; hydrocarbons such as 3-ethyl-1,5-octadiene isomers, a-copaene and a-farnesene; the esters (Z)-3-hexenyl acetate and hexyl acetate; as well as the aldehydes (E)-2-hexenal and (E)-2-decenal were identified only in the extra virgin olive oil-containing samples. The cake prepared with an extra virgin olive oil/margarine mixture was the most highly preferred by the panellists, whereas the sole use of virgin olive oil could decrease the panellist preference (Matsakidou et al. 2010).

Caponio et al. (2013) studied the type of fat used in *taralli*, a typical Italian bakery product. They made *taralli* using EVOO or olive oil, olive-pomace oil and palm oil. EVOO gave the lowest values of oxidation indices (except the peroxide value) and gave lower *trans* isomers. In addition, using EVOO instead of other fats added phenolic compounds to *taralli*, as well as characteristic sensory properties, with better visual appearance and odour.

It is worth mentioning that the use of EVOO to replace refined vegetable oils commonly used in bakery products led to a significant reduction of oxidative polymerisation, such as triacylglycerol oligopolymers and level of geometrical isomers which are nutritionally undesirable (Caponio et al. 2013).

9.8 Other Olive Products

Olive fruit has multiple uses. The present chapter has focused on olive oil, but exploitation of many other products obtained from the olive tree is possible. For example, the production of table olives, which have a quite big market worldwide, is carried out in a wide range of styles, giving different final products. Cultivars, processing of fruits for the preparation of the main commercial types and their bioactive constituents have been recently discussed by Boskou et al. (2015).

When olive oil is produced, and the fat has been extracted from the fruit, the solid parts of the olives can be dried and used for composting (pulp residues) or as a fuel (olive solid particles). Research is ongoing to recover phenolic compounds, especially from the olive mill wastewater (OMW) for a possible use in the cosmetic, pharmaceutical or food industries.

Caporaso et al. (2016a, b) used OMW phenolic extracts obtained by spray-drying of a concentrate recovered by subsequent membrane filtration. The extracts were coated using maltodextrins and added as an ingredient in olive oil-in-water (O/W) emulsions, where xanthan gum was used as a stabiliser, in combination with whey protein isolate (WPI). The authors showed that weak interactions between OMW phenolic extracts and WPI can be observed. OMW phenolics had a small effect in retarding lipid oxidation in emulsion. This is probably due to the coating material, which protects the phenolics from the external environment. During consumption, however, they might be released and exert their full potential. Functionalisation of food products by using olive mill wastewater biophenols has been recently reviewed by Caporaso et al. (2017).

Research like this can be positive to solve the issue of using food industry by-products, in particular olive mill by-products, by recovering bioactive compounds with interesting functional properties that can be further used by the food industry to produce new food products.

9.9 Future Challenges

Virgin olive oil, and especially extra virgin olive oil, has specific characteristics which make it unique among all other vegetable oils. The balanced fatty acid composition, the bitter-pungent sensory notes given by the phenolic compounds and the load of a wide range of volatile compounds are responsible for the hedonistic success of this product. Other categories of olive oils found in the market, olive oil (a mixture of virgin olive oil and refined olive oil) and olive-pomace oil (a mixture of virgin olive oil and refined pomace oil) have the same fatty acid composition but very little bioactive constituents and less intense organoleptic characteristics or even defective ones due to the absence of compulsory sensory analysis on these other categories.

The multiple culinary uses of olive oils have been broadly discussed mainly in the context of the traditional Mediterranean cuisine. There are, however, the literature that recently described new approaches to the use of food ingredients in cooking, including virgin olive oil and the other olive oil forms. These aim mainly at replacing other fats and at improving the nutritional profile in a series of food products. A greater awareness of the wide range of sensory characteristics and the presence of bioactive constituents by the consumers is expected to grow. This will lead to a better appreciation of this valuable natural product and a broadening of its culinary applications also in new countries that do not traditionally consume olive oils.

References

Ahmed CB, Rouina BB, Sensoy S, Boukhriss M (2009) Saline water effects on fruit development, quality and phenolic composition of virgin olive oils cv. *Chemlali*. J Agric Food Chem 57:2803–2811

Al-Saghir S, Thurner K, Wagner KH, Frisch G, Luf W, Razzazi-Fazeli E, Elmadfa I (2004) Effects of different cooking procedures on lipid quality and cholesterol oxidation of farmed salmon fish (Salmo salar). J Agric Food Chem 52(16):5290–5296

Andrewes P, Busch JL, de Joode T, Groenewegen A, Alexandre H (2003) Sensory properties of virgin olive oil polyphenols: identification of deacetoxy-ligstrosideaglycon as a key contributor to pungency. J Agric Food Chem 51(5):1415–1420

Andrikopoulos NK, Dedoussis GV, Falirea A, Kalogeropoulos N, Hatzinikola HS (2002) Deterioration of natural antioxidant species of vegetable edible oils during the domestic deep-frying and pan-frying of potatoes. Int J Food Sci Nutr 53(4):351–363

Angerosa F (2002) Influence of volatile compounds on virgin olive oil quality evaluated by analytical approaches and sensor panels. Eur J Lipid Sci Technol 104(9–10):639–660

Angerosa F, d'Alessandro N, Corana F, Mellerio G (1996) Characterization of phenolic and secoiridoid aglycons present in virgin olive oil by gas chromatography-chemical ionization mass spectrometry. J Chromatogr A 736(1–2):195–203

Angerosa F, Di Giacinto L, Solinas M (1992) Influence of Dacus Oleae infestation on flavor of oils, extracted from attacked olive fruits, by HPLC and HRGC analyses of volatile compounds. Grasas Aceites 43:134–142

Angerosa F, Di Giacinto L (1995) Caratteristiche di qualità dell'olio di oliva vergine in relazione ai metodi di frangitura. II. Rivista Italiana delle Sostanze Grasse 72(1):1–4

Angerosa F, Basti C, Vito R (1999) Virgin olive oil volatile compounds from lipoxygenase pathway and characterization of some Italian cultivars. J Agric Food Chem 47(3):836–839

Angerosa F, Servili M, Selvaggini R, Taticchi A, Esposto S, Montedoro G (2004) Volatile compounds in virgin olive oil: occurrence and their relationship with the quality. J Chromatogr A 1054(1):17–31

Angerosa F, Campestre C, Giansante L (2006) Analysis and authentication. In: Boskou D (ed) Olive oil: chemistry and technology. AOCS Press, Champaign, IL, pp 113–172

Angerosa F, Mostallino R, Basti C, Vito R (2000) Virgin olive oil odour notes: their relationships with volatile compounds from the lipoxygenase pathway and secoiridoid compounds. Food Chem 68(3):283–287

Antoun N, Tsimidou M (1997) Gourmet olive oils: stability and consumer acceptability studies. Food Res Int 30(2):131–136

Aparicio R, Harwood J (2013) Handbook of olive oil. Analysis and properties, 2nd edn. Springer, New York

Aparicio R, Luna G (2002) Characterisation of monovarietal virgin olive oils. Eur J Lipid Sci Technol 104(9–10):614–627

Aparicio R, Morales MT (1998) Characterization of olive ripeness by green aroma compounds of virgin olive oil. J Agric Food Chem 46(3):1116–1122

Aparicio R, Morales MT, Alonso V (1997) Authentication of European virgin olive oils by their chemical compounds, sensory attributes, and consumers' attitudes. J Agric Food Chem 45 (4):1076–1083

Baccouri O, Bendini A, Cerretani L, Guerfel M, Baccouri B, Lercker G, Zarrouk M, Miled DDB (2008) Comparative study on volatile compounds from Tunisian and Sicilian monovarietal virgin olive oils. Food Chem 111:322–328

Bazakos C, Spaniolas S, Kalaitzis P (2016) DNA-based approaches for traceability and authentication of olive oil. In: Boskou D, Clodoveo ML (eds) Products from olive tree. InTech, Rijeka, pp 115–133

Beltrán G, del Rio C, Sánchez S, Martínez L (2004) Influence of harvest date and crop yield on the fatty acid composition of virgin olive oils from cv. Picual. J Agric Food Chem 52 (11):3434–3440

Bonanome A, Visona A, Lusiani L, Beltramello G, Confortin L, Biffanti S, Pagnan A (1991) Carbohydrate and lipid metabolism in patients with non-insulin-dependent diabetes mellitus: effects of a low-fat, high-carbohydrate diet vs a diet high in monounsaturated fatty acids. Am J Clin Nutr 54(3):586–590

Bosetti C, Negri E, Franceschi S, Talamini R, Montella M, Conti E, La Vecchia C (2002) Olive oil, seed oils and other added fats in relation to ovarian cancer (Italy). Cancer Causes Control 13 (5):465–470

Boskou D (ed) (2006) Olive oil: chemistry and technology. AOCS Press, Urbana, IL

Boskou D (ed) (2008) Olive oil: minor constituents and health. CRC Press, Boca Raton, FL

Boskou D (ed) (2015) Olive and olive oil bioactive constituents. AOCS Press, Urbana IL

Boskou D, Blekas G, Tsimidou M (2006) Chapter 4. Olive oil composition. In: Boskou D (ed) Olive oil: chemistry and technology. AOCS Press, Champaign, IL

Boskou D, Camposeo S, Clodoveo ML (2015) Table olives as sources of bioactive compounds. In: Boskou D (ed) Olives and olive oil bioactive constituents. AOCS Press, Urbana, IL, pp 217–259

Caponio F, Giarnetti M, Paradiso VM, Summo C, Gomes T (2013) Potential use of extra virgin olive oil in bakery products rich in fats: a comparative study with refined oils. Int J Food Sci Technol 48(1):82–88

Caporale G, Policastro S, Monteleone E (2004) Bitterness enhancement induced by cut grass odorant (cis-3-hexen-1-ol) in a model olive oil. Food Qual Prefer 15(3):219–227

Caporaso N, Formisano D (2016) Developments, applications, and trends of molecular gastronomy among food scientists and innovative chefs. Food Rev Intl 32(4):417–435

Caporaso N, Paduano A, Nicoletti G, Sacchi R (2013) Capsaicinoids, antioxidant activity, and volatile compounds in olive oil flavored with dried chili pepper (Capsicum annuum). Eur J Lipid Sci Technol 115(12):1434–1442

Caporaso N, Panariello V, Sacchi R (2015a) The "True" Neapolitan pizza: assessing the influence of extra virgin olive oil on pizza volatile compounds and lipid oxidation. J Culinary Sci Technol 13(1):29–48

Caporaso N, Savarese M, Paduano A, Guidone G, De Marco E, Sacchi R (2015b) Nutritional quality assessment of extra virgin olive oil from the Italian retail market: do natural antioxidants satisfy EFSA health claims? J Food Compos Anal 40:154–162

Caporaso N, Genovese A, Burke R, Barry-Ryan C, Sacchi R (2016a) Physical and oxidative stability of functional olive oil-in-water emulsions formulated using olive mill wastewater biophenols and whey proteins. Food Funct 7(1):227–238

Caporaso N, Genovese A, Burke R, Barry-Ryan C, Sacchi R (2016b) Effect of olive mill wastewater phenolic extract, whey protein isolate and xanthan gum on the behaviour of olive O/W emulsions using response surface methodology. Food Hydrocoll 61:66–76

Caporaso N, Formisano D, Genovese A (2017) Use of phenolic compounds from olive mill wastewater as valuable ingredients for functional foods. Crit Rev Food Sci Nutr 58:2829–2841

Cardoso SM, Catarino MD, Semião MS, Pereira OR (2014) Virgin olive oil as a source of anti-inflammatory agents. In: de Leonardis A (ed) Virgin olive oil, production, composition, uses and benefits for man. Nova Science Publishers, Hauppauge, NY, pp 187–209

Clodoveo ML, Camposeo S, Amirante R, Dugo G, Cicero N, Boskou D (2015) Research and innovative approaches to obtain virgin olive oils with a higher level of bioactive constituents. In: Boskou D (ed) Olive and olive oil bioactive constituents. AOCS Press, Urbana IL, pp 179–215, Chapter 7

Codex Alimentarius (2003) Codex standard for olive oils, and olive pomace oils, CODEX STAN 33-1981. Codex Alimentarius, Roma, Itália, rev, 2

Conde C, Delrot S, Geros H (2008) Physiological, biochemical and molecular changes occuring during olive development and ripening. J Plant Physiol 165:1545–1562

Covas MI, Ruiz-Gutiérrez V, De La Torre R, Kafatos A, Lamuela-Raventós RM, Osada J, Visioli F (2006) Minor components of olive oil: evidence to date of health benefits in humans. Nutr Rev 64(suppl_4):S20–S30

Dhifi W, Hamrouni I, Ayachi S, Chahed T, Saidani M, Marzouk B (2004) Biochemical characterization of some tunisian olive oils. J Food Lipids 11:287–296

EC Regulation (1991) No. 2568/91/EEC, 1991. Off J Eur Commun L 248:1–83, July 11

EC Regulation (2002) No. 796/02. Off J Eur Commun L 123:8–28

EC Regulation (2012) No. 432/2012 of 16 May 2012 establishing a list of permitted health claims made on foods, other than those referring to the reduction of disease risk and to children's development and health. Off J Eur Union L 136:1–40

EC Regulation (2013) No. 1332/2013 amending Regulation (EU) No. 36/2012 concerning restrictive measures in view of the situation in Syria, 13 December 2013. OJ L, 335(14.12)

Féart C, Samieri C, Barberger-Gateau P (2010) Mediterranean diet and cognitive function in older adults. Curr Opin Clin Nutr Metab Care 13(1):14

Fini L, Hotchkiss E, Fogliano V, Graziani G, Romano M, De Vol EB, Ricciardiello L (2007) Chemopreventive properties of pinoresinol-rich olive oil involve a selective activation of the ATM–p53 cascade in colon cancer cell lines. Carcinogenesis 29(1):139–146
Galeone C, Talamini R, Levi F, Pelucchi C, Negri E, Giacosa A, Montella M, Franceschi S, La Vecchia C (2006) Fried foods, olive oil and colorectal cancer. Ann Oncol 18(1):36–39
Gambacorta G, Faccia M, Pati S, Lamacchia C, Baiano A, La Notte E (2007) Changes in the chemical and sensorial profile of extra virgin olive oils flavored with herbs and spices during storage. J Food Lipids 14(2):202–215
Garcia-Gonzales DL, Tena N, Aparicio R (2007) Characterization of olive paste volatiles to predict the sensory quality of virgin olive oil. Eur J Lipid Sci Technol 109:663–672
García-González DL, Tena N, Aparicio R (2010) Quality characterization of the new virgin olive oil var. Sikitita by phenols and volatile compounds. J Agric Food Chem 58(14):8357–8364
Garcia JM, Seller S, Perez-Camino MC (1996) Influence of fruit ripening on olive oil quality. J Agric Food Chem 44(11):3516–3520
Genovese A, Caporaso N, De Luca L, Paduano A, Sacchi R (2015) Influence of olive oil phenolic compounds on headspace aroma release by interaction with whey proteins. J Agric Food Chem 63(15):3838–3850
Gomes da Silva MDR, Costa Freitas AM, Cabrita MJB, Garcia R (2011) Olive oil composition: volatile compounds. In: Boskou D (ed) Olive oil—constituents, quality, health properties and bioconversions. InTech, Rijeka, pp 17–63
Gómez-Caravaca AM, Cerretani L, Bendini A, Segura-Carretero A, Fernández-Gutiérrez A, Del Carlo M, Compagnone D, Cichelli A (2008) Effects of fly attack (Bactrocera oleae) on the phenolic profile and selected chemical parameters of olive oil. J Agric Food Chem 56 (12):4577–4583
Gomez-Rico A, Salvador MD, Greca ML, Fregapane G (2006) Phenolic and volatile compounds of extra virgin olive oil (Olea europaea L. Cv Cornicabra) with regard to fruit ripening and irrigation management. J Agric Food Chem 54:7130–7136
Gomez-Rico A, Salvador MD, Fregapane G (2009) Virgin olive oil and olive fruit minor constituents as affected by irrigation management based on SWP and TDF as compared to ETc in medium-density young olive orchards (Olea europaea L. cv. Cornicabra and Morisca). Food Res Int 42:1067–1076
Gucci R, Lodolini EM, Rapoport HF (2007) Productivity of olive trees with different water status and crop load. J Hortic Sci Biotechnol 82(4):648–656
Guth H, Grosch W (1991) A comparative study of the potent odorants of different virgin olive oils. Eur J Lipid Sci Technol 93(9):335–339
Gutiérrez-Rosales F, Rios JJ, Gomez-Rey ML (2003) Main polyphenols in the bitter taste of virgin olive oil. Structural confirmation by on-line high-performance liquid chromatography electrospray ionization mass spectrometry. J Agric Food Chem 51(20):6021–6025
Hornero-Méndez D, Mínguez-Mosquera MI (2007) Bioaccessibility of carotenes from carrots: effect of cooking and addition of oil. Innov Food Sci Emerg Technol 8(3):407–412
International Olive Council (IOC) (2013) Trade standard applying to olive oil and olive-pomace oil. COI/T 15/NC No 3/Rev 7, May 2013
International Olive Council (IOC) (2015a) World olive oil production statistics
International Olive Council (IOC) (2015b) Sensory analysis of olive oil—Method for the organoleptic assessment of Virgin Olive Oil. 2015, COI/T.20/Doc. 15/ Rev. 8, November 2015
International Olive Council (IOC) (2015c) COI/T.20/Doc. 15/Rev. 7, February 2015. Method for the organoleptic assessment of Virgin Olive
Issaoui M, Jlamini G, Brahmi F, Dabbou S, Hassine KB, Taamali A, Chehab H, Ellouz M, Zarrouk M, Hammami M (2009) Effect of the growing area conditions on differentiation between Chemlali and Chetoui olive oils. Food Chem 119:220–225
Jiménez-Colmenero F, Herrero A, Pintado T, Solas MT, Ruiz-Capillas C (2010) Influence of emulsified olive oil stabilizing system used for pork backfat replacement in frankfurters. Food Res Int 43(8):2068–2076

Kalogeropoulos N, Kaliora AC (2015) Effect of fruit maturity on olive oil phenolic composition and antioxidant capacity. In: Boskou D (ed) Olive and olive oil bioactive constituents. AOCS Press, Urbana, IL, pp 123–147

Kalogeropoulos N, Chiou A, Mylona A, Ioannou MS, Andrikopoulos NK (2007) Recovery and distribution of natural antioxidants (α-tocopherol, polyphenols and terpenic acids) after pan-frying of Mediterranean finfish in virgin olive oil. Food Chem 100(2):509–517

Kalua CM, Bedgood DR, Bishop AG, Prenzler PD (2006) Changes in volatile and phenolic compounds with malaxation time and temperature during virgin olive oil production. J Agric Food Chem 54:7641–7651

Kalua CM, Allen MS, Bedgood DR, Bishop AG, Prenzler PD, Robards K (2007) Olive oil volatile compounds, flavour development and quality: a critical review. Food Chem 100(1):273–286

Keys A, Fidanza F, Scardi V, Bergami G, Keys MH, Di Lorenzo F (1954) Studies on serum cholesterol and other characteristics of clinically healthy men in Naples. AMA Arch Intern Med 93(3):328–336

Kotsiou K, Tasioula-Margari M (2016) Monitoring the phenolic compounds of Greek extra-virgin olive oils during storage. Food Chem 200:255–262

Lanza B (2012) Nutritional and sensory quality of table olives. In: Muzzalupo I (ed) Olive germplasm-the olive cultivation, table olive and olive oil industry in Italy. InTech, Rijeka, ISBN: 978-953-51-0883-2

Lercker G, Rodriguez-Estrada MT (2000) Chromatographic analysis of unsaponifiable compounds of olive oils and fat-containing foods. J Chromatogr A 881(1):105–129

López-López I, Cofrades S, Yakan A, Solas MT, Jiménez-Colmenero F (2010) Frozen storage characteristics of low-salt and low-fat beef patties as affected by Wakame addition and replacing pork backfat with olive oil-in-water emulsion. Food Res Int 43(5):1244–1254

Luna G (2002) Characterisation of monovarietal virgin olive oils. Eur J Lipid Sci Technol 104:614–627

Luna, G., Morales, M. T., & Aparicio, R. (2006). Characterisation of 39 varietal virgin olive oils by their volatile compositions. Food Chemistry, 98(2), 243-252

Malheiro R, Rodrigues N, Pereira JA (2015) Olive oil phenolic composition as affected by geographic origin, olive cultivar, and cultivation systems. In: Boskou D (ed) Olive and olive oil bioactive constituents. Academic Press and AOCS Press, Waltham, MA, pp 93–122

Mateos R, Cert A, Pérez-Camino MC, García JM (2004) Evaluation of virgin olive oil bitterness by quantification of secoiridoid derivatives. J Am Oil Chem Soc' 81(1):71–75

Matsakidou A, Blekas G, Paraskevopoulou A (2010) Aroma and physical characteristics of cakes prepared by replacing margarine with extra virgin olive oil. LWT-Food Sci Technol 43 (6):949–957

Mattson FH, Grundy SM (1985) Comparison of effects of dietary saturated, monounsaturated, and polyunsaturated fatty acids on plasma lipids and lipoproteins in man. J Lipid Res 26(2):194–202

Menendez JA, Vazquez-Martin A, Oliveras-Ferraros C, Garcia-Villalba R, Carrasco-Pancorbo A, Fernandez-Gutierrez A, Segura-Carretero A (2009) Extra-virgin olive oil polyphenols inhibit HER2 (erbB-2)-induced malignant transformation in human breast epithelial cells: relationship between the chemical structures of extra-virgin olive oil secoiridoids and lignans and their inhibitory activities on the tyrosine kinase activity of HER2. Int J Oncol 34(1):43–51

Montaño A, Sánchez AH, López-López A, de Castro A, Rejano L (2010) Chemical composition of fermented green olives: acidity, salt, moisture, fat, protein, ash, fiber, sugar, and polyphenol. In: Preedy VR, Watson RR (eds) Olives and olive oil in health and disease prevention. Academic Press, London, UK, pp 291–297

Montedoro G, Bertuccioli M, Anichini F (1978) Aroma analysis of virgin olive oil by head space (volatiles) and extraction (polyphenols) techniques. In: Charalambous G, Inglett GE (eds) Flavor of foods and beverages. Academic Press, New York, pp 247–281

Monti SM, Ritieni A, Sacchi R, Skog K, Borgen E, Fogliano V (2001) Characterization of phenolic compounds in virgin olive oil and their effect on the formation of carcinogenic/mutagenic heterocyclic amines in a model system. J Agric Food Chem 49(8):3969–3975

Morales MT, Rios JJ, Aparicio R (1997) Changes in the volatile composition of virgin olive oil during oxidation: flavors and off-flavors. J Agric Food Chem 45(7):2666–2673

Napolitano A, Morales F, Sacchi R, Fogliano V (2008) Relationship between virgin olive oil phenolic compounds and acrylamide formation in fried crisps. J Agric Food Chem 56 (6):2034–2040

Ninfali P, Bacchiocca M, Biagiotti E, Esposto S, Servili M, Rosati A, Montedoro F (2008) A 3-year study on quality, nutritional and organoleptic evaluation of organic and conventional extra-virgin olive oils. J Am Oil Chem Soc 85:151–158

Paraskevopoulou A, Boskou D, Kiosseoglou V (2005) Stabilization of olive oil–lemon juice emulsion with polysaccharides. Food Chem 90(4):627–634

Paraskevopoulou D, Boskou D, Paraskevopoulou A (2007) Oxidative stability of olive oil–lemon juice salad dressings stabilized with polysaccharides. Food Chem 101(3):1197–1204

Patumi M, d'Andria R, Fontanazza G, Morelli G, Giorgio P, Sorrentino G (1999) Yield and oil quality of intensively trained trees of three cultivars of olive (Olea europaea L.) under different irrigation regimes. J Hortic Sci Biotechnol 74(6):729–737

Pérez-Jiménez F, Ruano J, Perez-Martinez P, Lopez-Segura F, Lopez-Miranda J (2007) The influence of olive oil on human health: not a question of fat alone. Mol Nutr Food Res 51 (10):1199–1208

Pérez-López FR, Chedraui P, Haya J, Cuadros JL (2009) Effects of the Mediterranean diet on longevity and age-related morbid conditions. Maturitas 64(2):67–79

Peri C (ed) (2014) The extra-virgin olive oil handbook. Wiley, Chichester, West Sussex

Perona JS, Botham KM (2013) Olive oil as a functional food: nutritional and health benefits. In: Aparicio R, Harwood J (eds) Handbook of olive oil. Springer, New York, pp 677–714

Perona JS, Cabello-Moruno R, Ruiz-Gutierrez V (2006) The role of virgin olive oil components in the modulation of endothelial function. J Nutr Biochem 17(7):429–445

Piperno A, Toscano M, Uccella NA (2004) The Cannizzaro-like metabolites of secoiridoid glucosides in some olive cultivars. J Sci Food Agric 84:341–349

Piscopo A, Poiana M (2012) Packaging and storage of olive oil. In: Muzzalupo I (ed) Olive germplasm-the olive cultivation, table olive and olive oil industry in Italy. InTech, Rijeka, pp 291–222

Pitt J, Roth W, Lacor P, Smith AB III, Blankenship M, Velasco P, De Felice F, Breslin P, Klein WL (2009) Alzheimer's-associated Aβ oligomers show altered structure, immunoreactivity and synaptotoxicity with low doses of oleocanthal. Toxicol Appl Pharmacol 240(2):189–197

Preedy VR, Watson RR (eds) (2010) Olives and olive oil in health and disease prevention. Academic Press, London, UK

Raieta K, Muccillo L, Colantuoni V (2015) A novel reliable method of DNA extraction from olive oil suitable for molecular traceability. Food Chem 172:596–602

Ramírez MR, Estévez M, Morcuende D, Cava R (2004) Effect of the type of frying culinary fat on volatile compounds isolated in fried pork loin chops by using SPME-GC-MS. J Agric Food Chem 52(25):7637–7643

Ranalli A, Malfatti A, Cabras P (2001) Composition and quality of pressed virgin olive oils extracted with new enzyme processing aid. J Food Sci 66:592–600

Rasmussen OW, Thomsen C, Hansen KW, Vesterlund M, Winther E, Hermansen K (1993) Effects on blood pressure, glucose, and lipid levels of high-monounsaturated fat diet compared with a high-carbohydrate diet in NIDDM subjects. Diabetes Care 16(12):1565–1571

Rodis PS, Karathanos VT, Mantzavinou A (2002) Partitioning of olive oil antioxidants between oil and water phases. J Agric Food Chem 50(3):596–601

Roselli L, Clodoveo ML, Corbo F, De Gennaro B (2017) Are health claims a useful tool to segment the category of extra-virgin olive oil? Threats and opportunities for the Italian olive oil supply chain. Trends Food Sci Technol 68:176–181

Rotondi A, Bendini A, Cerretani L, Mari M, Lercker G, Toschi TG (2004) Effect of olive ripening degree on the oxidative stability and organoleptic properties of cv. Nostrana di Brisighella extra virgin olive oil. J Agric Food Chem 52(11):3649–3654

Sacchi R, Caporaso N, Paduano A, Genovese A (2015) Industrial-scale filtration affects volatile compounds in extra virgin olive oil cv. Ravece. Eur J Lipid Sci Technol 117(12):2007–2014
Sacchi R, Della Medaglia D, Paduano A, Caporaso N, Genovese A (2017) Characterisation of lemon-flavoured olive oils. LWT-Food Sci Technol 79:326–332
Sacchi R, Paduano A, Savarese M, Vitaglione P, Fogliano V (2014) Extra virgin olive oil: from composition to "molecular gastronomy". In: Advances in nutrition and cancer. Springer, Berlin, pp 325–338
Saghir S, Wagner KH, Elmadfa I (2005) Lipid oxidation of beef fillets during braising with different cooking oils. Meat Sci 71(3):440–445
Sakouhi F, Absalon C, Sebei K, Fouquet E, Boukhchina S, Kallel H (2009) Gas chromatographic–mass spectrometric characterisation of triterpene alcohols and monomethylsterols in developing Olea europaea L. fruits. Food Chem 116(1):345–350
Salas JJ, Sanchez J (1997) Biogenesis of alcohols present in the aroma of virgin olive oil. In: Williams JP, Khan MU, Lem NW (eds) Physiology, biochemistry and molecular biology of plant lipids. Springer, The Netherlands, pp 328–330
Sánchez J, Harwood JL (2002) Biosynthesis of triacylglycerols and volatiles in olives. Eur J Lipid Sci Technol 104(9–10):564–573
Sangiovanni E, Colombo E, Fumagalli M, Abbiati F, Caruso D, Dell'Agli M (2012) Inhibition of NF-κB activity by minor polar components of extra-virgin olive oil at gastric level. Phytother Res 26(10):1569–1571
Savarese M, Caporaso N, De Marco E, Sacchi R (2013) Extra virgin olive oil overall quality assessment during prolonged storage in PET containers. In: Proceedings in GV—the 1st global virtual conference, vol. I, issue 1. ISBN: 978-80-554-0649-7, ISSN: 1339-2778, p 670–674
Servili M, Selvaggini R, Taticchi A, Esposito S, Montedoro GF (2003) Air exposure time of olive pastes during the extraction process and phenolic composition of virgin olive oil. J Am Oil Chem Soc 80:685–695
Servili M, Taticchi A, Esposto S, Urbani S, Selvaggini R, Montedoro G (2008) Influence of the decrease in oxygen during malaxation of olive paste on the composition of volatiles and phenolic compounds in virgin olive oil. J Agric Food Chem 56(21):10048–10055
Servili M, Esposto S, Fabiani R, Urbani S, Taticchi A, Mariucci F et al (2009) Phenolic compounds in olive oil: antioxidant, health and organoleptic activities according to their chemical structure. Inflammopharmacology 17(2):76–84
Sirtori CR, Tremoli E, Gatti E, Montanari G, Sirtori M, Colli S et al (1986) Controlled evaluation of fat intake in the Mediterranean diet: comparative activities of olive oil and corn oil on plasma lipids and platelets in high-risk patients. Am J Clin Nutr 44(5):635–642
Škevin D, Rade D, Štrucelj D, Mokrovšak Ž, Neđeral S, Benčić Đ (2003) The influence of variety and harvest time on the bitterness and phenolic compounds of olive oil. Eur J Lipid Sci Technol 105(9):536–541
Spinelli S (2014) Investigating the culinary use of olive oils. In: Monteleone E, Langstaff S (eds) Olive oil sensory science. Wiley, Chichester, West Sussex, pp 195–225
Stefanoudaki E, Williams M, Chartzoulakis K, Harwood J (2009) Effect of irrigation on quality attributes of olive oil. J Agric Food Chem 57(15):7048–7055
Tamendjari A, Angerosa F, Mettouchi S, Bellal MM (2009) The effect of fly attack (Bactrocera oleae) on the quality and phenolic content of Chemlali olive oil. Grasas y Aceites 60(5):509–515
Tierney AC, Roche HM (2007) The potential role of olive oil-derived MUFA in insulin sensitivity. Mol Nutr Food Res 51(10):1235–1248
Tognetti R, D'Andria R, Sacchi R, Lavini A, Morelli G, Alvino A (2007) Deficit irrigation affects seasonal changes in leaf physiology and oil quality of Olea europaea (cultivars Frantobio and Leccino). Ann Appl Biol 150:169–186
Traynor MP, Burke R, O'Sullivan MG, Hannon JA, Barry-Ryan C (2013) Sensory and chemical interactions of food pairings (basmati rice, bacon and extra virgin olive oil) with banana. Food Res Int 54(1):569–577

Trichopoulou A, Vasilopoulou E (2000) Mediterranean diet and longevity. Br J Nutr 84(S2):S205–S209

Trichopoulou A, Costacou T, Bamia C, Trichopoulos D (2003) Adherence to a Mediterranean diet and survival in a Greek population. N Engl J Med 348(26):2599–2608

Tsimidou MZ, Georgiou A, Koidis A, Boskou D (2005) Loss of stability of "veiled" (cloudy) virgin olive oils in storage. Food Chem 93(3):377–383

Tura D, Failla O, Bassi D, Pedò S, Serraiocco A (2008) Cultivar influence on virgin olive (Olea europaea L.) oil flavor based on aromatic compounds and sensorial profile. Scientiahorticulturae 118(2):139–148

Vassiliou EK, Gonzalez A, Garcia C, Tadros JH, Chakraborty G, Toney JH (2009) Oleic acid and peanut oil high in oleic acid reverse the inhibitory effect of insulin production of the inflammatory cytokine TNF-α both in vitro and in vivo systems. Lipids Health Dis 8(1):25

Veneziani G, Sordini B, Taticchi A, Esposto S, Selvaggini R, Urbani S et al (2016) Improvement of olive oil mechanical extraction: new technologies, process efficiency, and extra virgin olive oil quality. In: Boskou D (ed) Products from olive tree. InTech, Rijeka

Vichi S, Pizzale L, Conte LS, Buxaderas S, Lopez-Tamames E (2003) Solid-phase microextraction in the analysis of virgin olive oil volatile fraction: modifications induced by oxidation and suitable markers of oxidative status. J Agric Food Chem 51:6564–6571

Visioli F, Poli A, Gall C (2002) Antioxidant and other biological activities of phenols from olives and olive oil. Med Res Rev 22(1):65–75

Vitaglione P, Savarese M, Paduano A, Scalfi L, Fogliano V, Sacchi R (2015) Healthy virgin olive oil: a matter of bitterness. Crit Rev Food Sci Nutr 55(13):1808–1818

Zullo BA, Ciafardini G (2017) Changes in physicochemical and microbiological parameters of short and long-lived veiled (cloudy) virgin olive oil upon storage in the dark. Eur J Lipid Sci Technol 120(1):1700309

Chapter 10
Flaxseed (*Linum usitatissimum*)

Sangita Ganguly, Narender Raju Panjagari, and Rakesh Kumar Raman

Abstract Flaxseed is an important crop being used for multiple purposes with increasing applications in functional food development because of its high content of bioactive fatty acids (omega-3) and other phytonutrients. It contains several biologically active components that prevent and cure several physiological conditions and noncommunicable diseases such as dyslipidemia, obesity, diabetes mellitus, several types of cancer, kidney and renal failure, bowel syndrome, immune functions, etc. Flaxseed products such as whole seed, meal, oil, or mucilage are predominantly used for functional food development. In the present chapter, the proximates composition, mechanisms involved in the health benefits, and the effect of flaxseed on the quality of different food products have been discussed.

Keywords Flaxseed oil · Linseed · Alpha-linolenic acid · Secoisolariciresinol diglucoside · Food applications · Omega-3 fatty acids

10.1 Introduction

Flaxseed (*Linum usitatissimum*), also known as linseed, is grown as rabi crop in India in winter season. The plant belonging to the Linaceae family is a self-pollinated herb with 60–100 cm length. There are major two types of cultivars, namely, fiber type and seed type. The fiber-type cultivars are slender and more branched, whereas the seed-type cultivars are shorter and more branched. It grows under the rainfall of 45–75 cm in sandy loam or clay loam soil of pH 5–7. Flaxseed is an important crop of which all parts of the plant can be utilized for industrial, food, as well as feed purposes. The stem yields good quality dietary fiber, while the seed provides oil rich in bioactive components such as α-linolenic acid (ALA, omega-3), phytosterols, etc. The whole grain approximately sized 2.5 × 5.0 × 1.5 mm is flat and oval with a color of dark to yellow (Freeman 1995). Like most other seeds, it contains a seed

S. Ganguly · N. R. Panjagari (✉) · R. K. Raman
Dairy Technology Division, ICAR-National Dairy Research Institute, Karnal, India

B. Tanwar, A. Goyal (eds.), *Oilseeds: Health Attributes and Food Applications*,
https://doi.org/10.1007/978-981-15-4194-0_10

coat and endosperm (36%), two embryos (55%), and an embryo axis (4%) (Bhatty 1995).

Although earlier considered as oilseed crop, flaxseed has been emerging as an attractive nutritional component due to its exceptionally high content of bioactive compounds (Kajla et al. 2015). Flaxseed is gaining additional focus among nutritionists and medical professionals due to its beneficial properties because of the presence of biologically active components providing prevention and cure in several physiological conditions and diseases (Al-Okbi 2005). Whole seed, meal oil, or mucilage of flaxseed are used to develop functional food products such as bread, muffin, chocolate, ice cream, etc. (Gowda et al. 2018; Kajla et al. 2015).

10.2 History of Flaxseed

Since ancient times, flaxseed has been cultivated and used in many aspects of everyday life (Goyal et al. 2014). Archeologically, its use is detectable in prehistoric times approximately 10,000 to 12,000 years ago (Fuller and Harvey 2006). It was used in southern Mesopotamia during 5200–4000 B.C., while Babylonians cultivated it as early as 3000 B.C. It was first identified in Eastern Turkey to make cloth named as linen during 6000 B.C. (Judd 1995). The botanical name of flaxseed, *Linum usitatissimum*, means "very useful" in Latin indicating its usefulness. Colonist was the first to introduce flaxseed or flax to the United States, primarily for fiber of clothing (Goyal et al. 2014). Flaxseed had been primarily used for fabrication of cloths and papers until the 1990s. Documented evidences are available for the presence of linen cloth in ancient Egyptian tombs and Babylonians burial chambers, garments made from flax worn by Jewish high priests (Madhusudhan 2009). The oil obtained from the seed was used in paint industry, and meal was used as feed for animals and fertilizer (Singh et al. 2011; Judd 1995).

Hippocrates, the father of medicine, recommended flaxseed for the relief of abdominal pain during 650 B.C., whereas Theophrastus advocated it for cold remedy (Goyal et al. 2014). Flax had been indicated in Ayurveda (1200 B.C.) for controlling mental and physical fatigue and aging process (Goyal et al. 2014). During middle ages, flaxseed oil was also used for the treatment of kidney disorders (Moghaddasi 2011). Apart from this, flax preparations were widely used as healing agent, treatment of gastrointestinal disorder, cough and cold remedy, and pain and anti-inflammatory agents (Ivanova et al. 2011; Moghaddasi 2011). In southern India, flaxseed is being consumed occasionally as *chutney* and also used as raw material for certain medicines (Faseehuddin and Madhusudhan 2007a, b). Although before World War II, flaxseed was commonly used as food, it was discontinued after that. However, it has been reintroduced as a functional food during the 1990s.

10.3 Production of Flaxseed

With an annual total annual production of about 2.92 million tons, flaxseed is produced in more than 50 countries (FAO 2016). The major flaxseed-producing countries are Canada, India, China, the USA, and Ethiopia (Singh et al. 2011) with Canada being the world's largest producer and exporter (Oomah 2001).

10.4 Composition of Flaxseed

Flaxseed contains approximately 41% fat, 20% protein, 28% dietary fiber, and 4% ash content (Table 10.1) (Morris 2007a; Gopalan et al. 2007). The potential functional components of flaxseed include alpha-linolenic acid, lignan, fiber, and phenolic components (Oomah 2001). The composition of flaxseed varies due to several factors such as growing environment, genetic, and processing conditions (Morris 2007a). The major vitamins of flaxseed are vitamins A, C, and E, whereas the major minerals are phosphorous, magnesium, potassium, sodium, iron, copper, manganese, and zinc.

10.4.1 Lipid

The major oil storage sites of flaxseed are the cotyledons containing 75% of oil (37–45%), while the remaining (22%) is present in seed coat and endosperm (Singh et al. 2011; Morris 2007a; Payne 2000). Flaxseed oil contains 98% triacylglycerol and phospholipid and small amounts of free fatty acids (0.1%) (Mueller et al. 2010). Flaxseed oil is rich in PUFA (polyunsaturated fatty acids) (73%) with lower amount of saturated fatty acids (9%) (Cunnane et al. 1993a, b). α-Linolenic acid (40–60%),

Table 10.1 Proximate composition of flaxseed

Components	Amount (g/100 g, db)
Moisture	8
Protein	20
Total fat	41
Saturated fatty acids	9
MUFA	18
PUFA	
ω-3 fatty acids	57
ω-6 fatty acids	16
Total dietary fiber	27
Mineral	4.0
Energy (kcal)	530.0

Source: Madhusudhan (2009)

linoleic acid (12–17%), oleic acid (13–19%), palmitic acid (5–8%), and stearic acid (2–4.5%) are the major fatty acids present in flaxseed oil. As an exclusive source of ω-3 fatty acids, flaxseed contributes alpha-linolenic acid (ALA) to vegetarian diets (Riediger et al. 2009). The ratio of $\omega_6:\omega_3$ in flaxseed is about 0.3 (Pellizzon et al. 2007). Monoglycerides, diaglycerides, tocopherols, and sterols and their esters, phospholipids, waxes, cyclolinopeptides (CL_s), free fatty acid (FFAs), carotenoids, chlorophyll, etc. are the minor lipid and lipid-soluble compounds present in flaxseed (Shim et al. 2014). Depending on the source, the bioavailability of ALA may vary. As compared to the whole seed, oil and milled seeds have shown higher bioavailability of ALA (Austria et al. 2008).

10.4.2 Protein

Protein content of flaxseed ranged between 20% and 30% (Hall et al. 2006), majority of which (56–70%) is found in cotyledons, whereas seed coat and endosperm contains the remaining 30% (Dev et al. 1986). Globulins (80%) containing linin and conlinin fractions are the major storage proteins of flaxseed (Madhusudhan 2009; Hall et al. 2006). Flaxseed protein is rich in amino acids such as arginine, aspartic acid, and glutamic acid, whereas the limiting amino acid is lysine (Table 10.2) (Singh et al. 2011). The presence of good quantities of cysteine and

Table 10.2 Approximate quantities of amino acids in flaxseed protein

Name of the amino acid	Quantity (g/100 g protein)
Alanine	4.4
Arginine	9.2
Aspartic acid	9.3
Cystine	1.1
Glutamic acid	19.6
Glycerine	5.8
Histidine	2.2
Isoleucine	4.0
Lysine	4.0
Methionine	1.5
Phenylalanine	4.6
Proline	3.5
Serine	4.5
Threonine	3.6
Tryptophan	1.8
Tyrosine	2.3
Valine	4.6

Source: Shim et al. (2014)

methionine is attributed to improved antioxidant ability, thus reducing the risk of cancer (Oomah 2001).

10.4.3 Carbohydrates and Dietary Fiber

Flaxseed contains very low level of carbohydrate (1 g/100 g) and thus contributes very less in total carbohydrate intake (Morris 2007a). Flaxseed contains high amount of dietary fiber, both soluble (9%) and insoluble (20%) fiber whose proportions vary in the range between 20:80 and 40:60 (Morris 2003; Cui 2001; Hadley et al. 1992). Cellulose and lignin form the insoluble fraction, while mucilage gums form the part of soluble fraction of flaxseed dietary fiber (Vaisey-Genser and Morris 2003). The soluble fiber portion is majorly present in the seed coat. Soluble fiber can be easily extracted with hot water (Cui et al. 1994) than cold water (Paynel et al. 2013). Acidic polysaccharides (L-rhamnose, L-galactose, L-fructose, and D-xylose) and neutral polysaccharides (L-arabinose and D-xylose/D-galactose) are the major components of soluble fiber (Anderson and Lowe 1947). Insoluble fiber contains cellulose, lignin, and acid detergent fiber, whereas the mucilage arabinogalactans are associated with protein (Cui et al. 1994; Ray et al. 2013). Insoluble fiber acts as laxative and helps in reduction of constipation (Greenwald et al. 2001), whereas soluble fiber helps in maintenance of blood glucose and blood cholesterol level (Kristensen et al. 2012).

10.4.4 Polyphenols and Lignan

Phenolic compounds are known for their anticancer and antioxidative properties. Phenolic acids, flavonoids, and lignans are the three major phenolic compound groups present in flaxseed. Ferulic acid, chlorogenic acid, and gallic acid are major phenolic acids present in defatted flaxseed, while p-Coumaric acid glucosides, hydroxycinnamic acid glucosides, and 4-hydroxybenzoic acid are present in minor quantities (Beejmohun et al. 2007; Mazza 2008). Flavone C and flavone O-glycosides are the major flavonoids present in flaxseed (Mazza 2008). Lignans can act as both antioxidants and phytoestrogens. Secoisolariciresinol diglucoside (SDG), matairesinol, lariciresinol, and pinoresinol are the major lignan compounds present in flaxseed (Tourre and Xueming 2010; Milder et al. 2005). The amount of SDG varies in defatted flaxseed flour (11.7–24.1 mg/g) and whole flaxseed (6.1–13.3 mg/g) (Johnsson et al. 2000). The SDG found in flax can provide antioxidant effects (Adlercreutz 2007) and can be helpful in controlling cancerous tumor growth (Tham et al. 1998).

10.4.5 Minerals and Vitamins

Flaxseed contains good quantity of minerals especially phosphorous, magnesium, and calcium (Table 10.3). It contains high amounts of potassium and is comparable to sources rich in potassium such as banana. However, it has very low quantities of sodium (Morris 2007a). Flaxseed oil has high levels of beta-tocopherol (200 ppm) than alpha (15–20 ppm) and gamma-tocopherol (5–7 ppm) (Nagaraj 2009). Higher levels of unsaturation and lower levels of alpha and gamma-tocopherols indicate that flaxseed oil is highly unstable with limited shelf life and bad frying quality. Flaxseed oil has about 23–27 ppm campesterol and contains about 70 ppm of other phytosterols.

10.4.6 Anti-Nutritional Factors of Flaxseed

Flaxseed, despite having several functional ingredients, contains certain anti-nutritional factors that have adverse influence on nutritional absorbance in humans. The major anti-nutritional compounds present in flaxseed are cyanogenic glycosides (250–550 mg/100 g) in which linustatin and neolinistatin are the major fractions, while linamarin is in minor quantities (Singh et al. 2011; Kajla et al. 2015). The level of these glycosides depends upon several factors such as the cultivar, location, environmental conditions, etc. (Oomah et al. 1992). In the intestine, cyanogenic

Table 10.3 Mineral and vitamin profile of flaxseed

Micronutrient	Quantity (mg/100 g)
Potassium	831
Calcium	236
Phosphorus	622
Iron	2.7
Copper	1
Magnesium	431
Manganese	3
Sodium	27
Zinc	4
Ascorbic acid	0.50
Thiamine	0.53
Riboflavin	0.23
Niacin	3.21
Pyridoxine	0.61
Pantothenic acid	0.57
Folic acid	112
Biotin	6
α-Tocopherol	7

Source: Morris (2007b)

glycosides release poisonous hydrogen cyanide, a potential respiratory inhibitor (Tanwar et al. 2018). Hydrogen cyanide can be converted to thiocyanates by intestinal β-glycosidase which can interfere with iodine uptake and in chronic cases may lead to iodine deficiency disorders, goiter, and cretinism. Several thermal and nonthermal processing techniques such as autoclaving, microwave roasting, pelleting, and soaking in saline water and enzymatic treatments by detoxifying enzymes such as β-glycosidases may destroy cyanogenic glycosides (Feng et al. 2003; Yamashita et al. 2007; Tanwar et al. 2018).

Another important antinutrient present in flaxseed is phytic acid (23–33 g/kg of flaxseed meal) (Oomah et al. 1996a, b). Phytic acid, a strong chelator, reduces the bioavailability of minerals and interferes with the absorption of several important minerals (Akande et al. 2010). Flaxseed antinutrients have lesser impact on human health as compared to that of soybean and canola (Ganorkar and Jain 2013). Flaxseed contains linatine (antipyrodoxidine factor) (10 mg/100 g) and may induce vitamin B_6 deficiency (Mazza 2008). Flaxseed also contains trypsin inhibitors, but their activity has been found to be not significant (Bhatty 1993).

10.5 Bioactive Compounds of Flaxseed

Linolenic acid, linoleic acid, lignans, cyclic peptides, etc. are the important bioactive compounds present in flaxseed (Shim et al. 2014). Flaxseed oil contains bioactive peptides that oxidize during the initial storage period (Bruhl et al. 2007). Secoisolariciresinol diglucoside (SDG), a precursor of phytoestrogen also has phytochemical and antioxidant properties (Prasad 2004). Flaxseed lignan complex, an oil-insoluble alcoholic extract of flaxseed containing SDG, phenylpropanoids, and hydroxymethyl glutaric acid, is reported to have multiple physiological effects in animals and humans (Shim et al. 2014).

Flaxseed proteins, rich in arginine, are associated with lower level of blood pressure (Udenigwe et al. 2012). It is reported that when hypertensive rats (SHR) were fed flaxseed protein isolates (FPI) (200 mg/kg bw), effectively lowered blood pressure was observed (Shim et al. 2014). Flaxseed proteins digested with flavourzyme-generated ACE inhibitory peptides having hydroxyl radicals scavenging activity (Marambe et al. 2008). Flaxseed proteins hydrolyzed either by thermolysin and pronase (Udenigwe and Aluko 2010) or alcalase (Omoni and Aluko 2006) yielded biologically active peptides that have ACE inhibitory properties.

Gorski et al. (2001) have established important biological activity of cyclolinopeptides isolated from flaxseed oil such as immunosuppressive activities in mice and downregulation of cholesterol levels in rats. However, limited data is available regarding the release of such peptides under gut conditions (Marambe et al. 2011).

10.6 Health Attributes

Studies have revealed that apart from providing nutrition, flaxseed can exert beneficial health effects in the prevention and treatment of various diseases and disorders. Flaxseed has been found useful in preventing heart diseases, dyslipidemia, diabetes mellitus, osteoporosis, rheumatoid arthritis, several kinds of cancer, and constipation and also favorably affects immunity (Singh et al. 2011). The therapeutic attributes of flaxseed are mainly because of the presence of three major bioactive compounds, namely, omega-3 fatty acids, dietary fibers (soluble and insoluble), and lignans (Goyal et al. 2014).

10.6.1 Reduction of Dyslipidemia and Cardiovascular Disease

Dyslipidemia and cardiovascular disease are the serious concerns among the noncommunicable diseases and are the causes of mortality of a large number of people. High cholesterol level is a major precursor for endothelial dysfunction, atherosclerosis, and cardiovascular disease. The effectiveness of flaxseed in lowering serum lipid or controlling hypercholesterolemic atherosclerosis is mainly due to its components and fatty acid profile as reported by several researchers (Kristensen et al. 2011; Khalesi et al. 2011; Leyvaa et al. 2011). Reducing serum cholesterol level, platelet aggregation, and inflammatory markers and acting as an antioxidant (Ridges et al. 2001; Bloedon and Szapary 2004; Dupasquier et al. 2007) are the mechanisms by which flaxseed protects against cardiovascular disease. The major lipid component of flaxseed oil is alpha-linolenic acid (ALA), which further metabolized by enzymes cyclooxygenase and lipoxygenase to ecosanoids, prostaglandins (E3 series), and leukotrienes having anti-inflammatory responses (James et al. 2000; Funk 2001; Kaur et al. 2012) leading to protection against CVDs.

10.6.1.1 Effect of Fiber

Soluble fiber reduces total cholesterol and LDL cholesterol by enhancing gastric emptying, altering transit time, interfering with bulk phase diffusion of fat, and increasing excretion of bile acids (Kritchevsky 1995).

10.6.1.2 Effects of Lignans

Modulation of enzymes such as 7α-hydroxylase and acyl Co-A cholesterol transferase involved in cholesterol metabolism, by lignan, may directly lower serum cholesterol (Prasad 1999). Studies have demonstrated the ability of SDG to scavenge hydroxyl-free radicals and serve as potential antioxidants (Toure and Xueming 2010). Human body produces free radicals during fat, protein, and carbohydrate oxidation, which damages tissues, membrane lipids, nucleic acids, etc. leading to various diseases like cancer, neurological disorder, aging, etc. Lignan can scavenge free radicals thus reducing oxidative stress (Hu et al. 2007). Under in vivo and in vitro conditions, SDG, enterodiol, and enterolactone act as antioxidants by inhibiting peroxidation of polyunsaturated fatty acids and decrease oxidation of LDL cholesterol (Kitts et al. 1999). Also, SDG has been found to be a platelet-activating factor antagonist (Hall et al. 1993).

10.6.1.3 Effects of Omega Fatty Acids

Through several mechanisms such as decreasing inflammatory response, inhibiting platelet aggregation and thrombosis, decreasing blood pressure, improving serum lipids, and preventing cardiac arrhythmias, omega-3 fatty acids prevent cardiovascular diseases (Lanzmann-Petithory et al. 2002). These fatty acids interfere with the production of pro-inflammatory and pro-aggregatory eicosanoids (prostaglandin E2, thromboxane Ax2, and leukotriene B4) and protect against cardiovascular diseases. Since ALA and linoleic acid (LA) are essential fatty acids, they need to be supplied through diet. Upon ingestion, LA and ALA yield different classes of the eicosanoids. Such eicosanoids have beneficial effects in case of inflammation, platelet aggregation, and vasoconstriction (Bloedon and Szapary 2004). Alternatively, omega-3 fatty acids are involved in alteration of enzyme synthesis, regulating gene transcription and expression, and modification of risk factors for coronary heart diseases (Chen et al. 2007).

One of the prime requirements for the development of atherosclerosis is the production of oxygen-free radicals. These free radicals are produced in the body due to several metabolic reactions causing endothelial dysfunction which is the causative agent for the development of hyperchoestrolemic atherosclerosis which causes ischemic heart disease, stroke, and peripheral vascular diseases (Prasad 2000). The effect of flaxseed consumption on serum lipid levels had been studied in several animal models and has been reported to have positive results. Similar studies have been conducted in humans where flaxseed has been found to have beneficial effect in CVDs and hypocholesterolemic activity.

Regular consumption of flaxseed has been found to decrease atherosclerosis (by 46%) and lowered the number of inflammatory polymorphonuclear leukocytes in rabbits (Prasad et al. 1998). Further, it was reported that when purified SDG (15 mg/kg) was added to an atherogenic diet in rabbits (for 8 weeks), levels of aortic malondialdehyde were also reduced (Prasad 1997). In a study involving 40 female weaning Wistar rats fed with 200 g flaxseed oil/kg body weight for 4 weeks, MacDonald-Wicks and Garg (2002) reported that flaxseed produced the lowest concentration of 8-iso-$PGF_{2\alpha}$ (an in vivo oxidative stress marker) when administered with an oxidative stress inducer (CCL4) as compared to rats that received either saturated fat or linoleic acid containing diets. SDG isolated from flaxseed reduced serum cholesterol, serum lipid profile, and hyperchoestrolemic atherosclerosis in white rabbits (Prasad 1999). It was observed that SDG reduced the development of hyperchoestrolemic atherosclerosis (73%) and total cholesterol (33%). In a study on antiatherogenic potential of flaxseed, Dupasquier et al. (2007) used a LDL receptor-deficient mouse (LDLrKO) model and supplemented the cholesterol-enriched diet with 10% ground flaxseed. It was reported that flaxseed was associated with lowering of plasma cholesterol and saturated fatty acids, increasing of plasma ALA levels, and reducing plaque formation in the aorta and aortic sinus. The effect of flaxseed on lipid profile in in vivo (animal and human) studies has been compiled and represented in Table 10.4.

10.6.2 Prevention of Diabetes Mellitus

Diabetes mellitus or hyperglycemia (fasting blood glucose level exceeding 126 mg/dL) is associated with abnormal metabolism of mainly carbohydrate and leading to the development of various secondary complications like CVDs, kidney failure, blindness, etc. (Mani et al. 2011). Scientific literature supported the positive correlations between the high blood glucose and triglyceride levels, obesity, and the risk of CVDs (Boden-Albala et al. 2008). Flaxseed components (dietary fiber, lignans, and ω-3 fatty acids) have protective effect in the management of diabetes. Data suggests that flaxseed may improve glucose homeostasis (Bloedon and Szapary 2004). Reduced glycemic response of flaxseed is attributed to the interaction of flaxseed protein with the gums that stimulates insulin secretion (Oomah 2001). Flaxseed fiber especially the insoluble fiber also plays a vital role in lowering the blood glucose levels (Kapoor et al. 2011). SDG, the flaxseed lignan, has been observed to suppress the expression of the phosphoenolpyruvate carboxykinase gene that codes for the enzyme responsible for the gluconeogenesis in the liver (Prasad 2002).

In a study conducted by Nazni et al. (2006), 25 diabetic subjects were fed with flaxseed powder supplemented bread. Authors observed a significant reduction ($p < 0.05$) in blood glucose levels at the end of 90 days. In another study on obese and diabetic tumor susceptible rats fed with CLA and flaxseed oil (0.5% each), Kelley et al. (2009) observed 20% reduction in glycemia. The effect of flaxseed powder on glucose level of female diabetic subjects was studied by Kapoor et al. (2011) who reported a decrease in the postprandial glucose levels by 7.9% and 19.1% when fed with 15 g and 20 g/day, respectively, at the end of 20 days. Similarly, Mani et al. (2011) reported that supplementation of flaxseed powder (10 g) to type 2 diabetic patients reduced fasting blood glucose level by 19.7% at the end of 1 month.

Flaxseed complex containing secoisolariciresinol diglucoside and its colonic metabolites enterodiol and enterolactone, cinnamic acids, and 3-hydroxy-3-methylglutaric acid were reported to have a role in the management of diabetes and related health issues (Barre et al. 2012). Studies were conducted on type

Table 10.4 Effect of flaxseed consumption on blood lipid profile of animal and human models

Model system	Diet particulars	Feeding particulars	Findings	Reference
Rat	Flaxseed powder- enriched diet	12 weeks	• Decreased level of total cholesterol • Decreased level of LDL cholesterol	Park and Velasquez (2012)
Hypercholesterolemic rat	Biscuit prepared with flaxseed oil	8 weeks	• Decreased levels of triglycerides • LDL and VLDL cholesterol decreased, while HDL cholesterol increased	Hassan et al. (2012)
Rats	Animals were fed with raw and heated flaxseed		• Total cholesterol level significantly reduced • HDL cholesterol level significantly increased • Significant reduction in LDL cholesterol in raw flaxseed group	Khalesi et al. (2011)
Hypercholesterolemic rabbit	Ground linseed	29 days	• Total cholesterol and LDL cholesterol level reduced	Prim et al. (2012)
Young healthy adult	Consumption of flax fiber daily at 5 g/day in the form of bread and drink	1 week	• Fecal excretion of fat increased by 50%. Total and LDL cholesterol reduced	Kristensen et al. (2012)
Hypercholesterolemic human	Isoenergetic diet, containing 36% energy from fat out of which about 70% provided by flaxseed oil	28 days	• Total, LDL, and HDL cholesterol level were reduced	Gillingham et al. (2011)
Hypercholesterolemic postmenopausal women	Flaxseed diet was provided at 30 g/day	3 months	• Total and LDL cholesterol level was lowered • HDL cholesterol and triglyceride levels remained unaltered	Patade et al. (2008)

2 diabetic patients who were given flaxseed complex (600 mg) for 3 months, and it was reported that flaxseed complex could reduce the hyperglycemia, dyslipidemia, blood pressure, central obesity, LDL oxidation, hypertension, inflammation, and prothrombotic state (Barre et al. 2012). Further, it was reported that due to the strong antioxidative properties, the flaxseed complex was effective in controlling diabetes mellitus (Pan et al. 2009). On the contrary, there are several studies in the literature indicating little or no effect of flaxseed supplementation on serum glucose and insulin level (Dodin et al. 2008; Barre et al. 2008).

10.6.3 Tumor and Cancer-Reducing Effects

Increasing evidences are being accumulated regarding the anti-cancerous effects of flaxseed (Truan et al. 2010). Flaxseed lignans are well-known to reduce the hormone-dependent breast and prostate cancers by influencing cell proliferation, protein synthesis, hormone metabolism, growth factors, etc. and also angiogenesis reducing the availability of estrogen (Branca and Lorenzetti 2005). Lignans are converted into enterolactone (EL) and enterodiol (ED) by gut microflora, which have potential antioxidant activity (Kitts et al. 1999; Prasad 2000). As they have structural similarity with that of human estrogen (17β-estradiol), they have binding affinity toward estrogen receptors (Penttinen et al. 2007). The principle mechanism of controlling hormone-dependent cancers involves increase in apoptosis through modulation of estrogen receptors as well as reduction in cell proliferation and angiogenesis (Penttinen et al. 2007). Hasler (1998) reported that ingestion of flaxseed (10 g) can reduce breast cancer risk. Thompson et al. (2000), in a placebo-controlled clinical study on breast cancer individuals, found that flaxseed and its component (lignan) reduced tumor growth. In an another study on preclinical athymic postmenopausal breast cancer mouse model, Power and Thompson (2007) found that tumor stimulatory effects were negated when the mouse was fed with a diet containing flaxseed and soy protein as compared to soy protein alone. Truan et al. (2010) studied the potential effect of flax lignans in breast cancer and hypothesized that its protective effect could be due to weak estrogenic and antioxidative activities. In a study on 25 men having prostate cancer who were fed with ground flaxseed (30 g/day), Demark-Wahnefried et al. (2001) observed decreased prostate cancer cell proliferation and increased apoptotic death of cancer cells at the end of 4 weeks.

Pisani (2008) reported that increasing risk of pancreatic and colorectal cancer has been associated with high level of blood insulin. By stimulating cell proliferation and increasing the survival rate of DNA damaged cells through antiapoptotic mechanism, insulin and insulin-like growth factor-1 increase cancer risk (Sturgeon et al. 2011). Antioxidative property of lignans and polyphenols present in flaxseed may be the main reason of anti-cancerous activity. Alternatively, high levels of sulfur containing amino acids (cysteine and methionine) in flaxseed are attributed to reduction of the risk of certain type of cancers by stabilizing the DNA during cell division (Oomah 2001). Further, the interaction between soluble polysaccharides and flaxseed protein can reduce the colon luminal ammonia, thereby protecting against colon cancer (Clinton 1992). Lignans may have partial chemoprotective effects due to their strong protein-binding properties (Adlercreutz et al. 1993). In a study on azoxymethane-induced colon cancer Fisher male rats, Williams et al. (2007) observed that dietary flaxseed meal (20%) and flaxseed oil (7%) reduced the aberrant crypt foci formation by 90% and 78%, respectively, as compared to control in the experimental animals. It was hypothesized that the ability of ω-3 fatty acids to suppress the inflammation by downregulating the cycloygenase-2 enzyme might be the mechanism behind the reduced crypt foci formation.

10.6.4 Role in Kidney and Renal Diseases

Chronic kidney disease is one of the serious health concerns among adult population that may lead to renal disease which need dialysis or transplantation for survival (Lauretani et al. 2009). Various studies have reported the useful effects of flaxseed ω-3 fatty acids in the improvement of renal health and reducing renal inflammation and fibrosis in animal models (Baggio et al. 2005; Ogborn et al. 2002, 2003). Extended supplementation of ω-3 fatty acids has been found to be related with significant reduction in systolic and diastolic blood pressure. Since hypertension is a risk factor and precursor of chronic kidney disease, this is considered as a potential mechanism for the protection of kidney disorders (Cicero et al. 2010). In a study on Lupus nephritis patients, Clark et al. (2001) observed the reno-protective effects of flaxseed which was associated with a decline in the serum creatinine and microalbumin levels during a period of 2 years. In an another study on male Han: SPRD-cy rats from the time of weaning till 8 weeks, Ogborn et al. (2002) reported that flaxseed ameliorates polycystic kidney disease through moderation of the associated chronic interstitial nephritis. Naqshbandi et al. (2013) studied the ameliorative effect of flaxseed oil in cisplatin-induced nephrotoxicity and reported that feeding of a diet rich in flaxseed oil markedly reduced nephrotoxicity.

10.6.5 Role in the Prevention of Obesity

Flaxseed fibers are effective hunger suppression agents due to the formation of highly viscous solution upon hydration (Kristensen et al. 2011; Wanders et al. 2011) and found to be useful in controlling obesity. Flaxseed mucilages are hydrophilic substances that form thick viscous solution on hydration that can delay gastric emptying and thereby can prevent obesity (Goyal et al. 2014). In an obese subject, downregulation of leptin receptor leads to underexpression of leptin by adipose tissue (Dubey et al. 2006) resulting in the lack of satiety, overeating, and finally obesity. McCullough et al. (2011) reported that consumption of flaxseed positively and significantly improved the leptin expression in male white rabbit model. Fukumitsu et al. (2008) studied the effect of flaxseed lignan SDG on the development of diet-induced obesity and adiponectin expression in mice and found that flaxseed regulates adipogenesis-related gene expressions. In another study, lignan-enriched flaxseed powder supplementation was found to be beneficial in the reduction of bodyweight and fat accumulation in rats (Park and Velasquez 2012). Also, lower fat deposition was reported in the livers of both lean and obese rats fed with flaxseed meal as compared to rats fed either on casein or soy protein diet (Bhathena et al. 2003).

10.6.6 Treatment of Bowel Syndrome

Intestinal bowel disorder is one of the common problems causing upset of gastrointestinal tract, abdominal distension, and changes in bowel habit leading to constipation and/or diarrhea. Dietary supplementation of flaxseed can combat the bowel syndrome. Dietary fibers are considered as a first line of treatment, and managing intestinal bowel syndrome and adequate amount of dietary fiber is essential in the prevention and treatment of constipation (Tarpila et al. 2005). Flaxseed is also one of the important sources of dietary fiber in which the proportion of soluble to insoluble dietary fiber is between 20 and 40:60 and 80 (Morris 2007a). The mechanism is likely that soluble and insoluble fiber of flaxseed promotes the healthy intestinal functions. Pentosans and mucilaginous matter are attributed for the laxative nature of flaxseed (Madhusudhan 2009). Dietary fiber of flaxseed is fermented by colonic microflora after reaching the large intestine and produces several metabolites. These include short chain fatty acids (SCFAs), biomass, and gases such as hydrogen, carbon dioxide, and methane all of which exhibit laxative effects (Kritchevsky 1979). These SCFA acts as major anions in the large intestine and contribute in regulation of various colonic motilities and decrease transit time by various mechanisms (Brownlee 2011; Xu et al. 2012). Dietary fibers have bulking effect resulting in increasing the contents of feces. Soluble fiber also increases microbial cell mass due to its water binding capacity (Malkki 2004).

In a study on constipated mice models, at the end of 14 days, Xu et al. (2012) observed that medium (5%) to high (10%) doses of defatted flaxseed meal supplemented diet increased the stool frequency and weight, whereas the low doses (2%) reduced the start time of defecation. Similarly, in another study, Tarpila et al. (2004) observed that defatted flaxseed significantly reduced the constipation and irritable bowel syndrome symptoms.

10.6.7 Role in Inflammatory Disease and Immune Function

Flaxseed, due to the presence of eicosanoids and cytokines, favorably affects the immunity. Javed (1999) reported that ALA and lignans in flaxseed have a beneficial role in the management of autoimmune diseases by modulating the immune responses. The conversion of fatty acids of flaxseed into prostaglandins may be helpful in regulation of inflammation (Ogborn et al. 2002; Fitzpatrick 2007). In a study on tumor necrosis factor-α (TNF-α) and interleukin-1β (IL-1β) in health subjects fed with flaxseed oil-based diet, Caughey et al. (1996) observed that incorporation of flaxseed oil (14 g) in domestic food preparation inhibited the production of mononuclear cell (TNF-α) and (IL-1β) nearly 30% at the end of 4 weeks. Roy et al. (2007) studied the anti-inflammatory effect of ALA in middle-aged men and observed that ALA supplementation significantly reduced inflammation markers as measured in terms of C-reactive protein (CRP).

10.6.8 Role in Reproduction and Fetal Development

Dietary phytoestrogen has a vital role in masculine reproductive health. Reports are available linking the role of lignans in combating against polycystic ovary syndrome in vulnerable women by reducing the levels of circulating testosterone (Nowak-Debra et al. 2007). The fetus and newborn are very much sensitive to environmental chemicals, and a small exposure can have important consequences (Jefferson et al. 2012). Detoxification process in the fetus and newborn is inefficient as the metabolites processing and elimination mechanisms are not fully developed (Chen et al. 2006). Further, several developmental processes in neonates are dependent on steroid hormones, and secreted proteins have the potential to be altered by phytoestrogens, which can interact with estrogen receptors such as ERα and Erβ (Jefferson et al. 2012). Fetal exposure to phytoestrogen can suppress testosterone synthesis, leading to cryptorchidism and adult testis dysfunction and also disrupts female reproductive tract development by altering expression of genes encoded signaling proteins (Clark and Cochrum 2007; Wohlfahrt-Veje et al. 2009; Ma 2009). These effects were found to have permanent consequences in both rodents and humans, which persists during postnatal life (Baird and Newbold 2005). Until the onset of puberty, many of the postnatal differentiation events are dependent on steroid hormone signaling (Gray et al. 2001). In several animal models, it was found that the brain development and sexual behavior are affected by exposure to phytoestrogen (Henry and Witt 2006; Sullivan et al. 2011). A recent study by Cardoso-Carraro et al. (2012) revealed that exposure of neonatal rats to estrogen results in the reduction of the spermatic concentration and plasmatic testosterone. In an another study on female Wistar rats, Assinder et al. (2007) observed that feeding of phytoestrogen prior to conception negatively affected the spermatogenesis of the hypothalamus-pituitary-testicular axis among weaned male pups.

10.6.9 Role in Menopause and Bone Metabolism

Menopause is experienced by every elderly woman (about 50 year of age) and is diagnosed by amenorrhea (Ghazanfarpour et al. 2016). It is associated with a cluster of symptoms including hot flashes, night sweating, vaginal atrophy, anxiety, nervousness, and reduced libido (Cramer et al. 2012). Because of structural and functional similarities with human estrogen (17β-estradiol), flaxseed lignans alter hormone metabolism (Bloedon and Szapary 2004). In the treatment of menopausal problems of women, several studies clearly indicated that flaxseed can serve as an alternative therapy to traditional estrogen (Bloedon and Szapary 2004). Further, among postmenopausal women, flaxseed may also reduce serum concentration of 17β-estradiol and estrone sulfate and increase prolactin levels. However, limited studies are available in the literature. Pruthi et al. (2007) also reported that hormonal modulation exerted beneficial effects such as decreased hot flashes and other symptoms among postmenopausal women due to weak estrogenic activity of lignans. Hutchins et al. (2000) reported that intake of flaxseed lignan (5–10 g/day) significantly increased urinary excretion of enterodiol, enterolactone, and total lignans. In another study on postmenopausal women, Haggans et al. (1999) reported that consumption of flaxseed increased urinary estrogen metabolites, namely, 2-hydroxy estrogen and 2/16 α-hydroxy estrogen excretion in a linear, dose-dependent manner.

Several studies have revealed that menopause is positively associated with the reduction of bone mineral density and the incidences of fractures (Borrelli and Ernst 2010). Lignanic compounds, which are estrogen like in nature, can exert effects on estrogen responsive tissues, including the bones. It has been observed that human osteoclasts and osteoblasts can express estrogen receptor to which phytoestrogens specifically bind (Mano et al. 1996; Arts et al. 1997; Setchell and Cassidy 1999). In spite of this, in several animal and human models, the intake of flax lignans resulted in little or no benefit to the bone (Lucas et al. 2002; Dodin et al. 2005; Cornish et al. 2009). However, Kim and Ilich (2011) reported that flaxseed intake along with estrogen therapy provided an extra benefit to the bones in animal models.

10.7 Adverse Effects of Flaxseed

Although individuals are not allergic to flaxseed, still in certain cases, potential problems with flax fiber and lignan components could still occur. Some flaxseed hypersensitivity- related anaphylactic cases have been reported (Gall 2000). Since flaxseed lignans have structural similarity with human estrogens, they may alter hormone metabolism. Although lignans are useful in the treatment of hormone-dependent cancers as well as menopausal symptoms, higher dose of lignan may lead to some adverse effects in menstruating women, which needs further investigations. Although in a study, flaxseed intake (10%) resulted in low birth weight and produced hormonal effects in both male and female rats (Tou et al. 1998), there are no published evidences of safety issues in pregnant and lactating women and children. The consumption of uncooked flaxseed could be toxic at higher doses due to the presence of cyanogenic glycosides. Morris (2000) reported that intake of uncooked flaxseed meal (more than 10 tablespoons/day) may increase the hydrogen cyanide content to potentially toxic levels (>50–60 mg) among adults. Other reports indicated that consumption of baked flaxseed powder (50 g/day) does not increase urinary thiocyanate levels (Cunnane et al. 1993a, b). Further, there is no report of association of processed flaxseed consumption to that of chronic or acute cyanide toxicity (Bloedon and Szapary 2004).

10.8 Food Applications

Before consumption, flaxseeds are processed by various processing methods, which involve multiple steps that lead to extraction of oil and various by-products. Flaxseed oil is known for its alpha-linolenic acid, which is a heat labile compound and destroyed at high-temperature treatment (Choo et al. 2007) and is also prone to autoxidation leading to quality deterioration. Hence, after cold extraction, it is incorporated with antioxidants and stored in dark packaging materials to prevent photo-induced deteriorative changes (Lukaszewicz et al. 2004). Flaxseed oil is generally extracted by mechanical methods such as screw press extraction (86–92% recovery) followed by sedimentation and filtration without any heat treatment or refining (Wiesenborn et al. 2005). Freshly extracted (unrefined) oil has a pleasant nutty flavor with attractive golden color. However, various pre-treatments prior to pressing results in significant improvement in oil recovery and quality. Solvent extraction using hexane at high temperatures yields high-quality oil with maximum recovery (Nash and Frankel 1986). Bioactive alpha-linolenic acid present in flaxseed oil can be degraded by exposure to high temperature; therefore, emerging techniques such as supercritical fluid extraction technique, which involves low-temperature processing (31 °C) could be exploited. Reports are emerging with regard to application of ultrasonic-assisted oil extraction techniques for speedy and enhanced oil recovery with lesser solvent (Zhang et al. 2008).

10.8.1 Food Applications of Flaxseed

Although flaxseed is well-known since ancient times, the concept of utilizing flaxseed in food formulations is comparatively recent. In the current functional food regime, with an array of health benefits, flaxseed has emerged as a potential functional ingredient. The nutritional and medicinal characteristics of flaxseed make it a potential source of functional ingredients in the development of functional foods (Rubilar et al. 2010). It can be utilized as a main food ingredient in order to enhance

Table 10.5 Food applications of flaxseed constituents

Flaxseed constituent	Functional property	Effect	Food products	References
Mucilage	Emulsification and stabilization of foam and emulsion	Improved body and texture	Sauces, sausages, meat emulsions, salad dressing	Stewart and Mazza (2000)
	Anti-staling property	Enhancement of keeping quality	Baked products	Lipilina and Ganji (2009)
	High water absorption, moisture binding capacity	Improved cooking quality	Noodles	Kishk et al. (2011)
	High water absorption	Shortening alternatives	Cookies, muffins, breads	Chetana et al. (2010)
Protein	Stabilization and emulsification	Improved body and texture	Ice cream, sausages, meat emulsion	Martinez-Flores et al. (2006)
	Viscoelastic texture	Texture improvement	Extruded snacks and breakfast cereals	Wu et al. (2010)
	Gluten free	Enhanced nutritional	Gluten-free food product	Gambus et al. (2009)
	Egg and gelatine substitution	Vegetarian product	Baked goods and ice cream	Shearer and Davies (2005)

functional and nutritional quality. Whole or ground flaxseed can be used in various food formulations to enhance omega-3 and other bioactive components especially to cater the vegetarian population. Apart from enhancing nutritional and therapeutic properties, flaxseed can contribute toward improving the physical properties of food products due to its protein and carbohydrate (gums) components. The effects of supplementation of flaxseed components in several food products are presented in Table 10.5. Inspite of several useful virtues, flaxseed incorporation may sometimes result in lower sensory acceptability, which needs to be carefully addressed. The characteristics of various flaxseed incorporated food products have been presented in Table 10.6.

10.8.1.1 Flaxseed in Bakery Products

Bakery products are ideal vehicles for fortification as well as incorporation of functional ingredients as they are widely and daily consumed across the world (Kadam and Prabhasankar 2010). Flaxseed components could be incorporated into bread to improve its nutritional quality (Hall et al. 2005). Composite flour-based bakery products could be manufactured by replacing some portion of refined wheat flour with that of flaxseed flour. Mentes et al. (2008) reported that loaf volume and specific loaf volume significantly ($P < 0.05$) increased, while rate of staling

Table 10.6 Characteristics of various food products supplemented with flaxseed

Product	Flaxseed/ingredient	Characteristics	Reference
Bakery products			
Yeast bread	Milled flaxseed flour (15%)	• Highly acceptable on sensory basis	Mentes et al. (2008)
Bread	Flaxseed flour and wheat flour	• Acceptable sensory quality • Decreased specific volume of bread with increased flaxseed levels • Enhanced alpha-linolenic acid	Lipilina and Ganji (2009)
	Flaxseed oil powder	• Water absorption capacity increased • No effect on sensory quality	Gokmen et al. (2011)
Brown bread	Deoiled cake (15%)	• Bread samples were acceptable to sensory panel	Ogunronbi et al. (2011)
Flat bread	Full fat (12%) and partially defatted flaxseed (16%) flour	• Unleavened bread samples were sensorily acceptable and had high nutritional quality	Hussain et al. (2012)
Muffins	Flaxseed powder	• Flaxseed muffins had lower acceptability than control	Ramcharitar et al. (2005)
	Flaxseed flour (raw and roasted) (20%) along with wheat flour	• Samples were soft and had enhanced nutritional quality	Chetana et al. (2010)
Muffins and snack bar	Flaxseed flour	• Products were least acceptable. However, incorporation of flavoring compounds increased sensory acceptability	Aliani et al. (2011)
Biscuits	Flaxseed oil	• Samples containing flaxseed oil (shortening) were observed to be similar to control	Hassan et al. (2012)
Cookies	Flaxseed flour (15%)	• Acceptable sensory and rheological properties • Enhanced omega-3 fatty acid (ALA) content	Rajiv et al. (2012)
	Flaxseed flour (12%)	• Acceptable by consumers • Flaxseed flour increased dough stickiness	Khouryieh and Aramouni (2012)
Extruded snacks and energy bar			
Extruded puffs	Flaxseed with corn meal	• Incorporation of 15% flaxseed resulted in a product with good puffing quality	Wu et al. (2007)
Energy bar	Roasted ground flaxseed	• Product having flaxseed at 15% level and 45% sweeteners resulted in acceptable sensory and nutritional quality	Mridula et al. (2013)

(continued)

Table 10.6 (continued)

Product	Flaxseed/ingredient	Characteristics	Reference
Milk and dairy products			
Milk	Flaxseed oil and DHA	• Rancid flavor after 3 days of storage	Divya et al. (2013)
	Flaxseed oil microcapsules (2 g/100 mL)	• Samples containing microcapsules were similar to that of control • Fortified milk remained stable up to 5 days	Goyal et al. (2017)
Milk and *dahi*	Flaxseed oil	• Lower flavor scores	Veena et al. (2017)
Dairy products	Lignan (SDG) isolated from flaxseed	• SDG remained stable during processing (pasteurization, fermentation and renneting) and storage	Hyvarinen et al. (2006)
Ice cream	Flaxseed oil	• Ice cream had soft creamy texture due to improved meltdown time	Goh et al. (2006)
		• Increased flaxseed oil levels decreased flavor scores	Lim et al. (2010)
Butter	Flaxseed additives	• Improved structure of butter	Ivanova et al. (2011)
Dairy beverage	Commercial flax lignan	• Improved oxidative stability of the product	Matumoto-Pintroa et al. (2011)
Cheese	Flaxseed oil	• Cheese retained high amounts of flaxseed oil • Shelf life was not affected	Aguirre and Canovas (2012)
Yoghurt	Flaxseed oil	• Product was acceptable by sensory panelists • Growth of the lactic acid bacteria was affected both during fermentation and storage	Bello et al. (2015)
Dahi	Flaxseed oil powder (microencapsulated)	• Product containing flaxseed oil powder (2%) was similar to control	Goyal et al. (2016)

decreased with the incorporation of ground flaxseed (10%). Hao and Beta (2012) reported that incorporation of flaxseed hull into Chinese steamed bread (CSB) resulted in enriched phytochemical profile with associated increase in the antioxidant activity. Lipilina and Ganji (2009) observed improved levels of α-linolenic acid, linoleic acid, and dietary fiber as compared to the control bread due to flaxseed flour (30%). However, in another study on bread, flaxseed flour (15%) incorporation resulted in musty aroma. Incorporation of full-fat flaxseed flour (20%) into cookies was found to be acceptable by the sensory panelists (Hussain et al. 2006). On the contrary, Ramcharitar et al. (2005) reported that muffins containing milled flaxseed (11.6%) were less acceptable than control. Similarly, incorporation of flaxseed (15%) into pretzel-type bread led to reduced flavor and overall acceptability scores (Alpaslan and Hayta 2006). However, incorporation of suitable flavoring agents could enhance the acceptability of the flaxseed fortified bakery products (Aliani et al. 2011).

10.8.1.2 Flaxseed in Dairy Products

As milk and milk products are consumed by all age groups, they have been proved to be ideal medium for incorporation of bioactive and functional ingredients. Flaxseed oil has been incorporated into ice cream and similar products (Hall and Schwarz 2002). It was used to replace up to 25% of the milk fat in ice cream formula, which exhibited an oil-like mouth feel that could not be detected by 60% of the informal sensory panelists. Hall III et al. (2004) studied the stability of lignan in yoghurt and reported that incorporation of flaxseed extract did not have any negative impact on the fermentation. On the other hand, Lim et al. (2010) reported decreased acceptability of ice cream due to the incorporation of flaxseed oil although it did not affect the physicochemical properties. In a study on the stability of flaxseed secoisolariciresinol diglucoside (SDG) in Edam cheese, Hyvärinen et al. (2006) reported that most of the added SDG was passed into the whey fraction, and only 6% was retained in the cheese curd. Further, it was reported that SDG was relatively stable at the end of 6 weeks of ripening of cheese stored at 9 °C. In a study on incorporation of flaxseed oil in ice cream, Goh et al. (2006) reported increased meltdown rate and decreased hardness and ascribed it to the melting point of flaxseed oil. Li et al. (2008) investigated the effect of flaxseed gum, guar gum, and denatured starch as stabilizer blends in stirred yoghurt and observed that the optimum proportion of the additives was in the ratio of 3:2:20, and the shelf life of yoghurt was prolonged beyond 27 day at 4 °C. Ivanova et al. (2011) reported that flaxseed additive containing butter had good spreadability with pure creamy flavor. Giroux et al. (2010) reported method of manufacture of linseed oil-enriched dairy-based beverages. Matumoto-Pintroa et al. (2011) reported that dairy beverages enriched with flaxseed lignan SDG extract increased oxidative stability of flaxseed oil.

10.9 Conclusion

Flaxseed is emerging as an attractive nutritional component due to its exceptionally high content of several bioactive compounds. With cultivation in more than 50 countries and proven health benefits, it has found applications in a variety of food products across the world. Bread and cookies among the bakery products and ice cream and yoghurt among the dairy products have gained commercial success with flaxseed incorporation. However, efforts are required to minimize the adverse sensory effects due to flaxseed incorporation among other food products.

References

Adlercreutz H (2007) Lignans and human health. Crit Rev Clin Lab Sci 44(5–6):483–525
Adlercreutz H, Fotsis T, Lampe T, Wahala K, Makela T, Brunnow G, Hase T (1993) Quantitative determination of lignans and isoflavonoids in plasma of omnivorous and vegetarian women by isotope dilution gas chromatography-mass spectrometry. Scand J Clin Lab Invest 53:5–18
Aguirre DB, Canovas GVB (2012) Fortification of queso fresco, cheddar and mozzarella cheese using selected sources of omega-3 and some nonthermal approaches. Food Chem 133 (3):787–797
Akande KE, Doma UD, Agu HO, Adamu HM (2010) Major anti nutrients found in plant protein sources: their effect on nutrition. Pak J Nutr 9:827–832
Aliani M, Ryland D, Pierce GN (2011) Effect of flax addition on the flavor profile of muffins and snack bars. Food Res Int 44:2489–2496
Al-Okbi SY (2005) Highlights on functional foods, with special reference to flaxseed. J Nat Fibers 2 (3):63–68
Alpaslan M, Hayta M (2006) The effects of flaxseed, soy and corn flours on the textural and sensory properties of a bakery product. J Food Qual 29:617–627
Anderson E, Lowe HJ (1947) The composition of flaxseed mucilage. J Biochem 168:289–297
Arts J, Kuiper GG, Janssen JM, Gustafsson JA, Lowik CW, Pols HA, van Leeuwen JP (1997) Differential expression of estrogen receptors α and β mRNA during differentiation of human osteoblast SV-HFO cells. Endocrinology 138:5067–5070
Assinder S, Davis R, Fenwick M, Glover A (2007) Adult-only exposure of male rats to a diet of high phytoestrogen content increases apoptosis of meiotic and post-meiotic germ cells. Reproduction 133:11–19
Austria JA, Richard MN, Chahine MN (2008) Bioavailability of alphalinolenic acid in subjects after ingestion of three different forms of flaxseed. J Am Coll Nutr 27:214–221
Baggio B, Musacchio E, Priante G (2005) Polyunsaturated fatty acids and renal fibrosis: pathophysiologic link and potential clinical implications. J Nephrol 18:362–367
Baird DD, Newbold R (2005) Prenatal diethylstilbestrol (DES) exposure is associated with uterine leiomyoma development. Reprod Toxicol 20:81–84
Barre DE, Mizier-Barre KA, Griscti O, Hafez K (2008) High dose flaxseed oil supplementation may affect fasting blood serum glucose management in human type 2 diabetics. J Oleo Sci 57 (5):269–273
Barre DE, Mizier-Barre KA, Stelmach E, Hobson J, Griscti O, Rudiuk A, Muthuthevar D (2012) Flaxseed lignan complex administration in older human type2 diabeticsmanages central obesity and prothrombosis-aninvitaton to further investigation into polypharmacy reduction. J Nutr Metab. Article ID 585170. https://doi.org/10.1155/2012/585170
Beejmohun V, Fliniaux O, Grand E, Lamblin F, Bensaddek L, Christen P, Kovensky J, Fliniaux MA, Mesnard F (2007) Microwave-assisted extraction of the main phenolic compounds in flaxseed. Phytochem Anal 18:275–282
Bello BD, Torri L, Piochi M, Zeppa G (2015) Healthy yogurt fortified with n-3 fatty acids from vegetable sources. J Dairy Sci 98(2):8375–8385
Bhathena SJ, Ali AA, Haudenschild C, Latham P, Ranich T, Mohamed AI, Hansen CT, Velasquez MT (2003) Dietary flaxseed meal is more protective than soy protein concentrate against hypertriglyceridemia and steatosis of the liver in an animal model of obesity. J Am Coll Nutr 22:157–164
Bhatty RS (1993) Further compositional analyses of flax: mucilage, trypsin inhibitors and hydrocyanic acid. J Am Oil Chem Soc 70:899–904
Bhatty RS (1995) Nutritional composition of whole flaxseed and flaxseed meal. In: Cunnane SC, Thompson LH (eds) Flaxseed in human nutrition. AOCS Press, Champaign, IL, pp 22–45
Bloedon LT, Szapary PO (2004) Flaxseed and cardiovascular risk. Nutr Rev 62(1):18–27

Boden-Albala B, Cammack S, Chong J, Wang C, Wright C, Rundek T, Elkind MS, Paik MC, Sacco RL (2008) Diabetes, fasting glucose levels, and risk of ischemic stroke and vascular events: findings from the Northern Manhattan study (NOMAS). Diabetes Care 31:1132–1137

Borrelli F, Ernst E (2010) Alternative and complementary therapies for the menopause. Maturitas 66(4):333–343

Branca F, Lorenzetti S (2005) Health effects of phytoestrogens. In: Elmadfa I (ed) Diet diversification and health promotion. Karger Publishers, Basel, Switzerland, pp 100–111

Brownlee IA (2011) The physiological roles of dietary fibre. Food Hydrocoll 25(2):238–250

Bruhl L, Matthaus B, Fehling E, Wiege B, Lehmann B, Luftmann H, Bergander K, Quiroga K, Scheipers A, Frank O, Hofmann T (2007) Identification of bitter off-taste compounds in the stored cold pressed linseed oil. J Agric Food Chem 55:7864–7868

Cardoso-Carraro JC, de Souza Dantas MI, Rocha-Espeschit AC, Duarte-Martino HS, Rocha-Ribeiro SM (2012) Flaxseed and human health: reviewing benefits and adverse effects. Food Rev Int 28:203–230

Caughey GE, Mantzioris E, Gibson RA, Cleland LG, James MJ (1996) The effect on human tumor necrosis factor alpha and interleukin 1 beta production of diets enriched in n-3 fatty acids from vegetable oil or fish oil. Am J Clin Nutr 63:116–122

Chen J, Power KA, Mann J, Cheng A, Thompson LU (2007) Flaxseed alone or in combination with tamoxifen inhibits mcf-7 breast tumor growth in ovariectomized athymic mice with high circulating levels of estrogen. Exp Biol Med 232:1071–1080

Chetana SML, Begum K, Ramasarma PR (2010) Nutritional characteristics of linseed/flaxseed (Linum usitatissimum) and its application in muffin making. J Texture Stud 41:563–578

Choo WS, Birch J, Dufour JP (2007) Physicochemical and quality characteristics of cold-pressed flaxseed oils. J Food Compos Anal 20(3):202–211

Cicero AF, Derosa G, Di G, Bove M, Gaddi AV, Borghi C (2010) Omega-3 polyunsaturated fatty acids supplementation and blood pressure levels in hypertriglyceridemic patients with untreated normal-high blood pressure and with or without metabolic syndrome: a retrospective study. Clin Exp Hypertens 32(2):137–144

Clark BJ, Cochrum RK (2007) The steroidogenic acute regulatory protein as a target of endocrine disruption in male reproduction. Drug Metab Rev 39:353–370

Clark WF, Kortas C, Heidenhiem AP, Garland J, Spanner E, Parbtani A (2001) Flaxseed in lupus nephritis: a two-year nonplacebo-controlled crossover study. J Am Coll Nutr 20(2):143–148

Clinton SK (1992) Dietary protein and the origins of human cancer. In: Liepa GU, Bietz DC, Beynen AC, Gorman MA (eds) Dietary proteins: how they alleviate disease and promote better health. American Oil Chemists' Society, Champaign, IL, pp 84–122

Cornish SM, Chilibeck PD, Paus-Jennsen L, Biem HJ, Khozani T, Senanayake V, Vatanparast H, Little JP, Whiting SJ, Pahwa P (2009) A randomized controlled trial of the effects of flaxseed lignan complex on metabolic syndrome composite score and bone mineral in older adults. Appl Physiol Nutr Metab 34(2):89–98

Cramer H, Lauche R, Langhorst J, Dobos G (2012) Effectiveness of yoga for menopausal symptoms: a systematic review and meta-analysis of randomized controlled trials. Evid Based Complement Alternat Med 2012:863905. https://doi.org/10.1155/2012/863905, p 1–11

Cui SW (2001) Polysaccharide gums from agricultural products: processing, structures and functionality. Technomic Publishing, Lancaster, PA

Cui W, Mazza G, Oomah BD, Biliaderis CG (1994) Optimization of an aqueous extraction process for flaxseed gum by response surface methodology. LWT Food Sci Technol 27:363–369

Cunnane SC, Ganguli S, Menard C, Liede AC, Hamadeh MJ, Chen ZY, Wolever TMS, Jenkins DJA (1993a) High α–linolenic acid flaxseed (Linum usitatissimum): some nutritional properties in human. Br J Nutr 69(2):443–453

Cunnane SC, Ganguli S, Menard C, Liede AC, Hamadeh MJ, Chen ZY, Wolever TM, Jenkins DJ (1993b) High α-linolenic acid flaxseed (Linumusitatissimum): some nutritional properties in humans. Br J Nutr 69(2):443–453

Demark-Wahnefried W, Price DT, Polascik TJ, Robertson CN, Anderson EE, Paulson DF, Walther PJ, Gannon M, Vollmer RT (2001) Pilot study of dietary fat restriction and flaxseed supplementation in men with prostate cancer before surgery: exploring the effects on hormonal levels, prostate-specific antigen, and histopathologic features. Urology 58:47–52

Dev DK, Quensel E, Hansen R (1986) Nitrogen extractability and buffer capacity of defatted linseed (Linum usitatissimum L.) flour. J Sci Food Agric 37:199–205

Divya KB, Kumar S, Mittal V, Sabikhi L (2013) Sensory attributes, fatty acid profile and oxidative stability of low fat buffalo milk supplemented with omega 3 fatty acids and dietary fiber. Indian J Dairy Sci 66(6):469–476

Dodin S, Lemay A, Jacques H, Légaré F, Forest JC, Mâsse B (2005) The effects of flaxseed dietary supplement on lipid profile, bone mineral density, and symptoms in menopausal women: a randomized, double-blind, wheat germ placebo-controlled clinical trial. J Clin Endocrinol Metab 90(3):1390–1397

Dodin S, Cunnane SC, Masse B, Lemay A, Jacques H, Asselin G, Tremblay-Mercier J, Marc I, Lamarche B, Legare F, Forest JC (2008) Flaxseed on cardiovascular disease markers in healthy menopausal women: a randomized, double-blind, placebo-controlled trial. Nutrition 24 (1):23–30

Dubey H, Das SK, Panda T (2006) Numerical simulation of a fully baffled biological reactor: the differential circumferential averaging mixing plane approach. Biotechnol Bioeng 95:754–766

Dupasquier CMC, Dibrov E, Kostenuk AL, Yeganeh B, Moghadasian MH, Pierce GN (2007) The beneficial effects of dietary flaxseed on atherosclerosis in the LDL receptor deficient mouse occur in part through anti-proliferative and anti-inflammatory actions. Am J Physiol Heart Circul Physiol 293:H2394–H2402

FAO (2016) FAOSTAT. Food and Agricultural Organization of the United Nations. http://faostat3.fao.org/home/index.html. Accessed 1 Dec 2017

Faseehuddin SKA, Madhusudhan B (2007a) Hypocholesterolemic and hepatoprotective eff ects of flaxseed chutney: evidence from animal studies. Ind J Clin Biochem 22:117–121

Faseehuddin SKA, Madhusudhan B (2007b) Effects of flaxseed (Linum usitatissimum) chutney powder on gamma-glutamyl transpeptidase and micronuclei profile in azoxymethane treated rats. Ind J Clin Biochem 22:129–131

Feng D, Shen Y, Chavez ER (2003) Effectiveness of different processing methods in reducing hydrogen cyanide content of flaxseed. J Sci Food Agric 83(8):836–841

Fitzpatrick K (2007) Innovation in Western Canadian functional food ingredients. Cereal Foods World 52:289–290

Freeman TP (1995) Structure of flaxseed. In: Cunnane SC, Thompson LH (eds) Flaxseed in human nutrition. AOCS Press, Champaign, IL, pp 11–21

Fukumitsu S, Aida K, Ueno N, Ozawa S, Takahashi Y, Kobori M (2008) Flaxseed lignan attenuates high-fat diet-induced fat accumulation and induces adiponectin expression in mice. Br J Nutr 100:669–676

Fuller DQ, Harvey E (2006) The archaeobotany of Indian pulses: identification, processing and evidence for cultivation. Environ Archaeol 11:241–268

Funk CD (2001) Prostaglandlins and leukotrienes: advances in eicasanoid biology. Science 294:1871–1875

Gall H (2000) Food dependent exercise induced anaphylaxis to flaxseed. Allergol Int 49 (3):219–221

Gambus H, Gambus F, Pastuszka D (2009) Quality of gluten-free supplemented cakes and biscuits. Int J Food Sci Nutr 60:31–50

Ganorkar PM, Jain RK (2013) Flaxseed: a nutritional punch. Int Food Res J 20:519–525

Ghazanfarpour M, Sadeghi R, Roudsari RL, Khadivzadeh T (2016) Effects of flaxseed and Hypericumperforatum on hot flash, vaginal atrophy and estrogen-dependent cancers in menopausal women: a systematic review and meta-analysis. Avicenna J Phytomed 6(3):273–283

Gillingham LG, Gustafson JA, Han SY, Jassal DS, Jones PJH (2011) High-oleic rapeseed (canola) and flaxseed oils modulate serum lipids and inflammatory biomarkers in hypercholesterolaemic subjects. Br J Nutr 105:417–427

Giroux HJ, Houde J, Britten M (2010) Use of heated milk protein-sugar blends as antioxidants in diary beverages enriched with linseed oil. LWT Food Sci Technol 43:1373–1376

Goh KKT, Ye A, Dale N (2006) Characterisation of ice cream containing flaxseed oil. Int J Food Sci Technol 41:946–953

Gokmen V, Magol BA, Lumaga RB, Fogliano-Vincezo F, Kaplun Z (2011) Development of functional bread containing nanoencapsulated omega-3 fatty acids. J Food Eng 105:585–591

Gopalan C, Ramasastri BV, Subramanian SC (2007) Nutritive value of Indian foods. National Institute of Nutrition, Hyderabad

Gorski A, Kasprzycka M, Nowaczyk M, Wieczoreck Z, Siemion IZ, Szelejewski W, Kutner A (2001) Cyclolinopeptide: a novel immunosuppressive agent with potential anti-lipemic activity. Transplant Proc 33:553

Gowda A, Sharma V, Goyal A, Singh AK, Arora S (2018) Process optimization and oxidative stability of omega-3 ice cream fortified with flaxseed oil microcapsules. J Food Sci Tech 55 (5):1705–1715

Goyal A, Sharma V, Upadhyay N, Gill S, Sihag M (2014) Flax and flaxseed oil: an ancient medicine & modern functional food. J Food Sci Technol 51(9):1633–1653

Goyal A, Sharma V, Sihag MV, Singh AK, Arora S, Sabikhi L (2016) Fortification of dahi (Indian yoghurt) with omega-3 fatty acids using microencapsulated flaxseed oil microcapsules. J Food Sci Technol 53(5):2422–2433

Goyal A, Sharma V, Sihag MV, Singh AK, Arora S, Sabikhi L (2017) Oxidative stability of alpha-linolenic acid (ω-3) in flaxseed oil microcapsules fortified market milk. Int J Dairy Technol 70 (2):188–196

Gray CA, Bartol FF, Tarleton BJ, Wiley AA, Johnson GA, Bazer FW, Spencer TE (2001) Developmental biology of uterine glands. Biol Reprod 65:1311–1323

Greenwald P, Clifford CK, Milner JA (2001) Diet and cancer prevention. Eur J Cancer 37:948–965

Hadley M, Lacher C, Mitchel-Fetch J (1992) Fibre in flaxseed. In: Proceedings of the 54th Flax Institute of United States, Flax Institute of United States, Fargo, ND, p 79–83

Haggans CJ, Hutchins AM, Olson BA, Thomas W, Martini MC, Slavin JL (1999) Effect of flaxseed consumption on urinary estrogen metabolites in postmenopausal women. Nutr Cancer 33 (2):188–195

Hall C III, Niehaus M, Wolf-Hall C, Bauer A, Auch C, Stoerzinger C (2004) Flaxseed in yoghurt: Lignan stability and effect on processing. In: Proceedings of the 60th Flax Institute of the United States, Fargo, ND, p 1–7

Hall C, Schwarz J (2002) Functionality of flaxseeds in frozen desserts-preliminary report. In: Proceedings of the 59th Flax Institute of the United States held at Fargo, ND, p 21–24

Hall AV, Parbtani A, Clark WF, Spanner E, Keeney M, Chin-Yee I, Philbrick DJ, Holub BJ (1993) Abrogation of MRL/lpr lupus nephritis by dietary flaxseed. Am J Kidney Dis 122:326–332

Hall AA, Manthey FA, Lee RE, Niehaus M (2005) Stability of α- linolenic acid and secoisolariciresinol diglusoside in flaxseed fortified macaroni. J Food Sci 70:C483–C489

Hall C, Tulbek MC, Xu Y (2006) Flaxseed. Adv Food Nutr Res 51:1–97

Hao M, Beta T (2012) Development of Chinese steamed bread enriched in bioactive compounds from barley hull and flaxseed hull extracts. Food Chem 133:1320–1325

Hasler CM (1998) Functional foods: their role in disease prevention and health promotion. Food Technol 52(11):63–70

Hassan AA, Rasmy NM, Foda MI, Bahgaat WK (2012) Production of functional biscuits for lowering blood lipids. World J Dairy Food Sci 7(1):1–20

Henry LA, Witt DM (2006) Effects of neonatal resveratrol exposure on adult male and female reproductive physiology and behavior. Dev Neurosci 28:186–195

Hu C, Yuan YV, Kitts DD (2007) Antioxidant activities of the flaxseed lignan secoisolariciresinol diglucoside, its aglycone secoisolariciresinol and the mammalian lignans enterodiol and enterolactone in vitro. Food Chem Toxicol 45:2219–2227

Hussain S, Anjum FM, Butt MS, Alamri MS, Khan MR (2012) Biochemical and nutritional evaluation of unleavened flat breads fortified with healthy flaxseed. Int J Agric Biol 14:190–196

Hussain S, Anjum FM, Butt MS, Khan MI, Asghar A (2006) Physical and sensoric attributes of flaxseed flour supplemeted cookies. Turk J Biol 30:87–92

Hutchins AM, Martini MC, Olson BA, Thomas W, Slavin JL (2000) Flaxseed influences urinary lignan excretion in a dose-dependent manner in postmenopausal women. Cancer Epidemiol Biomarkers Prev 9(10):1113–1118

Hyvarinen HK, Pihlava J, Hiidenhovi JA, Hietaniemi V, Korhonen HJT, Ryhanen E (2006) Effect of processing and storage on the stability of flaxseed lignan added to dairy products. J Agric Food Chem 54:8788–8792

Hyvärinen HK, Pihlava J, Hiidenhovi JA, Hietaniemi V, Korhonen HJT, Ryhänen E (2006) Effect of processing and storage on the stability of flaxseed lignan added to dairy products. J Agric Food Chem 54(23):8788–8792

Ivanova S, Rashevskaya T, Makhonina M (2011) Flaxseed additive application in dairy products production. Procedia Food Sci 1:275–280

James MJ, Gibson RA, Cleland LG (2000) Dietary polyunsaturated fatty acids and inflammatory mediator production. Am J Clin Nutr Suppl 71:343S–348S

Javed F (1999) Flaxseed a boon to our life. J Diab Assoc India 39:50–51

Jefferson WN, Patisaul HB, Williams CJ (2012) Reproductive consequences of developmental phytoestrogen exposure. Reproduction 143(3):247–260

Johnsson P, Kamal-Eldin A, Lundgren LN, Aaman P (2000) HPLC method for analysis of secoisolariciresinol diglucoside in flaxseeds. J Agric Food Chem 48(11):5216–5219

Judd A (1995) Flax-some historical perspective. In: Cunnane SC, Thompson LH (eds) Flaxseed in human nutrition. AOCS Press, Champaign, IL, pp 1–10

Kadam SU, Prabhasankar P (2010) Marine foods as functional ingredients in bakery and pasta products. Food Res Int 43:1975–1980

Kajla P, Sharma A, Sood DR (2015) Flaxseed—a potential functional food source. J Food Sci Technol 52(4):1857–1871

Kapoor S, Sachdeva R, Kochhar A (2011) Efficacy of flaxseed supplementation on nutrient intake and other lifestyle pattern in menopausal diabetic females. Ethnomedicine 5(3):153–160

Kaur N, Chugh V, Gupta AK (2012) Essential fatty acids as functional components of foods—a review. J Food Sci Technol 51:2289. https://doi.org/10.1007/s13197-012-0677-0

Kelley DS, Vemuri M, Adkins Y, Gill SHS, Fedor D, Mackey BE (2009) Flaxseed oil prevents trans-10, cis-12-conjugated linoleic acidinduced insulin resistance in mice. Br J Nutr 101:701–708

Khalesi S, Jamaluddina R, Ismailab A (2011) Effect of raw and heated flaxseed (Linum usitatissimum L.) on blood lipid profiles in rats. Int J Appl Sci Technol 1(4):84–89

Khouryieh H, Aramouni F (2012) Physical and sensory characteristics of cookies preparedwith flaxseed flour. J Sci Food Agric 92(11):2366–2372

Kim Y, Ilich JZ (2011) Implications of dietary α-linolenic acid in bone health. Nutrition 27 (11):1101–1107

Kishk YMK, Elsheshetawy HE, Mahmoud EAM (2011) Influence of isolated flaxseed mucilage as a non-starch polysaccharide on noodle quality. Int J Food Sci 46:661–668

Kitts DD, Yuan Y, Wijewickreme AN, Thompson LU (1999) Antioxidant activity of the flaxseed lignan secoisolariciresinol diglycoside and its mammalian lignin metabolites enterodiol and enterolactone. Mol Cell Biochem 202:91–100

Kristensen M, Savorani F, Christensena S, Engelsen SB, Bugelb S, Toubroc S, Tetensd I, Astrupa A (2011) Flaxseed dietary fibers suppress postprandial lipemia and appetite sensation in young men. Nutr Metab Cardiovasc Dis 23:136–143

Kristensen M, Jensen MG, Aarestrup J, Petersen KEN, Søndergaard L, Mikkelsen MS, Astrup A (2012) Flaxseed dietary fibers lower cholesterol and increase fecal fat excretion, but magnitude of effect depend on food type. Nutr Metab 9:8

Kritchevsky D (1979) Metabolic effects of dietary fiber (clinical nutrition symposium). West J Med 130:123–127

Kritchevsky D (1995) Fiber effects of hyperlipidemia. In: Cunnane SC, Thompson LUE (eds) Flaxseed in human nutrition. AOCS Press, Champaign, IL, pp 174–186

Lanzmann-Petithory D, Pueyo S, Renaud S (2002) Primary prevention of cardiovascular diseases by alphalinolenic acid. Am J Clin Nutr 76:1456–1457

Lauretani F, Maggio M, Pizzarelli F, Michelassi S, Ruggiero C, Ceda GP, Bandinelli L, Ferrucci L (2009) Omega-3 and renal function in older adults. Curr Pharm Design 15(36):4149–4156

Leyvaa DR, Zahradkab P, Ramjiawanc B, Guzmanc R, Alianid M, Piercea GN (2011) The effect of dietary flaxseed on improving symptoms of cardiovascular disease in patients with peripheral artery disease: rationale and design of the FLAX-PAD randomized controlled trial. Contemp Clin Trials 32(5):724–730

Li X, Lu J, Xia Z, Qiao C (2008) Application of flaxseed gum in stirred yoghurt production. Food Sci 29(12):331–335

Lim CW, Norziah MH, Lu HFS (2010) Effect of flaxseed oil towards physico-chemical and sensory characteristics of reduced fat ice-cream and its stability in ice cream upon storage. Int Food Res J 17:393–403

Lipilina E, Ganji V (2009) Incorporation of ground flaxseed into bakery products and its effect on sensory and nutritional characteristics-a pilot study. J Foodservice 20(1):52–59

Lucas EA, Wild RD, Hammond LJ, Khalil DA, Juma S, Daggy BP, Stoecker BJ, Arjmandi BH (2002) Flaxseed improves lipid profile without altering biomarkers of bone metabolism in postmenopausal women. J Clin Endocrinol Metab 87(4):1527–1532

Lukaszewicz M, Szopa J, Krasowska A (2004) Susceptibility of lipids from different flax cultivars to per oxidation and its lowering by added antioxidants. Food Chem 88:225–231

Ma L (2009) Endocrine disruptors in female reproductive tract development and carcinogenesis. Trends Endocrinol Metab 20:357–363

MacDonald-Wicks LK, Garg ML (2002) Modulation of carbon tetrachloride-induced oxidative stress by dietary fat in rats (open star). J Nutr Biochem 13:87–95

Madhusudhan B (2009) Potential benefits of flaxseed in health and disease-a perspective. Agric Conspec Sci 74(2):67–72

Malkki Y (2004) Trends in dietary fiber research and development: a review. Acta Aliment 33:39–62

Mani UV, Mani I, Biswas M, Kumar SN (2011) An open-label study on the effect of flax seed powder (Linumusitatissimum) supplementation in the management of diabetes mellitus. J Dietary Suppl 8(3):257–265

Mano H, Yuasa T, Kameda T, Miyazawa K, Nakamaru Y, Shiokawa M, Mori Y, Yamada T, Miyata K, Shindo H, Azuma H, Hakeda Y, Kumegawa M (1996) Mammalian mature osteoclasts as estrogen target cells. Biochem Biophys Res Commun 223:637–642

Marambe HK, Shand PJ, Wanasundara JPD (2008) An in vitro investigation of selected biological activities of hydrolysed flaxseed (Linum usitatissimum L.) proteins. J Am Oil Chem Soc 85:1155–1164

Marambe HK, Shand PJ, Wanasundara JPD (2011) Release of angiotensin I-converting enzyme inhibitory peptides from flaxseed (Linum usitatissimum L.) protein under simulated gastrointestinal digestion. J Agric Food Chem 59:9596–9604

Martinez-Flores H, Barrera E, Garnica-Romo M, Penagos C, Saavedra J, Macazaga-Alvarez R (2006) Functional characteristics of protein flaxseed concentrated obtained applying a response surface methodology. J Food Sci 71:495–498

Matumoto-Pintroa PT, Petita HV, Giroux HJ, Cortes C, Agnona NG, Britten M (2011) Effect of flaxseed lignans added to milk or fed to cows on oxidative degradation of dairy beverages enriched with polyunsaturated fatty acids. J Dairy Res 78(1):111–117

Mazza G (2008) Production, processing and uses of Canadian flax. In: First CGNA international workshop, Temuco, Chile, August 3–6

McCullough RS, Edel AL, Bassett CMC, LaVallee RK, Dibrov E, Blackwood DP, Ander BP, Pierce GN (2011) The alpha linolenic acid content of flaxseed is associated with an induction of adipose leptin expression. Lipids 46:1043–1052

Mentes O, Bakkalbassi E, Ercan R (2008) Effect of the use of ground flaxseed on quality and chemical composition of bread. Food Sci Technol Int 14(4):299–306

Milder IE, Arts IC, Van de Putte B, Venema DP, Hollman PC (2005) Lignan contents of Dutch plant foods: a database including lariciresinol, pinoresinol, secoisolariciresinol and matairesinol. Br J Nutr 93(3):393–402

Moghaddasi MS (2011) Linseed and usages in human life. Adv Environ Biol 5(6):1380–1392

Morris D (2000) Cyanogenic glycosides in flaxseed: implications for the health of north Americans. Flax Council of Canada, Winnepeg, MB

Morris DH (2003) Flax-a health and nutrition primer. Flax Council of Canada, Winnipeg, MB

Morris DH (2007a) Flax—a health and nutrition primer, 4th edn. Flax Council of Canada, Winnipeg, MB. http://www.flaxcouncil.ca

Morris DH (2007b) Flax-a health and nutrition primer (Chapter 1), 4th edn. Flax Council of Canada, Winnipeg, MB, pp 9–21

Mridula D, Singh KK, Barnwal P (2013) Development of omega-3 rich energy bar with flaxseed. J Food Sci Technol 50(5):950–957

Mueller K, Eisner P, Yoshie-Stark Y, Nakada R, Kirchhoff E (2010) Functional properties and chemical composition of fractionated brown and yellow linseed meal (Linum usitatissimum L.). J Food Eng 98:453–460

Nagaraj G (2009) Oilseeds-properties, processing, products and procedures. New India Publishing Agency, New Delhi, India, pp 123–132

Naqshbandi A, Rizwan S, Khan F (2013) Dietary supplementation of flaxseed oil ameliorates the effect of cisplatin on rat kidney. J Funct Foods 5(1):316–326

Nash AM, Frankel EM (1986) Limited extraction of soybeans with hexane. J Am Oil Chem Soc 63:244–246

Nazni P, Amrithaveni M, Poongodi VT (2006) Impact of flaxseed based therapeutic food on selected type II diabetic patients. Indian J Nutr Diet 43:141–145

Nowak-Debra A, Denise C, Snyder BAJ, Demark-Wahnefried W (2007) The effect of flaxseed supplementation on hormonal levels associated with polycystic ovarian syndrome: a case study. Curr Top Nutraceutical Res 5(4):177–181

Ogborn MR, Nitschmann E, Bankovic-Calic N, Weiler HA, Aukema H (2002) Dietary flax oil reduces renal injury, oxidized LDL content, and tissue n-6/n-3 FA ratio in experimental polycystic kidney disease. Lipids 37(11):1059–1065

Ogborn MR, Nitschmann E, Bankovic-Calic N, Weiler HA, Fitzpatrick-Wong S, Aukema HM (2003) Dietary conjugated linoleic acid reduces PGE2 release and interstitial injury in rat polycystic kidney disease. Kidney Int 64:1214–1221

Ogunronbi O, Jooste PJ, Abu JO, Vander-Merwe B (2011) Chemical composition, storage stability and effect of cold-pressed flaxseed oil cake inclusion on bread quality. J Food Process Preserv 35:64–79

Omoni AO, Aluko RE (2006) Effect of cationic flaxseed protein hydrolysate fractions on the in vitro structure and activity of calmodulin-dependent endothelial nitric oxide synthase. Mol Nutr Food Res 50:958–966

Oomah BD (2001) Flaxseed as a functional food source. J Sci Food Agric 81(9):889–894

Oomah BD, Mazza G, Kenaschuk EO (1992) Cyanogenic compounds in flaxseed. J Agric Food Chem 40:1346–1348

Oomah BD, Kenaschuk EO, Mazza G (1996a) Phytic acid content of flaxseed as influenced by cultivar, growing season and location. J Agric Food Chem 44:2663–2666

Oomah BD, Mazza G, Kenaschuk EO (1996b) Dehulling characteristics of flaxseed. Lebensm Wiss Technol 29:245–250

Pan A, Yu D, Demark-Wahnefried W, Franco OH, Lin X (2009) Meta-analysis of the effects of flaxseed interventions on blood lipids. Am J Clin Nutr 90(2):288–297

Park JB, Velasquez MT (2012) Potential effects of lignan-enriched flaxseed powder on bodyweight, visceral fat, lipid profile, and blood pressure in rats. Fitoterapia 83:941–946

Patade A, Devareddy L, Lucas EA, Korlagunta BPD, Arjmandi BH (2008) Flaxseed reduces total and LDL cholesterol concentrations in native American postmenopausal women. J Womens Health (Larchmt) 17:355–366

Payne TJ (2000) Promoting better health with flaxseed in bread. Cereal Foods World 45 (3):102–104

Paynel F, Pavlov A, Ancelin G, Rihouey C, Picton L, Lebrun L, Morvan C (2013) Polysaccharide hydrolases are released with mucilages after water hydration of flax seeds. Plant Physiol Biochem 62:54–62

Pellizzon M, Billheimer JT, Bloedon LT, Szapary PO, Rader DJ (2007) Flaxseed reduces plasma cholesterol levels in Hypercholesterolemic mouse models. J Am Coll Nutr 26(1):66–75

Penttinen P, Jaehrling J, Damdimopoulos AE, Inzunza J, Lemmen JG, van der Saag P, Pettersson K, Gauglitz G, Makela S, Pongratz I (2007) Diet-derived polyphenol metabolite enterolactone is a tissue-specific estrogen receptor activator. Endocrinology 148(10):4875–4886

Pisani P (2008) Hyper-insulinaemia and cancer, meta-analyses of epidemiological studies. Arch Physiol Biochem 114(1):63–70

Power KA, Thompson LU (2007) Can the combination of flaxseed and its lignans with soy and its isoflavones reduce the growth stimulatory effect of soy and its isoflavones on established breast cancer. Mol Nutr Food Res 51(7):845–856

Prasad K (1997) Hydroxyl radical–scavenging property of secoisolariciresinol diglycoside (SDG) isolated from flax-seed. Mol Cell Biochem 168(1–2):117–123

Prasad K (1999) Reduction of serum cholesterol and hypercholesterolemic atherosclerosis in rabbits by secoisolariciresinol diglucoside isolated from flaxseed. Circulation 99:1355–1362

Prasad K (2000) Flaxseed: a source of hypocholesterolemic and anti atherogenic agents. Drug News Perspect 13(2):99

Prasad K (2002) Suppression of phosphoenolpyruvatecarboxykinase gene expression by secoisolariciresinoldiglucoside (SDG), a new antidiabetic agent. Int J Angiol 11(2):107–109

Prasad K (2004) Antihypertensive activity of secoisolariciresinol diglucoside (SDG) isolated from flaxseed: role of guanylate cyclase. Int J Angiol 13:7–14

Prasad K, Mantha SV, Muir AD, Westcott ND (1998) Reduction of hypercholesterolemic atherosclerosis by CDC-flaxseed with very low alphalinolenic acid. Atherosclerosis 136:367–375

Prim CR, Baroncini LA, Precoma LB, Caron PH, Winter G, Poletti MO, Precoma DB (2012) Effects of linseed consumption for a short period of time on lipid profile and atherosclerotic lesions in rabbits fed a hypercholesterolaemic diet. Br J Nutr 107(5):660–664

Pruthi S, Thompson SL, Novotny PJ, Barton DL (2007) Pilot evaluation of flaxseed for the management of hot flashes. J Soc Integr Oncol 5(3):106–112

Rajiv J, Indrani D, Prabhasankar P, Rao GV (2012) Rheology, fatty acid profile and storage characteristics of cookies as influenced by flax seed (Linum usitatissimum). J Food Sci Technol 49(5):587–593

Ramcharitar N, Badrie M, Mattfeldt-Beman M, Matsuo H, Ridley C (2005) Consumer acceptability of muffins with flaxseed (Linum usitatissimum). J Food Sci 70(7):504–507

Ray S, Paynel F, Morvan C, Lerouge P, Driouich A, Ray B (2013) Characterization of mucilage polysaccharides, arabinogalactanproteins and cell-wall hemicellulosic polysaccharides isolated from flax seed meal: a wealth of structural moieties. Carbohydr Polym 93:651–660

Ridges L, Sunderland R, Moerman K, Meyer B, Astheimer L, Howe P (2001) Cholesterol lowering benefits of soy and linseed enriched foods. Asia Pac J Clin Nutr 10:204–211

Riediger ND, Othman R, Fitz E, Pierce GN, Suh M, Moghadasian MH (2009) Low n6:n3 fatty acid ratio, with fish or flaxseed oil, in high fat diet improves plasma lipids and beneficially alters tissue fatty acid composition in mice. Eur J Nutr 47:153–160

Roy H, Lundy S, Eriksen C, Kalicki B (2007) Flaxseed: a review of health benefits. Pennington Nutr Ser 11(10):3828–3825

Rubilar M, Gutiérrez C, Verdugo M, Shene C, Sineiro J (2010) Flaxseed as a source of functional ingredients. J Soil Sci Plant Nutr 10:373–377

Setchell KDR, Cassidy A (1999) Dietary isoflavones: biological effects and relevance to human health. J Nutr 129:758S–767S

Shearer AEH, Davies CGA (2005) Physicochemical properties of freshly baked and stored whole-wheat muffins with and without flaxseed meal. J Food Qual 28:137–153

Shim YY, Gui B, Arnison PG, Wang Y, Martin JTR (2014) Flaxseed (Linum usitatissimum L.) bioactive compounds and peptide nomenclature: a review. Trends Food Sci Technol 38(1):5–20

Singh KK, Mridula D, Rehal J, Barnwal P (2011) Flaxseed- a potential source of food, feed and fiber. Crit Rev Food Sci Nutr 51:210–222

Stewart S, Mazza G (2000) Effect of flaxseed gumon quality and stability of a model salad dressing. J Food Qual 23:373–390

Sturgeon SR, Volpe SL, Puleo E, Bertone-Johnson ER, Heersink J, Sabelawski S, Wahala K, Bigelow C, Kurzer MS (2011) Dietary intervention of flaxseed: effect on serum levels of IGF-1, IGF-BP3, and C-peptide. Nutr Cancer 63(3):376–380

Sullivan AW, Hamilton P, Patisaul HB (2011) Neonatal agonism of ERbeta impairs male reproductive behavior and attractiveness. Horm Behav 60:185–194

Tanwar B, Modgil R, Goyal A (2018) Antinutritional factors and hypocholesterolemic effect of wild apricot kernel (Prunus armeniaca L.) as affected by detoxification. Food Funct 9 (4):2121–2135

Tarpila S, Tarpila A, Grhn P, Silvennoinen T, Lindberg L (2004) Efficacy of ground flaxseed on constipation in patients with irritable bowel syndrome. Curr Topics Nutraceutical Res 2(2):119–125

Tarpila A, Wennberg T, Tarpila S (2005) Flaxseed as functional food. Curr Top Nutraceut R 3(3):167–188

Tham DM, Gardner CD, Haskell WL (1998) Potential health benefits of dietary phytoestrogens, a review of the clinical, epidemiological, and mechanistic evidence. J Clin Endocrinol Metab 83 (7):2223–2235

Thompson LU, Li T, Chen J, Goss PE (2000) Biological effects of dietary flaxseed in patients with breast cancer. Breast Cancer Res Treat 64(1):50

Tou JC, Chen J, Thompson LU (1998) Flaxseed and its lignan precursor, secoisolariciresinol diglycoside, affect pregnancy outcome and reproductive development in rats. J Nutr 128 (11):1861–1868

Toure A, Xueming X (2010) Flaxseed lignans: source, biosynthesis, metabolism, antioxidant activity, bio-active components and health benefits. Compr Rev Food Sci Food Saf 9:261–269

Tourre A, Xueming X (2010) Flaxseed lignans: source, biosynthesis, metabolism, antioxidant activity, bio-active components, and health benefits. Comp Rev Food Sci Food Saftey 9:261–269

Truan JS, Chen JM, Thompson LU (2010) Flaxseed oil reduces the growth of human breast tumors (MCF-7) at high levels of circulating estrogen. Mol Nutr Food Res 54(10):1414–1421

Udenigwe CC, Aluko RE (2010) Antioxidant and angiotensin converting enzyme-inhibitory properties of a flaxseed protein derived high Fischer ratio peptide mixture. J Agric Food Chem 58:4762–4768

Udenigwe CC, Adebiyi AP, Doyen A, Li H, Bazinet L, Aluko RE (2012) Low molecular weight flaxseed protein-derived arginine-containing peptides reduced blood pressure of spontaneously hypertensive rats faster than amino acid form of arginine and native flaxseed protein. Food Chem 132:468–475

Vaisey-Genser M, Morris DH (2003) Introduction, history of the cultivation and uses of flaxseed. In: Muir AD, Westcott ND (eds) Flax, the genus Linum. Taylor and Francis, London, pp 1–21

Veena N, Nath BS, Srinivas B, Balasubramanyam BV (2017) Quality attributes of dahi prepared from milk fortified with omega-3 fatty acids, phytosterols and polydetxrose. J Food Sci Technol 54(7):1765–1775

Wanders AJ, van den Borne JJ, de Graaf C, Hulshof T, Jonathan MC, Kristensen M, Mars M, Schols HA, Feskens EJ (2011) Effects of dietary fibre on subjective appetite, energy intake and body weight: a systematic review of randomized controlled trials. Obesity Rev 12(9):724–739

Wiesenborn D, Kangas N, Tostenson K, Hall C, Chang K (2005) Sensory and oxidative quality of screw-pressed flaxseed oil. J Am Oil Chem Soc 82:887–892

Williams D, Verghese M, Walker LT, Boateng J, Shackelford L, Chawan CB (2007) Flax seed oil and flax seed meal reduce the formation of aberrant crypt foci (ACF) in azoxymethane-induced colon cancer in fisher 344 male rats. Food Chem Toxicol 45(1):153–159

Wohlfahrt-Veje C, Main KM, Skakkebaek NE (2009) Testicular dysgenesis syndrome: foetal origin of adult reproductive problems. Clin Endocrinol 71:459–465

Wu W, Huff HE, Hsieh F (2007) Processing and properties of extruded flaxseed-corn puff. J Food Process Preserv 31:211–226

Wu M, Li D, Wang LJ, Ozkan N, Mao ZH (2010) Rheological properties of extruded dispersions of flaxseed-maize blend. J Food Eng 98:480–491

Xu J, Zhou X, Chen C, Deng Q, Huang Q, Yang JE, Yang N, Huang F (2012) Laxative effects of partially defatted flaxseed meal on normal and experimental constipated mice. BMC Complement Alternat Med 12(1):1–5

Yamashita T, Sano T, Hashimoto T, Kanazawa K (2007) Development of a method to remove cyanogen glycosides from flaxseed meal. Int J Food Sci Technol 42:70–75

Zhang ZS, Wang Li J, Li D, Jiao SS, Chen DX, Mao ZH (2008) Ultrasound-assisted extraction of oil from flaxseed. Sep Purif Technol 62:192–198

Chapter 11
Chia Seed (*Salvia hispanica*)

Sibele Santos Fernandes, Carlos Prentice,
and Myriam de las Mercedes Salas-Mellado

Abstract Chia (*Salvia hispanica* L.) is a herbaceous plant cultivated annually, being considered oleaginous due to the high oil content, which corresponds to 30–40% of the seed, being rich in polyunsaturated fatty acids, mainly omega-3 (linolenic acid, 54–67%), higher than in any known plant source, and omega-6 (linoleic acid, 12–21%) fatty acids. Consumption of the essential fatty acids present in chia seeds is associated with the prevention of various chronic diseases, such as obesity, cardiovascular disease, and cancer. This oil can be extracted by various techniques and can be encapsulated through micro- and nanoencapsulation to provide the incorporation and preservation of fatty acids in foods. However, commercially available foods and scientific studies concentrate on the addition of whole chia seed, mainly in bakery products. In addition to the high oil content, the chia seed presents high protein content and dietary fiber, evidencing its high potential as a functional food.

Keywords Chia oil · Essentials fatty acids · Omega-3 fatty acids · Cardiovascular diseases · Foods · Encapsulation

11.1 Origin and History

The chia seed (*Salvia hispanica* L.) is an annual herbaceous plant, belonging to the Lamiaceae family, being native to southern Mexico and northern Guatemala (Ixtaina et al. 2008; Capitani et al. 2012). In pre-Columbian times, besides being used as food along with corn, beans, and amaranth, it was also used for medical purposes (Coates 2011). Some historians suggest that *Salvia hispanica* was a staple more important than corn in some areas (Cahill 2003; Ayerza and Coates 2005).

S. S. Fernandes (✉) · C. Prentice · M. d. l. M. Salas-Mellado
Laboratory of Food Technology, School of Chemistry and Food, Federal University of Rio Grande, Carreiros, Brazil

B. Tanwar, A. Goyal (eds.), *Oilseeds: Health Attributes and Food Applications*,
https://doi.org/10.1007/978-981-15-4194-0_11

For centuries, chia seeds were used as food by the people of the west and of southern Mexico, besides being provided as a tribute in Mexico (Cahill 2003). For the Aztecs, chia was offered to the gods during religious ceremonies; however, this custom disappeared about 500 years after the conquest of the territory by the Spaniards, who eventually replaced chia with their favorite cultures, brought from Europe (Ayerza and Coates 2005).

The consumption of chia is considered limited and regional, due to the lack of information on the characteristics of the seed, high retail price, low availability, and consumption (Olivos-Lugo et al. 2010). In Brazil, the difficulties are concentrated by the producers due to the lack of knowledge about the harvesting of the seeds and their commercialization, which is a culture little explored in the country. Thus, the destination of the seeds ends up being the supply of markets and stores of natural products (Migliavacca et al. 2014).

The use of the chia seed in the religious ceremonies of the Aztecs and the Mayas provoked the wrath of the Spanish Catholics who saw a pagan ritual. With this, its cultivation was extinguished by centuries and was only resumed in the early 1990s. As of 2011, chia seed has been gaining popularity as a food, where more than 72 products based on this grain have reached the global market in the form of snacks, seasonings, yogurts, bakery products, etc. (Everitt 2016). According to one of the world's leading market research agencies, the consumption of chia seed has increased by 118% between 2012 and 2015, with a 70% increase in the launch of new products in the food and beverage sector (Benzi 2016).

11.2 Production

Chia seed is cultivated mainly in Mexico, Peru, Bolivia, Colombia, Ecuador, and Guatemala. In Argentina, mainly in the north of the country, in the provinces of Salta and Jujuy, this culture has become a very important economic activity (Martínez et al. 2012). The largest production center is located in Mexico, which currently exports seeds to several countries such as Japan, the USA, and Europe.

The average crop yields are around 500–600 kg ha^{-1} although some producers have obtained yields of up to 1200 kg ha^{-1} (Coates 2011). In Brazil, this seed is grown in the regions of western Paranaense and northwest of Rio Grande do Sul, with a yield of up to 800 kg ha^{-1} in May and 200–300 kg ha^{-1} in August (Migliavacca et al. 2014).

Baginsky et al. (2016) studied different regions of Chile for the cultivation of chia seed. These authors related that the Valle de Azapa (coastal deserts) and Canchones (normal desert climates) provide the best conditions for chia production under the conditions of the study. Among these two regions, the Valle de Azapa was more prominent for cultivation, due to the lower thermal oscillation, extreme temperatures, and freezing throughout the harvest, and it can produce, between mid-February and mid-March, yields over 2000 kg ha^{-1}, oil content above 550 L ha^{-1}, and α-linolenic acid content and linoleic acid above 350 and 90 L ha^{-1}, respectively.

11.3 Chemical Composition

Chia seed is rich in essential fatty acids, dietary fiber, and proteins (Capitani et al. 2012). The chemical composition of chia seeds varies depending on their geographical origin due to differences in the environment, climatic changes, nutrient availability, year of cultivation, or soil conditions. Various studies have reported the variations in the chemical compositions of chia seeds in different regions of Mexico (Ixtaina et al. 2010; Porras-Loaiza et al. 2014), Argentina (Ayerza and Coates 2011; Ixtaina et al. 2011), Bolivia and Ecuador (Ayerza and Coates 2011), and Brazil (da Silva et al. 2017).

Climatic parameters that influence the composition of the chia seed are mainly temperature and altitude. The increase in the temperature of the environment causes a decrease in polyunsaturated fatty acids (PUFAs) content (Ayerza and Coates 2005). Table 11.1 presents the nutritional composition of chia seed in different countries.

11.3.1 Lipids

Lipids in chia seed represent about 30–40% of total weight (Coorey et al. 2014), which is composed mainly of the polyunsaturated fatty acids (omega-3, α-linolenic acid, 54–67% and omega-6, linoleic acid, 12–21%). In addition to these fatty acids, oleic acid, palmitic acid, and stearic acid are also present in small quantities (Ixtaina

Table 11.1 Nutritional composition of chia seed (g/100 g) in different countries

	Brazil[a]	México[b]	Argentina[c]	Bolivia[c]	Ecuador[c]
Ash	4.56 ± 0.04	4.55 ± 0.29	–	–	–
Lipids	32.16 ± 0.29	23.66 ± 0.68	33.5	29.98	31.47
16:0	1.82 ± 0.12	6.22 ± 0.10	6.89	7.72	6.39
18:0	0.90 ± 0.07	2.95 ± 0.08	2.36	3.59	3.74
18:1 (Omega-9)	1.69 ± 0.10	7.05 ± 0.06	6.73	9.12	6.59
18:2 (Omega-6)	5.69 ± 0.42	20.12 ± 0.15	22.50	21.93	16.99
18:3 (Omega-3)	20.37 ± 1.38	62.78 ± 0.21	60.35	56.93	64.75
Saturated	2.88 ± 0.18	–	9.26	11.32	10.14
Polyunsaturated	27.75 ± 1.80	–	82.85	78.87	81.74
Protein	18.18 ± 1.20	22.03 ± 0.42	16.45	26.03	15.95
Total fiber	33.37 ± 0.26	33.47 ± 0.21	–	–	–
Carbohydrates	4.59 ± 0.34	9.05 ± 0.21	–	–	–

All the values of fatty acids are given in g/100 g of fatty acids except da Silva et al. (2017) where fatty acids are mentioned as g/100 g of chia seed

[a]da Silva et al. (2017)

[b]Porras-Loaiza et al. (2014)

[c]Ayerza and Coates (2011)

Table 11.2 Extraction method of chia oil

Extraction method	Details
Pressing	Cold-pressing technique and storage at low temperature (4 °C) in the dark using a Komet screw press (Ixtaina et al. 2011; Martínez et al. 2012; González et al. 2016; Bodoira et al. 2017) or using a Tamer hydraulic press (Rodea-González et al. 2012; Escalona-García et al. 2016)
Maceration	Using organic solvent (Guindani et al. 2016)
Ultrasound	Using ultrasound bath (Guindani et al. 2016)
Solvent	Hot extraction using the Soxhlet method (Ixtaina et al. 2011; Guindani et al. 2016) or cold extraction (Silva et al. 2016; Timilsena et al. 2016) using mainly hexane and other solvents
Supercritical fluids	Using carbon dioxide (Ixtaina et al. 2010; Uribe et al. 2011; Guindani et al. 2016; Scapin et al. 2017) or liquefied petroleum gas (LPG) (Scapin et al. 2017)

et al. 2011). Thus, this oil can be termed as "gourmet oil," since it is recognized as high-quality oil and appreciated for its taste, color, and healthy characteristics (Ixtaina et al. 2015).

An advantage associated with chia seed in comparison to the commercial products rich in omega-3 fatty acids derived from fish, fish oil and fish meal, is that it has no cholesterol (Tosco 2004) and free-form heavy metal contamination. Chia seed is mainly valued for its oil content. Several methods can be applied for the extraction of chia oil, but differences between the methods cause changes in oil yield, fatty acid quality, fatty acid content, and antioxidant activity. Table 11.2 summarizes the current methods used in extracting chia seed oil.

11.3.2 Carbohydrates

Total carbohydrate content of chia seeds ranges from 24.6 to 41.5% (European Union 2017). The carbohydrate content is represented mostly by 90% of fibers, of which the majority is soluble, and the rest by starch, with no sugar content (Segura-Campos et al. 2016). Among the soluble fibers, chia mucilage is formed by the hydration of chia seeds. According to Lin et al. (1994), the structure of chia mucilage is a tetrasaccharide with a main chain consisting the units of (1 → 4)-β-d-xylopyranosyl-(1 → 4)-α-d-glucopyranosyl-(1 → 4)-β-d-xylopyranosyl with 4-O-methyl-α-d-glucuronic acid ramifications in the O-2 position of β-d-xylopyranosyl in the main chain. The ratio of β-d-xylose to α-d-glucose monosaccharides to 4-O-methyl-α-d-glucuronic acid is 2:1:1.

Since it is gluten-free, the use of chia seed has an additional advantage to be used in the diet of patients with celiac disease (Vanesa Ixtaina 2010). This advantage has driven recent research into the development of new gluten-free products (Costantini et al. 2014; Steffolani et al. 2014; Huerta et al. 2016; Simurina et al. 2017).

11.3.3 Proteins

The protein content of chia seeds is 15–25% (Ali et al. 2012), similar to the content present in lentils (23%), peas (25%), and chickpeas (21%) (Olivos-Lugo et al. 2010). Among proteins, which are readily digestible, the globulins are the major components (52% of total protein fractions), and albumins, glutelins, and prolamins are present in almost equal proportion (Sandoval-Oliveros and Paredes-López 2013). Chia seeds contain all the essential amino acids, in particular glutamic acid, arginine, and aspartic acid (Sandoval-Oliveros and Paredes-López 2013), adding, by virtue of this, advantages in its use as a source of nutrients (Alvarado 2011).

11.3.4 Vitamins

Vitamins cannot be synthesized by the body and thus necessary to consume in diet from natural foods (Rendón-Villalobos et al. 2012). Chia seed is a good source of B-complex vitamins, such as B_6 with 0.1 mg/100 g, thiamine with 0.7 mg/100 g, riboflavin with 0.2 mg/100 g, and niacin with 7.2 mg/100 g of seed. Vitamin C is also present with 5.4 mg/100 g of seed (Craig and Sons 2004). da Silva et al. (2017) reported the average content of vitamin E in chia of 8.1 mg/100 g, with the appreciable quantities of γ-tocopherol.

11.3.5 Minerals

Minerals, like vitamins, are not synthesized by the body but are necessary for maintaining the body in optimal health. It is therefore necessary to use external sources such as food, nutritional supplements, and absorption through the skin, to ensure an adequate supply of minerals (Rendón-Villalobos et al. 2012). In chia seed, minerals, such as phosphorus, calcium, potassium, magnesium, iron, zinc, and copper, are present in adequate amounts along with safe levels of heavy metals for use in food (Migliavacca et al. 2014). Among the minerals, phosphorus, calcium, and magnesium are present in the highest proportions (Capitani et al. 2012). A significant difference between chia and other sources of omega-3 fatty acids is that it has a low sodium content in the seeds, which makes it an excellent food option for people suffering from hypertension and those who need a diet with low levels of sodium (Busilacchi et al. 2013).

11.4 Antinutritional Factors

There are no reports available in the literature till date describing allergic reactions due to the use of chia seeds, although there have been some cases of hypersensitivity to plants of the same family (Lamiaceae) (Coates 2011).

11.5 Bioactive Compounds

The composition and the concentration of bioactive compounds depend on several factors, such as climatic conditions, geographical origin, and the extraction methods (Ixtaina et al. 2011; Capitani et al. 2012). Chia seed extracts present caffeic acid, which is composed of a dihydroxyphenyl group linked with acrylic acid and plays an important role both chemically and biologically (De Falco et al. 2017). Moreover, in the metabolome of chia seeds, not only monomers of caffeic acid building block are present but also condensation products such as polymers ferulic acid (Martínez-Cruz and Paredes-López 2014), chlorogenic acid (Reyes-Caudillo et al. 2008; Coelho and de las Mercedes Salas-Mellado 2014), and rosmarinic acid (Martínez-Cruz and Paredes-López 2014). Caffeic acid dimers are also frequent in chia samples, and among them rosmarinic acid is the most abundant one (De Falco et al. 2017). Flavonoids identified in chia seed are myricetin (Taga et al. 1984), quercetin (Taga et al. 1984; Reyes-Caudillo et al. 2008; Coelho and de las Mercedes Salas-Mellado 2014), kaempferol (Taga et al. 1984; Reyes-Caudillo et al. 2008), and daidzin, glycitein, glycitin, genistein, and genistin (Martínez-cruz and Paredes-lópez 2014).

As previously mentioned, chia seed is rich in omega-3 and omega-6 bioactive compounds. Essential fatty acids are not synthesized by human metabolism, so they need to be ingested through food. The high omega-3 content, higher than any known plant source, refers to the use of chia seed as a source of omega-3 functional food (Ayerza and Coates 2011).

11.6 Health Attributes

α-Linolenic (ω-3) and linoleic (ω-6) fatty acids are fundamental in the prevention and treatment of cardiovascular diseases, as well as having other important roles in reducing the risks of hypertension, diabetes, arthritis, and autoimmune diseases (Carvalho et al. 2003; Goyal et al. 2014).

11.6.1 Chia Seeds and Cardiovascular Diseases

Although the dietary habits of the population are changing, still many consume high amounts of obesogenic foods, rich in sugars and saturated and trans fats, which promote the development of cardiovascular diseases (CVD), the main cause of death among individuals (Temple 2018). According to the World Health Organization, cardiovascular diseases is the world's number one cause of mortality (WHO 2017). Approximately 17.7 million people died from CVDs in 2015; of these deaths, an estimated 7.4 million were due to coronary heart disease, and 6.7 million were due to stroke.

The consumption of fatty acids of the omega-3 series, such as eicosapentaenoic acid (EPA, C20:5) and docosahexaenoic acid (DHA, C22:6), both mainly from fish, and α-linolenic (ALA, C18:3), derived primarily from plant sources act as major protectors of CVDs. However, the essential polyunsaturated fatty acids (PUFA) are then synthesized from linolenic and linoleic fatty acids, including arachidonic acid (AA) from linoleic acid and EPA and DHA from linolenic acid (El-Badry et al. 2007).

Studies have shown that supplementation of 50 g of chia seed per day significantly increases the plasma linolenic acid content (Nieman et al. 2009) and the ingestion of 25 g chia seed daily by healthy postmenopausal women significantly increases the levels of EPA and plasma linolenic acid (Jin et al. 2012). Although these studies have verified the increase of linolenic acid in humans by the consumption of chia seed cannot justify its protective effect against the CVD, and might only suggest long-term cardio-protective effects.

Sierra et al. (2015) have found that chia oil can protect vascular function against the deleterious effects of early hypercholesterolemia and may improve vascular relaxation even in the presence of elevated cholesterol levels. The ability to protect vascular function reinforces the idea that chia oil could be used as a functional food and be a potential approach to limit the progression of vascular dysfunction.

11.6.2 Antioxidative Effects

Chia seeds present high content of phenolic compounds which have been scientifically proven to exhibit antioxidative functions. Antioxidants and phenolic compounds have been found to have health-promoting properties and also confer protection from degenerative diseases such as cardiovascular diseases, cancer, diabetes, and diverticulosis (Guevara-Cruz et al. 2012). Chia oil contains a great quantity of γ-tocopherol and small amount of δ-tocopherol, as well as trace amounts of α-tocopherol (Bodoira et al. 2017). In chia oil, the main phenolic compounds are chlorogenic and caffeic acids, followed by myricetin, quercetin, and kaempferol, which are not present in other oilseeds (Ixtaina et al. 2011).

The extracts of water and methanol produced by extracting oil from chia seeds present strong antioxidant activity, mainly due to their chlorogenic acid, caffeic acid, and flavonol contents. These compounds assist in stabilizing the lipid composition of the seeds, which is why chia oil or flour does not require the addition of antioxidants for preservation (Segura-Campos et al. 2016).

There are several studies that have determined the antioxidant effect of seed and chia oil. The objective of one study was to compare the antioxidant activity of chia seeds in different types of regions, such as Jalisco and Sinaloa, Mexico (Reyes-Caudillo et al. 2008). These authors verified that the chia extracts of Jalisco and Sinaloa showed higher antioxidant activity than other *Salvia* species, such as *Salvia caespitosa* (55.9%), *Salvia candidissima* (62.3%), *Salvia hypargeia* (62.9%), *Salvia euphratica* (59.1%), *Salvia sclarea* (63.5%), and *Salvia aethiopis* (29.0%).

Other studies have focused on comparing the antioxidant activity of chia seed with other seeds, such as flax and perilla (Sargi et al. 2013). Among the flax and perilla species, the gold and white species had higher levels of omega-3 and omega-6, while both brown flax and perilla seeds showed higher antioxidant capacity, and chia showed a higher content of fatty acids and intermediate antioxidant capacity.

11.6.3 Diabetes

On ingestion, chia seeds form a gel with the acid present in the stomach. This gel acts as a physical barrier between consumed carbohydrates and digestive enzymes, which consequently slows carbohydrate digestion, leading to a more gradual and sustained conversion to glucose and, thus, avoids abrupt peaks in blood glucose after consuming the seeds (Reyes-Caudillo et al. 2008).

The soluble fiber present in the chia seed exerts an influence on the stabilization of blood glucose levels by regulating the rate at which complex carbohydrates are digested and assimilated in the body. It is proven that supplementing daily diets with 35–37 g of chia seeds for diabetics controls hyperglycemia (reduces blood glucose levels) and reduces systolic blood pressure (Vuksan et al. 2007; Toscano et al. 2014).

11.6.4 ACE-Inhibitory Activity

The elevation of blood pressure affects the risk of cardiovascular diseases such as arteriosclerosis, stroke, and myocardial infarction. To regulate blood pressure, the angiotensin I-converting enzyme (ACE, dipeptidylcarboxypeptidase, EC 3.4.15.1), through a rennin-angiotensin system, converts angiotensin I to angiotensin II and inactivates the vasodilator bradykinin. ACE inhibition mainly produces a hypotensive effect but can also influence the regulatory systems involved in immune defense and nervous system activity (Haque et al. 2009).

Segura-Campos et al. (2013) identified and quantified the ACE-inhibitory effect produced by hydrolysates from chia protein hydrolysates obtained with Alcalase®-Flavourzyme® sequential enzymatic system and evaluated the effect in a carrot cream. The chia protein hydrolysates produced with this sequential system exhibited ACE-inhibitory activity, suggesting that the peptides released from the proteins are the agents behind the inhibition. On the other hand, ACE-inhibitory activity in the carrot cream improved markedly with addition of the chia protein hydrolysates.

11.6.5 Anti-inflammatory Effects

Foods with anti-inflammatory components are responsible for reducing the production of substances that stimulate the inflammatory process, strengthening the immune system, making the body more resistant against respiratory diseases like flu and cold. For a food to be considered anti-inflammatory, it must be rich in substances such as allicin, omega-3 fatty acids, and vitamin C. Although there are no scientific studies that prove the anti-inflammatory action of chia seed, the high omega-3 content present in this seed has a great potential to show anti-inflammatory effects to the human body.

11.6.6 Inhibition of Melanin Hyperpigmentation

Diwakar et al. (2014) found that in addition to being essential fatty acids, omega-3 and omega-6 fatty acids play an important role in maintaining skin hydration and the stratum corneum epidermal barrier and influence melanogenesis in epidermal melanocytes. Consumption of chia seeds has shown the inhibition of melanin biosynthesis in melan-a cells. In addition, the authors combined chia seed extract with pomegranate fruit extract, providing inhibition of melanin production, possibly through downregulated expression of genes related to melanin synthesis (Diwakar et al. 2014).

11.6.7 Antimicrobial Effects

The rapid emergence of antibiotic-resistant bacterial pathogens is a serious public health concern, and considerable efforts have been focused on the development of new classes of antimicrobial agents, such as antimicrobial peptides. Segura-Campos et al. (2013) evaluated the antimicrobial activity of hydrolysates from chia protein-rich fraction against seven pathogens using the agar disk diffusion assay. The results obtained by the authors were that the chia protein hydrolysates did not inhibit the growth of Gram-negative (*E. coli*, *S. typhi*, *S. flexneri*) and Gram-positive bacteria

(*K. pneumonia*, *S. aureus*, *B. subtilis*, *S. agalactiae*), even at the lowest evaluated concentration (1 × 10^3 UFC/mL). Such a feat can be justified because the highest degree of hydrolysis of the chia protein indicates that they have higher yields of low-molecular-weight peptides, which ends up restricting their potential antimicrobial activity.

Guindani et al. (2016) studied the extraction of chia oil through several techniques from the industrially generated residues aiming to increase the yield of the oil. The authors characterized the oil extracted against various properties including antimicrobial activity. These authors verified that the extract did not show any inhibition against *E. coli* bacteria (Gram-negative), but against *B. cereus* (Gram-positive), a weak inhibitory activity was observed. In addition, the authors verified that extracts obtained by supercritical fluid extraction presented lower MIC (minimum inhibition concentration) values than extracts obtained by low-pressure extractions (Soxhlet extraction, maceration, and ultrasound-assisted extraction). The scarce studies that characterize chia seed and chia oil against the antimicrobial property show that this effect has not yet been proven, which means that more work in this area should be carried out.

11.6.8 Role in Obesity and Weight Loss

Although some authors mention that chia seed helps to control weight, research is still going on to establish this hypothesis. The existing studies evaluated the effect of chia seed supplementation on body weight. Outcomes showed no significant effect on weight loss by overweight individuals; however, a significant increase was observed in the levels of eicosapentaenoic acid (EPA) and plasma linolenic acid (Nieman et al. 2009, 2012; Jin et al. 2012).

11.7 Food Applications

Chia seed can be consumed in natura or as an ingredient of food. There are several studies indicating the use of chia seeds commercially. Chia seeds can be used whole, ground, in the form of flour, soaked in water or dried, or as a component extracted from them, such as mucilage and bioactive components. Food applications of chia seeds are discussed below.

11.7.1 In Bakery Products

The largest application of chia seed is in bakery products. According to the European Commission, initially chia seeds in bread and other bakery products were allowed subjected to the maximum amount of 5%, but later on, it was increased to 10% (European Union 2014).

In bakery products, studies focus on the use of whole seed as well as chia flour in bread for the replacement of part of the wheat flour (Justo et al. 2007; Segura-Campos et al. 2013; Costantini et al. 2014; Pizarro et al. 2014; Coelho and de las Mercedes Salas-Mellado 2015; Steffolani et al. 2015; Švec and Hrušková 2015; Verdú et al. 2015) aiming to improve the nutritional and technological properties of the product.

Recently, Fernandes and de las Mercedes Salas-Mellado (2017a) and Salgado-Cruz et al. (2017) used chia mucilage in the bread formulation. Salgado-Cruz et al. (2017) studied the changes in the microstructure of pita bread enriched with chia mucilage, while Fernandes and de las Mercedes Salas-Mellado (2017a) verified that breads with the substitution of 75% of fat replacement by chia mucilage dried at 50 °C presented a high index of sensorial acceptance, an improvement of the physical and technological characteristics, and a reduction of the lipid content by about 37%.

Chia seed does not contain gluten, so it is used in the development of gluten-free bread intended for people with celiac disease (Costantini et al. 2014; Steffolani et al. 2014; Huerta et al. 2016; Simurina et al. 2017). All authors working in this line of research sought to obtain better nutritional, technological, sensory, and microbiological characteristics.

As in bread, there are studies in relation to cakes aiming at the nutritional enhancement of the product. Pizarro et al. (2014) showed that up to 15 g of chia flour/100 g can be incorporated into cakes to improve nutritional properties and maintain sensory acceptance. Borneo et al. (2010) verified that chia gel can substitute up to 25% of the oil or eggs in cake formulations without affecting the functional and sensorial characteristics and increasing the content of α-linolenic acid and fibers.

Rendón-Villalobos et al. (2012) developed corn tortillas with 5, 10, 15, and 20% of chia flour, and all formulations presented higher fiber, protein, and lipid contents than control corn tortillas, being more significant with 15 and 20% of substitution. In contrast, Inglett et al. (2014) developed biscuits for the mixture of oat and chia seeds in order to ally the soluble fiber content of oats that is beneficial to the texture of foods and for the prevention of coronary diseases together with the health benefits of omega-3 present in chia seeds.

Oliveira et al. (2015) developed pastas with different percentages of chia flour to substitute wheat flour. The substitution enhanced the nutritional and technological properties of pasta. Menga et al. (2017) added chia seed or chia mucilage together with rice flour in order to develop gluten-free pasta and verified that a concentration of 10% of mucilage or seeds of chia entailed in nutritious and healthy pasta compared to commercial sample as confirmed by the high content of protein, dietary fiber and phenolic acids.

11.7.2 In Dairy Products

Campo et al. (2017) evaluated the potential of chia mucilage in the total substitution of emulsifiers and stabilizers in ice cream. The results obtained by the authors indicated that chia mucilage may replace emulsifiers and stabilizers in the formulation of ice cream while maintaining the quality of the product; however, the sensorial analysis shows significant differences between the samples of ice cream formulated with mucilage and the control ice cream in terms of color, and overall attributes, probably because of the dark mucilage color. The authors suggested that adding some food coloring to the formulation may be a great choice to improve acceptability.

Fernandes and de las Mercedes Salas-Mellado (2017b) studied chia mucilage to substitute oil and egg yolk in mayonnaise. The authors reported a reduction of 50% of lipid content in the mayonnaise with the substitution of 45% of oil with chia mucilage and a reduction of only 0.94% in the mayonnaise with substitution of 35% of egg yolk with chia mucilage, although higher evaluation scores in the sensorial analyses were observed.

Recently, a group of researchers used chia oil as an ingredient in the formulation of different products. Ullah et al. (2017) added olein fraction of chia oil in ice cream and reported significant increase in the concentration of omega-3 fatty acids, total phenolic content, total flavonoids, and improved DPPH free radical scavenging activity of the supplemented ice cream. Nadeem et al. (2017) found that the replacement of palm oil by chia oil at all levels (5, 10, 15, and 20%) enhanced the concentration of beneficial omega-3 and omega-6 fatty acids of margarine. Additionally, sensory characteristics of margarine supplemented with 15% chia oil were not different from the control.

11.7.3 In Meat and Meat Products

Scapin et al. (2015) studied the effect of the addition of chia seed extract on the physicochemical characteristics, microbiological stability, and sensory analysis of pork sausage during refrigerated storage at 4 °C. All the treatments showed a decrease in L∗ values, indicating darkening of the product and increased TBARS (thiobarbituric acid reactive substances) levels over time during storage. However, the extract of chia seeds was efficient in improving the oxidative stability of the pork sausage. In general, it was observed that the pork sausage containing 2% of chia seed extract presented better results in relation to the other concentrations in relation to lipid oxidation.

Pintado et al. (2016) used strategies to incorporate high concentrations of chia in reduced-fat frankfurters. The strategies were incorporation of chia flour (10%) in a meat matrix, with addition such as a component of an oil-in-water emulsion or an oil-in-water emulsion gel. The frankfurters added with chia presented better nutritional composition, with significantly greater amounts of proteins, insoluble dietary fiber, minerals (K, Mg, Ca, Fe, P, and Mn), and especially high levels of MUFAs and n-3 PUFAs (mainly linolenic acid). Although differences were detected in the sensory attributes of frankfurters reformulated with chia, these products were acceptable by judging panelists.

11.7.4 Other Products

Cereal bars and cookies are the most common products that are supplemented with chia seed and available in the market. Rupflin (2011) studied the inclusion of chia seeds as a source of protein, polyunsaturated fatty acids, and dietary fibers in food bars. Aiming to meet a specific niche of consumers having celiac disease, gluten-free chip-like snacks were developed that were high in protein and omega-3 fatty acids (Coorey et al. 2012). The substitution of 5% chia in chip was well accepted in the marketplace and provided almost half of the fiber intake recommended daily.

Battalwar and Shah (2015) verified that incorporation of chia seeds in fruit punch, kheer, and smoothie up to the level of 8, 4 ,and 4%, respectively, maintained the technological and sensorial characteristics of the original products, besides increasing the nutritional value of the products.

11.8 Alternative Applications

An alternative to the use of chia seed is the obtaining of protein concentrates and isolates and the extraction of the bioactive compounds from the seed. According to Segura-Campos et al. (2013), the chia protein hydrolysates with enhanced biological activity could prove to be an effective functional ingredient in a wide range of foods.

11.9 Future Challenges

Currently, the focus of studies on chia seed is concentrated on the encapsulation of chia oil. Although the fatty acid profile of chia seed oil is nutritionally favorable, the high degree of unsaturation renders it susceptible to oxidation so that its incorporation into food may lead the development of foreign/rancid flavors that affect the sensory properties of foods (Ixtaina et al. 2015). In this context, the techniques of microencapsulation and nanoencapsulation stand out. Most of the chia oil studies have focused on microencapsulation (by the spray-drying technique) using several wall materials (encapsulating agents) such as sodium caseinate and lactose (Ixtaina et al. 2015); hydroxymethyl cellulose and maltodextrin (Martínez et al. 2015); whey protein concentrate and mesquite gum (Rodea-González et al. 2012;

Escalona-García et al. 2016); isolated soy protein and maltodextrin (González et al. 2016); whey protein concentrate, pectin, maltodextrin, and modified starch (Noello et al. 2016); and chia seed protein isolate and chia mucilage (Timilsena et al. 2016).

There are few recent studies on nanoencapsulation of chia oil also. For example, Campo et al. (2017) have prepared chia oil nanoemulsion using mucilage extracted from the chia seed as the encapsulating material and obtained encapsulation efficiency of 82.8% with a particle size of 205 nm. Chia oil nanoemulsions were thermal stable with improved oxidative stability during storage. Teng et al. (2017) studied chia seed oil nanoemulsions by using microfluidization and spontaneous emulsification using polysorbate 80 (Tween 80) and sorbitan monooleate (Span 80), sodium caseinate, or sucrose monoesters as an emulsifier. An area not yet explored well is the use of proteins derived from chia seeds. As mentioned previously, chia protein hydrolysates can be used as a functional ingredient, but studies are limited.

11.10 Conclusion

Chia seed has many advantages to be used as a functional food. The major feature of chia seed is its high content of omega-3 fatty acids, which are related to the prevention and treatment of various inflammatory diseases, diabetes, CVD, autoimmune diseases, cancer, etc. In addition, the high protein and fiber content of chia seed is driving new studies in order to use them for diversified food applications. Although a number of foods containing chia seed are already available in the market, further studies should be conducted to prove the efficacy of chia seed consumption in humans.

References

Ali NM, Yeap S, YongHo W et al (2012) The promising future of chia, *Salvia hispanica* L. J Biomed Biotechnol 2012:1–9. https://doi.org/10.1155/2012/171956

Alvarado RDI (2011) Caracterización de la semilla del chan (*Salvia hispanica* L.) y diseño de un producto funcional que la contiene como ingrediente. Rev la Univ del Val Guatemala 23:43–49

Ayerza R, Coates W (2005) Ground chia seed and chia oil effects on plasma lipids and fatty acids in the rat. Nutr Res 25:995–1003. https://doi.org/10.1016/j.nutres.2005.09.013

Ayerza R, Coates W (2011) Protein content, oil content and fatty acid profiles as potential criteria to determine the origin of commercially grown chia (*Salvia hispanica* L.). Ind Crop Prod 34:1366–1371. https://doi.org/10.1016/j.indcrop.2010.12.007

Baginsky C, Arenas J, Escobar H, Garrido M, Valero N, Tello D, Pizarro L, Valenzuela A, Morale L, Silva H (2016) Growth and yield of chia (*Salvia hispanica* L.) in the Mediterranean and desert climates of Chile. Chil J Agric Res 76:255–264. https://doi.org/10.4067/S0718-58392016000300001

Battalwar R, Shah V (2015) Incorporation of chia seeds in Fruit Punch, Kheer, Smoothie and its sensory evaluation. Int J Food Nutr Sci 4:1–4

Benzi L (2016) Novos conceitos de super alimentos, baseados em sementes, ganham força no mercado global. http://www.alimentosebebidas.com.br/2016/12/22/novos-conceitos-de-super-alimentos-baseados-em-sementes-ganham-forca-no-mercado-global/. Accessed 16 Mar 2017

Bodoira RM, Penci MC, Ribotta PD, Martínez ML (2017) Chia (*Salvia hispanica* L.) oil stability: study of the effect of natural antioxidants. LWT Food Sci Technol 75:107–113. https://doi.org/10.1016/j.lwt.2016.08.031

Borneo R, Aguirre A, León AE (2010) Chia (*Salvia hispanica* L) gel can be used as egg or oil replacer in cake formulations. J Am Diet Assoc 110:946–949. https://doi.org/10.1016/j.jada.2010.03.011

Busilacchi H, Quiroga M, Bueno M et al (2013) Evaluación de Salvia hispanica L. cultivada em el Sur de Santa Fe (Republica Argentina). Cultiv Trop 34:55–59

Cahill JP (2003) Ethnobotany of Chia, *Salvia hispanica* L. (Lamiaceae). Econ Bot 57:604–618. https://doi.org/10.1663/0013-0001(2003)057[0604:eocshl]2.0.co;2

Campo C, dos Santos PP, Costa TMH, Paese K, Guterres SS, Rios AO, Flôres SH (2017) Nanoencapsulation of chia seed oil with chia mucilage (*Salvia hispanica* L.) as wall material: characterization and stability evaluation. Food Chem 234:1–9. https://doi.org/10.1016/j.foodchem.2017.04.153

Capitani MI, Spotorno V, Nolasco SM, Tomás MC (2012) Physicochemical and functional characterization of by-products from chia (*Salvia hispanica* L.) seeds of Argentina. LWT Food Sci Technol 45:94–102. https://doi.org/10.1016/j.lwt.2011.07.012

Carvalho PO, Campos PRB, Noffs MD et al (2003) Speaking figures. Quim Nova 181:344. https://doi.org/10.1093/nq/181.25.344-a

Coates W (2011) Whole and ground chia (*Salvia hispanica* L.) seeds, chia oil—effects on plasma lipids and fatty acids. Elsevier Inc.

Coelho MS, de las Mercedes Salas-Mellado M (2014) Chemical characterization of CHIA (*Salvia hispanica* L.) for use in food products. J Food Nutr Res 2:263–269. https://doi.org/10.12691/jfnr-2-5-9

Coelho MS, de las Mercedes Salas-Mellado M (2015) Effects of substituting chia (*Salvia hispanica* L.) flour or seeds for wheat flour on the quality of the bread. LWT Food Sci Technol 60:729–736. https://doi.org/10.1016/j.lwt.2014.10.033

Coorey R, Grant A, Jayasena V (2012) Effects of chia flour incorporation on the nutritive quality and consumer acceptance of chips. J Food Res 1:85. https://doi.org/10.5539/jfr.v1n4p85

Coorey R, Tjoe A, Jayasena V (2014) Gelling properties of chia seed and flour. J Food Sci 79. https://doi.org/10.1111/1750-3841.12444

Costantini L, Lukšič L, Molinari R et al (2014) Development of gluten-free bread using tartary buckwheat and chia flour rich in flavonoids and omega-3 fatty acids as ingredients. Food Chem 165:232–240. https://doi.org/10.1016/j.foodchem.2014.05.095

Craig R, Sons M (2004) Application for approval of whole chia (*Salvia hispanica* L.) seed and ground whole chia as novel food ingredients. Advis Comm Nov foods Process 1–29

da Silva BP, Anunciação PC, Matyelka JCDS et al (2017) Chemical composition of Brazilian chia seeds grown in different places. Food Chem 221:1709–1716. https://doi.org/10.1016/j.foodchem.2016.10.115

De Falco B, Amato M, Lanzotti V (2017) Chia seeds products: an overview. Phytochem Rev 16:745–760. https://doi.org/10.1007/s11101-017-9511-7

Diwakar G, Rana J, Saito L et al (2014) Inhibitory effect of a novel combination of *Salvia hispanica* (chia) seed and *Punica granatum* (pomegranate) fruit extracts on melanin production. Fitoterapia 97:164–171. https://doi.org/10.1016/j.fitote.2014.05.021

El-Badry AM, Graf R, Clavien PA (2007) Omega 3—omega 6: what is right for the liver? J Hepatol 47:718–725. https://doi.org/10.1016/j.jhep.2007.08.005

Escalona-García LA, Pedroza-Islas R, Natividad R et al (2016) Oxidation kinetics and thermodynamic analysis of chia oil microencapsulated in a whey protein concentrate-polysaccharide matrix. J Food Eng 175:93–103. https://doi.org/10.1016/j.jfoodeng.2015.12.009

European Union (2014). http://www.cadexco.bo/actual/boletin-inteligencia/Product-factsheet-chia-seeds-europe-vegetable-oils-oilseeds-2014.pdf

European Union (2017). https://eur-lex.europa.eu/legal-content/EN/TXT/PDF/?uri=CELEX:32017D2354&from=DA. Accessed 12 February 2019

Everitt L (2016) The chia craze. http://www.bbc.co.uk/news/magazine-1747669/ Accessed 16 Mar 2017

Fernandes SS, de las Mercedes Salas-Mellado M (2017a) Addition of chia seed mucilage for reduction of fat content in bread and cakes. Food Chem 227:237–244. https://doi.org/10.1016/j.foodchem.2017.01.075

Fernandes SS, de las Mercedes Salas-Mellado M (2017b) Development of mayonnaise with substitution of oil or egg yolk by the addition of chia *Salvia hispanica* L.) Mucilage. J Food Sci 1–10. https://doi.org/10.1111/1750-3841.13984

González A, Martínez ML, Paredes AJ et al (2016) Study of the preparation process and variation of wall components in chia (*Salvia hispanica* L.) oil microencapsulation. Powder Technol 301:868–875. https://doi.org/10.1016/j.powtec.2016.07.026

Goyal A, Sharma V, Upadhyay N, Gill S, Sihag M (2014) Flax and flaxseed oil: an ancient medicine & modern functional food. J Food Sci Technol 51:1633–1653. https://doi.org/10.1007/s13197-013-1247-9

Guevara-Cruz M, Tovar AR, Aguilar-Salinas CA et al (2012) A dietary pattern including nopal, chia seed, soy protein, and oat reduces serum triglycerides and glucose intolerance in patients with metabolic syndrome. J Nutr 142:64–69. https://doi.org/10.3945/jn.111.147447

Guindani C, Podestá R, Block JM et al (2016) Valorization of chia (*Salvia hispanica*) seed cake by means of supercritical fluid extraction. J Supercrit Fluids 112:67–75. https://doi.org/10.1016/j.supflu.2016.02.010

Haque E, Chand R, Kapila S (2009) Biofunctional properties of bioactive peptides of milk origin. Food Rev Int 25:28–43. https://doi.org/10.1080/87559120802458198

Huerta M, Alves S, Franco A, et al (2016) Sensory response and physical characteristics of gluten-free and gum-free bread with chia flour. Food Sci Technol Campinas 2–5. https://doi.org/10.1590/1678-457x.0032

Inglett GE, Chen D, Liu S (2014) Physical properties of sugar cookies containing chia-oat composites. J Sci Food Agric 94:3226–3233. https://doi.org/10.1002/jsfa.6674

Ixtaina V (2010) Caracterización de la semilla y el aceite de Chía (*Salvia hispanica* L.) obtenido mediante distintos procesos, aplicación en la tecnología de alimentos. J Chem Inf Model 53:1689–1699. https://doi.org/10.1017/CBO9781107415324.004

Ixtaina VY, Nolasco SM, Tomás MC (2008) Physical properties of chia (*Salvia hispanica* L.) seeds. Ind Crop Prod 28:286–293. https://doi.org/10.1016/j.indcrop.2008.03.009

Ixtaina VY, Vega A, Nolasco SM et al (2010) Supercritical carbon dioxide extraction of oil from Mexican chia seed (*Salvia hispanica* L.): characterization and process optimization. J Supercrit Fluids 55:192–199. https://doi.org/10.1016/j.supflu.2010.06.003

Ixtaina VY, Martínez ML, Spotorno V et al (2011) Characterization of chia seed oils obtained by pressing and solvent extraction. J Food Compos Anal 24:166–174. https://doi.org/10.1016/j.jfca.2010.08.006

Ixtaina VY, Julio LM, Wagner JR et al (2015) Physicochemical characterization and stability of chia oil microencapsulated with sodium caseinate and lactose by spray-drying. Powder Technol 271:26–34. https://doi.org/10.1016/j.powtec.2014.11.006

Jin F, Nieman DC, Sha W et al (2012) Supplementation of milled chia seeds increases plasma ALA and EPA in postmenopausal women. Plant Foods Hum Nutr 67:105–110. https://doi.org/10.1007/s11130-012-0286-0

Justo MB, Alfaro ADC, Aguilar EC et al (2007) Desarrollo de pan integral con soya, chía, linaza y ácido fólico como alimento funcional para la mujer. Arch Latinoam Nutr 57

Lin KY, Daniel JR, Whistler RL (1994) Structure of chia seed polysaccharide exudate. Carbohydr Polym 23:13–18. https://doi.org/10.1016/0144-8617(94)90085-X

Martínez ML, Marín MA, Salgado Faller CM et al (2012) Chia (*Salvia hispanica* L.) oil extraction: study of processing parameters. LWT Food Sci Technol 47:78–82. https://doi.org/10.1016/j.lwt.2011.12.032

Martínez ML, Curti MI, Roccia P et al (2015) Oxidative stability of walnut (*Juglans regia* L.) and chia (*Salvia hispanica* L.) oils microencapsulated by spray drying. Powder Technol 270:271–277. https://doi.org/10.1016/j.powtec.2014.10.031

Martínez-Cruz O, Paredes-López O (2014) Phytochemical profile and nutraceutical potential of chia seeds (*Salvia hispanica* L.) by ultra high performance liquid chromatography. J Chromatogr A 1346:43–48. https://doi.org/10.1016/j.chroma.2014.04.007

Menga V, Amato M, Phillips TD et al (2017) Gluten-free pasta incorporating chia (*Salvia hispanica* L.) as thickening agent: an approach to naturally improve the nutritional profile and the in vitro carbohydrate digestibility. Food Chem 221:1954–1961. https://doi.org/10.1016/j.foodchem.2016.11.151

Migliavacca RA, da Silva TRB, de Vasconcelos ALS et al (2014) desenvolvimento, sementes. Chia's cultivation in Brazil: future and perspectives. J Agron Sci 3:161–179

Nadeem M, Imran M, Taj I et al (2017) Omega-3 fatty acids, phenolic compounds and antioxidant characteristics of chia oil supplemented margarine. Lipids Health Dis 16:1–12. https://doi.org/10.1186/s12944-017-0490-x

Nieman DC, Cayea EJ, Austin MD et al (2009) Chia seed does not promote weight loss or alter disease risk factors in overweight adults. Nutr Res 29:414–418. https://doi.org/10.1016/j.nutres.2009.05.011

Nieman DC, Gillitt N, Jin F et al (2012) Chia seed supplementation and disease risk factors in overweight women: a metabolomics investigation. J Altern Complement Med 18:700–708. https://doi.org/10.1089/acm.2011.0443

Noello C, Carvalho AGS, Silva VM, Hubinger MD (2016) Spray dried microparticles of chia oil using emulsion stabilized by whey protein concentrate and pectin by electrostatic deposition. Food Res Int 89:549–557. https://doi.org/10.1016/j.foodres.2016.09.003

Oliveira MR, Novack ME, Santos CP et al (2015) Evaluation of replacing wheat flour with chia flour (*Salvia hispanica* L.) in pasta. Semin Agrar 36:2545–2553. https://doi.org/10.5433/1679-0359.2015v36n4p2545

Olivos-Lugo BL, Valdivia-López MÁ, Tecante A (2010) Thermal and physicochemical properties and nutritional value of the protein fraction of Mexican chia seed (*Salvia hispanica* L.). Food Sci Technol Int 16:89–96. https://doi.org/10.1177/1082013209353087

Pintado T, Herrero AM, Jiménez-Colmenero F, Ruiz-Capillas C (2016) Strategies for incorporation of chia (*Salvia hispanica* L.) in frankfurters as a health-promoting ingredient. Meat Sci 114:75–84. https://doi.org/10.1016/j.meatsci.2015.12.009

Pizarro PL, Almeida EL, Coelho AS et al (2014) Functional bread with n-3 alpha linolenic acid from whole chia (*Salvia hispanica* L.) flour. J Food Sci Technol 52:4475–4482. https://doi.org/10.1007/s13197-014-1477-5

Porras-Loaiza P, Jiménez T, Sosa-Morales ME et al (2014) Physical properties, chemical characterization and fatty acid composition of Mexican chia (*Salvia hispanica* L.) seeds. Int J Food Sci Technol 49:571–577. https://doi.org/10.1111/ijfs.12339

Rendón-Villalobos R, Ortíz-Sánchez A, Solorza-Feria J, Trujillo-Hernández CA (2012) Formulation, physicochemical, nutritional and sensorial evaluation of corn tortillas supplemented with chía seed (*Salvia hispanica* L.). Czech J Food Sci 30:118–125

Reyes-Caudillo E, Tecante A, Valdivia-López MA (2008) Dietary fibre content and antioxidant activity of phenolic compounds present in Mexican chia (*Salvia hispanica* L.) seeds. Food Chem 107:656–663. https://doi.org/10.1016/j.foodchem.2007.08.062

Rodea-González DA, Cruz-Olivares J, Román-Guerrero A et al (2012) Spray-dried encapsulation of chia essential oil (*Salvia hispanica* L.) in whey protein concentrate-polysaccharide matrices. J Food Eng 111:102–109. https://doi.org/10.1016/j.jfoodeng.2012.01.020

Rupflin DIA (2011) Caracterización de la semilla del chan (*Salvia hispanica* L.) y diseño de um producto funcional que la contiene como ingrediente. Rev la Univ del Val Guatemala 43:49

Salgado-Cruz MP, Ramírez-Miranda M, Díaz-Ramírez M et al (2017) Microstructural characterisation and glycemic index evaluation of pita bread enriched with chia mucilage. Food Hydrocoll 69:141–149. https://doi.org/10.1016/j.foodhyd.2017.01.027

Sandoval-Oliveros MR, Paredes-López O (2013) Isolation and characterization of proteins from chia seeds (*Salvia hispanica* L.). J Agric Food Chem 61:193–201. https://doi.org/10.1021/jf3034978

Sargi SC, Silva BC, Santos HMC et al (2013) Antioxidant capacity and chemical composition in seeds rich in omega-3: chia, flax, and perilla. Food Sci Technol 33:541–548. https://doi.org/10.1590/S0101-20612013005000057

Scapin G, Schimdt MM, Prestes RC et al (2015) Effect of extract of chia seed (*Salvia hispanica*) as an antioxidant in fresh pork sausage. Int Food Res J 22:1195–1202

Scapin G, Abaide ER, Nunes LF et al (2017) Effect of pressure and temperature on the quality of chia oil extracted using pressurized fluids. J Supercrit Fluids 127:90–96. https://doi.org/10.1016/j.supflu.2017.03.030

Segura-Campos MR, Salazar-Vega IM, Chel-Guerrero LA, Betancur-Ancona DA (2013) Biological potential of chia (*Salvia hispanica* L.) protein hydrolysates and their incorporation into functional foods. LWT Food Sci Technol 50:723–731. https://doi.org/10.1016/j.lwt.2012.07.017

Segura-Campos MR, Chel-Guerrero LA, Betancur-Ancona DA (2016) Salvia hispanica: nutritional and functional potential. In: Kristbergsson K, Ötles S (eds) Functional properties of traditional foods. Springer, New York, pp 115–118

Sierra L, Roco J, Alarcon G et al (2015) Dietary intervention with *Salvia hispanica* (Chia) oil improves vascular function in rabbits under hypercholesterolaemic conditions. J Funct Foods 14:641–649. https://doi.org/10.1016/j.jff.2015.02.042

Silva C, Garcia VAS, Zanette CM (2016) Chia (*Salvia hispanica* L.) oil extraction using different organic solvents: oil yield, fatty acids profile and technological analysis of defatted meal. Int Food Res J 23:998–1004

Simurina O, Radunovic A, Filipcev B et al (2017) Quality improvement of gluten-free bread based on soybean and enriched with sugar beet molasses. Food Feed Res 44:65–72. https://doi.org/10.5937/FFR1701065S

Steffolani E, de la Hera E, Pérez G, Gómez M (2014) Effect of chia (*Salvia hispanica* L.) addition on the quality of gluten-free bread. J Food Qual 37:309–317. https://doi.org/10.1111/jfq.12098

Steffolani E, Martinez MM, León AE, Gómez M (2015) Effect of pre-hydration of chia (*Salvia hispanica* L.), seeds and flour on the quality of wheat flour breads. LWT Food Sci Technol 61:401–406. https://doi.org/10.1016/j.lwt.2014.12.056

Švec I, Hrušková M (2015) Hydrated chia seed effect on wheat flour and bread technological quality. Agric Eng Int CIGR J 2015:259–263

Taga MS, Miller EE, Pratt DE (1984) Chia seeds as a source of natural lipid antioxidants. J Am Oil Chem Soc 61:928–931. https://doi.org/10.1007/BF02542169

Temple NJ (2018) Fat, sugar, whole grains and heart disease: 50 years of confusion. Nutrients 10:1–9. https://doi.org/10.3390/nu10010039

Teng J, Hu X, Wang M, Tao N (2017) Fabrication of chia (*Salvia hispanica* L.) seed oil nanoemulsions using different emulsifiers. J Food Process Preserv 1–9. https://doi.org/10.1111/jfpp.13416

Timilsena YP, Adhikari R, Barrow CJ, Adhikari B (2016) Microencapsulation of chia seed oil using chia seed protein isolate-chia seed gum complex coacervates. Int J Biol Macromol 91:347–357. https://doi.org/10.1016/j.ijbiomac.2016.05.058

Toscano LT, da Silva CSO, Toscano LT et al (2014) Chia flour supplementation reduces blood pressure in hypertensive subjects. Plant Foods Hum Nutr 69:392–398. https://doi.org/10.1007/s11130-014-0452-7

Tosco G (2004) Os benefícios da chia em humanos e animais. Atualidades ornitológicas 7

Ullah R, Nadeem M, Imran M (2017) Omega-3 fatty acids and oxidative stability of ice cream supplemented with olein fraction of chia (*Salvia hispanica* L.) oil. Lipids Health Dis 16:1–8. https://doi.org/10.1186/s12944-017-0420-y

Uribe JAR, Perez JIN, Kauil HC et al (2011) Extraction of oil from chia seeds with supercritical CO2. J Supercrit Fluids 56:174–178. https://doi.org/10.1016/j.supflu.2010.12.007

Verdú S, Vásquez F, Ivorra E et al (2015) Physicochemical effects of chia (*Salvia hispanica*) seed flour on each wheat bread-making process phase and product storage. J Cereal Sci 65:67–73. https://doi.org/10.1016/j.jcs.2015.05.011

Vuksan V, Whitham D, Sievenpiper JL et al (2007) Supplementation of conventional therapy with the novel grain salba (*Salvia hispanica* L.) improves major and emerging cardiovascular risk factors in type 2 diabetes. Diabetes Care 30:2804–2810. https://doi.org/10.2337/dc07-1144. Clinical

World Health Organization (WHO) (2017) Cardiovascular Diseases (CVDs). FactSheet no 317. http://www.who.int/mediacentre/factsheets/fs317/en/. Accessed 24 December 2017

Chapter 12
Sesame (*Sesamum indicum*) Seed

Loveleen Sharma, Charaniv Singh Saini, Sneh Punia, Vikash Nain, and Kawaljit Singh Sandhu

Abstract Sesame, an erect annual herb, is a member of the Pedaliaceae family and contains 45–65% oil. Tanzania is the largest producer of sesame oil followed by China, India, and Sudan. Apart from oil, sesame seeds are a rich source of protein and calcium. It is rich in methionine, valine, and tryptophan which are deficient in many other pulses and crops. Bioactive components such as phenolics, vitamins, phytosterols, and polyunsaturated fatty acids are present in sesame seeds which provide a beneficial effect on human health. Sesame has the potential to establish itself as a major oilseed crop, and its better performance, yield, oil quality, and quantity would also provide benefits to the farmers from the area where sesame is grown as a regular crop. Also, sesamin, a lipid-soluble lignan which is present in sesame has, been garnering considerable attention as an anticancer agent. Sesame seeds are reported to be a rich source of phytates compared to soya beans. Sesame oil, due to its excellent keeping quality, fatty acid composition, and good amount of highly active tocopherol, has found versatile use in the food processing industry as cooking oil, in salad dressing, soups, confectionery, etc. This chapter provides an overview origin, history, nutritional value, production status, and future aspects of sesame.

Keywords Sesame oil · Sesamin · Lignan · Anticancer · Sesame meal

---

L. Sharma (✉)
Amity Institute of Food Technology (AIFT), Amity University, Noida, Uttar Pradesh, India
e-mail: lsharma8@amity.edu

C. S. Saini
Department of Food Engineering and Technology, Sant Longowal Institute of Engineering and Technology, Longowal, Punjab, India

S. Punia · V. Nain
Department of Food Science and Technology, Chaudhary Devi Lal University, Sirsa, Haryana, India

K. S. Sandhu
Department of Food Science and Technology, Maharaja Ranjit Singh Punjab Technical University, Bathinda, Punjab, India

B. Tanwar, A. Goyal (eds.), *Oilseeds: Health Attributes and Food Applications*,
https://doi.org/10.1007/978-981-15-4194-0_12

12.1 Introduction

Sesame (*Sesamum indicum* L.) often called as "seed of immortality" is one of the earliest domesticated short-duration crop grown throughout the year and yields high amount of oil (45–50%). It was one of the first crops processed for oil and is also one of the earliest condiments used (De Carvalho et al. 2001). Currently, the largest commercial producers of sesame seeds include India (660,000 tons) and China (350,000 tons). It consists of 45–65% oil, 19–35% protein, 14–20% carbohydrates, and 15–20% hull material along with measurable quantities of oxalic acid, calcium, and other minerals. Sesame seeds are rich in saturated fatty acids such as palmitic (8.58%) and stearic acids (5.44%) and unsaturated fatty acids, mainly oleic (35–50%) and linoleic acids (46.26%). Unsaponifiable fractions in sesame oil consists of sesamin, sesamolin, sterols (59–60% β-sitosterol, 10–12% avenasterol 18–19% campesterol, 6–7% stigmasterol), and tocopherol. Apart from oil, sesame seeds are a rich source of protein and calcium. It is rich in methionine, valine, and tryptophan which are deficient in many other pulses and crops. Bioactive components such as phenolics, vitamins, phytosterols, and polyunsaturated fatty acids are present in sesame seeds which provide a beneficial effect on human health.

12.2 Origin and History

Sesame, an erect annual herb, is a member of the Pedaliaceae family and is commonly known as *til* (Hindi), *hu ma* (Chinese), *sesame* (French), *goma* (Japanese), *gergelim* (Portuguese), and *ajonjoli* (Spanish) (Anilakumar et al. 2010). The original area of domestication is obscure, but despite numerous claims, sesame is presumed to have originated in Africa and later spread to West Asia to India, China, and Japan (Alegbejo et al. 2003; Purseglove 1969). Archeological records indicate that it has been known and used in India for more than 5000 years, and it finds a mention in *Atharvaveda*, a holy scripture, as a food rather than an oil or medicine (I.26.15). Additionally, it is also recorded as a crop in Babylon and Assyria some 4000 years ago (Borchani et al. 2010), and the Assyrians used its oil for food, ointments, and medicine purposes (Purseglove 1969). Although referred to as "queen of oilseeds" and being cultivated in both tropical and temperate zones of Africa, Asia, Latin America, and some parts of southern United States, it is an orphan crop and hence not mandated by any International Crop Research Institute for semi-Arid Tropics (Islam et al. 2010).

12.3 Production

Until the mid-2000s, sesame production was around 35 million tons and thereafter steadily increased to nearly 40 million tons in 2014 (FAO 2014). Till 2009, China was the largest producer of sesame in the world accounting for more than 40% of the global production, followed by India with about 15% share, but after 2012, Tanzania became the largest producer followed by India, Sudan, and China. However, in 2014 according to the FAO data, with the annual yield of 14,658 hg/ha, China was the largest producer followed by Tanzania (12,000 hg/ha), Myanmar (5472 hg/ha), and India (4675 hg/ha). On the other hand, in terms of area, India is the largest cultivator and exporter of sesame. Major importing countries are Japan, the European Union, Vietnam, Indonesia, Mexico, and Russia (FAO 2014). Till 2012, Myanmar was the largest producer of sesame seed oil, but after 2012, Tanzania became the largest producer of oil followed by Myanmar, China, India, and Sudan. Tanzania processed 5,44,293 tons of sesame oil in 2014 (FAO 2014) (Figs. 12.1 and 12.2).

12.4 Composition

12.4.1 Lipids

Sesame oil is considered as a rich food as it has a high nutritive quality, is stable, and possesses numerous therapeutic properties, and hence, all these properties gives it an advantage over other vegetable oils (Prasad Nagendra et al. 2012). Sesame oil is rich in saturated fatty acids such as palmitic (9–10%) and stearic acids (5–7%) and unsaturated fatty acids, mainly oleic (35–50%) and linoleic acids (31–41%) (Ebuehi

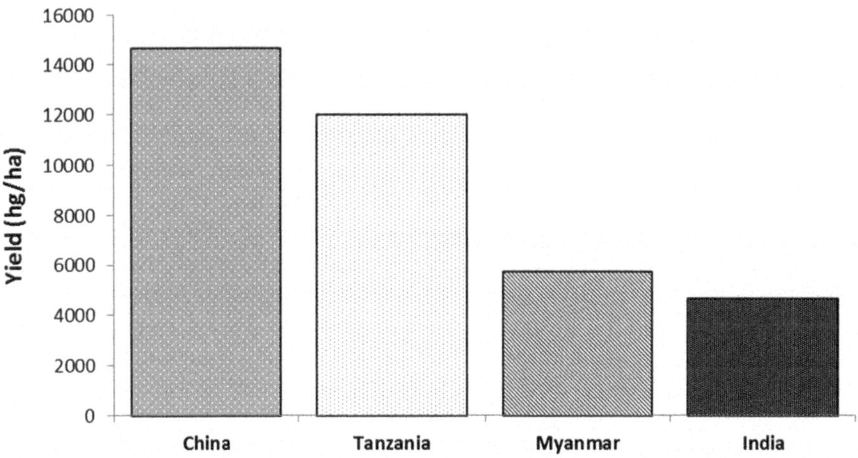

Fig. 12.1 Yield trend of sesame seeds in major countries (FAOSTAT 2017)

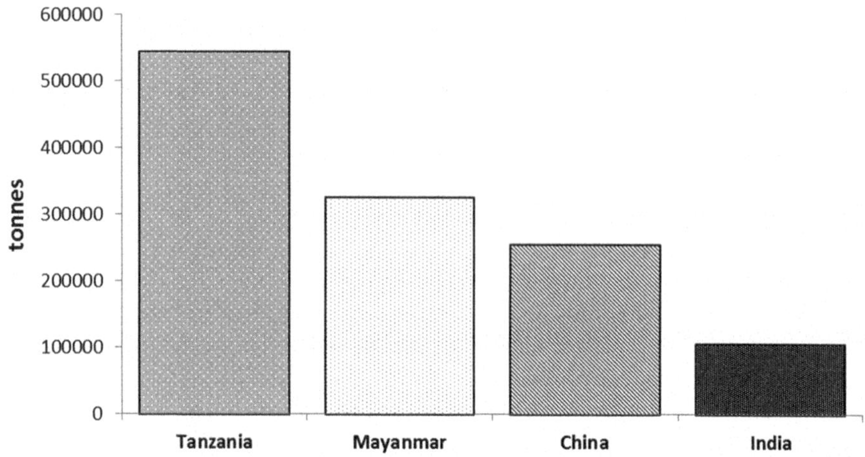

Fig. 12.2 Sesame seed oil production status of different countries (FAOSTAT 2017)

et al. 2006). Unsaponifiable fractions in sesame oil consist of sesamin (0.07–0.61%), sesamolin (0.2–0.48%), sterols (16–40%), and tocopherol (10–12%). Sesamin and sesamolin are the two natural antioxidants along with tocopherols. Low extent of unsaturated fats of high molecular weight was seen because of the low amount of saponification (1.87 mg KOH/g) and the iodine value (1.27 g T2/100g) suggesting that oils of this value are good for consumption (Mbaebie et al. 2010). Asghar and Majeed (2013) characterized the fatty acid profile of four sesame varieties (Till-90, P-37, Till-93, and S-17) grown in Pakistan for the variation in oil content, oil yield, chemical composition, and fatty acid composition. The oil content was reported to be ranging between 49.5 and 53.9%; caprylic acid, linoleic acid, palmitic, and stearic acids in the seed oil ranged from 16.9 to 21.8%, 4.7 to 12.5%, 16.0 to 19.3%, and 13.9 to 21.5%, respectively. Also, minor fatty acids, viz., oleic, linoleic acids, and palmitoleic acids which enhance the suitability of the sesame oil for human consumption, were reported. El-Khier et al. (2008) determined the chemical composition and oil qualities of ten different sesame seed cultivars (3015, Kenana1, local white, mixed, Aswad (Sudanese genotypes), and zirra2, zirra7, zirra9, hurria11, and hurria 49 developed in Sudan). The iodine values of the sesame seed cultivars were reported in the range of 101.52–114.85 g/100 g for the local cultivars and 97.70–111.30 g/100 g for the introduced cultivars, and all the values were consistent with those stated by FAO. Iodine value is used to determine the amount of unsaturation in fatty acids. Oil was extracted from sesame cultivars, mixed (local), zirra2, zirra7, and huria11, for the fatty acid composition. Oleic, linoleic, palmitic, and stearic acids were found in abundant amounts. Compared to huria11, the mixed cultivar had almost similar oleic acids (47.50 against 48.40%) and linoleic acids (36.40 against 35.80%). It was also observed that local cultivars had average high content of calcium but low protein content. Nzikou et al. (2009) studied the physicochemical properties of the sesame seed oil extracts and reported that the oil was liquid at room temperature (37 °C) and high levels of unsaturated fatty acids especially oleic (up to 38.84%), linoleic (up to 46.26%), and saturated fatty acids palmitic (8.58%) and stearic (5.44%) acids were found in oil.

12.4.2 Carbohydrates

Whole and dehulled white variety of sesame seeds contain about 14.90 and 14.70% carbohydrates, respectively, and whole and dehulled black variety contains 14.70 and 14.90% carbohydrates, respectively (Yasothai 2014). According to Zebib et al. (2015), carbohydrate content of three sesame varieties, Adi, Bawnji, and T-85, was reported in the range between 8.3 and 11.69%. On the other hand, Sudanese and US genotypes were observed to have much lower carbohydrate content (1.05–2.88%) (El-Khier et al. 2008). Highest carbohydrate content (27.90–45.15%) was reported in 13 sesame accession of Nigeria by Ogbonna and Ukaan (2013). Further, different fractions of carbohydrate of sesame are reported as 3.24% D-glucose, 0.06% D-galactose, 2.63% D-fructose, 0.17% sucrose, 0.24% raffinose, 0.23% stachyose, 0.59% planteose, 0.38% sesamose, 0.16% pentasaccharides, and 0.08% hexasaccharides on a moisture-free basis (Johnson et al. 1979).

12.4.3 Protein

Apart from being a prominent oilseed, sesame seeds are a rich source of protein (18–40%) (El-Khier et al. 2008; Onsaard et al. 2010; Ogbonna and Ukaan 2013) and consist of 8.6% albumin, 67.3% globulin, 1.4% prolamin, and 6.9% glutelin (Achouri et al. 2012). α-Globulin and β-globulin are the two major storage proteins and constitute 80–90% of the total seed protein (Tai et al. 2001). However, the presence of 95% 13S globulin which possesses hydrophobic properties limits the functionality of sesame proteins especially in beverage and fluid preparations (Sharma et al. 2016; Saini et al. 2018).

Sesame seed protein has a good combination of essential and nonessential amino acids. It is rich in sulfur-containing amino acids, valine and tryptophan, but is deficient in lysine (Achouri and Boye 2013). Mehta (2000) reported the presence of arginine (12–13 g), cysteine (2 g), histidine (2.4–2.8 g), isoleucine (3.3–4.9 g), leucine (6.5–8.9 g), lysine (2.5–3.5 g), methionine (2.0–4 g), phenylalanine (4.2–4.5 g), threonine (3.4–3.6 g), tryptophan (1.9–2.4 g), and valine (4.2–4.5 g) in dehulled sesame seeds. Maneemegalai and Prasad (2011) evaluated the amino acid composition and protein solubility profile of commercially available sesame meal and reported the presence of valine, isoleucine, arginine, aspartic acid, and glutamine in high concentration. The protein solubility of sesame protein in alkaline pH is high than in acidic pH in aqueous solution, and highest solubility was reported at pH 10.

12.4.4 Dietary Fiber

The total dietary fiber content in sesame seed coat was reported to be ranging between 31 and 42 g/100 g seed coat dry matter. The largest fraction was insoluble fiber (26–33 g/100 g seed coat dry matter), which is more when compared with the cereal derivatives such as corn bran (15–20 g/100), wheat bran (15 g–25/100 g), oat bran (10–15 g/100 g), and rice bran (16–20 g/100 g); soluble dietary fiber ranged between 5.5 and 8.6 g/100 g seed coat dry matter. Also, the dietary fiber contained high amounts of neutral sugars (15.11 g/100 g seed coat dry matter), insoluble uronic acids (10.52 g/100 g seed coat dry matter), and lignin (5.42 g/100 g seed coat dry matter) (Elleuch et al. 2012).

12.4.5 Minerals

Calcium (1200 mg/100 g) and copper (150 g/100 g) are present in higher amount in sesame seeds. It is also rich in phosphorous (700 mg/100 g), iron (9.3 mg/100 g), magnesium (521 mg/100 g), selenium (34.4 μg/100 g), manganese (2.5 mg/100 g), and zinc (3.8 mg/100 g). Daily value (DV) of a quarter cup of sesame seeds for copper is 74.0%, 31.6% DV for magnesium, and 35.1% DV for calcium, which showed that sesame seeds are the richest source of minerals (Jimoh et al. 2011; USDA 1999).

12.4.6 Vitamins

Sesame was reported as a rich source of thiamine (0.24 mg/100 g), riboflavin (0.20 mg/100 g), pantothenic (0.52 mg/100 g), niacin (6.7 mg/100 g), vitamin A (50 IU/100 g), and vitamin E (1.2 mg/100 g) present in sesame oil (USDA 1999). Vitamin and mineral composition of sesame seeds is presented in Table 12.1.

12.5 Phytochemicals/Antinutritional Factors

Secondary metabolites such as phytonutrients or antinutrients depending upon the concentration in the dietary intake have both beneficial and adverse health effects. Studies suggest that when present at a lower concentration in food, they confer health benefits such as reduced risk of cardiovascular diseases (CVDs), cancer, diabetes, hypertension, etc. and improved immunomodulatory functions, cognitive functions, etc. and are then termed "phytonutrients" (Shahidi 1997). However, at higher concentrations these compounds are termed as "antinutrients" as they might

Table 12.1 Vitamins and minerals per 100.00 gm of sesame seeds on dry weight basis

Vitamins	
A-carotenoid	0.97 RE
A-Retinol	0.00 RE
A-β-carotene	5.83 mcg
Riboflavin	0.21 mg
Thiamine	0.75 mg
Niacin	4.42 mg
Vitamin B6	0.75 mg
Biotin	10.69 mcg
Vitamin E	3.24 IU
Folate	93.96 mcg
Minerals	
Calcium	947 mg
Chloride	9.72 mg
Copper	3.9 mg
Iron	14.10 mg
Magnesium	340 mg
Manganese	2.37 mg
Molybdenum	28.72 mcg
Phosphorus	59.40 mg
Potassium	1263 mg
Selenium	6.48 mcg
Sodium	32.40 mg
Zinc	7.560 mg

Source: Shivhare and Satsangee (2012)
RE Retinol equivalent

reduce the digestibility and bioavailability of nutrients by inhibiting digestive enzymes and forming an insoluble mineral–protein complex in the gut (Soetan and Oyewole 2009).

Phytate, phenolic compounds, lectin, enzyme inhibitors, saponins, and oxalates are the anti-nutritional components present in abundant amount in sesame seeds and have an impact on the gastrointestinal tract and thus affect human metabolism (Nikmaram et al. 2017). Sesame seed coat contains oxalates (3–5%) and phytates (2.25–3.5%), which form insoluble complexes with minerals (calcium, magnesium, zinc, and iron) and ultimately interfere with their utilization (Adegunwa et al. 2012).

12.5.1 Phytates

Phytic acid is an important source of phosphorus and possesses a strong chelating ability to form protein complex, thereby contributing to anti-nutritional effects (Urbano et al. 2000). Phytates reduces bioavailability of minerals, impair protein digestibility by formation of phytic–protein complexes, and depress absorption of nutrients due to damage to the pyloric caeca region of the intestine (Francis et al. 2001). However, dietary phytates aid in lowering cholesterol level and also prevent cancer (Steer and Gibson 2002). Also, phytates help in reducing lipid peroxidation and are, thus, considered to be natural antioxidants (Serna-Saldivar et al. 2017).

Sesame seeds are reported to be a rich source of phytates (5.18%) as compared to soybean meal (1%) and soybean protein (1.5%) (De Boland et al. 1975). Toma et al. (1979) reported 4.7, 5.2, 4.7, and 5.1% phytin content in whole, dehusked, roasted dehusked, and roasted sesame seeds, respectively. Makinde and Akinoso (2013) reported 30 mg/100 g phytate in dehusked white cultivar and 25.07 mg/100 g in black cultivar of sesame seeds. The authors also observed that the high content of phytic acid and oxalic acid in sesame seed hinders the use of sesame protein as food.

12.5.2 Oxalates

Oxalic acid salts of calcium and magnesium are insoluble crystals and are not absorbed in the human body (Leeson and Summers 2001). Massey et al. (2001) reported that after absorption from diet, oxalate cannot be metabolized and is excreted by the kidney into urine, where it binds to calcium forming an insoluble salt that may precipitate to form kidney stones. Sesame seed contains 2.2% oxalic acid, and processing methods like pearling, soaking, steaming, oil expelling, and drying treatments aid in reducing the oxalate content of the seeds. Further, the authors reported that the combination of 20 min pearling, 15 min soaking, 15 min steaming at 100 kPa steam pressure, and drying at 50 °C reduced the oxalate content to a marked extent (0.43%) and, thus, can increase the functionality of the sesame seed protein to be used as a protein supplement (Manikantan et al. 2015).

12.5.3 Tannins

Tannin, a phenolic derivative of flavones, forms complexes with protein and reduces bioavailability of amino acids in the human body (Lampart-Szczapa et al. 2003) and is the chief ANF present in sesame meal (Mukhopadhyay and Bandyopadhyay 2003). Sesame seed was reported to contain 5.62 mg/100 g tannin, whereas it was also observed that cooking for 10, 20, and 30 min and toasting for 5, 10, and 15 min led to a marked decrease to 3.96, 1.54, and 0.49 mg/100 g and 2.19, 1.83, and 1.81 mg/100 g, respectively. The authors concluded from their study that moist heat treatment can be an appropriate method to reduce tannin content in sesame (Jimoh et al. 2011).

12.6 Phenolic Compounds

Phenolics like phenolic acid, sesamol, and chlorine containing naphthoquinone occur in sesame in small amounts (Lyon 1972; Dabrowski and Sosulski 1984; Shimoda et al. 1997).

Sesame seeds offer several health benefits like it aids in lowering cholesterol, reduces inflammation, reduces the risk of cardiovascular and oxidative stress-related diseases, and antimutagenic effects have also been reported (Chen et al. 2005; Lazarou et al. 2007; Gouveia et al. 2016). Increasing evidences suggests that these biological effects are due to the presence of sesamolin, sesaminol, and sesamolinol which are abundantly present in sesame seeds (Kochhar 2002; Rangkadilok et al. 2010). The presence of flavonoids in sesame seed oil helps in enhancing the antioxidant property which in turn helps in inhibiting the replication of human colon cancer cells (Sani et al. 2013).

12.6.1 Phenolic Acids

Sesame seeds, oil, and cake contain several phenolic compounds like ferulic, vanillic, cinnamic and p-coumaric acids, 4-hydroxybenzoic, protocatechuic acid, gallic acid, and sesamin (Mohdaly et al. 2013; Ben Othman et al. 2015) which have been reported to demonstrate important biological properties. Compounds like sesamol, sesamin, sesamolin, sesaminol glucosides, and tocopherol provide sesame oils protection against the autoxidation and therefore long shelf life (Chung 2004; Suja et al. 2004).

12.6.2 Lignans

Lignan, a constituent of lignin, is formed by the coupling of two p-hydroxyphenylpropane molecules by a bond between β-positions in the propane side chains. Lignans constitute a group of important plant phenolics and exist in two major groups, i.e., oil-soluble lignans and glycosylated water-soluble lignans. Sesamin (167–804 mg/100 g seeds), sesamolin (48–279 mg/100 g seeds), sesaminol (32–298 mg/100 g seeds), and sesamolinol (58 mg/100 g seeds) are the main oil-soluble lignans present in sesame (Moazzami et al. 2006; Pathak et al. 2014). However, higher amounts of sesamin (2.45 mg/g seed) and sesamolin (1.10 mg/g seed) were observed for Indian sesame species when compared with 65 sesame seeds harvested in Texas, USA (1.63 mg/g seed for sesamin and 1.01 mg/g seed for sesamolin), and 403 sesame landraces of Korea (2.09 mg/g for sesamin and 1.65 mg/g for sesamolin) (Kim et al. 2006; Moazzami et al. 2006; Pathak et al. 2014). The major glycosylated lignans are sesaminol triglucoside, pinoresinol triglucoside, sesaminol monoglucoside, pinoresinol monoglucoside, and two isomers of pinoresinol diglucoside and sesaminol diglucoside (Hemalatha 2004; Moazzami et al. 2006).

Lignans like sesamin and sesamolin exhibit antihypertensive, anticancerous, and hypocholesterolemic activities especially in humans and, thus, have become compounds of tremendous research interest in recent times. Moazzami et al. (2007) studied the total lignan content in 14 different sesame seeds and observed it to be ranging between 405 and 1178 mg/100 g of the seed. The major lignin observed in the present study was sesamin (167–804 mg/100 g seeds) followed by sesamolin (48–279 mg/100 g seeds) and sesaminol (32–298 mg/100 g seeds), while sesamolinol was found in much lower concentrations (58 mg/100 g seeds).

Shi et al. (2017) studied the concentration of lignans (sesamol, sesamin, sesamolin) in 100 sesame seeds and 56 sesame oils and reported total lignan content to be ranging between 2.52 to 12.76 mg/g and 3.38 to 11.53 mg/g, respectively. Additionally, it was also observed by the authors that color and processing affect the lignan content. Black sesame seeds had higher sesamin content (1.98–9.41 mg/g) than yellow, brown, and white sesame, seeds and hot-pressed and small mill sesame oils showed higher amounts of sesamol, sesamin, and total lignans as compared to cold-pressed and refined oil samples.

Ide et al. (2015) and Chu and Liu (2005) reported that sesame seed rich in lignan content more profoundly affects hepatic fatty acid oxidation and serum triacylglycerol level and possibly attenuates oxidative stress. These components not only help in lowering blood lipids (Hirata et al. 1996), arachidonic acid levels (Shimizu et al. 1991), and cholesterol levels by inhibiting absorption and synthesis of cholesterol (Hirose et al. 1991) but also possess anti-inflammatory (Hsu et al. 2005) and immunomodulatory activities (Nonaka et al. 1997) and, hence, promote health.

12.6.3 Tocopherols

Tocopherols belong to a class of plant phenolics and being a source of vitamin E not only possess nutritional potential but also prevent oxidation of oil, act as scavengers of free radicals, have strong antioxidant activity, and can prevent from lung and oral cancer, Alzheimer's disease, and diseases related to the nervous system (Brigelius-Flohe and Traber 1999; Brigelius-Flohé et al. 2002; Pasias et al. 2018). Various studies report that sesame seeds are a prominent source of γ-tocopherol (468.5–517.9 mg/kg lipid), while α- and δ-tocopherols are present in low amounts (Saldeen et al. 1999; Yoshida et al. 2007). Williamson et al. (2008) analyzed 11 genotypes of sesame seeds (US collection) and observed α-, δ-, and γ-tocopherols ranging between 0.034 and 0.175 μg/g, 0.44 and 3.05 μg/g, and 56.9 and 99.3 μg/g, respectively. Recently, Hajimahmoodi et al. (2010) reported 563–1095 mg/kg and 293–569 mg/kg of γ-tocopherol in sesame oil and seed, respectively. γ-Tocopherol is the most potent free-radical remover among the isomers of vitamin E, has strong anti-inflammatory activity, and aids in the inhibition of carcinogenesis (Ju et al. 2010). The alpha-tocopherol presents anti-inflammatory activity and modulates the expression of proteins involved in cholesterol metabolism (Wallert et al. 2014). Both α- and δ-tocopherols help in decreasing platelet aggregation and low-density lipid oxidation and delaying in thrombus formation and reduce photo carcinogenesis (Balan et al. 2009).

12.7 Health Attributes

12.7.1 In Diabetic Management

Diabetes is the most common metabolic syndrome in which the body does not produce or use insulin effectively. Sesame and its products can be efficient in the prevention of diabetes complications as they contain less carbohydrate, high dietary fiber, and high protein (Ley et al. 2014; Bigoniya et al. 2012). Various reports suggest that saturated fatty acid intake is one possible factor for diabetes risk, whereas edible oils rich in mono- and polyunsaturated fatty acids effectively reduce the risk of diabetes (Sankar et al. 2006; Riserus et al. 2009; Patel et al. 2010; Violi et al. 2015). Since sesame seed and sesame oil are a good source of mono- and polyunsaturated fatty acids, it can aid in the prevention of diabetes.

The influence of sesame oil (commercial diet containing 2% oil supplemented with 6% sesame oil) on blood glucose, lipid peroxidation, and status of antioxidants in normal and streptozotocin (STZ) diabetic rats was assessed for 42 days. The diabetic rats were observed to have elevated levels of blood glucose (322.61 ± 9.49 mg/dL), glycosylated hemoglobin, lipid hydroperoxides, thiobarbituric acid-reactive substances (TBARS), vitamin E and decreased levels of hemoglobin, vitamin C, and reduced glutathione (GSH). On the other hand, sesame oil consumption led to a significant reduction in the blood glucose (222.02 ± 8.27 mg/dL), glycosylated hemoglobin, lipid hydroperoxides, and TBARS and increased the hemoglobin, vitamin E, and GSH levels in the diabetic rats fed with sesame oil (Ramesh et al. 2005). Numerous studies suggest that sesamin, the major sesame lignan, is related to lipid metabolism through a series of biochemical actions in both humans and animals (Hirose et al. 1991; Matsumura et al. 1998; Chen et al. 2005). Mohammad Shahi et al. (2017) evaluated the effect of sesamin supplementation on 48 type 2 diabetic patients and reported that after 8 weeks of supplementation, the fasting blood sugar, glycated hemoglobin, tumor necrosis factor-alpha, interleukin-6, waist and hip circumference, and body adipose index levels reduced significantly ($p < 0.05$), whereas adiponectin levels increased when compared with the placebo. Thuy et al. (2017) investigated the effect of oral administration of sesamin for 4 weeks (100 and 200 mg/kg body weight) on cardiac function of diabetic rats (diabetes induced by streptozotocin). The results revealed marginal improvement in the blood glucose levels and body weight of the test group. However, significant improvement in the heart rate and blood pressure was observed when compared with the control rats. Both these studies suggest that sesamin could be used as a therapeutic approach in diabetes management.

Effects of sesame butter (1.25 g/kg) versus sesame oil (0.5 g/kg) (fed by oral gavage to diabetic rats for 6 weeks) on the serum levels of glucose, lipid profile, and oxidative stress biomarkers were examined, and significant reduction in the levels of glucose and increased levels of high-density lipoprotein were reported. The authors concluded that future intensive research on human diabetes should be undertaken to determine the exact dose and duration of supplementation of sesame butter for effective antihyperglycemic and partly lipid-reducing properties (Haidari et al. 2016). Bigoniya et al. (2012) observed that sesame meal-based products such as sesame meal-supplemented bread, *tahini* (sesame paste), and *halwa* (confectionary) were found suitable for diabetics as they contain less carbohydrate and high protein.

Aslam et al. (2018) investigated the impact of white sesame seed oil (WSSO) consumption on fasting blood glucose (GLU), insulin (INS), glycosylated hemoglobin (HbA1c), and hepatic antioxidant enzymes in 46 participants with type 2 diabetes at 0, 30, 60, and 90 days interval. The influence of WSSO on serum biochemistry, hepatic, cardiac, and renal functions was also studied. A significant reduction ($p < 0.05$) in GLU from mean 187.07 ± 5.63 mg/dl at 0 day to 137.83 ± 3.16 mg/dl at 90 day was observed in sesame oil-supplemented diabetic group, whereas the diabetic control group was reported to have 218.13 ± 5.92 mg/dl fasting blood glucose level. Furthermore, at the end of the study period, the HbA1c levels significantly reduced in the sesame oil-supplemented group when compared with control. INS increased from mean 12.12 ± 1.03 μU/ml at day 0 to 23.13 ± 1.15 μU/ml when compared with control (7.93 ± 0.38 μU/ml) at day 90. Additionally, the activities of superoxide dismutase, catalase, and glutathione peroxidase increased significantly, whereas in the control group, their activities decreased significantly ($p < 0.05$). The authors concluded that the white sesame seed oil can act as a functional food by improving the biomarkers of the liver, cardiac, and renal functions which improved significantly and, thus, can be used in diabetes management.

12.7.2 Hypertension

Owing to the rich content of polyunsaturated fatty acids, fiber, phytosterol, and lignans, sesame consumption may offer protection against hypertension and associated diseases like cardiovascular diseases, myocardial infarction, stroke, and renal failure (Khosravi-Boroujeni et al. 2017).

The antihypertensive effect of sesamin (1%w/w), a lignan from sesame oil, was observed using salt-loaded and salt-unloaded stroke-prone spontaneously hypertensive rats (SHRSP). Sesamin feeding significantly suppressed the development of hypertension in the salt-loaded groups, whereas a nonsignificant suppression was observed in the salt-unloaded group. In the histochemical studies of the aorta and superior mesenteric arteries, the authors observed a significant decrease in the wall thickness and wall area in the sesamin-fed salt-loaded group and hence concluded that sesamin holds the potential to be used as prophylactic treatment in hypertension (Matsumura et al. 1998).

Sankar et al. (2006) investigated the effect of sesame oil in 22 male and 18 female hypertensive diabetics patients aged 45–65 years old on atenolol (β-blocker) and glibenclamide (sulfonylurea) medications. The subjects were instructed to use sesame oil in place of other cooking oils for 45 days and then switch over to their regular oils for another 45 days. After 45 days of using sesame oil, the authors observed that the systolic and diastolic blood pressure decreased from 150 to 129.6 mmHg and 98.3 to 80 mmHg, respectively. However, when sesame oil substitution was withdrawn, at the end of the ninetieth day, the systolic and diastolic blood pressure increased to 140 and 90 mmHg, respectively. The authors reported that sesame oil substitution also brought a favorable change in the anthropometric measurements, lipid profile, lipid peroxidation, plasma glucose, and levels of electrolytes and antioxidants in the subjects.

A systematic review and meta-analysis of controlled trials were conducted by Khosravi-Boroujeni and others to ascertain the role of sesame in hypertension. Eight controlled trials (selected from PubMed (MEDLINE), Cumulative Index to Nursing and Allied Health Literature (CINAHL), and Cochrane Library (CENTRAL) databases) with a total of 843 participants were selected for the study, and after a random-effect meta-analysis, it was reported that sesame consumption can reduce systolic BP (−7.83 mmHg, 95% CI, −14.12, −1.54; $p < 0.05$, I2 = 99%) and diastolic BP (−5.83 mmHg, 95% CI, −9.58, −2.08; $p < 0.01$, I2 = 98%). Furthermore, for more accuracy only high methodology quality trials ($n = 4$) were analyzed, and it was observed that sesame consumption brought a significant reduction in systolic BP (−3.23 mmHg, 95% CI, −5.67, −0.79, I2 = 33%), whereas a nonsignificant reduction in diastolic BP (−2.08 mmHg, 95% CI, −4.85, 0.69, I2 = 62%) was observed (Khosravi-Boroujeni et al. 2017).

12.7.3 Cancer Prevention

Recently, sesame has been garnering considerable attention as an anticancer agent owing to the presence of sesamin. Harikumar et al. (2010) found that sesamin, a lipid-soluble lignan, inhibited the proliferation of tumor cells of lung, breast, pancreatic, colon, prostate cancer as well as of leukemia and multiple myeloma. It also potentiated tumor necrosis factor-alpha-induced apoptosis and the suppression of gene products linked to cell survival (e.g., Bcl-2 and survivin), proliferation (e.g., cyclin D1), inflammation (e.g., cyclooxygenase-2), invasion (e.g., matrix metalloproteinase-9, intercellular adhesion molecule 1), and angiogenesis (e.g., vascular endothelial growth factor). In another study, sesamin was reported to drastically inhibit the macrophage-enhanced proangiogenic activity, vascular endothelial growth factor (VEGF), and matrix metalloproteinase-9 (MMP-9) of breast cancer cell lines MCF-7 and MDA-MB-231 (Lee et al. 2011).

Truan et al. (2012) studied the comparative effect of flaxseed lignan, secoisolariciresinol diglucoside (SDG) and sesame seed lignan, and sesamin (SES) in human estrogen receptor-positive breast tumors (MCF-7) with high serum estrogen in athymic mice for 8 weeks. The authors reported that in mice fed with SES (1 g/kg), the tumor size reduced by 23% than in control; tumor cell proliferation also decreased along with an increase in apoptosis, whereas SDG-fed mice group neither differed from control in tumor size reduction nor was an increase in apoptosis was observed. Hence, the authors concluded that SES was found to be more successful in reducing the tumor size when compared with SDG at high serum estrogen levels.

In an in vitro study, Siao et al. (2015) observed that in the human breast cancer MCF-7 cell line, sesamin brought dose-dependent (1, 10, and 50 μM) reduction in the cell viability and increased lactic dehydrogenase (LDH) release and apoptosis. The authors suggested that as sesamin can modulate apoptotic signal pathways and also inhibit tumor cell growth, it can be used as a dietary supplement for prevention of breast cancer.

12.7.4 Bone Health

Phytoestrogens are naturally occurring plant compounds which consist of isoflavones, lignans, and coumestans. They are either structurally or functionally similar to β-estradiol (E2) and, thus, can play a prominent role in maintaining bone health in addition to their antioxidant activity (Hassan et al. 2013).

Boulbaroud et al. (2008) studied the effect of dietary flaxseed oil (7 and 10%) and sesame oil (7 and 10%) on biochemical and histological status of bone in 64 90-day-old female ovariectomized rats. After 4 weeks of treatment, the researchers observed that animal group fed with 10% flaxseed and 10% sesame seed oil had significantly reduced alkaline phosphatase activity (bone formation marker) and tartrate-resistant acid phosphatase activity (bone resorption marker) along with improved bone microarchitecture when compared with control. However, no significant decrease was observed in the groups which received 7% of the flaxseed or sesame seed oil-supplemented diet.

Hassan et al. (2013) studied the effect of soybean oil (15%)- and sesame oil (10%)-supplemented diets on bone mineral density, serum minerals (calcium, phosphorus, and magnesium), estrogen, serum lipid profile, enzymes (alkaline phosphatase and acid phosphatase activity), parathyroid hormone, osteocalcin, total protein, urea, and creatinine in ovariectomized rats for 2 months. The study showed that the supplementation of sesame and soybean oil in the diet of ovariectomized rats brought a significant improvement in all the abovementioned parameters and, thus, due to their high phytoestrogen content can act as natural anti-osteoporotic agents.

EL Wakf et al. (2014) also reported the osteoprotective effect of sesame oil- and soybean oil-supplemented diets (10% and 15%, respectively) on the ovariectomized (OVX) rats. The authors reported that after 2 months of supplementation, there was significant improvements in calcium and phosphorus level, the inflammatory indices (tumor necrosis factor-α, C-reactive protein levels, white blood cell counts, and acid phosphatase activity), antioxidant biomarkers (superoxide dismutase, catalase and reduced glutathione), and oxidative stress markers (malondialdehyde and protein carbonyl) in the sesame oil- and soybean oil-supplemented groups, when compared with that of the control. It was concluded that sesame and soybean oil supplementation might be useful for preventing bone loss caused by estrogen deficiency in OVX conditions.

12.8 Adverse Health Effects

Sesame and its products are increasingly popular in Western diet and is a common ingredient in a variety of foods. However, sesame allergy is a significant, serious, and growing problem. Persons allergic to various sesame seed products developed shock, lip stinging, palatial edema, and/or asthma on consumption of such products (Stutius et al. 2010). There has been a significant increase in the number of reports of hypersensitivity to sesame since the first report from the United States in 1950. Succeeding reports are commonly from developed countries, including the United States, Australia, many European countries, and Asia (Israel and Japan) (Gangur et al. 2005; Goto et al. 2008). The general prevalence of sesame allergy has been reported to about 0.1–0.2%, whereas in Israel it is the third most and second most common cause of IgE-mediated food reactions and food-induced anaphylaxis, respectively (Sicherer et al. 2003; Dalal et al. 2012). Consequently, the European Commission (EC) law (2003/89 EC) and Canadian Food Inspection Agency have listed sesame in the allergen labeling list; however, the Food and Drug Administration has not listed it under the allergen category (USFDA 1992, 2001; Goodwin 2004).

Studies have shown that protein and sesame oil components trigger the allergic response mediated by IgE. Eight sesame allergens (Ses i 1 to Ses i 8) have been characterized, wherein 11 s globulin (main protein component) comprises of Ses i 6 and Ses i 7, and oleoresins comprise Ses i 4 and Ses i 5 and cause severe allergic reactions like rhinoconjunctivitis (nasal congestion, sneezing and rhinorrhea, periocular pruritus, tearing, and conjunctival erythema) or asthma (wheezing, cough, dyspnea) (Beyer et al. 2007; Wolff et al. 2004; Permaul et al. 2009; Barbarroja-Escudero et al. 2015). Sesamin and sesamol in sesame oil have been reported to be associated with allergic contact dermatitis (Gangur et al. 2005).

12.9 Food Applications

12.9.1 As Cooking Oil

Sesame oil, due to its excellent keeping quality, fatty acid composition (43% oleic acid, 43% linoleic acid, 9% palmitic acid, and 4% stearic acid), good amount of highly active tocopherol, and stability, has found versatile use in food processing industry as cooking oil, in salad dressing, soups, confectionery, etc. and as a flavoring agent in the final stages of cooking (Sowmya et al. 2009; Asghar et al. 2014). Asia is the largest consumer of sesame and its products, mainly as cooking oil. In Japan, sesame seed is toasted prior to oil extraction which improves flavor and also improves the keeping quality of oil (Alegbejo et al. 2003).

12.9.2 Bakery Products

Various bakery products like cake have high amount of fats (more than 25%) mostly in the form of hydrogenated shortenings as it not only imparts the delicate and tender eating properties desired in cakes but also helps in entrapment of air during creaming process, physically meddling with the congruity of starch and protein particles and emulsification of the liquid in formulation (Anilakumar et al. 2010). However, due to the rising health-oriented concern in consumers, the application of hydrogenated fats is now discouraged. Sesame oil, as a replacement of hydrogenated fat, can play an important role in bakery shortenings as it has desirable characteristics along with numerous health benefits. Studies report that low-fat cakes can be prepared by partial (up to 50%) replacement of traditional fat with sesame oil without affecting the baking quality (Sowmya et al. 2009). Sesame meal has emerged as a popular source of protein as well as replacement of soy and legume proteins due to high levels of sulfur-containing amino acids, methionine, and cysteine (Evans and Bandemer 1967; Smith 1971), lack of trypsin-hindering factors, and a pleasant flavor (Boloorforooshan 1979; Brito and Núñez 1982). Some of the recently developed functional food products using sesame seed or its products as ingredients and their major findings have been summarized in Table 12.2.

12.9.3 Confectionary

In the Western world, sesame is essentially utilized as an ingredient in confectionary, as the seeds are incorporated into crackers, sesame sticks, and various other food products. Water-hulled, twofold washed, and dried sesame seeds are utilized on hamburger buns. Dehulled sesame seeds are sweet and oleaginous and are utilized specifically in various kinds of foods in different parts of the world, like in the

Table 12.2 Recent reports of various food products supplemented with sesame seed/products

Control product	Supplemented ingredient(s)	Quality attributes of sesame supplemented product	Reference
Wheat flour cookies	Germinated sesame flour (5, 10, and 15%)	Protein, fat, and ash content increased significantly ($p \leq 0.05$), whereas carbohydrate content decreased with increase in sesame supplementation. 15% sesame cookies had 18.80% protein, 25.02% fat, and 4.21% ash content. However, only 5% sesame-supplemented cookies matched with the control in terms of overall acceptability	Olagunju and Ifesan (2013)
	Unripe plantain and defatted sesame flours (100:0, 90:10, 80:20, and 70:30)	The diameter and thickness of unripe plantain and defatted sesame composite flour cookies were similar, whereas weight was significantly higher than control. The protein, fat, ash, iron, magnesium, crude fiber, and in vitro protein digestibility content of composite flour cookies was significantly ($p \leq 0.05$) higher than control. On the other hand, carbohydrate content, total sugar, sodium, and in vitro starch digestibility decreased when compared to control. Organoleptic evaluation results revealed no significant ($p \geq 0.05$) differences between composite cookies and control	Chinma et al. (2012)
	Dehulled sesame seed meal (10, 20, 30, and 40%)	Fortification with dehulled sesame seed meal did not result in any appreciable change in color, flavor, texture, or overall acceptability. However, with 40% substitution, the protein content increased from 12.52 to 16.86% and, thus, can be used to formulate high protein cookies	Inyang and Wayo (2005)
	Black sesame powder (BSP) (2, 4, 6, and 8%)	The spread ratio 2,2-Diphenyl-1-picrylhydrazyl and 2,2′-azino-bis (3-ethylbenzothiazoline-6-sulphonic acid) radical scavenging activities increased significantly ($P < 0.05$), whereas color characteristics, i.e., lightness (L∗), redness (a∗), yellowness (b∗), and hardness of cookies decreased with an increase in	Lim and Lee (2015)

(continued)

Table 12.2 (continued)

Control product	Supplemented ingredient(s)	Quality attributes of sesame supplemented product	Reference
		BSP in the cookies. The consumer acceptance test results revealed that 4% BSP incorporated cookies were preferred by the consumers when compared with the others	
Rice cookies based on 'Goami 2' rice	Sesame seeds (white and black) (substitution at 30% for butter)	The color characteristics, i.e., lightness (L∗), redness (a∗), and yellowness (b∗) values, were significantly different, and brittleness was significantly lower in the cookies made with black and white sesame seeds. Results of consumer test taken by 50 women students revealed that the Goami 2 rice cookies were the most preferred followed by black sesame cookies and white sesame cookies	Jung et al. (2007)
Millet flour biscuits	Defatted sesame seed flour (30, 40, and 50%)	Results revealed that the diameter and weight of biscuits reduced, while thickness, spread factor, and protein content increased ($p \leq 0.05$) with increasing level of sesame replacement. Defatted sesame seed flour-incorporated biscuits received good ratings for flavor and crispiness but were considered poor in color	Alobo (2001)
Wheat flour biscuits	Defatted sesame meal (5–8%)	Results revealed an increase in the protein content when compared to that of control and were found to be acceptable in the organoleptic evaluation	Gandhi and Taimini (2009)
Wheat flour bread	Black sesame powder (2.5, 5.0 and 7.5%)	The study revealed that with increase in the addition of black sesame powder, the weight of bread increased (522.0–532.0 g) than control (516.0 g) and the bread prepared with 7.5% black sesame powder addition was having highest hardness. Water-absorptive rate decreased along with lightness, redness, and yellowness values with increase in the addition of black sesame powder. The sensory evaluation	Choi and Chung (2005)

(continued)

Table 12.2 (continued)

Control product	Supplemented ingredient(s)	Quality attributes of sesame supplemented product	Reference
		showed that bread prepared with 205% black sesame powder incorporation was the best when compared with other incorporation levels and control	
	Defatted sesame meal (DSM) (4%)	The results revealed that the fortification of wheat flour with DSM at 4% level decreased improved essential amino acid scores (EAA) and overall nutritional value of the breads. However, in vitro protein digestibility was reportedly lesser in the fortified bread than the control	Saldivar et al. (1999)
Red wheat flour bread	Sesame products (sesame meal, roasted and autoclaved sesame meal, sesame protein isolate and concentrate) added to red wheat flour to produce blends at protein levels of 14, 16, 18, and 20%	Results of the study revealed that with an increase in the protein level of the blends, the water absorption, development time, and dough weakening were increased along with protein, minerals, and total essential amino acids, especially lysine and in vitro protein digestibility ($p < 0.05$); however, dough stability decreased. The study concluded that sesame protein isolate could be added up to 18% protein level to wheat flour and other sesame products could be added up to 16% protein level without affecting sensory properties	El-Adawy (1997)

manufacture of traditional confections such as *halva*, *laddu*, and *chikki* in India. They are also consumed as whole after roasting. A dessert called *laddu* is prepared from roasted groundnuts and sesame seeds by mixing with jaggery (ICMR 1977). The sweet has an extraordinary significance in numerous southern regions of India. It is distributed or exchanged between individuals to represent a great deal of sharing of good will (Mulky 1985). *Chikki* is another dessert prevalent in Maharashtra and other western parts of India.

12.9.4 As Ingredients

Numerous recipes contain sesame seeds as an ingredient, for example, sesame sprout, sesame spread, tangerine, sesame seed treats, hummus, sesame seed bagels, sesame granola, sesame broccoli rice, sesame mustard sauce, ginger sesame chicken, sesame cake, sesame seed sauce, and sesame green beans (Borchani et al. 2010; Noon 2003). In China sesame is sprinkled over rice and red beans and served at the exchange of wedding presents.

12.10 Alternative Applications

Different plant-based nutraceuticals are developed from sesame as it is reportedly rich in dietary and non-dietary phytochemicals and health. Sesame oil has nutritive, demulcent, and emollient properties and acts as laxative. It is also used as a solvent for intramuscular injections. From decades it is used to cure gum diseases, toothaches, headaches, blurred vision, and dizziness. Furthermore fatty acids obtained by the transesterification of triglycerides can be used as a substitute for fossil fuels. Also, sesame can be used for production of biodiesel by the use of methanol in the presence of sodium hydroxide as catalyst (Prasad Nagendra et al. 2012).

12.11 Conclusions

Sesame seeds are the richest source of protein apart from oil, lignans, antioxidants, tocopherols, calcium, vitamins, phosphorus, essential fatty acids, and others, but except sesame oil, very few value-added products are available in market. Sesame has inherent power to cure a number of diseases and contains bioactive components including phenolic compounds and phytosterols, which provide beneficial effect on human health and exhibit anticancer, antioxidative, anti-immunoregulation, and antihypersensitivity properties in addition to maintaining bone density and preventing bone loss during osteoporosis conditions. By-product like sesame meal obtained after oil extraction is also a valuable source of protein and holds the potential for developing various protein-based supplements. Sesame has the potential to establish itself as a major oilseed crop, and its better performance, yield, oil quality, and quantity would also provide benefits to the farmers from the area where sesame is grown as a regular crop. Sesame seed can be utilized as nutraceuticals to prevent malnutrition as well as global food security, and there is also enough scope for development of different value-added sesame products.

References

Achouri A, Boye JI (2013) Thermal processing, salt and high pressure treatment effects on molecular structure and antigenicity of sesame protein isolate. Food Res Int 53:240–251

Achouri A, Nail V, Boye JI (2012) Sesame protein isolate: fractionation, secondary structure and functional properties. Food Res Int 46:360–369

Adegunwa MO, Adebowale AA, Solano EO (2012) Effect of thermal processing on the biochemical composition, antinutritional factors and functional properties of beniseed (*Sesamum indicum*) flour. Am J Biochem Mol Biol 2:175–182

Alegbejo MD, Iwo GA, Abo EM, Idowu AA (2003) Sesame: a potential industrial and export oilseed crop in Nigeria. J Sustainable Agri 23:59–76

Alobo AP (2001) Effect of sesame seed flour on millet biscuit characteristics. Plant Foods Human Nutri 56:195–202

Anilakumar KR, Pal A, Khanum F, Bawa AS (2010) Nutritional, medicinal and industrial uses of sesame (*Sesamum indicum L.*) seeds-an overview. Agri Conspectus Scienti (ACS) 75:159–168

Asghar A, Majeed, MN (2013) Chemical characterization and fatty acid profile of different sesame verities in Pakistan. Am J Sci Indus Res:540–545

Asghar A, Majeed MN, Akhtar MN (2014) A review on the utilization of sesame as functional food. Am J Food Nutr 4:21–34

Aslam F, Iqbal S, Nasir M, Anjum AA (2018) White sesame seed oil mitigates blood glucose level, reduces oxidative stress, and improves biomarkers of hepatic and renal function in participants with type 2 diabetes mellitus. J Am College Nutri 26:1–2

Balan V, Rogers CA, Chundawat SP, da Costa SL, Slininger PJ, Gupta R, Dale BE (2009) Conversion of extracted oil cake fibers into bioethanol including DDGS, canola, sunflower, sesame, soy, and peanut for integrated biodiesel processing. J Am Oil Chem Soc 86:157–165

Barbarroja-Escudero J, Sanchez-Gonzalez MJ, Antolin-Amerigo D, Rodriguez-Rodriguez M, Pineda F, Alvarez-Mon M (2015) Diagnosis of IgE-mediated hypersensitivity to sesame seeds supplemented with lipid body proteins. Allergol Int 64:396–398

Ben Othman S, Katsuno N, Kanamaru Y, Yabe T (2015) Water-soluble extracts from defatted sesame seed flour show antioxidant activity in vitro. Food Chem 175:306–314

Beyer K, Grishina G, Bardina L, Sampson HA (2007) Identification of 2 new sesame seed allergens: Ses i 6 and Ses i 7. J Allergy Clin Immunol 119:1554–1556

Bigoniya P, Nishad R, Singh S (2012) Preventive effect of sesame seed cake on hyperglycemia and obesity against high fructose-diet induced Type 2 diabetes in rats. Food Chem 133 (4):1355–1361

Boloorforooshan MP (1979) Protein supplementation of navy beans with Sesame. J Food Sci 44:390–392

Borchani C, Besbes S, Blecker CH, Attia H (2010) Chemical characteristics and oxidative stability of sesame seed, sesame paste, and olive oils. J Agric Sci Technol 12:585–596

Boulbaroud S, Mesfioui A, Arfaoui A, Ouichou A, El Hessni A (2008) Preventive effects of flaxseed and sesame oil on bone loss in ovariectomized rats. Pak J Biol Sci 11:1696–1701

Brigelius-Flohe R, Traber MG (1999) Vitamin E: function and metabolism. The FASEB J 13:1145–1155

Brigelius-Flohé R, Kelly FJ, Salonen JT, Neuzil J, Zingg JM, Azzi A (2002) The European perspective on vitamin E: current knowledge and future research. Am J Clin Nutri 76:703–716

Brito OJ, Núñez N (1982) Evaluation of sesame flour as a complementary protein source for combinations with soy and corn flours. J Food Sci 47:457–461

Chen PR, Chien KL, Su TC, Chang CJ, Liu TL, Cheng H, Tsai C (2005) Dietary sesame reduces serum cholesterol and enhances antioxidant capacity in hypercholesterolemia. Nutri Res 25:559–556

Chinma CE, Igbabul BD, Omotayo OO (2012) Quality characteristics of cookies prepared from unripe plantain and defatted sesame flour blends. Am J Food Technol 7:398–408

Choi SN, Chung NY (2005) Quality characteristics of bread added with black sesame powder. Korean J Food Cookery Sci 21:655–661

Chu YF, Liu RH (2005) Cranberries inhibit LDL oxidation and induce LDL receptor expression in hepatocytes. Life Sci 77:1892–1901

Chung CH (2004). Molecular strategy for development of value-added sesame variety. Proceedings of international conference on sesame science, east asian society of dietary life, Seoul, pp 15–39
Dabrowski KJ, Sosulski FW (1984) Composition of free and hydrolyzable phenolic acids in defatted flours of ten oilseeds. J Agric Food Chem 32:128–130
Dalal I, Goldberg M, Katz Y (2012) Sesame seed food allergy. Current Allergy Asthma Rep 12:339–345
De Boland AR, Garner GB, O'Dell BL (1975) Identification and properties of phytate in cereal grains and oilseed products. J Agric Food Chem 23:1186–1189
De Carvalho PGB, Borgheetti F, Buckeridge MS, Morhy L, Filho EXF (2001) Temperature dependent germination and endo-b-mannanase activity in sesame seeds. R Bras Fisiol Veg 13:139–148
Ebuehi OAT, Umeh RA, Oletu F (2006) Physico-chemical and fatty acid composition of two common edible vegetable oils in Nigeria. Nigerian Food J 24:17–24
El Wakf AM, Hassan HA, Gharib NS (2014) Osteoprotective effect of soybean and sesame oils in ovariectomized rats via estrogen-like mechanism. Cytotechnology 66:335–343
El-Adawy TA (1997) Effect of sesame seed protein supplementation on the nutritional, physical, chemical and sensory properties of wheat flour bread. Food Chem 59:7–14
El-Khier KS, Ishag KEA, Yagoub AEA (2008) Chemical composition and oil characteristics of sesame seed varieties grown in Sudan. Res J Agric Biosci 14:1–6
Elleuch M, Bedigian D, Besbes S, Blecker C, Attia H (2012) Dietary fibre characteristics and antioxidant activity of sesame seed coats (testae). Int J Food Prop 15:25–37
Evans RJ, Bandemer SL (1967) Nutritive value of some oilseed proteins. Cereal Chem 44:417–426
FAO: Food and Agricultural Organization (2014) The State of Food Insecurity in the World. 6pp. www.fao.org/docrep/fao
FAOSTAT (2017) FAO. Food and Agriculture Organization of the United Nations. Rome, Italy
Francis G, Makkar HP, Becker K (2001) Antinutritional factors present in plant-derived alternate fish feed ingredients and their effects in fish. Aquaculture 199:197–192
Gandhi AP, Taimini V (2009) Organoleptic and nutritional assessment of sesame (*Sesame indicum*, L.) biscuits. Asian J Food Agroindust 2:87–92
Gangur V, Kelly C, Navuluri L (2005) Sesame allergy: a growing food allergy of global proportions? Ann Allergy Asthma Immunol 95:4–11
Goodwin P (2004) Food allergen testing. Food Quality:26–35
Goto S, Takaoka Y, Futamura M, Morishita M, Sakamoto T, Ito K (2008) Characteristics of sesame seed allergy and sesame-specific IgE test. J Allerg Clin Immunol 121:S247
Gouveia LA, Cardoso CA, de Oliveira GM, Rosa G, Moreira AS (2016) Effects of the intake of sesame seeds (*Sesamum indicum* L.) and derivatives on oxidative stress: a systematic review. J Med Food 19:337–345
Haidari F, Mohammadshahi M, Zarei M et al (2016) Effects of sesame butter (Ardeh) versus sesame oil on metabolic and oxidative stress markers in streptozotocin-induced diabetic rats. Iran J Med Sci 41(2):102–109. PMC4764959
Hajimahmoodi M, Oveisi MR, Sadeghi N, Jannat B, Bahaeddin Z, Mansoori S (2010) Gamma tocopherol content of Iranian sesame seeds. Iran J Pharm Res:135–139
Harikumar KB, Sung B, Tharakan ST, Pandey MK, Joy B, Guha S, Krishnan S, Aggarwal BB (2010) Sesamin manifests chemopreventive effects through the suppression of NF-κB–regulated cell survival, proliferation, invasion, and angiogenic gene products. Mol Cancer Res 11:1541–7786
Hassan HA, Wakf AME, Gharib NEE (2013) Role of phytoestrogenic oils in alleviating osteoporosis associated with ovariectomy in rats. Cytotechnology 65:609–619
Hemalatha S (2004) Lignans and tocopherols in Indian sesame cultivars. J Am Oil Chem Soc 81:467
Hirata F, Fujita K, Ishikura Y, Hosoda K, Ishikawa T, Nakamura H (1996) Hypocholesterolemic effect of sesame lignan in humans. Atherosclerosis 122:135–136

Hirose N, Inoue T, Nishihara K, Sugano M, Akimoto K, Shimizu S, Yamada H (1991) Inhibition of cholesterol absorption and synthesis in rats by sesamin. J Lipid Res 32:629–638
Hsu DZ, Su SB, Chien SP, Chiang PJ, Li YH, Lo YJ, Liu MY (2005) Effect of sesame oil on oxidative-stress-associated renal injury in endotoxemic rats: involvement of nitric oxide and proinflammatory cytokines. Shock 24:276–280
ICMR (1977) Studies on weaning and supplementary foods. Indian Council of Medical Research, Technical report series, No. 27 New Delhi
Ide T, Azechi A, Kitade S, Kunimatsu Y, Suzuki N, Nakajima C, Ogata N (2015) Comparative effects of sesame seeds differing in lignan contents and composition on fatty acid oxidation in rat liver. J Oleo Sci 64:211–222
Inyang UE, Wayo AU (2005) Fortification of cookies with dehulled sesame seed meal. Tropical Sci 45:103–105
Islam MB, Ali MY, Amin M, Zaman SM (2010) Climatic variations: farming systems and livelihoods in the high barind tract and coastal areas of Bangladesh. In: Climate change and food security in South Asia. Springer, Dordrecht, pp 477–497
Jimoh WA, Fagbenro OA, Adeparusi EO (2011) Effect of processing on some minerals, antinutrients and nutritional composition of sesame (*Sesamum indicum*) seed meals. Elect J Environ Agri Food Chem 10(1)
Johnson LA, Suleiman TM, Lusas EW (1979) Sesame protein: a review and prospectus. J Am Oil Chem Soc 56:463–468
Ju J, Picinich SC, Yang Z, Zhao Y, Suh N, Kong AN, Yang CS (2010) Cancer-preventive activities of tocopherols and tocotrienols. Carcinogenesis 31:533–542
Jung YJ, Seo HS, Myung JE, Shin JM, Lee EJ, Hwang IK (2007) Physicochemical and sensory characteristics of rice cookies based on Goami 2 with sesames (white and black) and perilla seeds. Korean J Food Cookery Sci 23:785–792
Khosravi-Boroujeni H, Nikbakht E, Natanelov E, Khalesi S (2017) Can sesame consumption improve blood pressure? A systematic review and meta-analysis of controlled trials. J Sci Food Agric 97:3087–3094
Kim KS, Lee JR, Lee JS (2006) Determination of sesamin and sesamolin in sesame (*Sesamum indicum L.*) seeds using UV spectrophotometer and HPLC. Korean J Crop Sci 51:95–100
Kochhar SP (2002) Sesame, rice-bran and flaxseed oils. In: Gunstone FD (ed) Vegetable oils in food technology. Blackwell Publishing, CRC Press, Oxford, pp 297–326
Lampart-Szczapa E, Siger A, Trojanowska K, Nogala-Kalucka M, Malecka M, Pacholek B (2003) Chemical composition and antibacterial activities of lupin seeds extracts. Nahrung/Food 47:286–229
Lazarou D, Grougnet R, Papadopoulos A (2007) Antimutagenic properties of a polyphenol-enriched extract derived from sesame seed perisperm. Mutation Res 634:163–171
Lee CC, Liu KJ, Wu YC, Lin SJ, Chang CC, Huang TS (2011) Sesamin inhibits macrophage-induced vascular endothelial growth factor and matrix metalloproteinase-9 expression and proangiogenic activity in breast cancer cells. Inflammation 34(3):209–221
Leeson S, Summers JD (2001) Nutrition of the chicken, 4th edn. University Books, Canada, pp 558–559
Ley SH, Hamdy O, Mohan V et al (2014) Prevention and management of type 2 diabetes: dietary components and nutritional strategies. Lancet 383:1999–2007
Lim JA, Lee JH (2015) Quality and antioxidant properties of cookies supplemented with black sesame powder. J Korean Soc Food Sci Nutri 44:1058–1063
Lyon CK (1972) Sesame: current knowledge of composition and use. J Am Oil Chem Soc 49:245–249
Makinde FM, Akinoso R (2013) Nutrient composition and effect of processing treatments on anti nutritional factors of Nigerian sesame (*Sesamum indicum* L) cultivars. Int Food Res J 20:2293–2300

Maneemegalai S, Prasad R (2011) Evaluation of amino acid composition and protein solubility profile of commercially available sesame and groundnut seed meal. Asian J Food Agroindus 4:161–166

Manikantan MR, Sharma R, Yadav DN, Gupta RK (2015) Selection of process parameters for producing high quality defatted sesame flour at pilot scale. J Food Sci Technol 52:1778–1783

Massey LK, Palmer RG, Horner HT (2001) Oxalate content of soybean seeds (Glycine max: Leguminosae), soyfoods, and other edible legumes. J Agric Food Chem 49:4262–4266

Matsumura Y, Kita S, Tanida Y, Taguchi Y, Morimoto S et al (1998) Antihypertensive effect of sesamin. III. Protection against development and maintenance of hypertension in stroke-prone spontaneously hypertensive rats. Biol Pharm Bull 21:469–473

Mbaebie B, Omosun G, Uti A, Oyedemi S (2010) Chemical composition of *Sesamum indicum* L. (Sesame) grown in Southeastern Nigeria and the physicochemical properties of the seed oil. Seed Sci Biotechnol 4:69–72

Mehta BV (2000) Sea millennium handbook 2000, 7th edn. The Solvent Extractors Association of India, Mumbai

Moazzami AA, Andersson RE, Kamal-Eldin A (2006) HPLC analysis of sesaminol glucosides in sesame seeds. J Agric Food Chem 54:633–638

Moazzami AA, Haese SL, Kamal Eldin A (2007) Lignan contents in sesame seeds and products. Eur J Lipid Sci Technol 109:1022–1027

Mohammad Shahi M, Zakerzadeh M, Zakerkish M, Zarei M, Saki A (2017) Effect of sesamin supplementation on glycemic status, inflammatory markers, and adiponectin levels in patients with type 2 diabetes mellitus. J Diet Suppl 14:65–75

Mohdaly AA, Ramadan-Hassanien MF, Mahmoud SMA, Smetanska I (2013) Phenolics extracted from potato, sugar beet, and sesame processing by-products. Int J Food Prop 16:1148–1168

Mukhopadhyay N, Bandyopadhyay S (2003) Extrusion cooking technology employed to reduce the anti-nutritional factor tannin in sesame (*Sesamum indicum*) meal. J Food Eng 56:201–202

Mulky MJ (1985) Utilization of oilseed meal for animal and human nutrition. In: Oilseeds production—constraints and opportunities. Oxford and IBH, New Delhi, pp 611–624

Nikmaram N, Leong SY, Koubaa M, Zhu Z, Barba FJ, Greiner R, Roohinejad S (2017) Effect of extrusion on the anti-nutritional factors of food products: an overview. Food Control

Nonaka M, Yamashita K, Iizuka Y, Namiki M, Sugano M (1997) Effects of dietary sesaminol and sesamin on eicosanoid production and immunoglobulin level in rats given ethanol. Biosci Biotechnol Biochem 61:836–839

Noon DD (2003) Sesame oil benefits blood pressure cooking oil switch may be high blood pressure treatment. Inter-American Society of Hypertension

Nzikou JM, Matos L, Bouanga-Kalou G, Ndangui CB, Pambou-Tobi NPG, Kimbonguila A, Silou T, Linder M, Desobry S (2009) Chemical composition on the seeds and oil of sesame (Sesamum indicum L.) grown in Congo-Brazzaville. Adv J Food Sci Technol 1:6–11

Ogbonna PE, Ukaan SI (2013) Chemical composition and oil quality of seeds of sesame accessions grown in the Nsukka Plains of South Eastern Nigeria. Afr J Agri Res 8:797–803

Olagunju AI, Ifesan BO (2013) Nutritional composition and acceptability of cookies made from wheat flour and germinated sesame (*Sesamum indicum*) flour blends. Br J App Sci Technol 3:702–706

Onsaard E, Pomsamud P, Audtum P (2010) Functional properties of sesame protein concentrates from sesame meal. Asian J Food Agroindus 3:420–431

Pasias IN, Kiriakou IK, Papakonstantinou L, Proestos C (2018) Determination of vitamin E in cereal products and biscuits by GC-FID. Foods 7:3

Patel PS, Sharp SJ, Jansen E et al (2010) Fatty acids measured in plasma and erythrocyte membrane phospholipids and derived by food-frequency questionnaire and the risk of new-onset type 2 diabetes: a pilot study in the European Prospective Investigation into Cancer and Nutrition (EPIC)-Norfolk cohort. Am J Clin Nutr 92:1214–1222

Pathak N, Rai AK, Kumari R, Bhat KV (2014) Value addition in sesame: a perspective on bioactive components for enhancing utility and profitability. Pharmacogn Rev 8:147

Permaul P, Stutius LM, Sheehan WJ, Rangsithienchai PA, Walter JE, Twarog F et al (2009) Characterizing sesame allergy: role of specific IgE and skin prick testing in predicting food challenge results. J Allergy Clin Immunol 123:24

Prasad Nagendra MN, Sanjay KR, Prasad DS, Vijay N, Kothari R, Nanjunda Swamy S (2012) A review on nutritional and nutraceutical properties of sesame. J Nutr Food Sci 2:2

Purseglove JW (1969) Tropical crops dicotyledons, vol 1. Longman, London

Ramesh B, Saravanan R, Pugalendi KV (2005) Influence of sesame oil on blood glucose, lipid peroxidation, and antioxidant status in streptozotocin diabetic rats. J Med Food 8:377–381

Rangkadilok N, Pholphana N, Mahidol C, Wongyai W, Saengsooksree K, Nookabkaew S, Satayavivad J (2010) Variation of sesamin, sesamolin and tocopherols in sesame (*Sesamum indicum L.*) seeds and oil products in Thailand. Food Chem 122:724–730

Riserus U, Willett WC, Hu FB (2009) Dietary fats and prevention of type 2 diabetes. Prog Lipid Res 48(1):44–51. https://doi.org/10.1016/j.plipres.2008.10.002

Saini CS, Sharma HK, Sharma (2018) Thermal, structural and rheological characterization of protein isolate from sesame meal. J Food Meas Charact 12:426–432

Saldeen T, Li D, Mehta JL (1999) Differential effects of alpha- and gamma-tocopherol on low-density lipoprotein oxidation, superoxide activity, platelet aggregation and arterial thrombogenesis. J Am Coll Cardiol 34:1208–1215

Saldivar SO, Abril-Dominguez JR, Lopez-Ahumada G, Ortega-Ramirez R (1999) Nutritional evaluation of table bread fortified with defated soybean and sesame meals. Arch Latinoam Nutr 49:260–264

Sani I, Sule FA, Warra AA, Bello F, Fakai IM, Abdulhamid A (2013) Phytochemicals and mineral elements composition of white *Sesamum indicum L.* seed oil. Int J Trad Nat Med 2(118):130

Sankar D, Rao MR, Sambandam G, Pugalendi KV (2006) A pilot study of open label sesame oil in hypertensive diabetics. J Med Food 9:408–412

Serna-Saldivar S O, Gutiérrez-Uribe J A, García-Lara S (2017) Phytochemical Profiles and Nutraceutical Properties of Corn and Wheat Tortillas. In *Tortillas* (pp 65–96)

Shahidi F (1997) Beneficial health effects and drawbacks of antinutrients and phytochemicals in foods: an overview. American Chemical Society

Sharma L, Singh C, Sharma HK (2016) Assessment of functionality of sesame meal and sesame protein isolate from Indian cultivar. J Food Meas Charact 10:520–526

Shi LK, Liu RJ, Jin QZ, Wang XG (2017) The contents of lignans in sesame seeds and commercial sesame oils of China. J Am Oil Chem Soc 94:1035–1104

Shimizu S, Akimoto K, Shinmen Y, Kawashima H, Sugano M, Yamada H (1991) Sesamin is a potent and specific inhibitor of Δ5 desaturase in polyunsaturated fatty acid biosynthesis. Lipids 26:512–516

Shimoda M, Nakada Y, Nakashima M, Osajima Y (1997) Quantitative comparison of volatile flavor compounds in deep-roasted and light-roasted sesame seed oil. J Agric Food Chem 45:3193–3196

Shivhare N, Satsangee N (2012) Wonders of sesame: nutraceutical uses and health benefits. In: Chemistry of phytopotentials: health, energy and environmental perspectives. Springer, Berlin Heidelberg, pp 63–68

Siao AC, Hou CW, Kao YH, Jeng KC (2015) Effect of sesamin on apoptosis and cell cycle arrest in human breast cancer mcf-7 cells. Asian Pac J Cancer Prev:163779–163783

Sicherer SH, Munoz-Furlong A, Sampson HA (2003) Prevalence of peanut and tree nut allergy in the United States determined by means of a random digit dial telephone survey: a 5-year follow-up study. J Allergy Clin Immunol 112:1203–1207

Smith KH (1971) Nutritional framework of oilseed proteins. J Am Oil Chem Soc 48:625–628

Soetan K, Oyewole O (2009) The need for adequate processing to reduce the anti-nutritional factors in plants used as human foods and animal feeds: a review. Afr J Food Sci 3:223–232

Sowmya T, Jeyarani R, Jyotsna D et al (2009) Effect of replacement of fat with sesame oil and additives on rheological, microstructural, quality characteristics and fatty acid profile of cakes. Food Hydrocoll 23:1827–1836

Steer TE, Gibson GR (2002) The microbiology of phytic acid metabolism by gut bacteria and relevance for bowel cancer. Int J Food Sci Technol 37:783–790

Stutius ML, Sheehan JW, Rangsithienchai P et al (2010) Characterizing the relationship between sesame, coconut, and nut allergy in children. Pediatr Allergy Immunol 21:1114–1118

Suja KP, Jayalekshmy A, Arumughan C (2004) Free radical scavenging behavior of antioxidant compounds of sesame (*Sesamum indicum L.*) in DPPH system. J Agric Food Chem 52:912–915

Tai SS, Lee TT, Tsai CC, Yiu TJ, Tzen JT (2001) Expression pattern and deposition of three storage proteins, 11S globulin, 2S albumin and 7S globulin in maturing sesame seeds. Plant Phys Biochem 39:981–992

Thuy TD, Phan NN, Wang CY, Yu HG, Wang SY, Huang PL, Do YY, Lin YC (2017) Novel therapeutic effects of sesamin on diabetes-induced cardiac dysfunction. Mol Med Rep 15:2949–2956

Toma RB, Tabekhia MM, Williams JD (1979). Phytate and oxalate contents in sesame seed (*Sesamum indicum L*). Nutr Rep Intern (USA)

Truan JS, Chen JM, Thompson LU (2012) Comparative effects of sesame seed lignan and flaxseed lignan in reducing the growth of human breast tumors (MCF-7) at high levels of circulating estrogen in athymic mice. Nutr Cancer 64:65–71

Urbano G, Lopez-Jurado M, Aranda P, Vidal-Valverde C, Tenorio E, Porres J (2000) The role of phytic acid in legumes: antinutrient or beneficial function? J Physiol Biochem 56:283–294

USDA, Nutrient Data Laboratory (U.S), & Consumer and Food Economics Institute (1999) USDA nutrient database for standard reference. Riverdal, MD, USDA, Agricultural Research Service

USFDA (Food and Drug Administration) (1992) A Statement of Policy. Rockville, MD: US Food and Drug Administration. pp 984–992

USFDA (Food and Drug Administration) (2001) CFSAN Program Priorities. Rockville, MD: US Food and Drug Administration, p 28

Violi F, Loffredo L, Pignatelli P et al (2015) Extra virgin olive oil use is associated with improved post-prandial blood glucose and LDL cholesterol in healthy subjects. Nutr Diabetes 5 (e172):1–7. https://doi.org/10.1038/nutd.2015.23

Wallert M, Schmölz L, Galli F, Birringer M, Lorkowski S (2014) Regulatory metabolites of vitamin E and their putative relevance for atherogenesis. Redox Biol 2:495–503

Williamson KS, Morris JB, Pye QN, Kamat CD, Hensley K (2008) A survey of sesamin and composition of tocopherol variability from seeds of eleven diverse sesame (*Sesamum indicum L*) genotypes using HPLC-PAD-ECD. Phytochem Anal 19:311–322

Wolff N, Yannai S, Karin N, Levy Y, Reifen R, Dalal I et al (2004) Identification and characterization of linear B-cell epitopes of β-globulin, a major allergen of sesame seeds. J Allergy Clinic Immunol 5:1151–1158

Yasothai R (2014) Chemical composition of sesame oil cake—review. Int J Sci Environ Technol 3:827–835

Yoshida H, Tanaka M, Tomiyama Y, Mizushina Y (2007) Antioxidant distributions and triacylglycerol molecular species of sesame seeds (*Sesamum indicum*). J Am Oil Chem Soc 84:165–172

Zebib H, Bultosa G, Abera S (2015) Physico-chemical properties of sesame (*Sesamum indicum L.*) varieties grown in northern area. Ethiopian Agric Sci 6:238

Chapter 13
Nigella (*Nigella sativa*) Seed

Paras Sharma and T. Longvah

Abstract There are several conventional oilseeds which are used to obtain edible oil such as soybean oil, sunflower oil, peanut oil, etc. However, few are nonconventional sources that play a crucial role in human health. *Nigella sativa* is one of the nonconventional oilseeds, which contains more than 30% (98.5% is fixed oil, 1.5% essential oil) oil. Fixed oil of nigella seed contains linoleic acid (48–62%) followed by oleic acid (19–25%) and limited amount of saturated fatty acids (arachidonic and eicosenoic acid), while the essential oil (volatile oil) contains a number of bioactive constitutes including thymoquinone (38.23%), p-cymene (28.61%), longifolene (5.4%), and 4-isopropyl-9-methoxy-1-methyl-1-cyclohexane (5.8%). Apart from oil, *Nigella sativa* seed also contains high levels of protein (up to 26%), dietary fiber, and micronutrients. A wide variation also exists in the nutritional value including minerals and vitamins of *Nigella* seeds owing to variation in growing conditions, region, climate, etc. Bioactive constituents of *Nigella* seeds exhibit strong health benefits including antidiabetic, anticancer, anti-inflammatory, antimicrobial, and antifungal activities. The present chapter covers the nutritional value, bioactive compounds, health benefits, and food uses of *Nigella sativa*.

Keywords *Nigella sativa* · Nutritional value · Fatty acids · Bioactive constituents · Micronutrients · Phytosterols

13.1 Introduction

Oilseeds play an important role in human nutrition since they provide oil; likewise, some of them are also consumed as a whole such as mustard seeds and *Nigella sativa* for culinary preparation. *Nigella sativa* (*kalonji* in Hindi) has been used since ancient time to cure a number of diseases and belongs to the family Ranunculaceae. *Nigella* seeds have been used as spice and additive in bread and cookies, as well as in other

P. Sharma (✉) · T. Longvah
Food Chemistry Division, ICMR-National Institute of Nutrition, Hyderabad, India

B. Tanwar, A. Goyal (eds.), *Oilseeds: Health Attributes and Food Applications*,
https://doi.org/10.1007/978-981-15-4194-0_13

dishes in various Asian and Eastern countries. *Nigella sativa* seed oil as such is not generally used for culinary preparation but known for its immense medicinal value. Nigella sativa is native to North Africa, Southern Europe, and Southwest Asia and is cultivated in many countries in the world including India, Pakistan, Syria, Turkey, and Saudi Arabia (Khare 2004).

Nigella seed has been reported in different traditional medicine systems including Unani and Ayurveda for remedy of numerous ailments. It is considered that *Nigella* seeds have the ability to cure almost all diseases except death (Muhammad 1994). *Nigella sativa* is known as the remedy of various diseases including asthma, diabetes, inflammation, cough, bronchitis, hypertension, headache, eczema, influenza, fever, and dizziness. Nigella seeds are angular, funnel-shaped, dark black or brown in color, and have a slightly bitter nutty–peppery taste and strong aroma.

Nutritionally, it contains high levels of oil (32–35%) which is recognized as medicinal food due to numerous health-promoting effects and therapeutic potential (Ramadan et al. 2003). Apart from oil, *Nigella* seed also contains high level of protein (26.7%), carbohydrate (25%), fiber (8.4%), and ash (4.8%) as well as micronutrients including vitamins, minerals, and numerous biological active compounds (Yarnell and Abascal 2011; Morsi 2000). Oil from *Nigella* seed contains fixed (stable) oil (98.2–99.9%) and volatile oil (essential) (0.11.8%) with plenty of bioactive constituents. Fixed oil is rich in linoleic acid (48–62%) followed by oleic acid (19–25%) as well as di-homo-γ-linoleic acid (strong antioxidant) and limited amount of saturated fatty acids (arachidonic and eicosenoic acid). This oil also contains fat-soluble vitamins (>0.2%) and a number of nutritionally important biological active constituents including phenolic compounds, pigments, carotenoids (β-carotene), tocopherol (α, β, γ), retinol, and a good amount of phytosterol particularly β-sitosterol, avenasterol, stigmasterol, campesterol, and lanosterol (Menounos et al. 1986), which have cholesterol-lowering effects. Besides the most valuable bioactive compound thymoquinone (38.23%), kalonji oil (KO) also contains other bioactive constituents including p-cymene (28.61%), longifolene (5.4%), and 4-isopropyl-9-methoxy-1-methyl-1-cyclohexane (5.8%) (Ramadan 2007).

Recently, kalonji oil has been introduced in several countries for edible purpose including Egypt, Southern Europe, Israel, and Lebanon as a novel source of edible oil. However, in India, its oil is not directly consumed, but seeds are widely used in food preparation. The present chapter will review the nutritional value, health attributes, and food application of *Nigella* seeds.

13.2 Nutritional Composition

Nigella seeds are reported to have high nutritive value including micronutrients such as fat-soluble vitamins and minerals. A large variation in different nutrients has been reported by Cheikh-Rouhou et al. (2007) among Tunisian- and Iranian-originated *Nigella* seed. They reported protein content between 22.6 and 26.7% and large variation in oil content (28.4–40.3%) as well. They observed total carbohydrate

Table 13.1 Nutritional composition of *Nigella* (*Nigella sativa*) seeds

Fat (%)	Protein (%)	Ash (%)	Total carbohydrate (%)	Fiber (%)	Sample collection country	Reference
28.48–40.35	22.6–26.7	4.41–4.86	32.7–40		Tunisian and Iranian	Cheikh-Rouhou et al. (2007)
31.72	23.07	5.29	34.91	–	Tehran (Iran)	Solati et al. (2014)
34.8	20.8	3.7	33.7	–	Egypt	Atta (2003)
31.16	22.80	4.20	–	–	Pakistan	Sultan et al. (2009)
38.2	20.9	404	31.9	–	Riyadh, Saudi Arabia	Al-Jassir (1992)
21.67	24.05	4.34	39.04	5.5[a]	Pakistan	Javed et al. (2012)
39.1–42.5	19.9–24.1	4.2–4.8	–	7.3–12.9[a]	India, Syria, and Turkey	Takruri and Dameh (1998)
36.8–38.4	19.1–20.3	3.8–4.6	31.2–33.1	TDF (26.5–36.8) IDF (20.5–27.1) SDF (6.5–8.9)	Yemen	Al-Naqeep et al. (2009)
37.33	20.2	6.72	30.52	–	Iran	Khoddami et al. (2011)
32.26	19.19	6.82	35.04	–	West Malaysia	Mohammed et al. (2016)

[a]Crude fiber; *TDF* total dietary fiber; *IDF* insoluble dietary fiber; *SDF* soluble dietary fiber

and ash content between 32.7 to 40% and 4.4 to 4.8%, respectively. The nutritional composition of *Nigella* seed is also affected by the growing environmental conditions, soil types, and agricultural practices (Atta 2003). Oil, ash, protein, and carbohydrate content in *Nigella* seed collected from the local market of Tehran (Iran) was found to be 31.7, 5.3, 23.0, and 34.9%, respectively (Solati et al. 2014). Sultan et al. (2009) also reported almost similar content of protein (22.8%), oil (31.1%), and ash (4.2%) in *Nigella* seeds. However, Malaysian *Nigella* seed samples exhibited 32.06% oil content, lower levels of protein (19.19%), high content of ash (6.82%), and 35% carbohydrate content (Mohammed et al. 2016). However, Khoddami et al. (2011) reported considerably different values of protein (20%), ash (6.27%), fat (37.3%), and total carbohydrate content (30.5%) in Nigella seeds (Table 13.1).

Similarly, Atta (2003) reported slightly different content of protein (20.8%), oil (34.8%), and ash (3.7%), while carbohydrate content (33.7%) was observed almost similar to other studies. *Nigella* sample collected from Saudi Arabia was evaluated for proximate composition and reported to contain lower level of protein (20.9%), high levels of oil (38.2%), carbohydrate (31.9%), crude fiber (7.9%), and ash content (4.4%). Overall, *Nigella* seed contains carbohydrate (39%) followed by protein (24%), fat (21.6%), fiber (5.5%), and ash content (4.3%) (Javed et al. 2012). Several *Nigella* seed samples were collected from few dominantly *Nigella*-producing countries and evaluated for proximate composition (Takruri and Dameh 1998), in which it was found that protein content varied between 19.9 and 24.4%, while oil, protein, ash, and crude fiber content was found between 39.7 and 42.5%, 19.9 and 22.5%, 4.2 and 4.8%, and 7.3 and 12.9%, respectively. A very high level of fat content (41%) and low content of total carbohydrate (17%) have been reported by Singh et al. (2014) in *Nigella* seeds grown in India.

Dietary fiber is also considered as biological active compounds due to their numerous health benefits. These are classified into three groups including total dietary fiber, soluble dietary fiber, and insoluble dietary fibers. Although most of the studies carried out on *Nigella* seed reported only crude fiber, nevertheless limited studies reported dietary fiber. It has been reported that total dietary fiber (TDF), soluble dietary fiber (SDF), and insoluble dietary fiber (IDF) varied between 6.5 and 8.9%, 20.56 and 27.1%, and 26.5 and 36.8%, respectively, among *Nigella* seeds collected from different regions of Yemen (Al-Naqeep et al. 2009) (Table 13.1).

Nigella seeds are also known as a good source of micronutrients including minerals and water-soluble vitamins. *Nigella* seed contains K (708–783 mg/kg), Mg (235–260 mg/kg), Ca (564–572 mg/kg), P (48.9–51.9 mg/kg), Fe (98.65–9.42 mg/kg), Zn (7–8.0 mg/kg), and Cu (1.48–1.65 mg/kg) (Cheikh-Rouhou et al. 2007) (Table 13.2). On the other hand, Al-Jassir (1992) reported the highest content of K (76 mg/kg) followed by P (18 mg/kg), Na (7.5 mg/kg), Fe (1.5 mg/kg), Zn (0.6 mg/kg), Ca (0.4 mg/kg), Mg (0.3 mg/kg), Mn (0.20 mg/kg), and Cu (0.20 mg/kg). Amin and Hosseinzadeh (2016) reported a wide range of minerals in *Nigella* seeds (3.7–7.0%). On the contrary, Ermumcu and Şanlıer (2017) observed a very narrow range of the same (1.76–3.44%). However, a study carried out by Babayan et al. (1978) on *Nigella* seed reported very high content of calcium (0.5–1.0%), phosphorus (0.6%), and potassium (0.6%) and low content of Na (0.1%).

Mineral profiling of *Nigella* seeds collected from different parts of Yemen exhibited a wide range of minerals including Fe (8.6–56.6 mg/100 g), Ca (544–811 mg/100 g), Mg (219–260 mg/100 g), Na (44–80 mg/100 g), K (447–563 mg/100 g), P (54–77 mg/100 g), Cu (1.30–1.6 mg/100 g), and Zn (1.84–2.5 mg/100 g) (Naqeep et al. 2009). Siong et al. (1989) reported Cu (1.8 mg/100 g), Fe (10.5 mg/100 g), zinc (6.0 mg/100 g), P (52.7 mg/100 g), and Ca content (186 mg/100 g). Mineral content of *Nigella* seeds also depends upon several factors including agricultural practices, soil composition, and on the method used for mineral quantification.

Nigella seed also contains water and fat-soluble vitamins. Among water-soluble vitamins, thiamin, niacin, folic acid, and pyridoxine were found abundantly, while a

Table 13.2 Mineral and water-soluble vitamin composition of *Nigella sativa* seeds

Micronutrient	References					
minerals (mg/kg)	Al-Jassir (1992)	Cheikh-Rouhou et al. (2007)	Takruri and Dameh (1998)	Al-Naqeep et al. (2009)	Abdel-Aal and Attia (1993)	Nargiz and Otles (1993)
Potassium	760	708–783	4423–5623	4473–5630	5606	11,800
Phosphorus	180	48.9–51.9	5043–5769	542–774	5699	–
Sodium	75	18.5–20.8	419–550	440–807	535	853
Iron	15.0	8.65–9.42	91–130	86–566	93	575
Zinc	6.0	7.03–8.4	56–66	19–25	59	–
Calcium	0.4	564–572	1544–2005	5440–8110	2000	1885
Magnesium	3.0	235–260	–	2190–2600	–	–
Manganese	2.0	3.37–4.43	–	–	–	–
Copper	2.0	1.48–1.65	15–24	13–16	17	–
Vitamins (μg/100 g)						
Thiamine (μg/100 g)	–	–	1300–1800	–	–	831
Riboflavin (μg/100 g)	–	–	–	–	–	63
Niacin(μg/100 g)	–	–	3300–9700	–	–	6311
Pyridoxin (μg/100 g)	–	–	400–600	–	–	789
Folic acid (mg/100 g)	–	–	40–87	–	–	42

small quantity of ascorbic acid has also been reported (Takruri and Dameh 1998) (Table 13.2). Among the fat-soluble vitamin, *Nigella* seed is an abundant source of vitamin E including α-, β-, and γ-tocopherol. Apart from tocopherol, it also contains retinol and carotenoids (β-carotene). Overall, fat-soluble vitamin comprises more than 0.2% of total oil content (Al-Saleh et al. 2006; Muhammad 1994), while altogether all vitamins range between 1 and 4% depending upon crop growing condition, genetic variation among seeds, and agricultural practices. Takruri and Dameh (1998) evaluated *Nigella* seeds collected from different countries including India, Syria, Turkey, and Jordan and reported that thiamin, pyridoxine, niacin, and folic acid content varied between 13 and 18, 04 and 15, 33 and 97, and 470 and 870 mg/kg, respectively.

Tocopherols are known to have potential antioxidant activity, and these effects are strongly associated with γ-tocopherol content. The total tocopherol content of the *Nigella* seeds are reported between 9.15 mg/100 g and 27.92 mg/100 g (Matthaus and Özcan 2011). Fat-soluble vitamins including α-tocopherols (10.19 μg/g), δ-tocopherol (2.28 μg/g), retinol (0.18 μg/g), vitamin D_2 (1.38 μg/g), phylloquinones (1.85 μg/g), and menaquinone-4 content (2.15 μg/g) were also reported by Vatansev et al. (2013).

Phytosterols are important bioactive compounds and have been shown numerous health benefits including cholesterol-lowering effects (Iriti and Varoni 2017). Phytosterols have been proved as strong antioxidant due to the formation of allylic free radical and its conversion into stable free radicals (Wang et al. 2002). Apart from the antioxidants, plant sterols also contribute in the prevention of colon cancer (Awad et al. 2000).

Nigella seeds contain significant quantities of phytosterols including β-sitosterol, campesterol, avenasterol, stigmasterol, and lanosterol (Amin and Hosseinzadeh 2016). However, very limited studies are available on phytosterol content in *Nigella* seed. β-Sitosterol was found to be the highest (1135–1182 μg/g) followed by D5-avenasterol (925–1025 μg/g) and D7-avenasterol (615–809 μg/g). Other sterols including stigmasterol, campesterol, and lanosterol were noticed in lower amounts (Ramadan and Mörsel 2003). *Nigella* seed contains total phytosterol of 3.66 g/kg (Ramadan et al. 2003). β-Sitosterol was found to be the chief phytosterol which accounts for 44 and 54% of the total sterols in Tunisian and Iranian varieties of black seed oils, respectively (Matthaus and Ozcan 2011). Ying et al. (2018) recorded slightly different content of phytosterol in *Nigella* seed oil including campesterol (6.7 mg/100 g), stigmasterol (6.1 mg/100 g), β-sitosterol (19.7 mg/100 g), avenasterol (2.2 mg/100 g), and cycloartenol (4.5 mg/100 g). It is also observed that *Nigella* seed oils extracted using cold press method and solvent extraction method did not have significant difference in their phytosterol content except D7-Avenasterol content (Gharby et al. 2015).

Nigella seed contains high level of oil which has medicinal value and protective effects against different diseases. *Nigella* seed contains fixed (stable) (32–40%) and essential (volatile) oil (0.40–0.50%), which are responsible for several health benefits. Fixed oil contains the highest amount of polyunsaturated fatty acid (PUFA) followed by monounsaturated fatty acids (MUFA) and saturated fatty acids (SFA) (Cheikh-Rouhou et al. 2007; Ramadan and Mörsel 2004). Linoleic acid (50–58%) and oleic acid (20–28%) are the highest among the PUFA and MUFA, respectively (Table 13.3). A small amount of α-linolenic acid is also found in *Nigella* seed oil. Among the total fat, linoleic acid is the highest (50.31%) existing fatty acid followed by oleic (25%) and palmitic acid (17.2%) (Cheikh-Rouhou et al. 2007). The total SFA, MUFA, and PUFA were recorded 22.7, 26.6, and 50.7%, respectively. On the other hand, higher content of linoleic acid (59%) and low content of palmitic acid were recorded by Solati et al. (2014). Most of the *Nigella* seed oil contains major proportions of unsaturated fatty acid (linoleic acid) between 50 and 60% followed by oleic acid (20%) and dihomolinoleic acid (10%) and eicosadienoic acid (3%) by linoleic acid (Karna 2013).

It is reported that oil extraction methods including Soxhlet and supercritical fluid extraction did not show significant effect on the fatty acid composition of *Nigella* seed (Solati et al. 2014). Nutritional composition including fatty acid composition is affected by the agricultural practices, soil composition, and environmental conditions (Asdadi et al. 2014). They reported a very high content of linoleic acid of 58.5%, followed by oleic acid (23.8%), palmitic acid (13.1%), and very high content of total PUFA of 83%. Similarly, Al-Jassir (1992) also reported the high content of

Table 13.3 Fatty acid composition of *Nigella* seeds (% of total fatty acid methyl esters)

Myristic s (C14, 0)	Myristoleic (C14, 1)	Palmitic (C16, 0)	Palmitoleic (C16, 1)	Margaric (C17, 0)	Stearic (C18, 0)	Oleic (C18, 1)	Linoleic (C18, 2)	Linolenic (18, 3n)	Arachidic (C20, 0)	Eicosenoic (C20, 1n9)	Behenic (C22, 0)	Lignoceric (C24, 0)	Reference
–	–	13.49	–	–	2.26	23.52	59.70	0.77	–	–	–	–	Solati et al. (2014)
0.21	–	12.84	0.28	–	3.38	22.58	58.23	–	0.22	0.38	0.035	0.041	Hosseini et al. (2013)
0.35–0.41	–	17.2–18.4	0.78–1.15	–	2.8–3.69	23.7–25	49.1–50.3	0.32–0.34	0.14–0.22	0.32–0.34	1.98–2.6	–	Cheikh-Rouhou et al. (2007)
1.0	–	13.1	0.2		2.3	23.8	58.5	0.4	0.5	–	–	–	Asdadi et al. (2014)
0.90	0.18	11.9	0.30	–	2.28	23.58	59.34	0.30	0.14	–	–	1.08	Al-Jassir (1992)
–	–	13	–	–	3.16	24.1	57.3	–	–	–	–	–	Ramadan et al. (2003)
–	–	12.3–14.8	–	–	3.2–4.6	14.2–18.4	60.5–61.7	0.4–0.7	–	–	–	–	Edris (2010)
0.13	–	12.01	0.25	0.06	2.77	23.95	57.49	0.25	0.15	0.27	–	0.31	Kiralan et al. (2014)
1.02–1.30	–	14.11–15.0	ND	–	3.0–3.04	21.5–21.7	55.65	2.24	–	–	–	–	Khoddami et al. (2011)

Table 13.4 Amino acid composition of *Nigella sativa* seeds

Amino acid	References		
	Al-Jassir (1992) (mg/100 g protein)	Babayan et al. (1978)	Al-Okbi et al. (2015) (mg/100 g protein)
Leucine	665 (5.82%)#	10.88%*	721.0
Valine	527 (4.61%)#	3.07%*	434.4
Lysine	462 (4.04%)#	7.62%*	475.8
Threonine	417 (3.65%)#	2.61%*	432.6
Phenylalanine	413 (3.61%)#	7.93%*	306.3
Isoleucine	395 (3.46%)#	4.03%*	119.5
Histidine	383 (3.35%)#	–	777.3
Methionine	188 (1.65%)#	6.16%*	60.5
Total essential amino acids	3450 (30.19%)#	–	4635.0
Glutamic acid	2829 (24.74%)#	13.21%*	9620.0
Arginine	2051 (9.19%)#	19.52%*	1308.4
Aspartic acid	1022 (8.94%)#	5.02%*	1277.6
Glycine	642 (5.61%)#	4.17%*	893.0
Proline	560 (4.90%)#	5.34%*	NR
Serine	493 (4.31%)#	1.98%*	926.7
Alanine	427 (3.73%)#	3.77%*	2233
Tyrosine	411 (3.59%)#	6.08%*	79.7
Ammonium	325 (2.84%)#	–	NR
Cystine	224 (1.96%)#	–	NR
Proline	NR	NR	234
Total nonessential amino acids	7984 (69.81%)	–	15,264

Values in the parenthesis # and * show the percent contribution of a particular amino acid to the total protein
NR Not reported

PUFA (83.7%) in nigella seed oil. Overall, oleic acid (20%), linoleic acid (50–60%), eicosadienoic acid (3%), and dihomolinoleic acid (10%) are found as chief unsaturated fatty acids, while palmitic acid and stearic acid are the major saturated fatty acids along with α-sitosterol (44–54%) and stigmasterol (6.57–20.92% of total unsaponifiable matter) (Islam et al. 2017).

Amino acids are a crucial component of any food which provides information about protein quality. Amino acid composition of *Nigella* seed's protein hydrolysates has been analyzed by Babayan et al. (1978) using gas chromatography and indicated the presence of 15 amino acids including all essential amino acids. Among the essential amino acids, leucine was found to be maximum (665 mg/100 g protein), while glutamic acid was found to be the highest among nonessential amino acids (Al-Jassir 1992). The total content of essential and nonessential amino acid was found to be 3450 and 7984 mg/100 g protein, respectively (Table 13.4).

13.3 Bioactive Compounds

Besides the important nutrients, *Nigella* also contains numerous biological active compounds which improve the human health (Amin and Hosseinzadeh 2016). Various bioactive compounds from *Nigella* seeds have been isolated and characterized, and their health effects have been evaluated. Thymoquinone is identified as chief bioactive compound with total concentration of 30–48% of volatile oil fraction of *Nigella* seeds, while p-cymene, thymohydroquinone, and dithymoquinone contributes 7–15% followed by carvacrol (6–12%), 4-terpineol (2–7%), and t-anethol (1–4%). Apart from these compounds, small quantity of longifolene and sesquiterpene (1–8%) α-pinene and thymol are also recorded. *Nigella* seed contains different alkaloids including isoquinolin, nigellicimine, nigellicimine-N-oxide, and indazole or pyrazol alkaloid ring bearing alkaloids including nigellidine and nigellicine. *Nigella* seeds also contain alpha-hederin which is a pentacyclic triterpene, and saponin, which are water-soluble compounds and have disease prevention potential (Al-Jassir 1992).

Nigella seeds contain trace amount of other biological active compound including limonene, carvone, and citronellol. Most of the health benefits of *Nigella* seeds are attributed to the quinine compounds (Ahmad et al. 2013). A number of active compounds have been isolated and identified which are nigellone, avenasterol-5-ene, avenasterol-7-ene, 24-methylene-cycloartanol, taraxerol, tirucallol, 3-O-[β-D-xylopyranosyl (1 → 3)-α-L-rhamnopyranosyl (1 → 2)-α-L-arabino-pyranosyl]-28-O-[α-L-rhamnopyranosyl(1 → 4)-β-D-glucopyranosyl(1 → 6)-β-D-glucopyranosyl]-hederagenin), 3-O-[α-L-rhamnopyranosyl-(1 → 2)-α-L-arabinopyranosyl]28-O-[α-L-rhamnopyranosyl-(1 → 4)-β-D-glucopyranosyl-(1 → 6)-β-D-glucopyranosyl]-hederagenin, and 3-O-[β-D-xylopyranosyl-(1 → 3)-α-L-rhamnopyranosyl-(1 → 2)-α-L-arabinopyranosyl]–hederagenin (Taşkin et al. 2005). In addition, aliphatic alcohol, β-unsaturated hydroxy ketone, hederagenin glycoside, melanthin, melanthigenin, 3-O-[β-D-xylopyranosyl-(1 → 2)-α-L-rhamno-pyranosyl-(1 → 2)-β-D-glucopyranosyl]-11-methoxy-16, 23-dihydroxy-28-methy-lolean-12-enoate,stigma-5, 22-dien-3-β-D-gluco-pyranoside, cycloart-23-methyl-7, 20, 22-triene-3β, 25-diol, and nigellidine-4-O-sulfite have also been reported in *Nigella* seed oil (Nickavar et al. 2003; Morikawa et al. 2004; Mehta et al. 2009).

13.4 Health Benefits

Numerous health-protecting effects including antioxidant activity, antidiabetic, anticancer, anti-inflammatory, nephroprotective activity, antibacterial, and antifungal properties have been shown by the bioactive constituents of *Nigella*, *Nigella* oil, and *Nigella* seed extract (Table 13.5).

Table 13.5 Health benefits of bioactive constituents of *Nigella* seeds bioactive constituents

Bioactive component	Health effects	Reference
Ether extract	Antibacterial effects (Gram +ve and −ve) both	Hanafy and Hatem (1991)
Methanolic, aqueous, chloroform extract and oil	Antibacterial effects (*Staphylococcus aureus* and *Escherichia coli*)	El-Nagerabia et al. (2012); Sunita and Meenakshi (2013)
Ethanolic extract	Antibacterial effect (*Staphylococcus aureus*, methicillin-resistant)	Hannan et al. (2008)
Thymoquinone (TQ) and thymohydroquinone (THQ)	Antibacterial activity against *E. coli*, *Pseudomonas aeruginosa*; *Staphylococcus aureus*; *Shigella flexneri*; *Sal. Typhimurium*, *Salmonella enteritidis*	Halawani (2009)
Aqueous extract	Antifungal activity (*Candida albicans*)	Develi et al. (2014); Suguna et al. (2013)
Oil	Antifungal activity (*Aspergillus flavus* and *Aspergillus parasiticus*) on the production of *aflatoxin*, *Chrysosporium evolceanui*; *Chrysosporium tropicum*; *Trichophyton simii*; *Trichophyton rubrum*, and *Microsporum gypseum*	Bita et al. (2012); Aljabre et al. (2005); Halamova et al. (2010)
TQ	*Candida albicans*	Nadaf et al. (2015); Aljabre et al. (2015)
Nanoparticles	Effective against lung cancer	Manju et al. (2016)
TQ	Brest cancer; anti-colon cancer activity	Woo et al. (2011); Salim (2010)
TQ	Polycystic ovary remedy	Arif et al. (2016)
TQ	Inhibition of induced superoxide production (antioxidant activity)	Boudiaf et al. (2016)
TQ	Antiangiogenic activity (positive effects in osteosarcoma)	Peng et al. (2013)
Ethanolic extract (seeds)	Nephroprotective activity	Canayakin et al. (2016); Saleem et al. (2012); Ulu et al. (2012); Abul-Nasr et al. (2001); Ali (2004)
TQ	Anticancer activity	Khalife et al. (2016)
Oil and TQ	Oral health	Al-Attass et al. (2016)
Oil and seed extract	Anti-schistosomiasis activity	Mohamed et al. (2005); Shenawy et al. (2008); Umar et al. (2012)
Nigella seeds and oil	Antidiabetic activity and cardiac protective activity	Salama (2011); Abdelmeguid et al. (2010); Kanter et al. (2009); Pari and Sankaranarayanan (2009)
TQ	Antidiabetic activity	Najmi et al. (2008)
Nigella seeds	Antidiabetic activity (in human)	Bamosa et al. (2010)
TQ	Anticancer and infectious disease; effective against pancreatic cancer	Torres et al. (2010)

(continued)

Table 13.5 (continued)

Bioactive component	Health effects	Reference
Aqueous extract of *Nigella* seeds	Effectiveness against YAC tumor cell; Anti-inflammatory; anti-analgesic activity	Majdalawieh et al. (2010); Alemi et al. (2013)
TQ and seed extract	Osteoporosis prevention activity	Shuid et al. (2012)
Aqueous extract; methanolic extract	Immunomodulatory activity	Ghonime et al. (2011); Mohamed et al. (2009)
TQ, oil, seeds; extracts	Cardiovascular activity	Nemmar et al. (2011)
Seed extract	Thyroid hormone regulation	Abdel-Sater (2009)
TQ, Oil	Gastro-protective activity	Magdy et al. (2012); Al-Mofleh et al. (2008); El-Abhar et al. (2003); Tayman et al. (2012)
TQ, *Nigella* seeds	Pulmonary-protective and anti-asthmatic effects	Wienkotter et al. (2008); Boskabady et al. (2007); Hossein et al. (2008)
Defatted seed, TQ	Neuroprotective activity	Perveen et al. (2009); Gilhotra and Dhingra (2011)

13.4.1 Cardiovascular Health Benefits

Cardiovascular disease is growing rapidly across the globe and thus is a major concern toward public health. The active principal thymoquinone (TQ) of *Nigella* seeds lowers the total cholesterol, thyroglobulin (TG), and low-density lipoprotein (LDL) and at the same time improves the high-density lipoprotein (HDL) cholesterol (Sahebkar et al. 2016). It is demonstrated that *Nigella* seeds also decrease motor fuel-induced systolic blood pressure, IL-6, leukocytes, and plasma superoxide dismutase activity (Ahmad et al. 2013). TQ isolated from *Nigella* seeds has been evaluated for its protective effects against diesel exhaust particles (DEP) on cardiopulmonary parameters in vivo, and it was concluded that TQ significantly prevented the adverse effects of DEP.

13.4.2 Antioxidant Activity

Nigella seed extracts have strong antioxidant potential even higher than synthetic antioxidant (Ermumcu and Şanlıer 2017). Most of the antioxidant compounds are contributed from volatile oil fraction of the seeds; however, other compounds including tocopherols, phenolic acids, and flavonoids also contribute to the antioxidant potential of *Nigella* seeds (Sultan et al. 2009). It is also proposed that *Nigella* seed extract could be used as a natural antioxidant in food to protect from oxidation by scavenging free radicals (Bourgou et al. 2012). Besides these, it has been proved that the antihyperglycemia and antihyperlipedimia effects of the TQ and other bioactive constituents of *Nigella* seed extract are also obtained through antioxidant mechanism. Several studies reported that *Nigella* prevents the lipid peroxidation and improves the antioxidant enzymes (Al-Mahasneh et al. 2008; Bamosa et al. 2010; Ragheb et al. 2008).

Studies carried out on rat model showed that injecting TQ and *Nigella* seed oil has protective effects on lipid peroxidation during ischemia–reperfusion injury (IRI) in rat hippocampus (Hosseinzadeh et al. 2007). Another study carried out on broiler chicks showed that feeding *Nigella* seeds prevents the liver from oxidative stress through increasing the few enzymes including glutathione-S-transferase, myeloperoxidase, and adenosine deaminase and by lowering hepatic lipid peroxidation (Sogut et al. 2008). Overall, a number of research have concluded that *Nigella* seed extract, seed oil, and TQ have potential radical scavenging activity by efficiently changing certain parameters including catalase (CAT), adenosine deaminase (ADA), myeloperoxidase (MPO), lipid peroxidase (LPO), reduced glutathione (GSH), superoxide dismutase (SOD) glutathione-S-transferase (GSH-ST), glutathione peroxidase (GPx), and nitric oxide (NO).

13.4.3 Antidiabetic Activity

Many studies have been carried out on antidiabetic potential of *Nigella* seeds and found its effectiveness against diabetes by reducing blood glucose levels through several mechanisms. *Nigella* extract which is rich in TQ reduces the tissue malondialdehyde (MDA) levels, prevents the DNA damage, vacuolization/fragmentation of mitochondria, as well as maintains the integrity of β-cells. TQ is also involved in improving insulin level thus significantly lowers the glucose and glycosylated hemoglobin levels in blood. Studies on rat model showed that the combination of three different compounds including α-lipoic acid (α-LA), *Nigella sativa*, and L-carnitine as well as individually significantly controls the blood glucose levels (Salama 2011). Similarly, *Nigella* seed extracts also have a synergistic effect along with parathyroid hormone (PTH) and improve bone mass, biomechanical behavior, and connectivity; consequently, it improves the diabetic conditions in type II diabetic rat model (Heshmati and Namazi 2015). Heshmati et al. (2015) suggested that feeding of 3 g *Nigella* seed oil three times a day significantly improved the glycemic status of diabetic patients.

13.4.4 Anticancer Activity

Bioactive principals of *Nigella* seed including TQ have shown anticancer activity against proliferation of cancer cells (Shafiq et al. 2014). Many studies have demonstrated that several bioactive compounds of *Nigella* have potential antioxidant activity which reduce the burden of free radicals, thus preventing cancer (Randhawa and Alghamdi 2011; Khan et al. 2011; Alenzi et al. 2010). In additions, these bioactive compounds also promote the activity of certain antioxidant enzymes including dismutase, catalase, superoxide, and glutathione peroxidase which also act as antioxidants and consequently maintain cell health (Badary et al. 2003; Bourgou et al. 2008; Randhawa and Alghamdi 2011).

13.4.5 Anti-inflammatory Activity

Various studies carried out on animal model shows that *Nigella* seed extract, TQ, and *Nigella* seed oil are effective against the inflammation (Ahmad et al. 2013; Chehl et al. 2009; Aljabre et al. 2015). This activity is due to the decline in production of NO, cyclooxygenase-1 and cyclooxygenase-2 (COX-1 and COX-2), and histone deacetylase (HDAC) along with other pro-inflammatory mediators including interleukin (IL)-1β, IL-6, tumor necrosis factor (TNF)-α, interferon (IFN)-γ, and prostaglandin (PGE2) (Chehl et al. 2009). Application of TQ at topical areas stimulates the expression of (NAD (P)H)-quinoneoxidoreductase-1, hemeoxygenase-1, glutathione S-transferases (GSH-ST), and glutamate cysteine ligase in mice. On the other hand, *Nigella* seed oil acts through the arachidonate metabolism by inhibition of 5-LPO and COXs (Aljabre et al. 2015). In the mice model, it has been shown that TQ reduces the DNA binding of nuclear factor kappa B through the obstruction of phosphorylation and further degradation of IκBα, which is a member of cellular protein (Ahmad et al. 2013).

13.4.6 Other Health-Promoting Activity

Apart from above-discussed positive health effects, *Nigella* essential oil and its extract also have the ability to improve nerve function (Ahlatci et al. 2014) and lower the blood pressure (Hosseini et al. 2013). In addition, *Nigella* seeds also have gastro-protective activity (Al-Mofleh et al. 2008; Kanter et al. 2005, El-Dakhakhny et al. 2000), pulmonary protective activity (Boskabady et al. 2007; Kanter et al. 2009), immunomodulatory activity (Islam et al. 2017; Swamy and Tan 2000), nephroprotective activity (Yaman and Balikci 2010), antibacterial activity (Kokoska et al. 2008), and antifungal activity (Khan et al. 2003).

13.5 Food Applications

Nigella seeds are pungent–bitter in taste and have strong aroma. Seeds are used in the Middle East and India as a spice and condiment and sporadically in Europe as a substitute of pepper as well as spices (Srinivasan 2018; Sharma et al. 2005). It is widely used as a minor ingredient in food preparation including pickles, naan (a type of Indian flat bread), and bakery products to give a characteristic flavor. In the Middle Eastern countries, *Nigella* seeds are used in bread dough and are a vital component of *choereg* rolls. Roasted *Nigella* seeds are widely used to flavor the curries, *dhal*, and *chutney* and several other culinary preparations including mildly braised lamb dishes such as korma. *Nigella* seeds, seed meal, and oil are also used as adjuncts to give characteristic flavor.

Nigella is also used as an ingredient in a mixture of spices (garam masala) and is a crucial part of one of the five spices in *panch phoran* which is a combination of five different whole spices including *Nigella* seed, fenugreek seed, black mustard seed, cumin seed, and fennel seed. The *panch phoran* originated from the Indian subcontinent and widely used in Bangladesh, Eastern India, and Southern Nepal cuisines (Malhotra 2012). Since *Nigella* seeds have a strong flavor and aroma, they are also used as insect repellent to preserve the grains. Conventionally, *Nigella* seed oil is not used for edible purpose; however, it is widely utilized for medicinal purpose due its numerous health benefits including antifungal, antibacterial, antidiabetic, antioxidant, anticancer, and cholesterol-lowering activities. Looking in to *Nigella* seed oil's health benefits, recently, it has been introduced as an edible oil in several countries including Egypt, Southern Europe, Israel, and Lebanon.

13.6 Conclusions

Nigella seeds are widely used as part of food recipe; however, it is not utilized as a major ingredient in food preparation due to its pungent flavor which comes from its essential oil. *Nigella* seed contains all nutrients including fat and water-soluble vitamins as well as minerals. *Nigella* seeds have balanced proteins and contain all essential amino acids which may play an important role in combating protein malnutrition. The high oil content of *Nigella* seeds makes it a perfect source of edible oil which contains numerous biological active compounds and may help in reducing the burden of noncommunicable disease. Presently, *Nigella* seeds have limited food application, and further efforts are required to develop food recipes where it should be used as key functional ingredient.

References

Abdelmeguid NE, Fakhoury R, Kamal SM, Al Wafai RJ (2010) Effects of *Nigella sativa* and thymoquinone on biochemical and subcellular changes in pancreatic β-cells of streptozotocin-induced diabetic rats. J Diabetes 2(4):256–266

Abdel-Aal ES, Attia RS (1993). Characterization of black cumin (*Nigella sativa*) seeds 1-chemical composition and lipids. Alexandria Science Exchange, 14:467-467

Abdel-Sater KA (2009) Gastroprotective effects of *Nigella sativa* oil on the formation of stress gastritis in hypothyroidal rats. Int J Phys Pathophys Pharm 1(2):143

Ahlatci A, Kuzhan A, Taysi S, Demirtas OC, Alkis HE, Tarakcioglu M, Demirci A, Caglayan D, Saricicek E, Cinar K (2014) Radiation-modifying abilities of *Nigella sativa* and thymoquinone on radiation-induced nitrosative stress in the brain tissue. Phytomedicine 21(5):740–744

Ahmad A, Husain A, Mujeeb M, Khan SA, Najmi AK, Siddique NA, Damanhouri ZA, Anwar F (2013) A review on therapeutic potential of *Nigella sativa*: a miracle herb. Asian Pac J Trop Biomed 3(5):337–352

Al-Mofleh IA, Alhaider AA, Mossa JS, Al-Sohaibani MO, Al-Yahya MA, Rafatullah S, Shaik SA (2008) Gastroprotective effect of an aqueous suspension of black cumin *Nigella sativa* on necrotizing agents-induced gastric injury in experimental animals. Saudi J Gastroenterol 14:128–134

Al-Attass SA, Zahran FM, Turkistany SA (2016) *Nigella sativa* and its active constituent thymoquinone in oral health. Saudi Med J 37:235–244

Alemi M, Sabouni F, Sanjarian F, Haghbeen K, Ansari S (2013) Anti-inflammatory effect of seeds and callus of *Nigella sativa* L. extracts on mix glial cells with regard to their thymoquinone content. AAPS PharmSciTech 14(1):160–167

Alenzi FQ, El-Bolkiny YES, Salem ML (2010) Protective effects of *Nigella sativa* oil and thymoquinone against toxicity induced by the anticancer drug cyclophosphamide. Br J Biomed Sci 67(1):20–28

Ali BH (2004) The effect of *Nigella sativa* oil on gentamicin nephrotoxicity in rats. Am J Chin Med 32(01):49–55

Aljabre SH, Randhawa MA, Akhtar N, Alakloby OM, Alqurashi AM, Aldossary A (2005) Antidermatophyte activity of ether extract of *Nigella sativa* and its active principle, thymoquinone. J Ethnopharmacol 101(1–3):116–119

Aljabre SH, Alakloby OM, Randhawa MA (2015) Dermatological effects of *Nigella sativa*. J Dermatol Dermatol Surg 19(2):92–98

Al-Jassir MS (1992) Chemical composition and microflora of black cumin (*Nigella sativa* L.) seeds growing in Saudi Arabia. Food Chem 45(4):239–242

Al-Mahasneh MA, Ababneh HA, Rababah T (2008) Some engineering and thermal properties of black cumin (*Nigella sativa* L.) seeds. Int J Food Sci Technol 43(6):1047–1052

Al-Naqeep GN, Ismail MM, Al-Zubairi AS, Esa NM (2009) Nutrients composition and minerals content of three different samples of *Nigella sativa* L. cultivated in Yemen. Asian J Biol Sci 2 (2):43–48

Al-Okbi SY, Mohamed DA, Hamed TE, El-Sayed EM, Mohamed MS, Mabrok HB (2015) Protective role of *Nigella sativa* seed meal and its alcohol extract in hepatorenal syndrome model in rats. Research J Pharm Biol Chem Sci 6(6):1355–1363

Al-Saleh IA, Billedo G, El-Doush II (2006) Levels of selenium, dl-α-tocopherol, dl-γ-tocopherol, all-trans-retinol, thymoquinone and thymol in different brands of *Nigella sativa* seeds. J Food Compos Anal 19(2–3):167–175

Amin B, Hosseinzadeh H (2016) Black cumin (*Nigella sativa*) and its active constituent, thymoquinone: an overview on the analgesic and anti-inflammatory effects. Planta Med 82 (01/02):8–16

Arif M, Thakur SC, Datta K (2016) Implication of thymoquinone as a remedy for polycystic ovary in rat. Pharm Biol 54(4):674–685

Asdadi A, Harhar H, Gharby S, Bouzoubaâ Z, Yadini AE, Moutaj R, Hadek ME, Chebli B, Hassani LM (2014) Chemical composition and antifungal activity of *Nigella sativa* L. oil seed cultivated in Morocco. Int J Pharm Sci Invent 3:9–15

Atta MB (2003) Some characteristics of nigella (*Nigella sativa* L.) seed cultivated in Egypt and its lipid profile. Food Chem 83(1):63–68

Awad AB, Downie A, Fink CS, Kim U (2000) Dietary phytosterol inhibits the growth and metastasis of MDA-MB-231 human breast cancer cells grown in SCID mice. Anticancer Res 20(2A):821–824

Babayan VK, Koottungal D, Halaby GA (1978) *Proximate analysis, fatty acid and amino acid composition of Nigella sativa* L. seeds. J Food Sci 43(4):1314–1315

Badary OA, Taha RA, Gamal El-Din AM, Abdel-Wahab MH (2003) Thymoquinone is a potent superoxide anion scavenger. Drug Chem Toxicol 26(2):87–98

Bamosa AO, Kaatabi H, Lebdaa FM, Elq AM, Al-Sultanb A (2010) Effect of *Nigella sativa* seeds on the glycemic control of patients with type 2 diabetes mellitus. Indian J Physiol Pharmacol 54 (4):344–354

Bita A, Rosu AF, Calina D, Rosu L, Zlatian O, Dindere C, Simionescu A (2012) An alternative treatment for Candida infections with *Nigella sativa* extracts. Eur J Hospital Pharma Science and Pract 19(2):162

Boskabady MH, Javan H, Sajady M, Rakhshandeh H (2007). The possible prophylactic effect of *Nigella sativa* seed extract in asthmatic patients. Fundamental & Clinical Pharmacology, 21 (5):559–566

Boudiaf K, Hurtado-Nedelec M, Belambri SA, Marie JC, Derradji Y, Benboubetra M, El-Benna J, Dang PM (2016) Thymoquinone strongly inhibits fMLF-induced neutrophil functions and exhibits anti-inflammatory properties in vivo. Biochem Pharmacol 104:62–73

Bourgou S, Ksouri R, Bellila A, Skandrani I, Falleh H, Marzouk B (2008) Phenolic composition and biological activities of Tunisian *Nigella sativa* L. shoots and roots. C R Biol 331(1):48–55

Bourgou S, Pichette A, Marzouk B, Legault J (2012) Antioxidant, anti-inflammatory, anticancer and antibacterial activities of extracts from *Nigella sativa* (black cumin) plant parts. J Food Biochem 36(5):539–546

Canayakin D, Bayir Y, Kilic Baygutalp N, Sezen Karaoglan E, Atmaca HT, Kocak Ozgeris FB, Keles MS, Halici Z (2016) Paracetamol-induced nephrotoxicity and oxidative stress in rats: the protective role of *Nigella sativa*. Pharm Biol 54(10):2082–2091

Chehl N, Chipitsyna G, Gong Q, Yeo CJ, Arafat HA (2009) Anti-inflammatory effects of the *Nigella sativa* seed extract, thymoquinone, in pancreatic cancer cells. HPB 11(5):373–381

Cheikh-Rouhou S, Besbes S, Hentati B, Blecker C, Deroanne C, Attia H (2007) *Nigella sativa* L.: chemical composition and physicochemical characteristics of lipid fraction. Food Chem 101 (2):673–681

Develi S, Evran B, Kalaz EB, Koçak-Toker N, Erata GÖ (2014) Protective effect of *Nigella sativa* oil against binge ethanol-induced oxidative stress and liver injury in rats. Chin J Nat Med 12 (7):495–499

Edris AE (2010) Evaluation of the volatile oils from different local cultivars of *Nigella sativa* L. grown in Egypt with emphasis on the effect of extraction method on thymoquinone. J Essent Oil Bear Plants 13(2):154–164

El-Abhar HS, Abdallah DM, Saleh S (2003) Gastroprotective activity of *Nigella sativa* oil and its constituent, thymoquinone, against gastric mucosal injury induced by ischaemia/reperfusion in rats. J Ethnopharmacol 84(2–3):251–258

El-Dakhakhny M, Mady NI, Halim MA (2000) *Nigella sativa* L. oil protects against induced hepatotoxicity and improves serum lipid profile in rats. Arzneimittelforschung 50(09):832–836

El-Nagerabia SA, Al-Bahryb SN, Elshafieb AE, AlHilalib S (2012) Effect of *Hibiscus sabdariffa* extract and *Nigella sativa* oil on the growth and aflatoxin B1 production of *Aspergillus flavus* and *Aspergillus parasiticus* strains. Food Control 25:59–63

Ermumcu MŞ, Şanlıer N (2017) black cumin (*Nigella sativa*) and its active component of thymoquinone: effects on health. J Food Health Sci 3(4):170–183

Gharby S, Harhar H, Guillaume D, Roudani A, Boulbaroud S, Ibrahimi M, Charrouf, Z (2015). Chemical investigation of *Nigella sativa L.* seed oil produced in Morocco. Journal of the Saudi Society of Agricultural Sciences, 14(2):172–177

Ghonime M, Eldomany R, Abdelaziz A, Soliman H (2011) Evaluation of immunomodulatory effect of three herbal plants growing in Egypt. Immunopharmacol Immunotoxicol 33(1):141–145

Gilhotra N, Dhingra D (2011) Thymoquinone produced antianxiety-like effects in mice through modulation of GABA and NO levels. Pharmacol Rep 63(3):660–669

Halamova K, Kokoska L, Flesar J, Sklenickova O, Svobodova B, Marsik P (2010) In vitro antifungal effect of black cumin seed quinones against dairy spoilage yeasts at different acidity levels. J Food Prot 73(12):2291–2295

Halawani E (2009) Antibacterial activity of thymoquinone and thymohydroquinone of *Nigella sativa* L. and their interaction with some antibiotics. Adv Biol Res 3(5–6):148–152

Hanafy MS, Hatem ME (1991) Studies on the antimicrobial activity of *Nigella sativa* seed (black cumin). J Ethnopharmacol 34(2–3):275–278

Hannan A, Saleem S, Chaudhary S, Barkaat M, Arshad MU (2008) Anti bacterial activity of *Nigella sativa* against clinical isolates of methicillin resistant *Staphylococcus aureus*. J Ayub Med Coll Abbottabad 20(3):72–74

Heshmati J, Namazi N (2015) Effects of black seed (*Nigella sativa*) on metabolic parameters in diabetes mellitus: a systematic review. Complement Ther Med 23(2):275–282

Heshmati J, Namazi N, Memarzadeh MR, Taghizadeh M, Kolahdooz F (2015) *Nigella sativa* oil affects glucose metabolism and lipid concentrations in patients with type 2 diabetes: a randomized, double-blind, placebo-controlled trial. Food Res Int 70:87–93

Hossein BM, Nasim V, Sediqa A (2008). The protective effect of nigella sativa on lung injury of sulfur mustard–exposed guinea pigs. Experimental Lung Research, 34(4):183-194

Hosseini MS, Mirkarimi SA, Amini M, Mohtashami R, Kianbakht S, Fallah HH (2013) Effects of *Nigella sativa* L. seed oil in type II diabetic Patients: a randomized, double-blind, placebo-controlled clinical trial. J Med Plants 3(47):93–99

Hosseinzadeh H, Fazly Bazzaz BS, Haghi MM (2007) Antibacterial activity of total extracts and essential oil of *Nigella sativa* L. seeds in mice. Pharmacology 2:429–435

Iriti M, Varoni EM (2017) Pulses, healthy, and sustainable food sources for feeding the planet. Int J Mol Sci 18(3):640

Islam MT, Guha B, Hosen S, Riaz TA, Shahadat S (2017) Nigellalogy: a review on *Nigella sativa*. MOJ Bioequiv Availab 3(6):00056

Javed S, Shahid AA, Haider MS, Umeera A, Ahmad R, Mushtaq S (2012) Nutritional, phytochemical potential and pharmacological evaluation of *Nigella sativa* (Kalonji) and *Trachyspermum ammi* (Ajwain). J Med Plants Res 6(5):768–775

Kanter M, Coskun O, Budancamanak M (2005) Hepatoprotective effects of *Nigella sativa* L and *Urtica dioica* L on lipid peroxidation, antioxidant enzyme systems and liver enzymes in carbon tetrachloride-treated rats. World J Gastroenterol: WJG 11(42):6684

Kanter M, Akpolat M, Aktas C (2009) Protective effects of the volatile oil of *Nigella sativa* seeds on β-cell damage in streptozotocin-induced diabetic rats: a light and electron microscopic study. J Mol Histol 40(5–6):379–385

Karna SKL (2013). Phytochemical screening and gas chromatography mass spectrometry and analysis of seed extract of *Nigella sativa* Linn. Int J Chem Studies, 1(4):183–188

Khalife R, Hodroj MH, Fakhoury R, Rizk S (2016) Thymoquinone from *Nigella sativa* seeds promotes the antitumor activity of noncytotoxic doses of topotecan in human colorectal cancer cells in vitro. Planta Med 82:312–321

Khan MAU, Ashfaq MK, Zuberi HS, Mahmood MS, Gilani AH (2003) The in vivo antifungal activity of the aqueous extract from *Nigella sativa* seeds. Phytother Res 17(2):183–186

Khan A, Chen HC, Tania M, Zhang DZ (2011) Anticancer activities of *Nigella sativa* (black cumin). Afr J Tradit Complement Altern Med 8(5S):226–232

Khare CP (2004) Encyclopedia of Indian medicinal plants. Springer-Verlag Berlin Heidelberg, New York

Khoddami A, Ghazali HM, Yassoralipour A, Ramakrishnan Y, Ganjloo A (2011) Physicochemical characteristics of nigella seed (*Nigella sativa* L.) oil as affected by different extraction methods. J Am Oil Chem Soc 88(4):533–540

Kiralan M, Özkan G, Bayrak A, Ramadan MF (2014) Physicochemical properties and stability of black cumin (*Nigella sativa*) seed oil as affected by different extraction methods. Ind Crop Prod 57:52–58

Kokoska L, Havlik J, Valterova I, Sovova H, Sajfrtova M, Jankovska I (2008) Comparison of chemical composition and antibacterial activity of *Nigella sativa* seed essential oils obtained by different extraction methods. J Food Prot 71(12):2475–2480

Magdy MA, Hanan EA, Nabila EM (2012) Thymoquinone: novel gastroprotective mechanisms. Eur J Pharmacol 697(1–3):126–131

Majdalawieh AF, Hmaidan R, Carr RI (2010) *Nigella sativa* modulates splenocyte proliferation, Th1/Th2 cytokine profile, macrophage function and NK anti-tumor activity. J Ethnopharmacol 131(2):268–275

Malhotra SK (2012) *Nigella*. In: Handbook of herbs and spices, vol 2, 7nd edn, pp 391–416

Manju S, Malaikozhundan B, Vijayakumar S, Shanthi S, Jaishabanu A, Ekambaram P, Vaseeharan B (2016) Antibacterial, antibiofilm and cytotoxic effects of *Nigella sativa* essential oil coated gold nanoparticles. Microb Pathog 91:129–135

Matthaus B, Özcan MM (2011) Fatty acids, tocopherol, and sterol contents of some *Nigella* species seed oil. Czech J Food Sci 29(2):145–150

Mehta BK, Pandit V, Gupta M (2009) New principles from seeds of *Nigella sativa*. Nat Prod Res 23 (2):138–148

Menounos P, Staphylakis K, Gegiou D (1986) The sterols of *Nigella sativa* seed oil. Phytochemistry 25(3):761–763

Mohamed AM, Metwally NM, Mahmoud SS (2005). Sativa seeds against Schistosoma mansoni different stages. Memórias do Instituto Oswaldo Cruz, 100(2):205–211

Mohamed A, Waris HM, Ramadan H, Quereshi M, Kalra J (2009). Amelioration of chronic relapsing experimental autoimmune encephalomyelitis (cr-eae) using thymoquinone-biomed 2009. Biomedical sciences instrumentation, 45:274–279

Mohammed NK, Manap A, Yazid M, Tan CP, Muhialdin BJ, Alhelli AM, Hussin M, Shobirin A (2016) The effects of different extraction methods on antioxidant properties, chemical composition, and thermal behavior of black seed (*Nigella sativa* L.) oil. Evid Based Complement Alternat Med:2016

Morikawa T, Xu F, Ninomiya K, Matsuda H, Yoshikawa M (2004) Nigellamines A3, A4, A5, and C, new dolabellane-type diterpene alkaloids, with lipid metabolism-promoting activities from the Egyptian medicinal food black cumin. Chem Pharm Bull 52(4):494–497

Morsi NM (2000) Antimicrobial effect of crude extracts of *Nigella sativa* on multiple antibiotics-resistant bacteria. Acta Microbiol Pol 49(1):63–74

Muhammad AA (1994) This black seed is healing for all diseases except as-saam (death). Translator: Saheeh al-Bukhaaree. In: Khan MM (ed) Book of medicine (English translation). Maktaba Dar-us-Salaam, Riyadh, p 939

Nadaf NH, Gawade SS, Muniv AS, Waghmare SR, Jadhav DB, Sonawane KD (2015) Exploring anti-yeast activity of *Nigella sativa* seed extracts. Ind Crop Prod 77:624–630

Najmi A, Haque SF, Naseeruddin M, Khan RA (2008) Effect of *Nigella sativa* oil on various clinical and biochemical parameters of metabolic syndrome. Int J Diabetes Dev Countries 16:85–87

Nargiz C, Otles S (1993) Chemical composition of Nigella sativa seeds. Food Chemistry 48(3):259-261

Nemmar A, Al-Salam S, Zia S, Marzouqi F, Al-Dhaheri A, Subramaniyan D, Dhanasekaran S, Yasin J, Ali BH, Kazzam EE (2011) Contrasting actions of diesel exhaust particles on the pulmonary and cardiovascular systems and the effects of thymoquinone. Br J Pharmacol 164 (7):1871–1882

Nickavar B, Mojab F, Javidnia K, Amoli MA (2003) Chemical composition of the fixed and volatile oils of *Nigella sativa* L. from Iran. Zeitschrift für Naturforschung C 58(9–10):629–631

Pari L, Sankaranarayanan C (2009) Beneficial effects of thymoquinone on hepatic key enzymes in streptozotocin–nicotinamide induced diabetic rats. Life Sci 85(23–26):830–834

Peng L, Liu A, Shen Y, Xu HZ, Yang SZ, Ying XZ, Liao W, Liu HX, Lin ZQ, Chen QY, Cheng SW (2013) Antitumor and anti-angiogenesis effects of thymoquinone on osteosarcoma through the NF-κB pathway. Oncol Rep 29(2):571–578

Perveen T, Haider S, Kanwal S, Haleem DJ (2009) Repeated administration of *Nigella sativa* decreases 5-HT turnover and produces anxiolytic effects in rats. Pak J Pharm Sci 22(2)

Ragheb A, Elbarbry F, Prasad K, Mohamed A, Ahmed MS, Shoker A (2008) Attenuation of the development of hypercholesterolemic atherosclerosis by thymoquinone. Int J Angiol 17(4):186

Ramadan MF (2007) Nutritional value, functional properties and nutraceutical applications of black cumin (*Nigella sativa* L.): an overview. Int J Food Sci Technol 42(10):1208–1218

Ramadan MF, Mörsel JT (2003) Analysis of glycolipids from black cumin (*Nigella sativa* L.), coriander (*Coriandrum sativum* L.) and niger (*Guizotia abyssinica* Cass.) oilseeds. Food Chem 80(2):197–204

Ramadan MF, Mörsel JT (2004) Oxidative stability of black cumin (*Nigella sativa* L.), coriander (*Coriandrum sativum* L.) and niger (*Guizotia abyssinica* Cass.) crude seed oils upon stripping. Eur J Lipid Sci Technol 106(1):35–43

Ramadan MF, Kroh LW, Mörsel JT (2003) Radical scavenging activity of black cumin (*Nigella sativa* L.), coriander (*Coriandrum sativum* L.), and niger (*Guizotia abyssinica* Cass.) crude seed oils and oil fractions. J Agric Food Chem 51(24):6961–6969

Randhawa MA, Alghamdi MS (2011) Anticancer activity of *Nigella sativa* (black seed)—a review. Am J Chin Med 39(06):1075–1091

Sahebkar A, Soranna D, Liu X, Thomopoulos C, Simental-Mendia LE, Derosa G, Maffioli P, Parati G (2016) A systematic review and meta-analysis of randomized controlled trials investigating the effects of supplementation with *Nigella sativa* (black seed) on blood pressure. J Hypertens 34(11):2127–2135

Salama RH (2011) Hypoglycemic effect of lipoic acid, carnitine and *Nigella sativa* in diabetic rat model. Int J Health Sci 5(2):126

Saleem U, Ahmad B, Rehman K, Mahmood S, Alam M, Erum A (2012) Nephro-protective effect of vitamin C and *Nigella sativa* oil on gentamicin associated nephrotoxicity in rabbits. Pak J Pharm Sci 25(4):727–730

Shafiq HAA, Masud T, Kaleem M (2014). Cardio-protective and anti-cancer therapeutic potential of *Nigella sativa*. Iranian journal of basic medical sciences, 17(12):967

Sharma PC, Yelne MB, Dennis TJ (2005) Database on medicinal plants used in Ayurveda. New Delhi, pp 420–440

Shenawy EL, Nahla S, Soliman MF, Reyad SI (2008) The effect of antioxidant properties of aqueous garlic extract and *Nigella sativa* as anti-schistosomiasis agents in mice. Rev Inst Med Trop Sao Paulo 50(1):29–36

Shuid AN, Mohamed N, Mohamed IN, Othman F, Suhaimi F, Mohd Ramli ES, Muhammad N, Soelaiman IN (2012) *Nigella sativa*: a potential antiosteoporotic agent. Evid Based Complement Alternat Med 2012

Siong TE, Choo KS, Shahid, SM (1989) Determination of iron in foods by atomic absorption spectrophotometric and colorimetric methods. Pertanika 12:313–322

Singh S, Das SS, Singh G, Schuff C, de Lampasona MP, Catalán CA (2014) Composition, in vitro antioxidant and antimicrobial activities of essential oil and oleoresins obtained from black cumin seeds (*Nigella sativa* L.). Biomed Res Int:2014

Sogut B, Celik I, Tuluce Y (2008) The effects of diet supplemented with the black Cumin (*Nigella sativa* L.) upon immune potential and antioxidant marker enzymes and lipid peroxidation in broiler chicks. J Anim Vet Adv 7(10):1196–1199

Solati Z, Baharin BS, Bagheri H (2014) Antioxidant property, thymoquinone content and chemical characteristics of different extracts from *Nigella sativa* L. seeds. J Am Oil Chem Soc 91 (2):295–300

Srinivasan K (2018) Cumin (*Cuminum cyminum*) and black cumin (*Nigella sativa*) seeds: traditional uses, chemical constituents, and nutraceutical effects. Food Qual Saf 2(1):1–6

Suguna P, Geetha A, Aruna R, Siva GV (2013) Effect of thymoquinone on ethanol and high fat diet induced chronic pancreatitis—a dose response study in rats. Indian J Exp Biol 51(4):292–302

Sultan MT, Butt MS, Anjum FM, Jamil A, Akhtar S, Nasir M (2009) Nutritional profile of indigenous cultivar of black cumin seeds and antioxidant potential of its fixed and essential oil. Pak J Bot 41(3):1321–1330

Sunita M, Meenakshi S (2013) Chemical composition and antidermatophytic activity of *Nigella sativa* essential oil. Afr J Pharm Pharmacol 7(20):1286–1292

Swamy SM, Tan BK (2000) Cytotoxic and immunopotentiating effects of ethanolic extract of *Nigella sativa* L. seeds. J Ethnopharmacol 70(1):1–7

Takruri HR, Dameh MA (1998) Study of the nutritional value of black cumin seeds (Nigella sativaL). J Sci Food Agric 76(3):404–410

Taşkin MK, ÇALIŞKAN ÖA, Anil H, Abou-Gazar H, Khan IA, BEDİR E (2005) Triterpene saponins from *Nigella sativa* L. Turk J Chem 29(5):561–569

Tayman C, Cekmez F, Kafa IM, Canpolat FE, Cetinkaya M, Uysal S, Tunc T, Sarıcı SU (2012) Beneficial effects of *Nigella sativa* oil on intestinal damage in necrotizing enterocolitis. J Investig Surg 25(5):286–294

Torres MP, Ponnusamy MP, Chakraborty S, Smith LM, Das S, Arafat HA, Batra SK (2010) Effects of thymoquinone in the expression of mucin 4 in pancreatic cancer cells: implications for the development of novel cancer therapies. Mol Cancer Ther 27:1535–7163

Ulu R, Dogukan A, Tuzcu M, Gencoglu H, Ulas M, İlhan N, Muqbil I, Mohammad RM, Kucuk O, Sahin K (2012) Regulation of renal organic anion and cation transporters by thymoquinone in cisplatin induced kidney injury. Food Chem Toxicol 50(5):1675–1679

Umar S, Zargan J, Umar K, Ahmad S, Katiyar CK, Khan HA (2012) Modulation of the oxidative stress and inflammatory cytokine response by thymoquinone in the collagen induced arthritis in Wistar rats. Chem Biol Interact 197(1):40–46

Vatansev H, Ciftci H, Ozkaya A, Ozturk B, Evliyaoglu N, Kiyici A (2013) Chemical composition of *Nigella sativa* L. seeds used as a medical aromatic plant from East Anatolia Region, Turkey. Asian J Chem 25(10):5490–5492

Wang T, Hicks KB, Moreau R (2002) Antioxidant activity of phytosterols, oryzanol, and other phytosterol conjugates. Journal of the American Oil Chemists' Society 79(12):1201–1206

Wienkötter N, Höpner D, Schütte U, Bauer K, Begrow F, El-Dakhakhny M, Verspohl EJ (2008). The effect of nigellone and thymoquinone on inhibiting trachea contraction and mucociliary clearance. Planta medica, 74(02):105-108

Woo CC, Loo SY, Gee V, Yap CW, Sethi G, Kumar AP, Tan KH (2011) Anticancer activity of thymoquinone in breast cancer cells: possible involvement of PPAR-γ pathway. Biochem Pharmacol 82(5):464–475

Yaman İ, Balikci E (2010) Protective effects of *Nigella sativa* against gentamicin-induced nephrotoxicity in rats. Exp Toxicol Pathol 62(2):183–190

Yarnell E, Abascal K (2011) *Nigella sativa*: holy herb of the middle East. Altern Comp Ther 17 (2):99–105

Ying Q, Wojciechowska P, Siger A, Kaczmarek A, Rudzinska M (2018) Phytochemical content, oxidative stability, and nutritional properties of unconventional cold-pressed edible oils. J Food Nutr Res 6:476–485

Chapter 14
Borage (*Borago officinalis*) Seed

Beenu Tanwar, Ankit Goyal, Vikas Kumar, Prasad Rasane, and Manvesh Kumar Sihag

Abstract Borage (*Borago officinalis* L.) seed is the richest source of γ-linolenic acid (GLA), which has been recognised to possess therapeutic potential against chronic inflammatory diseases such as atopic dermatitis (eczema), rheumatoid arthritis, cancer, atherosclerosis, etc. Under physiological conditions, the conversion of GLA into dihomo-γ-linolenic acid (DGLA) not only acts as the precursor of anti-inflammatory compounds but also inhibits the formation of pro-inflammatory eicosanoids from arachidonic acid (AA). Several animal and clinical trials have suggested the protective role of borage oil and GLA in the treatment of non-communicable diseases (NCDs) possibly due to their antioxidative and anti-inflammatory activities. The other parts of borage herb such as leaves and flowers are edible and consumed as raw and cooked in the form of salad and beverages. However, owing to the medicinal value of GLA, borage is commercially grown for its oil worldwide. In addition to GLA, borage oil contains minor quantity of tocopherols, phytosterols and squalene and, thus, used frequently in cosmetic products also.

Keywords Borage seed oil · *Borago officinalis* · Gamma-linolenic acid · GLA · Atopic dermatitis (eczema) · Starflower · Pyrrolizidine alkaloids · Rheumatoid arthritis

---

B. Tanwar
Department of Dairy Technology, Mansinhbhai Institute of Dairy and Food Technology, Mehsana, Gujarat, India

A. Goyal (✉)
Department of Dairy Chemistry, Mansinhbhai Institute of Dairy and Food Technology, Mehsana, Gujarat, India

V. Kumar · P. Rasane
Department of Food Science and Technology, College of Agriculture, Punjab Agricultural University, Ludhiana, Punjab, India

M. K. Sihag
Department of Dairy Chemistry, College of Dairy Science & Technology, Guru Angad Dev Veterinary and Animal Sciences University, Ludhiana, Punjab, India

B. Tanwar, A. Goyal (eds.), *Oilseeds: Health Attributes and Food Applications*,
https://doi.org/10.1007/978-981-15-4194-0_14

14.1 History and Introduction

Borage, also known as burrage, boretsch (Greek), bourrache (French), bugloss, borragine (Italy), bee bread, bee plant, burage, ox's tongue, cool tankard or star-flower, is botanically equivalent to *Borago officinalis* L. belonging to the family Boraginaceae (De Smet 1993; Osborne 1999). It is an annual herb believed to be originated in Aleppo (Syria) but now widely distributed in Europe, North Africa, minor Asia and several Mediterranean countries. The word 'borage' is thought to be originated from the Latin *borago*, which is a corruption of the word *corrago*, indicating *cor*, heart and *ago*, I lead, because of its 'cordial' effect (strengthening effect on the heart) (Mifsud 2004). The species *officinalis* is supposed to be derived from its 'official' use in pharmaceutical and healing applications (Mifsud 2004). In the past, this epithet was given to the plants with well-known medicinal properties. There are several other theories also for the origin of the name of borage. Italian *borra*, French *bourra* and Latin *burra* signify 'hair' or 'wool', as the plant is characterised to have numerous white, stiff and prickly hairs all over the stem and leaves (Fig. 14.1).

It is noteworthy that the family Boraginaceae comprises a total five species: *B. officinalis*, *B. trabutii*, *B. longifolia*, *B. pygmaea* and *B. morisiana* (Fabrikov et al. 2019). However, currently, the only species which is grown worldwide at commercial level is *B. officinalis*, and others are endemics and threatened for extinction (Selvi et al. 2006). In India, *Plectranthus amboinicus* [Lour.] Spreng, belonging to the family Lamiaceae, is known as Indian borage and/or country borage (Bhatt and Negi 2012). Borage plant is short in height (60–100 cm) having hollow and succulent stem, alternate wrinkled and pointy leaves and star-shaped flower with white, pink, purplish or most commonly blue-coloured petals (Fig. 14.1). The flower also contains sepals alternate to each petal which become dark red to brown in colour

Fig. 14.1 Borage (*Borago officinalis* L.) plant, flower and seeds

from green with time. After completion of the life span of the flower, petals shed off, and sepals close up to form a pot-shaped structure, covering and protecting the fruit inside. The fruit simply consists of four bare seeds known as nutlets (Mifsud 2004). When the nutlets are ripe, they become dark brown to black in colour and simply fall off without any mechanical dispersion.

Traditionally, borage has been grown for its therapeutic, nutritional and medicinal values. Its fresh leaves and flowers give cucumber-like flavour and are consumed raw and cooked in the form of salads as well as in the preparation of beverages. The borage leaves and flower were also used for the preparation of decoctions and tea by fusion, culinary and decorative purposes as well as in traditional medicines such as for the treatment of inflammation, cough, skin rashes, irritation, etc. (Roberts 2000; Rio-Celestino et al. 2008). The infusion or extract of fresh borage leaves has also been reported to show diuretic, diaphoretic, febrifuge (antipyretic), galactagogue (which promotes lactation), calmative, blood-purifying and vitalizing properties (De Smet 1993). Commercially, borage is cultivated worldwide for its oil content (30–40%), which has received considerable interest due to the presence of high amount of γ-linolenic acid (GLA) (18–30% of total oil) (Beaubaire and Simon 1987). GLA has shown antioxidative, anti-inflammatory, antiarthritic, antimutagenic, immunomodulatory and antiobesity activities in human and animals, which will be discussed in further sections. In addition to GLA, borage oil contains minor quantity of tocopherols, phytosterols and squalene and, thus, used frequently in cosmetic products also.

14.2 Market Status

As discussed previously, borage oil has nutritional, pharmaceutical as well as therapeutic potential owing to the presence of GLA (an essential omega-6 fatty acid), tocopherols and other nonsaponifiable phytonutrients. Owing to its bioactivity and functional properties, GLA has witnessed enormous demand in global market. According to a report, global borage oil market was valued to 33.91 million USD in 2015, which is expected to grow to 54.9 million USD in 2024 with a CAGR of 5.5% (Grand View Research 2016). The increased demand of natural bio-based cosmetic products, supplements and medicines is the key factor behind the massive growth of borage oil market worldwide. Currently, borage is cultivated throughout the world for its oil content. Europe, Canada and New Zealand are the main producers of borage oil probably due to the high altitude and favourable climatic conditions. It is reported that the Asia-Pacific region accounted approximately 30% of the total market demand of borage oil in 2015. The developing countries such as China, India and Indonesia are identified as a hub by major cosmetics and pharmaceutical industries for borage oil production. The oil's quality is directly associated to the GLA content, which ranged from 18 to 35% depending on the growing conditions. The quality of borage seeds grown in Central and South America is comparatively low due to low GLA content (<20%), which could be due to less favourable agroclimatic conditions. Europe has been witnessed to provide highest quality of borage oil (>25% GLA) and, therefore, has huge demand by leading cosmetic and food supplement industries such as L'Oréal and Unilever (Grand View Research 2016).

14.3 Proximate Composition

Till date, no literature is available on the proximate composition of borage seed except for the oil content. It has been reported that borage seeds contain approximately 30–40% oil (Gomez and de la Ossa 2002; Namal Senanayake and Shahidi 2000). It is noteworthy that being rich in antioxidants and phytonutrients, borage's leaves and flowers (petals) are widely used for edible purpose in raw and/or cooked form. Therefore, several researchers have reported the proximate composition of borage leaves (USDA 2019; Pereira et al. 2011; Pilerood and Prakash 2014), which is presented in Table 14.1. Here, the constituents and nutrients of borage leaf and/or flower will not be discussed as these do not lie within the scope of the chapter. To study the same, readers can refer the paper authored by Pereira et al. (2011).

14.4 Lipids

Borage seed oil is a yellow-coloured oil comprising triglycerides majorly of oleic acid (18–20%), linoleic acid (LA, 37.00–39.57%) and γ-linolenic acid (GLA: 21–23%) (De Smet 1993; Gomez and de la Ossa 2002; Mhamdi et al. 2009; Tasset-Cuevas et al. 2013; Szterk et al. 2010). The molecular structures of LA, α-linolenic acid (ALA) and GLA are shown in Fig. 14.2. On average, borage seed oil contains approximately 15–16% and 84–85% saturated and unsaturated fatty acids, respectively. Along with borage seed oil, GLA (omega-6) is also found in some

Table 14.1 Proximate composition of borage leaves

	Amount (g/100 g)		
Parameter	(on wet basis)[a]	(On dry basis)[b]	(On dry basis)[c]
Total lipids	0.70	1.25	1.20
Crude protein	1.80	8.93	7.30
Carbohydrates	3.06	71.94	NR
– Soluble fibre	NR	NR	45.00
– Insoluble fibre	NR	NR	38.00
Ash	1.44	17.88	NR

NR not reported
[a]USDA (2019)
[b]Pereira et al. (2011)
[c]Pilerood and Prakash (2014)

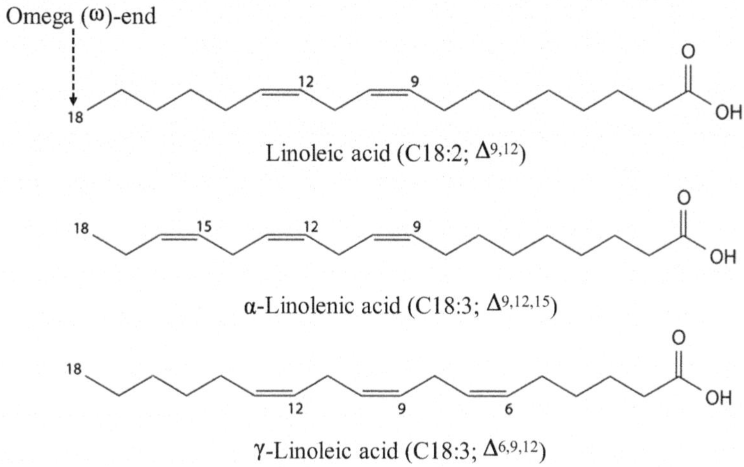

Fig. 14.2 Molecular structures of linoleic acid (LA), α-linolenic acid (ALA) and γ-linolenic acid (GLA)

specialty oil such as evening primrose oil (10–11%) and black currant oil (12.9–17.0%) (Sergeant et al. 2016; Sovova et al. 2001; Dubois et al. 2007).

Though the content of GLA in evening primrose oil (10–11%) is approximately half to that of borage seed oil (21–23%) (Redden et al. 1995), majority of the cosmetic applications use evening primrose oil as a source of GLA (Grand View Research 2016). However, contradictory results have been reported by Szterk et al. (2010), who evaluated the fatty acid composition of selected plant oils and reported 23.1% and 0.1% GLA in crude borage and evening primrose oil, respectively. Namal Senanayake and Shahidi (2000) estimated the lipid profile of borage seeds and reported triglycerides, diglycerides, monoglycerides, free fatty acids and sterols to the value of 99.1%, 0.06%, 0.02%, 0.91% and 0.02% of total lipids. The majority of the lipids were neutral lipids (95.7 ± 1.2%) followed by phospholipids (2.3 ± 1.7%) and glycolipids (2.0 ± 1.5%) (Namal Senanayake and Shahidi 2000). Fatty acid profile of borage seed oil reported by different researchers is presented in Table 14.2.

14.5 Phytochemicals and Minor Components

Apart from the GLA, borage seed oil contains sufficient amount of bioactive components such as tocopherols, phytosterols, squalene, phospholipids, phenolic acids, etc. Along with the free fatty acids and/or other components which reduce the shelf life of the oil, the common vegetable oil-refining process causes the loss of bioactive components also. As a few speciality oils such as borage seed oil, evening primrose oil and/or black currant oil are not used as conventional/cooking oils, they

Table 14.2 Fatty acid composition of borage seed oil (% of total fatty acids)

	Amount (% of total fatty acids)				
Fatty acid	a	b	c	d	e
Palmitic (C16:0)	13.28–13.32	10.20	10.70	5.7	10.4–10.9
Palmitoleic (C16:1)	0.18–0.19	0.10	NR	NR	0–0.2
Stearic (C18:0)	4.58–5.06	5.60	6.40	1.4	4.6–4.9
Oleic (C18:1)	19.78–20.77	24.20	18.50	16.1	15.1–17.8
Linoleic (C18:2)	37.00–39.57	35.40	36.60	46.3	34.0–37.2
α-Linolenic acid (C18:3; $\Delta^{9,12,15}$)	0.27–0.65	0.60	NR	0.1	0–2.6
γ-Linolenic acid (C18:3; $\Delta^{6,9,12}$)	21.04–22.29	20.40	21.10	23.1	20.4–21.9
Arachidic (C20:0)	NR	0.20	NR	0.1	NR
Eicosanoic (C20:1)	NR	3.30	4.20	3.6	4.1–4.4
Erucic acid (C22:1)	NR	NR	2.30	2.0	2.4–2.7
Total SFAs	NR	16.00	NR	NR	15.1–15.8
Total UFAs	NR	84.00	NR	NR	83.4–83.8

SFAs saturated fatty acids, *UFAs* unsaturated fatty acids, *NR* not reported
[a]Gomez and de la Ossa (2002); [b] Mhamdi et al. (2009); [c] Tasset-Cuevas et al. (2013); [d] Szterk et al. (2010) (in crude oil); [e]Namal Senanayake and Shahidi (2000)

are usually extracted using cold-pressed methods (Czaplicki et al. 2011), in order to retain maximum bioactive components in oils.

14.6 Tocopherols

Tocopherols are natural antioxidants present in all the vegetable oils. Czaplicki et al. (2011) characterised the bioactive components in various nonconventional oils and reported the highest amount of tocopherols (1603 mg/kg of oil) in borage oil. It is noteworthy that α- and β-tocopherols have not been detected in borage oil (Czaplicki et al. 2011; Fabrikov et al. 2019), whereas δ-tocopherol has been observed at a very high concentration, when compared to other nonconventional oils (Czaplicki et al. 2011). The content of different tocopherols in borage oil is represented in Table 14.3. Similar results have been observed by Szterk et al. (2010), who analyzed the chemical composition of selected plant oils (rapeseed oil, linseed oils, crude *Camelina sativa* oil, crude primrose oil, crude borage oil, pumpkin seed oil and crude amaranth oil) and detected highest level of tocopherols (including δ-tocopherol) in borage oil. Tocopherols show antioxidant activity by free radical mechanism. Mortensen and Skibsted (1997) demonstrated that among all the tocopherols, δ-tocopherol shows the highest activity against free radicals, which could be the possible reason behind the highest oxidative stability observed for borage oil among different unconventional plant seed oils (Czaplicki et al. 2011; Szterk et al. 2010). Recently, Fabrikov et al. (2019) evaluated the phytochemicals in farmed (*Borago officinalis* L.) and endemic-wild *Borago* species (*Borago officinalis*

Table 14.3 Minor/bioactive components in borage seed oil

Component	Czaplicki et al. (2011)	Fabrikov et al. (2019)	Szterk et al. (2010)
Squalene (mg/g oil)	0.22	0.19	NR
Tocopherols (mg/kg oil)			
– α-Tocopherol	ND	ND	ND
– β-Tocopherol	ND	ND	–
– γ-Tocopherol	171 ± 15.4	331.2	200.0[a]
– δ-Tocopherol	1432 ± 119	2068.8	3000.0
Phytosterols (mg/100 g oil)	196 ± 11.6	NR	365.5
– 24-Methylcholesta-5,23-dienol	21 ± 1.1	NR	NR
– Cholesterol	ND	NR	ND
– Stigmasterol	ND	NR	NR
– Campesterol	43.8 ± 2.1	140.2	82.6
– β-Sitosterol	56.6 ± 2.3	62.3	80.7
– Δ5-Avenasterol	43.6 ± 4.9	151.5	87.1
– Gramisterol	11.2 ± 0.1	NR	NR
– Citrostadienol	3.2 ± 0.4	NR	16.5
– Cycloartenol	17.2 ± 0.7	NR	35.2

NR not reported, *ND* not detected
[a]Indicates β + γ-tocopherol

L., *Borago morisiana Bigazzi* et Ricceri, *Borago pygmaea* (DC.) Chater & Greuter, *Borago longifolia* Poir, *Borago trabutii* Maire) and observed the highest amount of tocopherols in *B. pygmaea* and *B. morisiana* with 514 and 296 mg/100 g oil, respectively.

14.7 Phytosterols

Phytosterols are a class of natural, cholesterol-like compounds present in most of the crude vegetable oils. Phytosterols have been reported to inhibit dietary cholesterol absorption through the gut, thus reducing the risk of atherosclerosis and heart diseases. Borage oil contains total phytosterols in the range of 196.0–365.5 mg/100 g oil (Czaplicki et al. 2011; Szterk et al. 2010). The major phytosterols are campesterol (43.8–140.2 mg/100 g), β-sitosterol (56.6–80.7 mg/100 g oil) and Δ5-avenasterol (43.6–87.1 mg/100 g oil). Czaplicki et al. (2011) studied the detailed profile of phytosterols in borage oil and detected small fraction of gramisterol, citrostadienol and cycloartenol also (Table 14.3). Recently, Fabrikov et al. (2019) detected higher amount of squalene in wild variety of *Borago officinalis* L. (27.4 ± 0.6 mg/100 g oil) than that of farmed variety (19.7 ± 0.3 mg/100 g oil). When crude borage oil was compared with crude primrose oil, the former showed approximately half (365.5 mg/100 g oil) the content of phytosterols present in the latter (720.4 mg/100 g oil) (Szterk et al. 2010).

14.8 Phenolic Acids

Phenolic acids are the secondary metabolites in plants, which have shown antioxidative, antimutagenic, anti-inflammatory, antibacterial and antiviral activities (Mhamdi et al. 2009; Robbins 2003; Zadernowski et al. 2002; Godos et al. 2017). Wettasinghe et al. (2001) evaluated the phenolic acids in borage seed meal using thin-layer chromatography and observed 0.383, 0.113 and 0.121% rosmarinic acid, syringic acid and sinapic acid, respectively. Rosmarinic acid is a phenolic compound and ester of caffeic acid. It is commonly found in species of the Boraginaceae and Lamiaceae. Rosmarinic acid has been reported to show antiallergic, antioxidative and anti-inflammatory activities (Rocha et al. 2015; Osakabe et al. 2004). On the other hand, Zadernowski et al. (2002) reported hydroxycaffeic acid as a major phenolic acid (65.43 ± 3.95) in borage seed meal with 185.10 ± 5.85 mg total phenolic acids per kg. In another study, Mhamdi et al. (2009) reported total phenolic compounds to the value of 6.39 mg GAE (gallic acid equivalent)/g dry weight of defatted borage seed extract. The major phenolic acids reported by several authors are presented in Table 14.4.

Table 14.4 Major phenolic acids and their content in whole borage seed and defatted meal

	Amount		
Phenolic acid	g/100 g (in defatted meal)[a]	mg/100 g dry weight (in whole seed)[b]	mg/100 g (In defatted meal)[c]
Rosmarinic acid	0.383	165.64	NR
Syringic acid	0.113	23.93	NR
Sinapic acid	0.121	NR	0.23
Gallic acid	NR	22.07	0.51
p-Coumaric acid	NR	91.17	0.16
Trans-cinnamic acid	NR	49.64	NR
Chlorogenic acid	NR	54.26	NR
p-hydroxybenzoic acid	NR	NR	1.45
Hydroxycaffeic acid	NR	NR	6.54
Ferulic acid	NR	NR	5.08

NR not reported
[a]Wettasinghe et al. (2001)
[b]Mhamdi et al. (2009)
[c]Zadernowski et al. (2002)

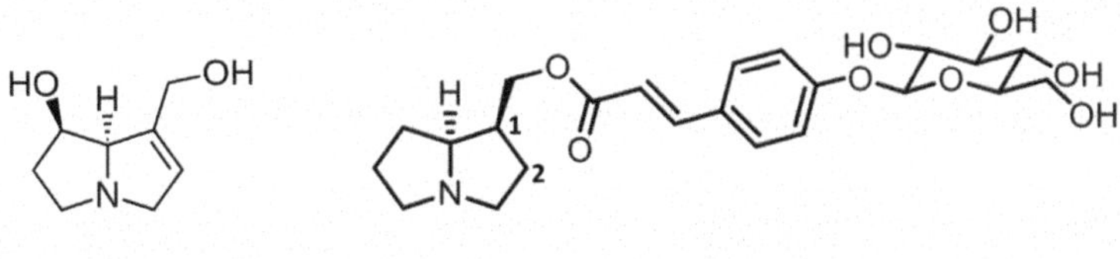

Fig. 14.3 Structure of necine base (**a**) and (+)-thesinine-4′-O-β-D-glucoside (**b**), the major alkaloid in borage seed

14.9 Antinutrients/Toxic Components

The members of Boraginaceae family are well known for their content of pyrrolizidine alkaloids (PAs), which are plants' secondary metabolites (Herrmann et al. 2002). Dodson and Stermitz (1986) have reported 0.03% crude alkaloids with thesinine [(+)-thesinine-4′-O-β-D-glucoside] (0.02%) and amabiline (0.002%) as major and minor alkaloids, respectively. The researchers also analyzed the borage seed oil samples, but no alkaloids were detected at any concentration when estimated by TLC and/or capillary GLC technique (Dodson and Stermitz 1986). The toxic effects of PAs depend on the structure and/or position of unsaturation. PAs having one double bond at 1–2 position in necine base are considered as hepatotoxic and carcinogenic (Tamariz et al. 2018). Thesinine is a saturated alkaloid (Fig. 14.3) and is not known to cause any toxic effect till date (Tamariz et al. 2018).

In another report, borage seed oil has been reported to contain unsaturated PAs down to the level of 5 μg/g (De Smet 1993). As we know that borage seed oil is used as a dietary supplement and/or nutraceutical, it means if consumers take two to four capsules per day containing 500 mg borage oil per capsule, they will be consuming 5–10 μg unsaturated PAs per day. However, as per the German Federal Health Agency, the daily intake of unsaturated PAs should not be more than 1 μg/day (De Smet 1993; Wretensjö and Karlberg 2003). Wretensjö and Karlberg (2003) evaluated the crude and refined borage oil for the presence of pyrrolizidine alkaloid content using GC-MS. In the study, no PAs were detected in borage seed oil above detection limit of 20 ppb. They also observed that pyrrolizidine content in crude borage oil was reduced overall by a factor of about 30,000 in the refining process (Wretensjö and Karlberg 2003). Recently, Vacillotto et al. (2013) examined the PAs level in *Borago officinalis* seed oil by advanced electrospray ionization method and reported PAs level lower than 200 ppt, if present. Overall, it can be concluded that refined borage oil seems to be safe to consume under proper supervision (FDA 2016). It should be noted that the antinutrients and PAs present in borage leaves and flowers are not discussed here and due consideration should be given before their consumption. The alkaloids present in borage leaves and flowers and their toxicity concerns have been discussed by Larson et al. (1984), De Smet (1993) and Herrmann et al. (2002) in detail.

14.10 Health Effects

14.10.1 Anti-inflammatory Effects

Borage oil has gained considerable interest owing to the presence of GLA, which has been associated to attenuate the signs and symptoms of chronic inflammation. Several clinical and animal studies have confirmed the anti-inflammatory effects of GLA metabolites. GLA, di-homo γ-linolenic acid (DGLA) and arachidonic acid (AA) are the intermediates of the metabolism of dietary linoleic acid (C18:2; $\Delta^{9,12}$) which is an essential fatty acid. The metabolism of dietary LA and the production of eicosanoids compounds are presented in Fig. 14.4. In the presence of Δ6-desaturase and elongase, dietary LA is converted into GLA and DGLA, respectively. A large number of in vivo and in vitro studies have demonstrated that human neutrophils have elongase but not Δ5-desaturase activity (Chapkin and Coble 1991; Navarette et al. 1992; Ziboh et al. 2000). Furthermore, it is worthy to note that the skin, murine peritoneal and platelets also appear to have high elongase activity than Δ5-desaturase activity; resulting in more synthesis of DGLA as compared to AA (Luthria and Sprecher 1994; Pawlosky et al. 1994; Sergeant et al. 2016). In normal situation, where Δ5-desaturase and elongase both have equal activities (e.g. in the liver, kidney, testes, brain, intestine), DGLA is further converted into arachidonic acid (AA) in the presence of cyclooxygenase and lipoxygenase and produces 2-series prostaglandins (PGE2) and thromboxanes (TX2) and 4-series leukotrienes (LT), respectively. These two-series eicosanoid components are potent pro-inflammatory and known to increase the risk of chronic inflammation, mutagenic activity, platelet aggregation, rheumatoid arthritis, atherosclerosis, obesity, etc. (Fan and Chapkin 1998; Kapoor and Huang 2006; Sergeant et al. 2016).

Contrarily, DGLA, in the presence of the same enzymes (cyclooxygenase and lipoxygenase), produces one-series eicosanoid compounds such as prostaglandins (PGE1), thromoxanes and 15-hydroxydihomo-γ-linolenic acid (Wang et al. 2012; Sergeant et al. 2016), which act as potential anti-inflammatory compounds. It is supposed that DGLA competes with AA for the same enzymes, i.e. cyclooxygenase and lipoxygenases, and favours lesser production of pro-inflammatory AA metabolites and more production of one-series PGE1 and PGH1 molecules (Belch and Hill 2000; Wang et al. 2012) (Fig. 14.4). It is interesting to note that DGLA itself cannot be converted into LTs but can form 15-hydroxy-derivatives that supress the formation of LTs (LT4) from AA (Belch and Hill 2000). One-series metabolites of DGLA have been reported to supress inflammation, promote vasodilation, lower blood pressure, inhibit smooth muscle cell proliferation and exert antineoplastic activities (Sergeant et al. 2016; Wang et al. 2012; Wang et al. 2015). The metabolism of GLA and the detailed mechanism of pro- and anti-inflammatory compounds have been explained by Sergeant et al. (2016).

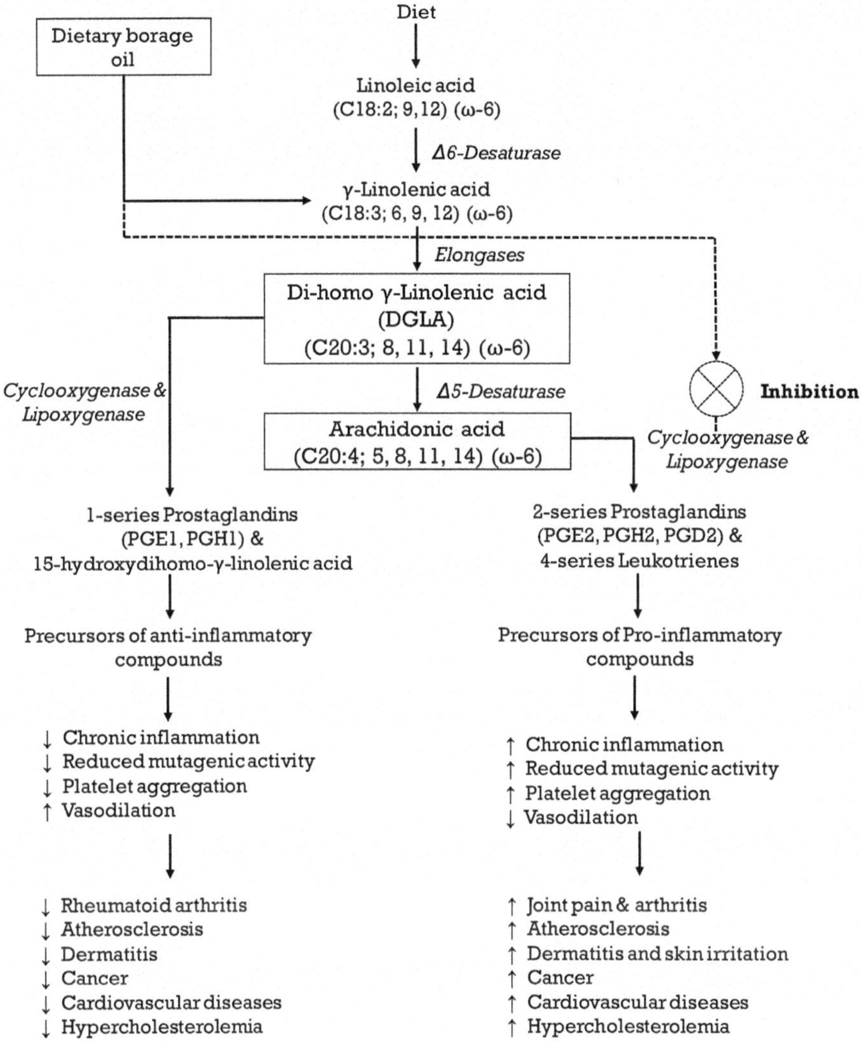

Fig. 14.4 Conversion of dietary linoleic acid (LA) to arachidonic acid (AA), synthesis of eicosanoid compounds and their health effects

14.10.2 Anticarcinogenic Activity

Cancer is the second leading cause of death globally and is responsible for an estimated 9.6 million deaths in 2018 (WHO 2018). Borage oil has been reported to exert anticarcinogenic effects in several in vivo and in vitro animal models (Jiang et al. 2000; Zhang et al. 2015; Wauquier et al. 2012). A few studies have been conducted to know the effect of dietary supplementation of borage oil and/or GLA only, whereas most of the studies have been conducted with GLA in addition to omega-3-rich oils (aicosapentaenoic acid, EPA and docosahexaenoic acid, DHA) (Chas et al. 2019; Wauquier et al. 2012; Camargo et al. 2018). The anticarcinogenic activity of GLA could be attributed to the increased generation of reactive oxygen species (ROS), accumulation of intracellular Ca^{2+}, decrease in mitochondrial membrane potential and ATP level which further causes the dysfunctioning of mitochondrial membrane of tumour cells and upregulation of gene expression for antioxidant proteins (Zhang et al. 2015; Tasset-Cuevas et al. 2013). Another possible reason could be ascribed to the enhanced activation of caspases, which are cysteine–aspartic proteases or a family of protease enzymes and essential for apoptosis (Zhang et al. 2015). Tasset-Cuevas et al. (2013) conducted an interesting study on the toxicity, genotoxicity, antigenotoxicity and cytotoxicity effects of borage seed oil and GLA individually. In the study, the life span of *Drosophila melanogaster* as affected by low doses of GLA and borage seed oil was also evaluated. The results demonstrated that borage seed oil and GLA both were DNA safe (non-genotoxic) and antimutagenic. Borage oil was non-toxic to *D. melanogaster* at lower concentrations (<125 μl/ml) and increased the health span portion $\geq 75\%$ of the life span curves. However, GLA did not show any positive effects on health span of *D. melanogaster*. It should also be noted that borage seed oil and GLA both showed cytotoxic activity at low doses (IC50 of 1 μl/ml and 0.087 mM, respectively) (Tasset-Cuevas et al. 2013). In another study, Jiang et al. (2000) studied the effect of GLA on breast cancer cell lines (MCF-7 and MDA MB 231) and observed reduced growth of cancerous cells due to decreased expression of peroxisome proliferator activator receptors (PPARs). PPARs are key components of cellular responses to PUFAs. Normally, cancerous cells express PPARs, particularly PPAR-γ. On interacting with GLA, PPAR-γ gets phosphorylated and translocated from cytoplasm to nucleus, and thus the growth of breast cancer cells is inhibited (Jiang et al. 2000). Zhang et al. (2015) studied the effect of omega-3 fatty acids (ALA, EPA and DHA) and omega-6 fatty acids (GLA and AA) on colon cancer cells LoVo and RKO. In the study, it was observed that all the PUFAs (at concentrations above 120 μM) significantly supressed the growth of cancer cells probably through the inhibition of ATP synthesis and dysfunctioning of mitochondria.

14.10.3 In Atopic Dermatitis (Eczema)

The GLA supplementation was first introduced to attenuate the signs and symptoms of atopic dermatitis preceded by the oral dose of linoleic acid-rich oils (Sergeant et al. 2016). Atopic dermatitis is a chronic inflammatory skin condition characterised by red, itchy rashes normally on the cheeks, arms and legs. An estimated 10% of all people worldwide are affected by atopic dermatitis at some point in their life (National Eczema Association 2019). A growing literature has demonstrated the beneficial effects of dietary GLA or GLA-rich oils on relieving the symptoms of chronic dermatitis (Kawamura et al. 2011; Chung et al. 2002; Jung et al. 2014; Fujii et al. 2013; Foster et al. 2010; Brosche and Platt 2000; Simon et al. 2014; Tanaka et al. 2015). One-series prostaglandins (PGE1) and 15-hydroxy-dihomo-γ-linolenic acid derived from GLA are known to inhibit pro-inflammatory four-series thromboxane (TX4) and exert the anti-proliferative effect on cells, which is further associated with increased ceramide synthesis and improved barrier function of epidermis (Chung et al. 2002). Kawamura et al. (2011) studied the effect of dietary supplementation of GLA (extracted from fungus *Mucor circinelloides*, 28.7% GLA) on patients ($n = 130$) with dry skin and mild-to-moderate atopic dermatitis. In the study, results showed that transepidermal water loss (TEWL) (which remains high in dry skin condition) reduced from -1.47 ± 5.25 to -3.24 ± 5.44 and -0.90 ± 3.23 to -2.03 ± 3.77 g/m^2/h in cheeks and forearm skin, respectively, and, thus, improved skin barrier function and reduced epidermal hyperproliferation. In GLA-supplemented group, authors associated improved skin barrier functions with anti-inflammatory compounds generated through GLA metabolism. Similarly, Brosche and Platt (2000) studied the effect of GLA supplementation (360–720 mg GLA/day for 2 months) in elderly people ($n = 13$) and reported the changes in fatty acid metabolism and a significant improvement in TEWL.

In a randomised, double-blind, controlled trial, Jung et al. (2014) studied the effect of omega-3 fatty acids (EPA + DHA, 2000 mg/day) and GLA (400 mg) (in the form of borage oil) supplementation in 45 patients having mild-to-moderate acne (acne vulgaris). The authors observed a significant decrease in inflammatory and noninflammatory acne lesions and interleukin-8 (IL-8) after 10 weeks. Interleukin-8 is a chemotactic and inflammatory cytokine associated with epidermal hyperplasia, follicular hyperkeratosis and acne inflammation (Abd El All et al. 2007). Contrary to the previous positive results, Henz et al. (1999) conducted a double-blind study over 24 weeks to test the effectiveness of borage oil in patients ($n = 160$) with moderate atopic eczema and found highly disappointing results. In the study, patients consumed three borage oil capsules (500 mg oil/capsule) twice daily. Though there was a significant increase in GLA metabolites and decrease in IgE levels in serum of the borage oil group, overall response could not reach to the statistical significance (Henz et al. 1999). Similarly, in an intervention review written by Bamford et al. (2013), 27 studies (1596 participants) were assessed for the effect of oral supplementation of evening primrose and borage oil and found no significant ($P < 0.05$) improvement in the sign and symptoms of eczema. However, it is also important to note that several clinical studies involving healthy and health-compromised (atopic dermatitis, asthma and atopic eczema) infants (<12 months), children (8–26 months) and adults consuming 0.03–6.00 g GLA/day for 6–18 months did not show any adverse effects in subjects, which have been compiled and reviewed by the FDA (2016).

14.10.4 In Rheumatoid Arthritis (RA)

Rheumatoid arthritis is an autoimmune chronic inflammatory disease characterised by long-lasting pain and swelling in joints, connective tissues, muscle, tendons and fibrous tissue. It is surprising to note that the people who have obesity, heart diseases, diabetes or other chronic diseases are at a higher risk of developing arthritis. In RA, the inflammation is mainly caused by the overproduction of AA metabolites such as PGE2, PGH2, interleukins-2, TNF-α, cytokines and other pro-inflammatory compounds. The overproduction of these pro-inflammatory eicosanoids could be due to excess consumption of LA-rich oils (regular vegetable oils), suboptimal diet including processed foods and *trans*-fats, stress, inadequate physical activity, etc. A number of clinical and animal studies have demonstrated that regular consumption of omega-3 fatty acid-rich oils (marine and fish, flaxseed, chia seed and walnut) and/or GLA-rich oils (borage, evening primrose, black currant) decreases the prevalence of developing rheumatoid arthritis by producing anti-inflammatory compounds (Dawczynski et al. 2011; Kast 2001; Reed et al. 2014; Belch and Hill 2000; Veselinovic et al. 2017; Vasiljevic et al. 2016; Labrousse et al. 2018; Lindqvist et al. 2019). Dawczynski et al. (2011) conducted a double-blind, placebo-controlled parallel designed trial in which the patients having rheumatoid ($n = 54$) and/or psoriatic ($n = 6$) arthritis were divided into four groups and given omega-3 LC-PUFA (3 g) (group 1), 3 g GLA (group 2), 1.6 g omega-3 + 1.8 g GLA (group 3) or 3 g olive oil (group 4: control) per day in the form of capsules for 12 weeks. Results displayed no significant ($P < 0.05$) change in AA/EPA ratio in groups 2–4. However, group 1 showed a considerable ($p \leq 0.001$) decrease in AA/EPA ratio from 6.5 ± 3.7 to 2.7 ± 2.1 in plasma lipids and from 25.1 ± 10.1 to 7.2 ± 4.7 in erythrocyte membranes, indicating a positive effect in arthritis patients. Nonetheless, the other groups (2–4) exhibited no significant ($P < 0.05$) increase in pro-inflammatory AA concentration in plasma lipids, cholesteryl esters and erythrocyte membranes, suggesting a possible positive effect PUFAs in chronic inflammatory diseases. Kast (2001) suggested the possible mechanism behind the reduction of RA by GLA and/or borage oil. The dietary intake of GLA increases the PE1 levels in blood, which in turn increases cAMP levels. cAMP is inversely correlated with TNFα, a potent inflammatory metabolite (Kast 2001). In an 18-month randomised and double-blind trial, the efficacy of fish oil, borage oil and combination of both was compared in RA patients. Though all the groups significantly decreased the disease activity in patients, there was no significant difference in changes in diseases activity among all the three groups (Reed et al. 2014). To study the effects of different drugs or nutraceuticals on RA, there are several tests that should be examined such as Disease Activity Score 28 (DAS28), visual analogue scale (VAS) (pain), erythrocyte sedimentation rate (ESR), C-reactive protein (CRP) and other pro-inflammatory biomarkers. The DAS28 is a measure of disease activity in RA, where 28 refers to the '28 joints' that are examined in this assessment (Vander Cruyssen et al. 2005). On the other hand, visual analogue scale is a unidimensional measure of pain intensity in which

0 indicates 'no pain' and 100 indicates 'worst imaginable pain' on a 100 mm scale (Jensen et al. 1986). In a recent study, Veselinovic et al. (2017) evaluated the effect of supplementation of fish oil alone and in combination with evening primrose oil in RA patients and observed a significant ($P < 0.001$) decrease in Disease Activity Score 28 (DAS28) from 4.76–4.99 to 3.79–3.91 and the number of tender joints and visual analogue scale (VAS) score from 55.7–59.0 to 46.7–50.5 after 12 weeks in both groups.

14.10.5 In Neurodevelopment and Obesity

Long-chain PUFAs such as DHA and AA are critically important for the optimal development of brain, visual and cognitive functions (Belkind-Gerson et al. 2008; Ryan et al. 2010; Wang et al. 2016). As the gestation progresses particularly during the last trimester, DHA and AA accrete in the brain and continue during the first postnatal year in human and animal infants (Makrides et al. 2010). Several clinical manifestations such as dyspraxia, dyslexia, autism or other cognitive disorders have been associated with the deficiency of LC-PUFAs (Wang et al. 2016; Belkind-Gerson et al. 2008). However, numerous researchers have reported that such clinical manifestations could be relieved to some extent with the supplementation of long-chain omega-3 fatty acids during early life (Ward 2000; Makrides et al. 2009; Ryan et al. 2010). Fewtrell et al. (2004) conducted a randomised, double-blind trial on preterm (<35 weeks) infants ($n = 238$) to test the efficacy and safety of LC-PUFAs (EPA, DHA and GLA)-supplemented formula for 9 months after birth. In the study, no significant different was observed in neurodevelopment between the supplemented group and control. However, LC-PUFAs-supplemented group showed significantly higher Bayley Mental Developmental Index (MDI), Motor (Psychomotor) Developmental Index (PDI), greater weight and length gain than that of control group. MDI and PDI are the Bayley Scales of Infant Development (BSID), which are widely used to monitor neurodevelopmental outcomes in young children up to 3 years of age (Bos 2013).

A few researchers have linked elevated levels of inflammatory biomarkers such as C-reactive protein (CRP) with increased prevalence of obesity (Visser et al. 1999; Schirmer and Phinney 2007; Horvei et al. 2016; Ellulu et al. 2017). Schirmer and Phinney (2007) evaluated the efficacy of GLA (5 g/day borage oil equivalent to 890 mg GLA/day) in obese people ($n = 50$) and observed no significant change in weight gain and fat regain in borage oil group after 1 year when compared with control (olive oil group) and concluded that GLA reduced weight gain followed by major weight loss in obese people. The clear mechanism behind the suppression of weight gain by GLA is not clear, but it could possibly be due to enhanced arachidonate synthesis, which in turn might improve peripheral glucose disposal via increased insulin sensitivity, downregulation of lipogenesis, increased lipid oxidation, and enhanced leptin secretion (Schirmer and Phinney 2007). In other studies, conducted on Zucker obese rats, supplementation of GLA supressed food intake and weight gain in the obese animals, whereas no effect was observed in lean genotypes (Phinney et al. 1993; Thurmond et al. 1993). Earlier, Takahashi et al. (2000) studied the effect of dietary GLA (from borage oil) on body fat accumulation in male Sprague-Dawley rats. It was observed that in spite of having high fat (20%), GLA-rich diet enhanced the gene expression for lipoprotein lipase and hepatic mitochondrial fatty acid oxidation rate, which in turn decreased the body fat mass in animals when compared to control having low-fat diet (2% safflower oil).

14.11 Food Applications

A large number of studies are available on the health effects of borage oil and GLA. However, no literature is available wherein borage oil as such is supplemented or fortified in food products in order to improve their physico-chemical, techno-functional and/or sensory properties. The other parts of borage herb such as leaves and flowers are edible, rich in antioxidants and consumed raw and/or cooked in the form of salads and beverages. Several researchers have worked on the addition of borage leaves' extract in meat (Miceli et al. 2014; Gupta and Negi 2016), fish and vegetables (Miceli et al. 2014) and pasta (Miceli et al. 2015) owing to its antioxidative and antimicrobial properties. Globally, borage oil is available as dietary supplements in the form of capsules and microencapsulated powder (dietary supplements or nutraceuticals). In a study, Puch et al. (2008) supplemented borage oil (300 mg GLA) along with green tea (47 mg catechins) and vitamin E (2 mg) in a fermented milk beverage, which was consumed by healthy women ($n = 36$) for 6 months. Data showed a significant increase in stratum corneum barrier function and decrease in transepidermal water loss in test group than that of control.

14.12 Conclusion

Borage oil, an important plant source of γ-linolenic acid, is recognised for its anti-inflammatory effects, which, in turn, improve the skin health and reduce the sign and symptoms of atopic dermatitis and other chronic inflammatory diseases. In some areas (such as anticarcinogenic, eczema, rheumatoid arthritis), the mechanisms for the positive effects are well-understood. However, in other areas (like obesity, respiratory problems, etc.), more clinical studies are required to establish the clear molecular mechanisms and safety levels of borage oil for human consumption.

References

Abd El All HS, Shoukry NS, El Maged RA, Ayada MM (2007) Immunohistochemical expression of interleukin 8 in skin biopsies from patients with inflammatory acne vulgaris. Diagn Pathol 2:1–6. https://doi.org/10.1186/1746-1596-2-4

Bamford JT, Ray S, Musekiwa A, Van Gool C, Humphreys R, Ernst E (2013) Oral evening primrose oil and borage oil for eczema. Cochrane Database Syst Rev 4:1–91. https://doi.org/10.1002/14651858.CD004416.pub2

Beaubaire N, Simon JE (1987) Production of *Borago officinalis* L. Acta Hortic 208:101–114. https://doi.org/10.17660/ActaHortic.1987.208.12

Belch JJ, Hill A (2000) Evening primrose oil and borage oil in rheumatologic conditions. Am J Clin Nutr 71(1):352s–356s. https://doi.org/10.1093/ajcn/71.1.352s

Belkind-Gerson J, Carreón-Rodríguez A, Contreras-Ochoa CO, Estrada-Mondaca S, Parra-Cabrera MS (2008) Fatty acids and neurodevelopment. J Pediatr Gastroenterol Nutr 1(47):S7–S9

Bhatt P, Negi PS (2012) Antioxidant and antibacterial activities in the leaf extracts of Indian borage (*Plectranthus amboinicus*). Food Nutr Sci 3:146–152. https://doi.org/10.4236/fns.2012.32022

Bos AF (2013) Bayley-II or Bayley-III: what do the scores tell us? Dev Med Child Neurol 55 (11):978–979

Brosche T, Platt D (2000) Effect of borage oil consumption on fatty acid metabolism, transepidermal water loss and skin parameters in elderly people. Arch Gerontol Geriatr 30:139–150

Camargo CQ, Brunetta HS, Nunes EA (2018) Effects of cotreatment with omega-3 polyunsaturated fatty acids and anticancer agents on oxidative stress parameters: a systematic review of in vitro, animal, and human studies. Nutr Rev 76(10):765–777

Chapkin RS, Coble KJ (1991) Utilization of gamma-linolenic acid by mouse peritoneal macrophages. Biochim Biophys Acta 1085:365–370

Chas M, Goupille C, Arbion F, Bougnoux P, Pinault M, Jourdan ML, Chevalier S, Ouldamer L (2019) Low eicosapentaenoic acid and gamma-linolenic acid levels in breast adipose tissue are associated with inflammatory breast cancer. Breast 45:113–117

Chung S, Kong S, Seong K, Cho Y (2002) Gamma-linolenic acid in borage oil reverses epidermal hyperproliferation in guinea pigs. J Nutr 132:3090–3097

Czaplicki S, Ogrodowska D, Derewiaka D, Tańska M, Zadernowski R (2011) Bioactive compounds in unsaponifiable fraction of oils from unconventional sources. Eur J Lipid Sci Technol 113(12):1456–1464

Dawczynski C, Hackermeier U, Viehweger M, Stange R, Springer M, Jahreis G (2011) Incorporation of n-3 PUFA and γ-linolenic acid in blood lipids and red blood cell lipids together with their influence on disease activity in patients with chronic inflammatory arthritis-a randomized controlled human intervention trial. Lipids Health Dis 10(1):1–12. https://doi.org/10.1186/1476-511X-10-130

De Smet PAGM (1993) *Borago officinalis*. In: Adverse effects of herbal drugs 2. Springer, Berlin, Heidelberg, pp 147–152

Dodson CD, Stermitz FR (1986) Pyrrolizidine alkaloids from borage (*Borago officinalis*) seeds and flowers. J Nat Prod 49(4):727–728

Dubois V, Breton S, Linder M, Fanni J, Parmentier M (2007) Fatty acid profiles of 80 vegetable oils with regard to their nutritional potential. Eur J Lipid Sci Technol 109(7):710–732

Ellulu MS, Patimah I, Khaza'ai H, Rahmat A, Abed Y (2017) Obesity and inflammation: the linking mechanism and the complications. Arch Med Sci 13(4):851–863

Fabrikov D, Guil-Guerrero JL, González-Fernández MJ, Rodríguez-García I, Gómez-Mercado F, Urrestarazu M, Lao MT, Rincón-Cervera MÁ, Álvaro JE, Lyashenko S (2019) Borage oil: tocopherols, sterols and squalene in farmed and endemic-wild *Borago* species. J Food Compos Anal 83:103299. https://doi.org/10.1016/j.jfca.2019.103299

Fan YY, Chapkin RS (1998) Importance of dietary γ-linolenic acid in human health and nutrition. J Nutr 128(9):1411–1414

FDA (2016). https://www.fda.gov/media/99226/download. Accessed 15 October 2019

Fewtrell MS, Abbott RA, Kennedy K, Singhal A, Morley R, Caine E, Jamieson C, Cockburn F, Lucas A (2004) Randomized, double-blind trial of long-chain polyunsaturated fatty acid supplementation with fish oil and borage oil in preterm infants. J Pediatr 144(4):471–479

Foster RH, Hardy G, Alany RG (2010) Borage oil in the treatment of atopic dermatitis. Nutrition 26 (7–8):708–718

Fujii M, Nakashima H, Tomozawa J, Shimazaki Y, Ohyanagi C, Kawaguchi N, Ohya S, Kohno S, Nabe T (2013) Deficiency of n-6 polyunsaturated fatty acids is mainly responsible for atopic dermatitis-like pruritic skin inflammation in special diet-fed hairless mice. Exp Dermatol 22 (4):272–277

Godos J, Sinatra D, Blanco I, Mulè S, La Verde M, Marranzano M (2017) Association between dietary phenolic acids and hypertension in a Mediterranean cohort. Nutrients 9(10):1–13. https://doi.org/10.3390/nu9101069

Gomez AM, De la Ossa EM (2002) Quality of borage seed oil extracted by liquid and supercritical carbon dioxide. Chem Eng J 88(1–3):103–109

Grand View Research (2016). Borage oil market size, share & trend analysis report by application (cosmetics, medical, dietary supplements) by region and segment forecast, 2016–2022. https://www.grandviewresearch.com/industry-analysis/borage-oil-market. Accessed 10 October 2019

Gupta SK, Negi SP (2016) Antibacterial activity of Indian borage (*Plectranthus amboinicus* Benth) leaf extracts in food systems and against natural microflora in chicken meat. Food Technol Biotechnol 54(1):90–96

Henz BM, Jablonska S, Stingl G, Blaszczyk M, Vandervalk PG, Veenhuizen R, Muggli R, Raederstorff D (1999) Double-blind, multicentre analysis of the efficacy of borage oil in patients with atopic eczema. Br J Dermatol 140(4):685–688

Herrmann M, Joppe H, Schmaus G (2002) Thesinine-4′-O-β-D-glucoside the first glycosylated plant pyrrolizidine alkaloid from *Borago officinalis*. Phytochemistry 60(4):399–402

Horvei LD, Grimnes G, Hindberg K, Mathiesen EB, Njølstad I, Wilsgaard T, Brox J, Braekkan SK, Hansen JB (2016) C-reactive protein, obesity, and the risk of arterial and venous thrombosis. J Thromb Haemost 14(8):1561–1571

Jensen MP, Karoly P, Braver S (1986) The measurement of clinical pain intensity: a comparison of six methods. Pain 27:117–126

Jiang WG, Redfern A, Bryce RP, Mansel RE (2000) Peroxisome proliferator activated receptor-γ (PPAR-γ) mediates the action of gamma linolenic acid in breast cancer cells. Prostaglandins Leukot Essent Fatty Acids 62(2):119–127

Jung JY, Kwon HH, Hong JS, Yoon JY, Park MS, Jang MY, Suh DH (2014) Effect of dietary supplementation with omega-3 fatty acid and gamma-linolenic acid on acne vulgaris: a randomised, double-blind, controlled trial. Acta Derm Venereol 94(5):521–526

Kapoor R, Huang YS (2006) Gamma linolenic acid: an antiinflammatory omega-6 fatty acid. Curr Pharm Biotechnol 7(6):531–534

Kast RE (2001) Borage oil reduction of rheumatoid arthritis activity may be mediated by increased cAMP that suppresses tumor necrosis factor-alpha. Int Immunopharmacol 1(12):2197–2199

Kawamura A, Ooyama K, Kojima K, Kachi H, Abe T, Amano K, Aoyama T (2011) Dietary supplementation of gamma-linolenic acid improves skin parameters in subjects with dry skin and mild atopic dermatitis. J Oleo Sci 60(12):597–607

Labrousse VF, Leyrolle Q, Amadieu C, Aubert A, Sere A, Coutureau E, Grégoire S, Bretillon L, Pallet V, Gressens P, Joffre C (2018) Dietary omega-3 deficiency exacerbates inflammation and reveals spatial memory deficits in mice exposed to lipopolysaccharide during gestation. Brain Behav Immun 73:427–440

Larson KM, Roby MR, Stermitz FR (1984) Unsaturated pyrrolizidines from borage (*Borago officinalis*), a common garden herb. J Nat Prod 47(4):747–748

Lindqvist HM, Gjertsson I, Andersson S, Calder PC, Bärebring L (2019) Influence of blue mussel (*Mytilus edulis*) intake on fatty acid composition in erythrocytes and plasma phospholipids and

serum metabolites in women with rheumatoid arthritis. Prostaglandins Leukot Essent Fat Acids 150:7–15
Luthria DL, Sprecher H (1994) A comparison of the specific activities of linoleate and arachidonate in liver, heart and kidney phospholipids after feeding rats ethyl linoleate-9,10,12,13-d4. Biochim Biophys Acta 1213:1–4
Makrides M, Gibson RA, McPhee AJ, Collins CT, Davis PG, Doyle LW, Simmer K, Colditz PB, Morris S, Smithers LG, Willson K (2009) Neurodevelopmental outcomes of preterm infants fed high-dose docosahexaenoic acid: a randomized controlled trial. JAMA 301(2):175–182
Makrides M, Smithers LG, Gibson RA (2010) Role of long-chain polyunsaturated fatty acids in neurodevelopment and growth. In: Importance of growth for health and development, vol 65. Karger Publishers, pp 123–136
Mhamdi B, Wannes WA, Bourgou S, Marzouk B (2009) Biochemical characterization of borage (*Borago officinalis* L.) seeds. J Food Biochem 33(3):331–341
Miceli A, Aleo A, Corona O, Sardina MT, Mammina C, Settanni L (2014) Antibacterial activity of Borago officinalis and *Brassica juncea* aqueous extracts evaluated in vitro and in situ using different food model systems. Food Control 40:157–164
Miceli A, Francesca N, Moschetti G, Settanni L (2015) The influence of addition of *Borago officinalis* with antibacterial activity on the sensory quality of fresh pasta. Int J Gastron Food Sci 2(2):93–97
Mifsud S (2004) *Borago officinalis* (common Borage). http://www.maltawildplants.com/BORG/Borago_officinalis.php. Accessed 16 October 2019
Mortensen A, Skibsted LH (1997) Real time detection of reactions between radicals of lycopene and tocopherol homologues. Free Radic Res 27(2):229–234
Namal Senanayake SP, Shahidi F (2000) Lipid components of borage (*Borago officinalis* L.) seeds and their changes during germination. J Am Oil Chem Soc 77(1):55–61
National Eczema Association (2019). https://nationaleczema.org/eczema/types-of-eczema/atopic-dermatitis/. Accessed 10 October 2019
Navarette R, Tang W, Ziboh VA (1992) Dietary intake of concentrated gamma-linolenic acid (GLA)-enriched oil suppresses cutaneous level of dihomo-gamma-linolenic acid (DGLA): possible in vivo inhibition of microsomal elongation of GLA to DGLA. Prostaglandins Leukot Essent Fatty Acids 46:139–144
Osakabe N, Takano H, Sanbongi C, Yasuda A, Yanagisawa R, Inoue KI, Yoshikawa T (2004) Anti-inflammatory and anti-allergic effect of rosmarinic acid (RA); inhibition of seasonal allergic rhinoconjunctivitis (SAR) and its mechanism. Biofactors 21(1–4):127–131
Osborne JL (1999) Borage. Bee World 80(1):33–36
Pawlosky R, Barnes A, Salem N Jr (1994) Essential fatty acid metabolism in the feline: relationship between liver and brain production of long-chain polyunsaturated fatty acids. J Lipid Res 35:2032–2040
Pereira C, Barros L, Carvalho AM, Ferreira IC (2011) Nutritional composition and bioactive properties of commonly consumed wild greens: potential sources for new trends in modern diets. Food Res Int 44(9):2634–2640
Phinney SD, Tang AB, Thurmond DC, Nakamura MT, Stern JS (1993) Abnormal polyunsaturated lipid metabolism in the obese Zucker rat, with partial metabolic correction by g-linolenic acid administration. Metabolism 42:1127–1140
Pilerood SA, Prakash J (2014) Evaluation of nutritional composition and antioxidant activity of Borage (*Echium amoenum*) and Valerian (*Valerian officinalis*). J Food Sci Technol 51 (5):845–854
Puch F, Samson-Villeger S, Guyonnet D, Blachon JL, Rawlings AV, Lassel T (2008) Consumption of functional fermented milk containing borage oil, green tea and vitamin E enhances skin barrier function. Exp Dermatol 17(8):668–674
Redden PR, Huang YS, Lin X, Horrobin DF (1995) Separation and quantification of the triacylglycerols in evening primrose and borage oils by reversed-phase high-performance liquid chromatography. J Chromatogr A 694(2):381–389

Reed GW, Leung K, Rossetti RG, VanBuskirk S, Sharp JT, Zurier RB (2014) Treatment of rheumatoid arthritis with marine and botanical oils: an 18-month, randomized, and double-blind trial. Evid Based Complement Alternat Med 2014:1–9. https://doi.org/10.1155/2014/857456

Rio-Celestino MD, Font R, de Haro-Bailón A (2008) Distribution of fatty acids in edible organs and seed fractions of borage (*Borago officinalis* L.). J Sci Food Agric 88(2):248–255

Robbins RJ (2003) Phenolic acids in foods: an overview of analytical methodology. J Agric Food Chem 51(10):2866–2887

Roberts MJ (2000). Edible & medicinal flowers. New Africa Books. p 8. ISBN: 9780864864673

Rocha J, Eduardo-Figueira M, Barateiro A, Fernandes A, Brites D, Bronze R, Duarte CM, Serra AT, Pinto R, Freitas M, Fernandes E (2015) Anti-inflammatory effect of rosmarinic acid and an extract of *Rosmarinus officinalis* in rat models of local and systemic inflammation. Basic Clin Pharmacol Toxicol 116(5):398–413

Ryan AS, Astwood JD, Gautier S, Kuratko CN, Nelson EB, Salem N Jr (2010) Effects of long-chain polyunsaturated fatty acid supplementation on neurodevelopment in childhood: a review of human studies. Prostaglandins Leukot Essent Fatty Acids 82(4–6):305–314

Schirmer MA, Phinney SD (2007) γ-linolenate reduces weight regain in formerly obese humans. J Nutr 137(6):1430–1435

Selvi F, Coppi A, Bigazzi M (2006) Karyotype variation, evolution and phylogeny in Borago (Boraginaceae), with emphasis on subgenus Buglossites in the Corso-Sardinian system. Ann Bot 98:857–868

Sergeant S, Rahbar E, Chilton FH (2016) Gamma-linolenic acid, dihommo-gamma linolenic, eicosanoids and inflammatory processes. Eur J Pharmacol 785:77–86

Simon D, Eng PA, Borelli S, Kägi R, Zimmermann C, Zahner C, Drewe J, Hess L, Ferrari G, Lautenschlager S, Wüthrich B (2014) Gamma-linolenic acid levels correlate with clinical efficacy of evening primrose oil in patients with atopic dermatitis. Adv Ther 31(2):180–188

Sovova H, Zarevucka M, Vacek M, Stránský K (2001) Solubility of two vegetable oils in supercritical CO2. J Supercrit Fluids 20(1):15–28

Szterk A, Roszko M, Sosińska E, Derewiaka D, Lewicki PP (2010) Chemical composition and oxidative stability of selected plant oils. J Am Oil Chem Soc 87(6):637–645

Takahashi Y, Ide T, Fujita H (2000) Dietary gamma-linolenic acid in the form of borage oil causes less body fat accumulation accompanying an increase in uncoupling protein 1 mRNA level in brown adipose tissue. Comp Biochem Physiol B: Biochem Mol Biol 127(2):213–222

Tamariz J, Burgueño-Tapia E, Vázquez MA, Delgado F (2018) Pyrrolizidine alkaloids. In: The alkaloids: chemistry and biology, vol 80. Academic Press, pp 1–314

Tanaka A, Amagai Y, Oida K, Matsuda H (2015) Dihomo-γ-linolenic acid prevents the development of atopic dermatitis through prostaglandin D1 production in NC/Tnd mice. Allergy Eur JAllergy Clin Immunol 70:30–37. https://doi.org/10.1016/j.jdermsci.2015.03.010

Tasset-Cuevas I, Fernández-Bedmar Z, Lozano-Baena MD, Campos-Sánchez J, de Haro-Bailón A, Muñoz-Serrano A, Alonso-Moraga Á (2013) Protective effect of borage seed oil and gamma linolenic acid on DNA: in vivo and in vitro studies. PLoS One 8(2):1–9. https://doi.org/10.1371/journal.pone.0056986

Thurmond DC, Tang AB, Nakamura MT, Stern JS, Phinney SD (1993) Time dependent effects of progressive g-linolenate feeding on hyperphagia, weight gain, and erythrocyte fatty acid composition during growth of Zucker obese rats. Obes Res 1:118–125

USDA (2019). https://fdc.nal.usda.gov/fdc-app.html#/food-details/170481/nutrients. Accessed 10 October 2019

Vacillotto G, Favretto D, Seraglia R, Pagiotti R, Traldi P, Mattoli L (2013) A rapid and highly specific method to evaluate the presence of pyrrolizidine alkaloids in *Borago officinalis* seed oil. J Mass Spectrom 48(10):1078–1082

Vander Cruyssen B, Van Looy S, Wyns B, Westhovens R, Durez P, Van den Bosch F, Veys EM, Mielants H, De Clerck L, Peretz A, Malaise M, Verbruggen L, Vastesaeger N, Geldhof A, Boullart L, De Keyser F (2005) DAS28 best reflects the physician's clinical judgment of response to infliximab therapy in rheumatoid arthritis patients: validation of the DAS28 score in patients under infliximab treatment. Arthritis Res Ther 7:R1063–R1071. https://doi.org/10.1186/ar1787

Vasiljevic D, Veselinovic M, Jovanovic M, Jeremic N, Arsic A, Vucic V, Lucic-Tomic A, Zivanovic S, Djuric D, Jakovljevic V (2016) Evaluation of the effects of different supplementation on oxidative status in patients with rheumatoid arthritis. Clin Rheumatol 35(8):1909–1915

Veselinovic M, Vasiljevic D, Vucic V, Arsic A, Petrovic S, Tomic-Lucic A, Savic M, Zivanovic S, Stojic V, Jakovljevic V (2017) Clinical benefits of n-3 PUFA and ɣ-linolenic acid in patients with rheumatoid arthritis. Nutrients 9(4):1–11. https://doi.org/10.3390/nu9040325

Visser M, Bouter LM, McQuillan GM, Wener MH, Harris TB (1999) Elevated C-reactive protein levels in overweight and obese adults. JAMA 282:2131–2135

Wang X, Lin H, Gu Y (2012) Multiple roles of dihomo-γ-linolenic acid against proliferation diseases. Lipids Health Dis 11(1):25

Wang F, Miao M, Chen B, Wang R, Sun B, Ren D, Lu J (2015) Antineoplastic activity of γ-linolenic acid extract from *Spirulina platensis* on HepG2 cells and its inhibition effect on platelet aggregation. Food Agric Immunol 26(1):97–108

Wang Q, Cui Q, Yan C (2016) The effect of supplementation of long-chain polyunsaturated fatty acids during lactation on neurodevelopmental outcomes of preterm infant from infancy to school age: a systematic review and meta-analysis. Pediatr Neurol 59:54–61

Ward PE (2000) Potential diagnostic aids for abnormal fatty acid metabolism in a range of neurodevelopmental disorders. Prostaglandins Leukot Essent Fatty Acids 63:65–68

Wauquier F, Barquissau V, Léotoing L, Davicco MJ, Lebecque P, Mercier S, Philippe C, Miot-Noirault E, Chardigny JM, Morio B, Wittrant Y (2012) Borage and fish oils lifelong supplementation decreases inflammation and improves bone health in a murine model of senile osteoporosis. Bone 50(2):553–561

Wettasinghe M, Shahidi F, Amarowicz R, Abou-Zaid MM (2001) Phenolic acids in defatted seeds of borage (*Borago officinalis* L.). Food Chem 75(1):49–56

WHO (2018). https://www.who.int/news-room/fact-sheets/detail/cancer. Accessed 10 October 2019

Wretensjö I, Karlberg B (2003) Pyrrolizidine alkaloid content in crude and processed borage oil from different processing stages. J Am Oil Chem Soc 80(10):963–970

Zadernowski R, Naczk M, Nowak-Polakowska H (2002) Phenolic acids of borage (*Borago officinalis* L.) and evening primrose (*Oenothera biennis* L.). J Am Oil Chem Soc 79(4):335–338

Zhang C, Yu H, Shen Y, Ni X, Shen S, Das UN (2015) Polyunsaturated fatty acids trigger apoptosis of colon cancer cells through a mitochondrial pathway. Arch Med Sci 11(5):1081–1094. https://doi.org/10.5114/aoms.2015.54865

Ziboh VA, Miller CC, Cho Y (2000) Metabolism of polyunsaturated fatty acids by skin epidermal enzymes: generation of antiinflammatory and antiproliferative metabolites. Am J Clin Nutr 71:361s–366s

Chapter 15
Hempseed (*Cannabis sativa*)

Irshaan Syed, Sourav Garg, Saumya Agarwal, and Nupur Mohapatra

Abstract Hempseed is a well-known source of highly digestible protein (edestin and albumin). It has significant amounts of all essential amino acids, bioactive compounds along with dietary fibre, vitamins and minerals. The hempseed oil extracted from the achenes of cannabis usually has more than 80% of polyunsaturated fatty acids. Particularly, it is a rich source of α-linolenic acid and linoleic acid and has been used as a replacement for fish oil. Moreover, various clinical trials carried out on hempseed oil have signified its importance as a functional food and are often used for the treatment of disorders. The moisture content present in the seed at various stages such as harvest and pressing is typically around 15 and 10%, respectively. The presence of excess moisture content during processing would favour mould growth and is a possible reason for reduced shelf life. This article covers the challenges and adverse effects associated with the use of hempseed that include long-term stability issues and toxicity. One among the significant adverse effects includes the presence of a psychoactive compound such as Δ-9-tetrahydrocannabinol (THC). Thus, the scope of this review is to highlight and focus on the significant findings that will contribute to broadening the application of hempseed for food applications and related health benefits.

Keywords Hempseed · Health benefits · Nutritional composition · Toxic effects · Hypercholesterolemia

15.1 Origin and History

The use of hempseed (*Cannabis sativa*) as a source of food, fibre and medicine has been a subject of extensive research. Secondly, the use of whole seed as food and condiment persists by various consumer groups across the world, especially in Asia

I. Syed (✉) · S. Garg · S. Agarwal · N. Mohapatra
Department of Food Process Engineering, National Institute of Technology Rourkela, Rourkela, Odisha, India
e-mail: 514fp1004@nitrkl.ac.in

B. Tanwar, A. Goyal (eds.), *Oilseeds: Health Attributes and Food Applications*,
https://doi.org/10.1007/978-981-15-4194-0_15

(Girgih et al. 2011a, 2014; Salentijn et al. 2015). Botanically it is distinct from all other plants and is readily recognised. However, the botanical classification of hemp is still a controversial topic (Allegret 2013; Chabbert et al. 2013).

Hemp is one of the anciently cultivated crops, with its initial cultivation dating back to 4000–6000 years ago. Although there is no written record, it is believed that hemp might have been grown in China 20,000 years ago (Przybylski 2006). Initially, hemp was valued as a source of fibre, but by the sixteenth century, it became an important cash crop in Europe and was widely utilised for both its fibre and its seeds (Johnson 1999). Hemp was introduced to America and New England, in 1645. However, in the late 1800s, hemp-based industries in the United States began to decline because of the technical advancements in cotton ginning, leading to reduced labour and consequent replacement of hemp-derived products with cotton products (Johnson 1999).

Cannabis is believed to be an Asiatic plant. There is no concerted agreement among botanists as to where the wild plant originally grew and where its cultivation first began. However, evidence of the use of hemp fibres has been found in Neolithic records in northern China (Lu and Clarke 1995). The oldest existing documents that describe the use of hempseed as both food and medicine are from China, where *Cannabis* stalks, leaves and seeds were found in tombs that are over 4500 years old (Callaway 2004).

The earliest or the primary use of the plant was probably for its fibres. It was the only fibre plant known to ancient people in northern China, north-eastern China and eastern Siberia. In addition, it is reported that patterns of woven material (possibly *Cannabis* fibres) in the form of nets were used for trapping small animals. The tremendous cultural importance attached to hemp as a textile fibre is indicated by the practice of wearing hemp fabric clothes while mourning the death of a parent or parents, since Confucian times (Mwaikambo and Ansell 2002).

There is no easy way to distinguish between wild and spontaneous or adventitious and semi-cultivated or cultivated plants. Therefore, further progress in the determination of the geographical origin of the plant is yet to be carried out. The influence of man must be considered along with the botanical facts to unveil the complex nature of the plant (Mwaikambo and Ansell 2002). By providing a nutritious food from its seed, a durable fibre from its stalk and efficacious medicine from its flower and leaves, *Cannabis* has assisted in the human development like no other plant species (Zuardi 2006).

15.2 Production

The inception of hempseed oil production begins with low temperature (<25 °C) air-drying of fresh and well-cleaned seeds. This drying unit operation is monitored over several days so as to reduce the moisture content of the seeds to 10% before proceeding for storage or processing (Callaway 2004; Callaway and Pate 2009). However, mould growth between the time of harvest and drying is a primary

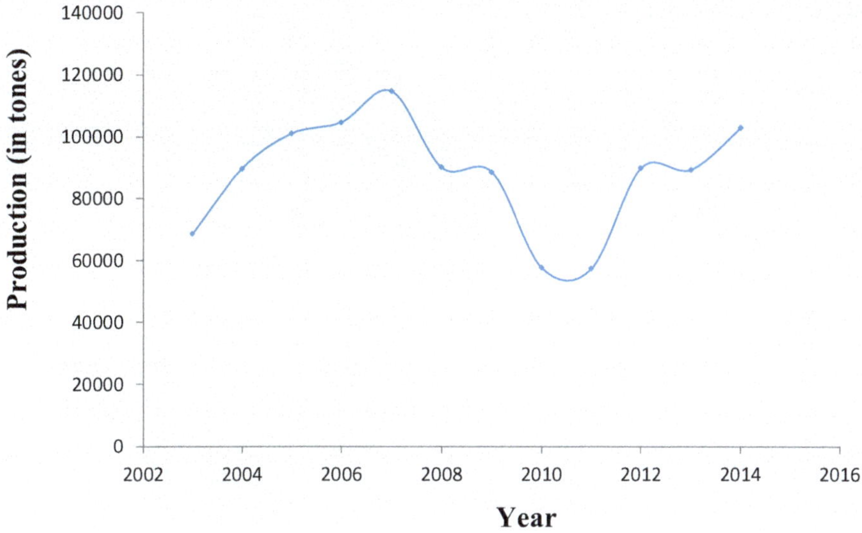

Fig. 15.1 World annual production of hempseed oil

concern. Various safety measures are undertaken for maintaining the quality of the oil produced (Oomah et al. 2002).

The consumable hempseed oil is extracted by cold-pressing the seeds using small-scale screw presses. Further, the oil is stored in vessels made up of ceramic, glass or glazed steel. In addition, it is allowed to settle for 1 or 2 weeks following which decantation is carried out into small containers which are then transported to retail stores around the globe.

Industrial processing demands the use of larger screw presses and advanced filtering operations that conveniently segregate the final sediments. Similarly, monitoring suitable inert atmosphere during the processing of the oil is one among the prerequisites for maintaining the quality of oil (Callaway and Pate 2009; Kostić et al. 2013). Figure 15.1 illustrates the variation in annual hempseed oil production over the years, while Fig. 15.2 represents the share of different regions in the production of hempseed (Callaway and Pate 2009).

Hydraulic processing could also substitute the cold-pressing technique used for extracting the oil (Callaway and Pate 2009; Kostić et al. 2013). However, the latter is not profoundly put into use owing to the low market demand for food-grade hempseed oil. However, industrial refining or bleaching tends to infringe the specific attributes (taste, nutrients and useful components) of the oil (Callaway and Pate 2009). Bleaching with clays is commonly practiced in the edible oil-refining industries to reduce the plant-based pigments. (Zschau 2001). Specific clays like bentonite, sepiolite and palygorskite are generally used in bleaching of vegetable oils (Didi et al. 2009). In the recent past, application of ultrasound bleaching technique has been introduced (Aachary et al. 2016). It is reported that the use of ultrasound bleaching technique reduces the quantity of the pigments such as chlorophyll and

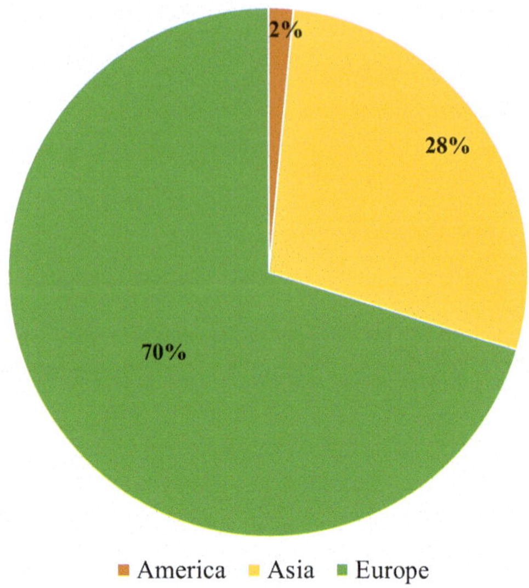

Fig. 15.2 Production share of hempseed by region

carotenoids of cold-pressed hempseed oil (Aachary et al. 2016). The presence of chlorophyll is known to trigger lipid oxidation that deteriorates the quality of hempseed oil during storage conditions. Simultaneous application of ultrasound was studied on cold-processed hempseed oil and compared with conventional bleaching techniques (Aachary et al. 2016). Conventional methods (clay bleaching) were observed to possess drawbacks such as increased treatment time and higher amount of clay used. These existing drawbacks of bleaching clay necessitated devising an alternate approach which is more sustainable and eco-friendly. Hence, ultrasonic treatment was brought in use, although its use in bleaching of vegetable oil is limited (Abedi et al. 2015).

15.3 Chemical Composition

The superior-quality hempseed oil obtained through cold pressing is clear and olive green in colour (Callaway and Pate 2009). It is observed to possess a characteristic fresh nutty taste (Blade et al. 2006) and is examined as a predominant source of polyunsaturated fatty acids with its concentration ranging from 75 to 85% of the total oil content (Blade et al. 2006; Matthäus et al. 2006).

15.3.1 Carbohydrates

Whole hempseeds consist of about 20–30% carbohydrates (Deferne and Pate 1996). Finola variety of hempseed was found to have an average carbohydrate content of 27.6% (Leizer et al. 2000) with glucose and fructose being the predominant ones at amounts of 0.30 g/100 g and 0.45 g/100 g, respectively. However, other carbohydrates such as lactose and maltose were also reported to be present but in small quantities (<0.1 g/100 g) (Leyva et al. 2011).

15.3.2 Proteins

Hempseed, also known as the seed of nondrug cultivars of industrial hemp, is often considered as an underexploited nonlegume protein-rich seed (Aiello et al. 2016). Hempseed is an excellent source of digestible protein with protein content varying from 20 to 25%. It is an abundant source of all the amino acids which are essential for the human body (Deferne and Pate 1996; Erasmus 1999). The amino acid composition of hempseeds is depicted in Table 15.1. Due to the absence of antinutritional trypsin-inhibiting factors, proteins found in hempseed are easier to digest and are more readily available for absorption. Moreover, it is an excellent source of arginine and some sulphur-containing amino acids, i.e. methionine and cystine (Callaway 2004). Just like other vegetable-based protein sources, it contains

Table 15.1 Amino acid profile of hempseed

Amino Acid	Quantity (in g) per 100 g
Tryptophan	0.20
Threonine	0.88
Isoleucine	0.98
Leucine	1.72
Lysine	1.03
Methionine	0.58
Cystine	0.41
Phenylalanine	1.17
Tyrosine	0.82
Valine	1.28
Arginine	3.10
Histidine	0.71
Alanine	1.28
Aspartic acid	2.78
Glutamic acid	4.57
Glycine	1.14
Proline	1.15
Serine	1.27

Adapted from Clifford Hall et al. (2015)

a relatively lower proportion of lysine. However, it cannot be considered as the only source of dietary protein for children (less than 10 years) as per WHO essential amino acid requirements (Nishida and Uauy 2009).

The major drawback with hempseed is that a significant fraction of total proteins is accounted by edestin (60–80%), while albumin balances the rest proportion. Edestin is a hexamer which is made up of six identical AB protein subunits and can serve as a potential biopolymer for the formation of cast films for its use as biodegradable and even edible food packaging (Odani and Odani 1998). In the recent past, hempseed protein finds its potential application in the development of nutraceuticals and functional foods (Aiello et al. 2016). Adequate treatment of hempseed proteins with suitable enzymes is known to produce hydrolyzed proteins that benefit the health as hypotensive agents and antioxidants (Aiello et al. 2016).

15.3.3 Fats

Hempseed oil is an exceptionally high source of unsaturated fatty acids with linoleic acid and α-linolenic acid being the predominant fatty acids (Leizer et al. 2000). The concentration of linoleic acid and α-linolenic acid in total fatty acid was observed to range in between 52–62% and 12–23%, respectively. This variation in fatty acid profile is generally observed due to the natural difference in samples, different processing conditions and storage methods adopted (Leizer et al. 2000). In addition, it is considered as a brilliant source of omega-3 polyunsaturated fatty acids (PUFA) and known to possess the corresponding omega-6 to omega-3 PUFA ratio of 3:1 (linoleic acid to α-linolenic acid) (Erasmus 1999). Table 15.2 describes the profile of fatty acids present in hempseed oil.

On the other hand, hempseed oil was found to possess lower proportion of saturated fatty acids (9 g/100 g) such as stearic acid and palmitic acid with palmitic acid being the predominant saturated fatty acid (Ally and Horrobin 1980; Brodt-Eppley and Myatt 1998). It also contains a significant amount of γ-linolenic acid and

Table 15.2 Fatty acid profile of hempseed oil

Name of fatty acid	Quantity (% of total fats)
Saturated fat	3.3
• C16:0	3.44
• C18:0	1.46
• C20:0	0.28
Monounsaturated fats	5.8
• C18:1 (omega-9)	9
Total polyunsaturated	36.2
• C18:2 (omega-6)	56
• C18:3 (omega-6)	4
• C18:3 (omega-3)	22

Adapted from Callaway (2004)

stearidonic acid which act as precursor for long-chain fatty acids and eicosanoids in the human body. Eicosanoids are less stable, hormonelike substances that play a crucial role in the control of inflammation, vascular tone and initiation of contractions during delivery (Ally and Horrobin 1980; Brodt-Eppley and Myatt 1998). The concentration of γ-linolenic acid varies from 0.34 to 6.8 g/100 g. However, a higher concentration of γ-linolenic acid is exhibited in evening primrose (*Oenothera macrocarpa*) (8.5 g/100 g) and borage (*Borago officinalis*) (16.3 and 20.5 g/100 g, respectively) (Pina et al. 1984; Senanayake and Shahidi 2000). The process of conversion of linoleic acid to γ-linolenic acid becomes slow in mammals due to stress, hypertension, ageing and various other diseases. So, in case of disturbed conversion, a dietary supplement of γ-linolenic acid is found to be beneficial (Clifford Hall et al. 2015).

Studies conducted on fatty acids had reported that the stearidonic acid concentration in hempseed oil to be up to 2.5 g/100 g (Mölleken and Theimer 1997). Conversely, Matthäus et al. (2006) reported it to range between 0.4 and 1.5 g/100 g. Similarly, the amount of monounsaturated fatty acid such as oleic acid was observed to range between 7 and 16 g/100 g (Deferne and Pate 1996; Matthäus et al. 2006). Stearidonic acid was found to increase the concentration of eicosapentaenoic acid more efficiently than α-linolenic acid. Moreover, it is demonstrated to be beneficial for patients suffering from a deficit in their D-6-desaturase function. Therefore, additional supplementation of stearidonic acid can improve the formation of eicosanoids in their body (Erasmus 1993).

As per the nutrition societies of Germany, Austria and Switzerland (D-A-CH recommendations for nutrition), the ratio of omega-6 to omega-3 polyunsaturated fatty acids should be 4:1 to 5:1. However, hempseed oil was examined to possess omega-6 to omega-3 polyunsaturated fatty acids in the ratio of 3:1 (für die Nährstoffzufuhr 2000). A lower ratio of omega-6 to omega-3 polyunsaturated fatty acids was highly desirable as it reduces the occurrence of many chronic diseases (Simopoulos 2008).

However, linolenic acid is highly susceptible to oxidation and can result in the formation of oxidised LDL (low-density lipoprotein) cholesterol which is responsible for the advancement of arteriosclerosis. In addition to this, oxidation products of polyunsaturated fatty acids also play a crucial role in the development of various types of tumours and can also cause premature ageing process (für die Nährstoffzufuhr 2000; Simopoulos 2008). Hence, it is highly suggested that the consumption of hempseed oil should be in moderate quantity.

15.3.4 Dietary Fibres

Dietary fibre has always been an essential part of the diet and linked with improved digestion. Table 15.3 illustrates the quantities of significant fibres present in hempseed. The fibres are broadly divided into two categories:

Table 15.3 Quantification of natural blast fibres in hempseed

Component	Quantity (g/1000 g)
Cellulose	67–78
Hemicellulose	16–18
Pectin	0.8
Lignin	3.5–5.5
Fat/wax	0.7

Adapted from Zimniewska and Batog (2012)

Table 15.4 Characterisation of vitamins in hempseed

Type of vitamin	Unit	Value per 100 g
Vitamin C	mg	1.0
Thiamine	mg	0.4
Riboflavin	mg	0.11
Niacin	mg	2.8
Vitamin B-6	mg	0.12
Vitamin A	IU	3800
Vitamin D	IU	2277.5
Vitamin E	mg	90.00

Adapted from Callaway (2004)

(a) Soluble fibre: Constituting around 20% of the fibres in hempseed, these are soluble in water and can form a gel-like substance in the gut. These are a valuable source of nutrients for the beneficial digestive bacteria and may also reduce spikes in blood sugar and regulate cholesterol levels (Matthäus and Brühl 2008; Matthäus et al. 2006).

(b) Insoluble fibre: Constituting around 80% of the fibres, they cannot be dissolved and instead adds bulk to faecal matter and may help food and waste pass through the gut. Consuming insoluble fibre has also been linked with a reduced risk of diabetes (de Munter et al. 2007; Weickert and Pfeiffer 2008). However, due to the removal of the fibre-rich shell in de-hulled or shelled hempseeds, the fibre content is substantially reduced (Callaway 2004).

15.3.5 Vitamins

Hempseed has received considerable attention as an excellent source of vitamins, with vitamin E being the predominant one (quantitative data regarding the presence of different types of vitamins in hempseed is reported in Table 15.4). Generally, tocopherols are a class of vitamin E compounds and exist in four different forms designated as α, β, δ, and γ. Supplementation of tocopherols is focused on its α form, and among all forms, hempseed oil is demonstrated to contain a relatively high proportion of γ-tocopherol. The amount of α-tocopherols and γ-tocopherol were found to be in amounts of 7–80 ppm and 710–870 ppm, respectively (Deferne and

Pate 1996). α-Tocopherol is primarily secreted into blood plasma, while γ-tocopherol is located in the intestine. Biologically, α-tocopherol is preferable than γ-tocopherol due to its better biological activity. However, this does not mean that α-tocopherol is a better antioxidant than γ-tocopherol. The inhibitory action of γ-tocopherol against the formation of phosphatidylcholine hydroperoxide is more profound at low peroxynitrite concentrations than that of α-tocopherol (Leizer et al. 2000). On the other hand, in fedora strain, α-tocopherol was found to be in traces, while the quantity of γ-tocopherol was found to be 468 mg/L (Leizer et al. 2000). However, both of them have their respective health benefits and roles as antioxidants.

15.3.6 Minerals

The total ash (inorganic matter) content of hempseed was found to be 5.6 g/100 g which depicts that it is highly rich in minerals. The composition of some important minerals in hempseed is depicted in Table 15.5 with phosphorus and potassium being the major ones (Callaway 2004).

15.3.7 Other Natural Compounds

Hempseed oil is a natural and abundant source of various classes of natural compounds mainly cannabidiol, β-caryophyllene, myrcene, β-sitosterol, α/γ-tocopherol and methyl salicylate which have numerous health benefits.

Table 15.5 Composition of minerals in hempseed

Mineral	Quantity (mg/100 g)
Calcium	145
Iron	14
Magnesium	483
Phosphorous	1160
Potassium	859
Sodium	12
Zinc	7
Copper	2
Manganese	7
Selenium	<0.02

Adapted from Callaway (2004)

15.3.7.1 Cannabidiol

Three types of cannabinoids are found in the hempseeds which include THC (tetrahydrocannabinol), CBD (cannabidiol) and CBN (cannabinol), the most important one being the THC (Petrović et al. 2015). Hempseeds were found to contain 10 mg/kg of cannabidiol which accounts for many of its health attributes (Petrović et al. 2015).

In the patients suffering from dystonic movement disorders, consumption of cannabidiol at the dosage of 100–600 mg/day has been reported to reduce tremor by 20–25% (Consroe et al. 1986; Formukong et al. 1988). Cannabidiol has also been reported to possess anticonvulsant, anti-epileptic, analgesic, anti-inflammatory and antimicrobial properties (Formukong et al. 1988; Karler and Turkanis1981). It can efficiently inhibit induction of both phenyl benzoquinone-induced writhing and tetradecanoyl phorbol-acetate-induced erythema. The antimicrobial activity is specifically found to be against the growth of Gram-positive bacteria mainly *Streptomyces griseus* and *Staphylococcus aureus* even at a very low concentration of 5 ppm (Formukong et al. 1988).

15.3.7.2 β-Sitosterol

β-Sitosterol is another bioactive compound found in hempseed oil that possesses several health benefits such as reduction in hypercholesterolemia, anti-inflammatory and antifungal properties (Malini and Vanithakumari 1990). The concentration of β-sitosterol in hempseed oil was found to be 100–148 g/L. Mattson et al. (1982) reported that certain plant sterols like β-sitosterol can block absorption of cholesterol in plasma either by its co-precipitation or by its crystallisation. The action is primarily effective for inhibition of cholesterol delivered through dietary fats. Isolated ethanolic extracts of *Hedychium spicatum* having the appropriate amount of β-sitosterol have been found to possess anti-inflammatory activity (Sharma et al. 1975).

15.3.7.3 Terpenes

The most abundant terpenes in hempseed oils are β-caryophyllene and myrcene whose concentrations are 740 mg/L and 160 mg/L, respectively. These are mostly found in the essential oil of hempseed, rather than directly in the seed oil (Malingre et al. 1975). They impart the characteristic aroma and various functional properties to the hempseed oil. The seed oil has highly active properties of β-caryophyllene such as anti-inflammatory and cytoprotective activities, while myrcene possesses excellent antioxidant properties (Duke and Beckstrom-Sternberg 1994).

15.3.7.4 Methyl Salicylate (Oil of Wintergreen)

Plant salicylates have been well known for centuries for their numerous health benefits. Aspirin or acetylsalicylic acid, which is one of the most commonly used drugs for its antipyretic, anti-inflammatory and analgesic uses, is closely associated and can be derived from methyl salicylate. Upon injection into the human body, it quickly hydrolyzes to form salicylic acid. Hence, methyl salicylate can be used as a potential replacer of aspirin. Although present in trace amounts, it is an important beneficial component in hempseed oil (Leizer et al. 2000).

15.4 Health Attributes

Various compounds found in hempseed are known to provide some general health benefits that may not necessarily be disease-specific. Consumption of cannabis as a nutraceutical can therefore provide a general improvement in health. For instance, certain cannabinoids and terpenes have been shown to have antioxidant and neuroprotective properties. They are found to be effective against inflammation, which is an underlying factor in many diseases. Such compounds may also provide beneficial effects on metabolic pathways, which may be useful for conditions related to diabetes, metabolic syndrome and obesity. In recent studies, it has been proved that peptides that are derived from the gastrointestinal simulated digestion of hempseed proteins are characterised by antioxidant and antihypertensive properties (Aiello et al. 2016). *Cannabis sativa* L., an annual plant in Cannabaceae family, has been a prime source of food, fibre, psychoactive/religious drug and as a medicine for ages. It is necessary to distinguish among the types of *Cannabis Sativa* L. such as drug type and nondrug type:

(a) Drug type: It is commonly known as marijuana, hashish or *Cannabis* tincture which contains enough Δ9-Tetrahydrocannabinol (THC) to exhibit psychoactivity.
(b) Nondrug type: It is the industrial hemp having THC concentration low enough to show any psychoactive properties.

In the United States, it is illegal to cultivate hempseed owing to its similarity with marijuana. It is believed that if industrial hemp is legalised, then it may also lead to the legalisation of marijuana. Governments of other countries have accepted the distinction between the above two types of *Cannabis sativa* L. because of its excellent nutritional values.

15.4.1 In Reducing the Risk of Heart Disease

Hemp constitutes almost all of the essential amino acids and amazingly high levels of the amino acid arginine, which is a metabolic precursor for the production of nitric oxide. Nitric oxide is known to act as a pivotal signalling messenger in the cardiovascular system that engages in the control of homeostasis, fibrinolysis, platelet and leukocyte interactions with the arterial wall, regulation of vascular tone, proliferation of vascular smooth muscle cells and homeostasis of blood pressure. Dietary hempseed is rich in omega-6 fatty acid linolenic acid and omega-3 fatty acid α-linolenic acid with ideal proportion (between 2:1 and 3:1 ratio) for a healthy diet and improving cardiovascular health (Rodriguez-Leyva and Pierce 2010).

15.4.2 In Skin-Related Disorders

A comparative study of dietary hempseed oil and olive oil in a 20-week randomised, single-blind crossover study was performed with the atopic patients in which fatty acid profiles were measured in plasma triglyceride, cholesterol and phospholipid fractions. Statistical information regarding the effects of hempseed oil based on skin dryness, itchiness, and usage of dermal medications was collected using a patient questionnaire. Skin transepidermal water loss was also calculated for every patient. The results depicted improvement in clinical symptoms of atopic dermatitis by the consumption of dietary hempseed oil because of a balanced supply of polyunsaturated fatty acids through hempseed oil. Hempseed oil also helped in relieving dry skin, improved itchiness and reduced the need for skin medication (Callaway et al. 2005).

15.4.3 In Reducing the Symptoms of Premenstrual Syndrome (PMS) and Menopause

Premenstrual emotional and physical changes affect up to 80% of women of reproductive age, and 20–40% of them suffer significant changes in the work, lifestyle and their relationships (Morino et al. 2016). These symptoms have been likely to be caused due to the increased production of the hormone prolactin. The production of prostaglandin E1 from γ-linolenic acid can reduce the effects of prolactin. In a study, it was found that consumption of 1 gram of essential fatty acids (γ-linolenic acid, 210 mg; oleic acid, 175 mg; linoleic acid, 345 mg; other polyunsaturated acids, 250 mg; and vitamin E, 20 mg) per day can result in a noticeable reduction in such symptoms (Rocha Filho et al. 2011). Oils which are rich in γ-linolenic acid can also be highly efficient for women who did not get any significant benefit from other therapies. Hempseed being rich in γ-linolenic acid is also the reason that studies are going on to suggest benefits of hempseed in hormone imbalances and inflammation associated with menopause (Rocha Filho et al. 2011).

15.4.4 In Hypercholesterolemia

In hypercholesterolemia, the aggregation of platelets is greatly enhanced which can lead to the development of various cardiovascular diseases (Prociuk et al. 2008). Platelet aggregation can be induced by production of eicosanoids (arachidonic acid) by metabolism of omega-6 fatty acids (Simopoulos 2008). In recent years, various research groups have worked to study the effect of hempseed on platelet aggregation in various animals. Hypercholesterolemic study of hempseed on male, white rabbits of New Zealand proved that a 10% supplementation of hempseed could reduce the platelet aggregation (caused by hypercholesterolemia) to normal levels within 8 weeks (Prociuk et al. 2008). They conferred that the effect was due to a higher amount of plasma γ-linolenic acid in hempseed and not due to a decrease in the level of cholesterol in plasma. A study on male Sprague-Dawley rats revealed that the addition of 5–10% of hempseed could reduce the rate of platelet aggregation within 12 weeks (Richard et al. 2007). The positive results were attributed to the fatty acid composition of hempseed, particularly omega-3 α-linolenic acid which can be metabolised to form anti-aggregatory eicosanoids. Since most of the cardiovascular diseases are caused mainly due to platelet aggregation, hempseed oil can be used as a potential preventive against such diseases.

15.4.5 In Maintaining Blood Pressure

Increased blood pressure or hypertension is one of the significant problems linked with the development of cardiovascular disease events, mainly heart failure, myocardial infarction and end-stage diabetes (Hong et al. 2008). Renin and angiotensin are the major enzymes which regulate the blood pressure levels by modulating the renin–angiotensin system. Hempseed protein hydrolysate and its peptide fractions have been studied for their potential application in the reduction of hypertension (Girgih et al. 2011b). In the experiment, peptide fractions were prepared by the consecutive action of pepsin and pancreatin and followed by membrane ultrafiltration. The prepared hempseed protein hydrolysate was found to inhibit the activities of angiotensin I-converting enzyme and renin and thereby reducing the chances of abnormal increase in blood pressure level. It is reported that oral intake of hempseed protein hydrolysate (200 mg/kg body weight) could reduce the blood pressure level of 30 mmHg within 8 h, while hempseed peptide fractions could balance the hypotension at the level of 15 mmHg after 6–8 h of oral intake. Thus, these results suggest for the potential application of hempseed proteins for the low-cost treatment of both hypertension and hypotension in human beings.

15.4.6 Antitumour Activity

Owing to its perfect fatty acid profile of omega-3 fats and γ-linolenic acid, hempseed naturally helps to balance inflammation levels and strengthen the immune system. Studies have reported that the presence of various cannabinoids, especially Δ9-tetrahydrocannabinol, in hempseeds can inhibit tumour growth (Guzman et al. 2006). Cannabinoids have demonstrated to inhibit human breast cancer cell proliferation through modulating the extracellular signal-regulated kinase and reactive oxygen species pathways (McAllister et al. 2007, 2011).

15.4.7 Effects on Atherosclerosis

In many countries, cardiovascular diseases account for the majority of deaths, principally caused by atherosclerosis, long-term inflammatory disease of the arteries which ends up in the formation of blood clots and hardening of arteries (Dahlöf 2010; Hansson 2005). Past theories on this subject have suggested that such diseases can be prevented by reducing the intake of omega-6 fatty acids (Matthäus and Brühl 2008). The anti-atherosclerotic activity of hempseed water extract (HWE) was studied in apolipoprotein E knockout mice (Seo et al. 2012). A reduction in total plasma cholesterol, LDL-cholesterol, atherosclerotic index and cardiac risk factor was reported in the HWE group along with a subsequent reduction of plaque lesion areas in aortic sinus and more weight gain. The results proved the potential use of HWE both as a nutritional resource and for its anti-atherosclerotic activity in apolipoprotein E knockout mice. In the future, studies can be done to assess the pharmacological mechanisms underlying these effects.

15.4.8 Improved Immunomodulatory Functions

Immunomodulatory functions involve provocation or curtailment of the immune system in the body. Various peptides can enhance immunomodulatory functions by regulating the expression of cytokines or by producing antibodies (Hartmann and Meisel 2007). Initial studies were carried on mice to examine the effects of hempseed protein on fatigue and immunomodulation defect (Li et al. 2008). The results revealed that hempseed protein could improve the swimming time of mice and contents of liver heparin distinctly and can reduce the contents of lactic acid in the blood. However, further research is required to confirm similar results of the consumption of hempseed on humans as models. The presence of omega-3-polyunsaturated fatty acids in the diet can help in the annihilation of interferon-γ production in patients suffering from immunological disorders such as multiple sclerosis (Weinstock-Guttman et al. 2005). A hot-nature diet with added hempseed and evening primrose oils was also found to be effective in patients who have multiple sclerosis (Rezapour-Firouzi et al. 2013).

15.5 Adverse Effects and Individual Concerns

The countries having industrial production of hemp have a regulation for the maximum level of tetrahydrocannabinol of 0.3% so that the psychoactive effects can be eliminated even after consumption of large quantities (Small and Marcus 2003). Owing to side effects of tetrahydrocannabinol, particularly its sleep-causing characteristics, certain regulations were mandated in Switzerland to control the consumption of hemp (Sarmento et al. 2015). Due to the association of hempseed with platelet aggregation, it can also result in prolonged bleeding in case of surgical interventions and other injuries. Also, the interaction of bioactive compounds in hempseed with another type of drug has not been completely studied. Intake of hempseed has been found to cause an adverse effect in individuals suffering from diarrhoea. However, no cases of toxicological issues have been reported yet with consumption of hempseed products (Leizer et al. 2000). Although there is a reported case of allergic reactions caused by consumption of de-hulled hempseed meal, no such evidences for intake of hempseed oil have been reported so far.

15.6 Food Applications

Hempseed enzymatic hydrolysates have proven effective during both in vitro and in vivo tests as an antioxidant and antihypertensive agents. Thus, hempseed proteins and hydrolysates have great potential to be used as prime ingredients in formulating functional foods which can be easily incorporated in the diet. In addition, it has considerate levels of arginine along with sulphur-rich proteins which helps in further enhancing the nutritional value of foods (Aluko 2017). Partially defatted hempseed flour is a valuable source of nutritional components mainly fibre content, total unsaturated fatty acids, bioactive carbohydrates, bioactive proteins and minerals and was found to enrich normal wheat flour with these compounds (Apostol et al. 2015a). Modern technologies have enabled the usage of waste produced from processing of raw materials. Exploiting food industry by-products has been proved to be beneficial with slight demerits which can be done away with additional processing methods. The utilization of pressed hemp cake to produce highly valuable bakery products and snacks is one such example. Mechanically pressed hempseed oil is used as salad dressing; however, due to high unsaturated fatty acid content, it is not suitable for at high-temperature applications.

In a study, nine recipes were studied in which hempseed oil, hemp grits, hemp flour and hemp protein added to wheat flour, or combinations thereof, were considered during evaluation (Mullerova et al. 2016). The baking test confirmed a higher dough yield and volume yield compared to the control after the addition of hemp oil. On addition of hemp flour, up to 10% reduction was reported in the baking loss along with an increased bread yield. Incorporation of hemp grits mostly affected bread arching and index number. Hemp oil was found to be the most suitable ingredient (according to the results of sensory analysis), which improved the crust colour, crumb elasticity, ease of bite, crumb moisture, consistency and overall impression when compared to the control variant. Also, addition of hemp grit elevated the quality of bread arching, index number and aroma. However, during processing (baking), it encountered few drawbacks such as occurrence of green colour and typical strong odour. Nevertheless, on combining the results of Rapid-Mix Test (bakery experiment test) and sensory analysis, it was concluded that hemp oil could be used for ideal addition both in terms of technology and nutrition (Apostol et al. 2015b). The nutritional content of butter extracted from hempseeds was found to be much more than that of peanut butter (Callaway 2004).

Soups, porridge and gruels made out of hempseed flour had sufficed the needs of thousands during famines. Ground hempseeds and flour have been used for making cakes, bread and granola bars. Results of the bakery test reported that flour from the wheat variety Scorpion had the highest specific volume and lowest convexity of the pastries. The highest ratio number was determined by the recipe which contained Rosso wheat flour and hemp grits (Ruban et al. 2016). Hempseed cake attained after oil pressing was processed in the laboratory conditions to obtain hemp flour. Its incorporation to wheat dough had influenced the dough rheological characteristics. The hemp flour affects water absorption and dough development time and consequently bread volume, colour, structural and textural properties of the breadcrumb (Pojić et al. 2014). Incorporation of hemp flour beyond a specific limit (15%) was found to decrease dough stability and dough strength. The study revealed that addition of hemp flour to bread elevated the amounts of important nutrients such as proteins and iron (Pojić et al. 2014). Moreover, the fatty acid composition and the antioxidant properties of the oilseed cakes support for its potential usage as nutritional supplements. Hemp fibre can be termed as an efficacious by-product of hempseed which can be used in industries for manufacturing durable fabrics and papers.

In a similar study, addition of hemp protein concentrates along with hemp flour was carried out for preparation of gluten-free starch bread (Korus et al. 2017b). In the case of formulation involving hemp protein, a significant increase in bread volume was observed with comparatively smaller change in bread crumb structure. On the other hand, formulations based on hemp flour weakened the dough structure which therefore made the dough prone to easy deformation. Hence, such supplementations of starch bread with hemp controlled hardening of the crumb, which is the primary cause of bread staling. Besides increasing the sensory acceptance, formulations incorporated with hemp were also demonstrated to increase the nutritional value. This is because hemp-based formulations were found to be rich in protein content, fats, minerals and dietary fibre. Therefore, it is worthy to note that such gluten-free bread products will have higher nutritional value when compared to its wheat and rye bread counterpart.

On the other hand, they simultaneously limit the ageing and staling of bread (Korus et al. 2017b). However, for industrial purposes, it is mandated that hemp flour content in case of wheat bread should not exceed 30% (Mikulec et al. 2018). In another study, incorporation of lacto-fermented hulled hempseed was found to be responsible for preparation of bread with good texture and higher overall quality. These findings indicate the use of *Pediococcus* species for fermentation process (Bartkiene et al. 2016).

In another similar study, addition of barley flour had demonstrated to reduce the specific volume of bread. Incorporation of wheat–barley blends in ratios of 70:30 and 50:50 was exhibited to reduce the specific volume of bread by 30% and 43%, respectively. Furthermore, within a group of bakery products containing 50% of barley flour, hulled hemp wholemeal was found to partially suppress the negative effect of barley flour, i.e. the specific volume of bread was found to increase by about 15% (Hrušková and Švec 2016).

Cookies are considered to be concentrated foods due to the high content of carbohydrates, fats and low moisture. Manufacturing cookies, biscuits and cones using hempseeds was initiated to enhance the taste and produce desirable flavours along with supplementing them with nutrients that control the blood sugar level and high blood cholesterol. Milk powder, egg yolk and hempseeds used in the cookie batter served multifaceted benefits. Although addition of hemp flour to the cookies increased the protein content, it lowered the quality. Even in the case of biscuits, it was noticed that a 20% addition of hemp flour increased the protein content (16–27%) as compared to control biscuits made out of regular corn flour. Addition of hemp flour to biscuits also elevated their mineral content by a certain level. Similarly, the total dietary fibre content received a massive boost upon addition of hemp flour to the normal flour used in the biscuit manufacturing process. Furthermore, it was observed that the addition of hemp flour changed the hydration properties of mixtures and the oil absorption capacity. However, the biscuits made with hemp flour scored less upon sensory evaluation and underwent visible colour changes (Korus et al. 2017a).

15.7 Alternative Applications

Hempseeds have been put into a wide spectrum of applications over the years that ranged from the manufacture of paints and varnishes to formulations of hempseed oil. Initially, hemp oil was used as lighting oil, but with the advent of hydrocarbons, this usage of hemp oil became extinct. Production of soap and linoleum also witnessed the usage of hempseed oil. The fibre varieties of hempseed are used as animal fodder, especially as bird feed. The medical applications of hempseed oil include its encapsulation for treatment of stomach and ear pain, cough and other diseases (Grigoriev 2002). The high unsaturated fatty acid content in the oil makes it unsuitable for high-temperature applications, but the hempseed oil finds utility in a broad horizon of products stretching from salad dressings to cosmetics like creams, lotions and lip balms (Rongrong et al. 2010). Gelatin-edible films have captured the limelight for a quite a short while now due to their efficient lipid barrier properties and film-forming capacity. However, the inclusion of bioactive compounds is necessitated by the high water-binding properties of the films which in turn cause swelling. Hemp oil has also been used in gelatin-edible films in order to increase their antimicrobial strength. Incorporation of hemp oil in gelatin-based films had exhibited to increase its solubility and light barrier property and subsequently decrease film thickness (Cozmuta et al. 2015).

15.8 By-product Applications

The hemp hurd is a by-product of the hemp plant obtained during hemp fibre separation. Generally, hemp hurd has been used as animal bedding (Khan et al. 2015). In order to produce hemp powder, hemp hurd is milled to different particle size distributions. Hemp hurd powder serves as a good filler for plastic-reinforced composites. In addition, it has been incorporated in the filament material used for 3D printing that plays a crucial role in developing a 3D food printing arena. On the other hand, hemp powder is used for the production of high specific surface area-activated carbon (Khan et al. 2015). Recent studies have revealed that industrial hemp can produce by-products such as hemp hurds and hemp dust which can prove to be a source of high value-added hydroxycinnamic acids (HCA), namely, ferulic and p-coumaric acids (Candy et al. 2017).

The hemp fibre has demonstrated to possess excellent antimicrobial properties and is therefore a major component of functional textiles. Moreover, the hemp hurd and powder are also examined to possess antibacterial properties. They have been used to eliminate the occurrence of *Escherichia coli*. Incorporating hemp hurd powder into the composites is certain to yield a lighter product with enhanced antimicrobial activities. Hemp hurd powder can be used to produce eco-friendly food packaging composites (Khan et al. 2015).

15.9 Future Challenges

The efficiency of the oil recovery from hempseeds can be enhanced by advancement in design and development sector of processing equipment. Currently, fewer studies are undertaken for examination of quality and grade of the oils available in the market. Poorly processed hempseed was investigated to possess psychoactive compound, Δ-9-tetrahydrocannabinol (THC). In Australia and New Zealand, consumption of hemp-derivative food products containing a trace level of THC is proposed. However, studies revealed that on consumption of low-content THC oil (5 mL bearer sesame oil containing 10 mg/kg THC), there was negligible detection of THC in oral fluid, blood or urine samples (Hayley et al. 2018).

15.10 Conclusion

Hempseed oil was observed to have potential health benefits as nutraceutical product possessing high nutritional value and as an excellent source of essential fatty acids, omega-3 and omega-6. Other than oil, tremendous insights on dried flower products and related extracts derived from a plant of *Cannabis sativa* L. have been found to have therapeutic applications globally. The fatty acid composition of hempseed, particularly omega-3 α-linolenic acid, can be metabolised to form anti-aggregatory eicosanoids and thus can prevent the occurrence of cardiovascular diseases. Hemp flour prepared from hempseed cake could influence the rheological characteristics of the wheat dough for preparation of bread and cookies. However, high unsaturated fatty acid content in the oil makes it unsuitable for high-temperature applications, but the hempseed oil finds utility in a broad horizon of products stretching from salad dressings to cosmetics like creams, lotions and lip balms.

References

Aachary AA, Liang J, Hydamaka A, Eskin NM, Thiyam-Holländer U (2016) A new ultrasound-assisted bleaching technique for impacting chlorophyll content of cold-pressed hempseed oil. LWT Food Sci Technol 72:439–446

Abedi E, Sahari MA, Barzegar M, Azizi MH (2015) Optimisation of soya bean oil bleaching by ultrasonic processing and investigate the physico-chemical properties of bleached soya bean oil. Int J Food Sci Technol 50:857–863

Aiello G, Fasoli E, Boschin G, Lammi C, Zanoni C, Citterio A, Arnoldi A (2016) Proteomic characterization of hempseed (*Cannabis sativa* L.). J Proteome 147:187–196

Allegret S (2013) The history of hemp Hemp: industrial production and uses CABI. Wallingford:4–26

Ally A, Horrobin D (1980) Thromboxane A2 in blood vessel walls and its physiological significance: relevance to thrombosis and hypertension. Prostaglandins Med 4:431–438

Aluko RE (2017) Chapter 7—hemp seed (*Cannabis sativa* L.) proteins: composition, structure, enzymatic modification, and functional or bioactive properties. In: Nadathur SR, Wanasundara JPD, Scanlin L (eds) Sustainable protein sources. Academic Press, San Diego, pp 121–132. https://doi.org/10.1016/B978-0-12-802778-3.00007-X

Apostol L, Popa M, Mustatea G (2015a) *Cannabis sativa* L. partially skimmed flour as source of biocompounds in the bakery industry. Romanian Biotechnol Lett 20:10843–10852

Apostol L, Popa ME, Mustatea G (2015b) Development of some innovative bakery products based on wheat flour enriched in bioactive compounds with functional role. Sci Bull Ser F Biotechnol 19:117–120

Bartkiene E et al (2016) Development and quality evaluation of lacto-fermented product based on hulled and not hulled hempseed (*Cannabis sativa* L.). LWT Food Sci Technol 72:544–551

Blade SF, Ampong-Nyarko K, Przybylski R (2006) Fatty acid and tocopherol profiles of industrial hemp cultivars grown in the high latitude prairie region of Canada. J Indust Hemp 10:33–43

Brodt-Eppley J, Myatt L (1998) Changes in expression of contractile FP and relaxatory EP2 receptors in pregnant rat myometrium during late gestation, at labor, and postpartum. Biol Reprod 59:878–883

Callaway J (2004) Hempseed as a nutritional resource: an overview. Euphytica 140:65–72
Callaway JC, Pate DW (2009) Hempseed oil. In: Gourmet and health-promoting specialty oils. Elsevier, pp 185–213
Callaway J, Schwab U, Harvima I, Halonen P, Mykkänen O, Hyvönen P, Järvinen T (2005) Efficacy of dietary hempseed oil in patients with atopic dermatitis. J Dermatol Treat 16:87–94
Candy L, Bassil S, Rigal L, Simon V, Raynaud C (2017) Thermo-mechano-chemical extraction of hydroxycinnamic acids from industrial hemp by-products using a twin-screw extruder. Ind Crop Prod 109:335–345
Chabbert B, Kurek B, Beherec O (2013) Physiology and botany of industrial hemp V: Bouloc P, Allegret S, Laurent A(ur) Hemp: industrial production and uses CABI. Wallingford, Boston, pp 27–47
Clifford Hall I, Fitzpatrick KC, Kamal-Eldin A (2015) Flax, perilla, and camelina seed oils: a-linolenic acid-rich oils gourmet and health-promoting specialty oils:151
Consroe P, Sandyk R, Snider SR (1986) Open label evaluation of cannabidiol in dystonic movement disorders. Int J Neurosci 30:277–282
Cozmuta AM et al (2015) Preparation and characterization of improved gelatin films incorporating hemp and sage oils. Food Hydrocoll 49:144–155
Dahlöf B (2010) Cardiovascular disease risk factors: epidemiology and risk assessment. Am J Cardiol 105:3A–9A
de Munter JS, Hu FB, Spiegelman D, Franz M, van Dam RM (2007) Whole grain, bran, and germ intake and risk of type 2 diabetes: a prospective cohort study and systematic review. PLoS Med 4:e261
Deferne J-L, Pate DW (1996) Hemp seed oil: a source of valuable essential fatty acids. J Int Hemp Assoc 3:4–7
Didi M, Makhoukhi B, Azzouz A, Villemin D (2009) Colza oil bleaching through optimized acid activation of bentonite. A comparative study. Appl Clay Sci 42:336–344
Duke JA, Beckstrom-Sternberg SM (1994) Dr. Duke's phytochemical and ethnobotanical databases
Erasmus U (1993) Fats that heal, fats that kill: the complete guide to fats, oils, cholesterol, and human health. Book Publishing Company
Erasmus U (1999) Fats that heal, fats that kill (1993). Alive books
Formukong E, Evans A, Evans F (1988) Analgesic and antiinflammatory activity of constituents ofCannabis sativa L. Inflammation 12:361–371
für die Nährstoffzufuhr DCR (2000) Deutsche Gesellschaft für Ernährung, Österreichische Gesellschaft für Ernährung, Schweizerische Gesellschaft für Ernährungsforschung, Schweizerische Vereinigung für Ernährung Umschau Braus GmbH, Frankfurt am Main
Girgih AT, Udenigwe CC, Aluko RE (2011a) In vitro antioxidant properties of hemp seed (*Cannabis sativa* L.) protein hydrolysate fractions. J Am Oil Chem Soc 88:381–389
Girgih AT, Udenigwe CC, Li H, Adebiyi AP, Aluko RE (2011b) Kinetics of enzyme inhibition and antihypertensive effects of hemp seed (*Cannabis sativa* L.) protein hydrolysates. J Am Oil Chem Soc 88:1767–1774
Girgih AT, Alashi A, He R, Malomo S, Aluko RE (2014) Preventive and treatment effects of a hemp seed (*Cannabis sativa*). Eur J Nutr 53:1237–1246
Grigoriev OV (2002) Application of hempseed (*Cannabis sativa* L.) oil in the treatment of ear, nose and throat (ENT) disorders. J Indust Hemp 7:5–15
Guzman M et al (2006) A pilot clinical study of Δ9-tetrahydrocannabinol in patients with recurrent glioblastoma multiforme. Br J Cancer 95:197–203
Hansson GK (2005) Inflammation, atherosclerosis, and coronary artery disease. N Engl J Med 352:1685–1695
Hartmann R, Meisel H (2007) Food-derived peptides with biological activity: from research to food applications. Curr Opin Biotechnol 18:163–169

Hayley AC et al (2018) Detection of delta-9-tetrahydrocannabinol (THC) in oral fluid, blood and urine following oral consumption of low-content THC hemp oil. Forensic Sci Int 284:101–106
Hong F, Ming L, Yi S, Zhanxia L, Yongquan W, Chi L (2008) The antihypertensive effect of peptides: a novel alternative to drugs? Peptides 29:1062–1071
Hrušková M, Švec I (2016) Fermented dough characteristics of wheat-barley-hemp composites. Comparison of two dosages of barley and hemp wholemeal/flour. Acta Universitatis Agriculturae et Silviculturae Mendelianae Brunensis 64:803–812
Johnson P (1999) Industrial hemp: a critical review of claimed potentials for *Cannabis sativa*. TAPPI J 82:113–123
Karler R, Turkanis SA (1981) The cannabinoids as potential antiepileptics. J Clin Pharmacol:21
Khan BA, Wang J, Warner P, Wang H (2015) Antibacterial properties of hemp hurd powder against *E. coli*. J Appl Polym Sci 132(10)
Korus A, Gumul D, Krystyjan M, Juszczak L, Korus J (2017a) Evaluation of the quality, nutritional value and antioxidant activity of gluten-free biscuits made from corn-acorn flour or corn-hemp flour composites. Eur Food Res Technol 243:1429–1438
Korus J, Witczak M, Ziobro R, Juszczak L (2017b) Hemp (*Cannabis sativa* subsp. *sativa*) flour and protein preparation as natural nutrients and structure forming agents in starch based gluten-free bread. LWT Food Sci Technol 84:143–150
Kostić MD, Joković NM, Stamenković OS, Rajković KM, Milić PS, Veljković VB (2013) Optimization of hempseed oil extraction by n-hexane. Ind Crop Prod 48:133–143
Leizer C, Ribnicky D, Poulev A, Dushenkov S, Raskin I (2000) The composition of hemp seed oil and its potential as an important source of nutrition. J Nutraceuticals Funct Med Foods 2:35–53
Leyva DR, McCullough RS, Pierce GN (2011) Medicinal use of hempseeds (*Cannabis sativa* L.): effects on platelet aggregation. In: Nuts and seeds in health and disease prevention. Elsevier, pp 637–646
Li Y, Yang R, Hu X, Long Z (2008) Initial study of Hemp seeds protein on antifatigue and the immunomodulation effects in mice. J Hyg Res 37:175–178
Lu X, Clarke RC (1995) The cultivation and use of hemp (*Cannabis sativa* L.) in ancient China. J Int Hemp Assoc 2:26–30
Malingre T, Hendriks H, Batterman S, Bos R, Visser J (1975) The essential oil of *Cannabis sativa*. Planta Med 28:56–61
Malini T, Vanithakumari G (1990) Rat toxicity studies with β-sitosterol. J Ethnopharmacol 28:221–234
Matthäus B, Brühl L (2008) Virgin hemp seed oil: an interesting niche product. Eur J Lipid Sci Technol 110:655–661
Matthäus B, Schumann E, Brühl L, Kriese U (2006) Hempseed oil—influence of the genotype on the composition in a two-year study. J Indust Hemp 10:45–65
Mattson FH, Grundy SM, Crouse JR (1982) Optimizing the effect of plant sterols on cholesterol absorption in man. Am J Clin Nutr 35:697–700
McAllister SD, Christian RT, Horowitz MP, Garcia A, Desprez P-Y (2007) Cannabidiol as a novel inhibitor of Id-1 gene expression in aggressive breast cancer cells. Mol Cancer Ther 6:2921–2927
McAllister SD et al (2011) Pathways mediating the effects of cannabidiol on the reduction of breast cancer cell proliferation, invasion, and metastasis. Breast Cancer Res Treat 129:37–47
Mikulec A, Kowalski S, Sabat R, Skoczylas Ł, Tabaszewska M, Wywrocka-Gurgul A (2018) Hemp flour as a valuable component for enriching physicochemical and antioxidant properties of wheat bread LWT
Mölleken H, Theimer R (1997) Evaluierung von C. sativa Saatgutherkünften im Hinblick auf eine verbesserte Ölqualität. In: Bioresource Hemp 97, Proceedings of the Symposium, Frankfurt, Germany. nova-Institute, Köln, Germany, pp 485–499
Morino S, Egawa M, Hirata H, Nishimura H, Aoyama T (2016) Association between premenstrual syndrome and daily physical activity levels. J Womens Health Issues Care 5:2

Mullerova M et al (2016) Use of hemp raw materials in common bakery product recipes. MendelNet 23:610–615
Mwaikambo LY, Ansell MP (2002) Chemical modification of hemp, sisal, jute, and kapok fibers by alkalization. J Appl Polym Sci 84:2222–2234
Nishida C, Uauy R (2009) WHO Scientific Update on health consequences of trans fatty acids: introduction. Nature Publishing Group
Odani S, Odani S (1998) Isolation and primary structure of a methionine-and cystine-rich seed protein of *Cannabis sativa*. Biosci Biotechnol Biochem 62:650–654
Oomah BD, Busson M, Godfrey DV, Drover JC (2002) Characteristics of hemp (*Cannabis sativa* L.) seed oil. Food Chem 76:33–43
Petrović M, Debeljak Ž, Kezić N, Džidara P (2015) Relationship between cannabinoids content and composition of fatty acids in hempseed oils. Food Chem 170:218–225
Pina M, Graille J, Grignac P, Lacombe A, Quenot O, Garnier P (1984) Recherche d'Oenothères riches en acide gammalinolénique. Oléagineux 39:593–596
Pojić M, Mišan A, Sakač M, Dapčević Hadnađev T, Šarić B, Milovanović I, Hadnađev M (2014) Characterization of byproducts originating from hemp oil processing. J Agric Food Chem 62:12436–12442
Prociuk M, Edel A, Richard M, Gavel N, Ander B, Dupasquier C, Pierce G (2008) Cholesterol-induced stimulation of platelet aggregation is prevented by a hempseed-enriched diet This article is one of a selection of papers published in the special issue Bridging the Gap: where progress in Cardiovascular and Neurophysiologic Research Meet. Can J Physiol Pharmacol 86:153–159
Przybylski R (2006) Hemp Revival of a forgotten oilseed crop. University of Lethbridge, Canada
Rezapour-Firouzi S et al (2013) Immunomodulatory and therapeutic effects of Hot-nature diet and co-supplemented hemp seed, evening primrose oils intervention in multiple sclerosis patients. Complement Ther Med 21:473–480
Richard M, Ganguly R, Steigerwald S, Al-Khalifa A, Pierce G (2007) Dietary hempseed reduces platelet aggregation. J Thromb Haemost 5:424–425
Rocha Filho EA, Lima JC, Neto JSP, Montarroyos U (2011) Essential fatty acids for premenstrual syndrome and their effect on prolactin and total cholesterol levels: a randomized, double blind, placebo-controlled study. Reprod Health 8:2
Rodriguez-Leyva D, Pierce GN (2010) The cardiac and haemostatic effects of dietary hempseed. Nutr Metab 7:32
Rongrong HJCTL, Ping LGCMQ, Chaojin CSYZZ (2010) Study on the Comprehensive Utilization and Industrialization of Hempseed (Fructus Cannabis)[J]. J Chin Inst Food Sci Technol 3:018
Ruban A et al (2016) The use of hemp and color wheat flour as baking ingredients. In: Proceedings of international PhD students conference, (MENDELNET 2016). Mendelova univerzita v Brně
Salentijn EM, Zhang Q, Amaducci S, Yang M, Trindade LM (2015) New developments in fiber hemp (Cannabis sativa L.) breeding. Ind Crop Prod 68:32–41
Sarmento L, Carus M, Grotenhermen F, Kruse D, Brenneisen R, Grassi G, Knapsack C (2015) Scientifically sound guidelines for THC in food in Europe Report of nova-Institut GmbH
Senanayake SN, Shahidi F (2000) Lipid components of borage (*Borago officinalis* L.) seeds and their changes during germination. J Am Oil Chem Soc 77:55–61
Seo J-H et al (2012) Hempseed water extract ameliorates atherosclerosis in apolipoprotein E knockout mice. Food Sci Biotechnol 21:927–932
Sharma SC, Shukla Y, Tandon J (1975) Alkaloids and terpenoids of *Ancistrocladus heyneanus*, *Sagittaria sagitifolia*, *Lyonia formosa* and *Hedychium spicatum*. Phytochemistry 14:578–579
Simopoulos AP (2008) The importance of the omega-6/omega-3 fatty acid ratio in cardiovascular disease and other chronic diseases. Exp Biol Med 233:674–688
Small E, Marcus D (2003) Tetrahydrocannabinol levels in hemp (*Cannabis sativa*) germplasm resources. Econ Bot 57:545–558
Weickert MO, Pfeiffer AF (2008) Metabolic effects of dietary fiber consumption and prevention of diabetes. J Nutr 138:439–442

Weinstock-Guttman B et al (2005) Low fat dietary intervention with ω-3 fatty acid supplementation in multiple sclerosis patients. Prostaglandins Leukot Essent Fat Acids 73:397–404
Zimniewska M, Batog J (2012) Ultraviolet-blocking properties of natural fibres. In: Handbook of natural fibres: processing and applications. Elsevier, pp 141–167
Zschau W (2001) Bleaching of edible fats and oils. Eur J Lipid Sci Technol 103:505–551
Zuardi AW (2006) History of cannabis as a medicine: a review. Revista Brasileira de Psiquiatría 28:153–157

Chapter 16
Rice (*Oryza sativa*) Bran

Upasana Yadav, Shalini Arora, and Isha Kaushik

Abstract Rice bran oil (RBO) is named as wonder oil or "heart oil" in many Asian countries and is considered an imminent functional food ingredient in western countries. High amounts of unsaturated fatty acids namely oleic and linoleic, 3–4% wax, 0.8% glycolipids, 1–2% phospholipids, and bioactive compounds such as γ-oryzanol (1.2–1.7%), phytosterols, tocopherols (0.02–0.08%), and tocotrienols (0.025–0.17%) are present in crude RBO. These bioactive compounds in RBO demonstrate antioxidative, antidermatitic, antidiabetic, chemopreventive, and cholesterol-lowering properties which makes it an ideal oil to be used in pharmaceutical, chemical, and food industries. In food industries owing to its palatability, plasticity, and spreadability it finds application in margarine and shortenings manufacturing and can also be incorporated in products like bread, cakes, noodles, pasta, and ice creams to increase their nutritive value without affecting textural and functional properties.

Keywords Rice bran oil · Rice bran · Oryzanol · Health benefits · Functional food

16.1 Introduction

Rice (*Oryza sativa* L.) is the primary staple for around half of the world's populace, principally Asia (Mingyai et al. 2017) and after corn (maize) the second largest crop grown in the world (Inform article 2014). Over the past years the global per capita

U. Yadav (✉)
Department of Nutrition and Dietetics, Manav Rachna International Institute of Research and Studies, Faridabad, Haryana, India

S. Arora
College of Dairy Science and Technology, Lala Lajpat Rai University of Veterinary and Animal Sciences, Hisar, Haryana, India

I. Kaushik
Department of Food Science & Technology, Chaudhary Charan Singh Haryana Agricultural University, Hisar, Haryana, India

B. Tanwar, A. Goyal (eds.), *Oilseeds: Health Attributes and Food Applications*,
https://doi.org/10.1007/978-981-15-4194-0_16

consumption of rice has increased from 50 to 65 kg per annum and the worldwide rice consumption during the year 2015–2016 has been estimated to be 478.441 million metric tons, whereas as estimated by United States Department of Agriculture, world's rice production during the year 2016–2017 was 481.5 million metric tons (Statistica 2016). China and India are the top producers of rice in the world and contribute over 25% and 23%, respectively, to the world's rice production (Ali and Devarajan 2017).

Rice is an annual plant which can rise to the height of 1–1.6 m. It has branched arching to pendulous inflorescence 32–48 cm long and can be grown anywhere, even on mountain area with the use of water-controlling terrace systems or on a steep hill (Orthoefer 2005; Sanghi and Tiwle 2015). The paddy grain comprises the husk, bran, germ, and endosperm. Rice bran is obtained by removing the bran layer which represents 8% of rice paddy by-products during the milling process (Sharif et al. 2014).

Annual rice bran production is up to 610 million metric tons (Gul et al. 2015). The bran removed during polishing of rice comprises of 18–22% oil, fatty acids, protein, dietary fiber, minerals (phosphorus, iron, and magnesium), and functional compounds such as unsaponifiable lipids like γ-oryzanols, polycosanols, phytosterols, squalene, tocopherols, tocotrienols, and carotenoids (Oliveira et al. 2011; Sookwong et al. 2007; Zhang et al. 2010). In Asian countries like Indonesia and India, rice bran is largely being fed to cattle. However, since ages, Japan contemplated rice bran as a valuable source of oil with therapeutic benefits and is known as "Heart Oil" in Japan (Pal and Pratap 2017). In western countries also, it has acquired the status of a "Functional Food" or a "Health Food". Now, in several Asian countries, RBO is emerging as a popular cooking oil, mainly for deep and shallow frying applications (CAC 2003). Although whole rice bran in itself does not have anti-cholesterol properties, its oil offers nutritional benefits. It contains vitamin E (tocopherol and tocotrienol) and γ-oryzanol which, like antioxidants, eliminate or prevent free radicals' formation which may cause damages to the body, and also inhibit proliferation of certain diseases such as some cancers and cardiovascular diseases. It has a good composition of triglycerides, diglycerides, monoglycerides, fatty acids, phospholipids, and unsaponifiable lipids (Wilson et al. 2000, 2007; Most et al. 2005).

Due to the medicinal importance, RBO is nowadays extensively used in pharmaceutical, chemical, and food industries. It is an ideal oil for making margarine and shortenings with good palatability, smooth plasticity, and spreading qualities (Gosh 2007; Liang et al. 2014). Not only RBO, rather rice bran is also used to improve nutritive value in various food formulations like bread, cakes, noodles, pasta, and ice creams without having any detrimental effect on their functional and textural properties (Sookwong and Mahatheeranont 2017).

16.2 Production

Global market size of RBO in 2015 was evaluated over 1.2 million tons and India holds immense potential in the production of rice bran oil (RBO), due to increased paddy production at 150 million tons over the last decade. Currently, the RBO production in India is 9 lakh tons out of which only 3 lakh tons is consumed while rest 6 lakh tons is used by Vanaspati industry or blended with other oils. India is the largest producer of RBO followed by Japan, Thailand, and China (Nayik et al. 2015). As per The Solvent Extractors Association (SEA) data, RBO production in India is remarkably rising by 50,000 tons annually (Nayik et al. 2015). The total revenue collected from the export of oils by India during the year 2015–2016 was nearly 250 crores. Nearly 24,000 metric tons of RBO was exported, having value 22.25 crores INR. Thus, it can be concluded that RBO serves the 10% of the total revenue obtained from the export of oils, which is very notable in increasing India's GDP growth (Pal and Pratap 2017). The global market share of RBO is segmented and ruled by firms such as Ricela Health Foods, 3F Industries Ltd., Sethia, BCL Industries & Infrastructure, and A.P. Refineries Pvt. Ltd. Refined or blended RBO industry also consists of leading marketers such as Adani Wilmar, Marico Ltd., ITC Ltd., and N.K. Proteins (www.gminsights.com/industry-analysis/rice-bran-oil-market).

16.3 Chemical Composition

Crude RBO is rich in unsaturated oleic and linoleic fatty acids, and bioactive compounds, for example, γ-oryzanol (1.2–1.7%), phytosterols, tocopherols (0.02–0.08%), and tocotrienols (0.025–0.17%). The chief component of rice bran oil is the glyceride component which makes 80% of the crude RBO in which 80.5% triacylglycerols, 4.8% diacylglycerol, and 1.8% monoacylglycerols. More than 90% of the fatty acid portion of the glycerides constitute three fatty acids: palmitic, oleic and linoleic acid. Crude RBO contains 3–4% wax, 0.8% glycolipids, 1–2% phospholipids (PL), and 4% unsaponifiable lipids (Gosh 2007; Lu et al. 1991). The fatty acid composition of RBO is comparable to that of groundnut oil, i.e., caprylic acid, capric acid, lauric acid, and myristic acid are not detected in groundnut oil however (0.1–0.5%) is found in RBO. An average 15.4% palmitic and 43.2% oleic acid are detected in both oils (Pal and Pratap 2017; White 2008). Both the oils contain 35% linoleic acid but RBO has a high unsaponifiable lipid fraction (4.2%), which is higher than the level recommended by the Central Committee of Food Standards (Gingras 2002). The shelf life of RBO is higher compared to other cooking oils due to the presence of good amounts of antioxidants. Less oil is absorbed during cooking in RBO due to its low viscosity, thereby reducing overall calorie intake (Gosh 2007).

16.4 Bioactive Components

Apart from an ample amount of wax (1.5–4%) and phosphatides (0.5–1.5%), RBO also contains bioactive compounds such as phytosterols, γ-oryzanol, tocopherols, and tocotrienols. The free fatty acid content (59.19%) of RBO is higher than other plant oils such as groundnut oil, in which about 70% is oleic and linoleic acid, while 22% is palmitic acid (Kaneto et al. 2010; Afinisha and Arumughan 2012; Gopala et al. 2006). A major bioactive component, γ-Oryzanol consists of a group of compounds containing 4-hydroxy-3-methoxycinnamic acid (ferulate) esters of cycloartenol, 24-methylenecycloartanol, β-sitosterol, and campesterol (Fig. 16.1). Crude RBO contains approximately 1.5% γ-oryzanol, which is a component unique to rice bran. γ-oryzanol is marketed with health claims as depicted in Fig. 16.2 for enhancing energy, improving muscle condition and some of the reported physiological effects related to γ-oryzanol and other components include the ability to decrease cholesterol absorption, reduces serum cholesterol and triglyceride concentration, maintain plasma glucose concentration, inhibit early atherosclerosis and platelet aggregation, and enhance excretion of fecal bile acids (Ryan 2011; Friedman 2013; Pali 2013; Pal and Pratap 2017).

It has been reported by Liang et al. (2014) that oryzanol has been used to increase the muscle mass, treat menopausal disorders and hyperlipidemia. Oryzanol present in RBO can increase the release of cholesterol and its metabolites and thereby,

Names	Chemical Structure	Molecular formula
Cycloartenyl ferulate Oryzanol A	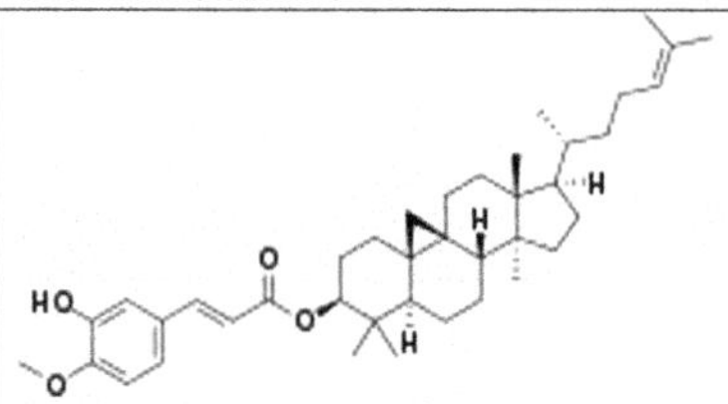	$C_{40}H_{58}O_4$
24-Methylenecycloartanyl ferulate Oryzanol C	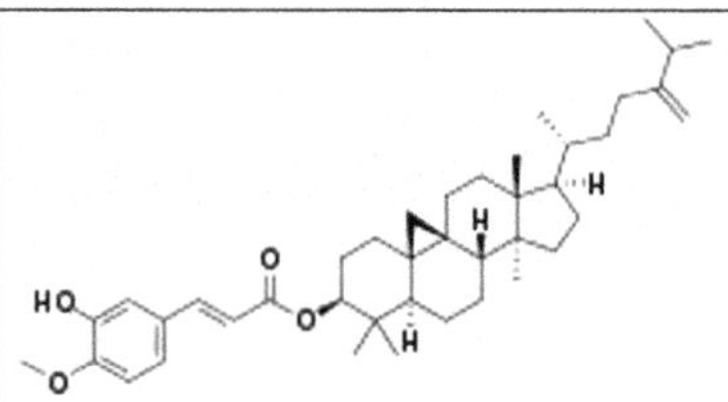	$C_{41}H_{60}O_4$
Campesteryl ferulate	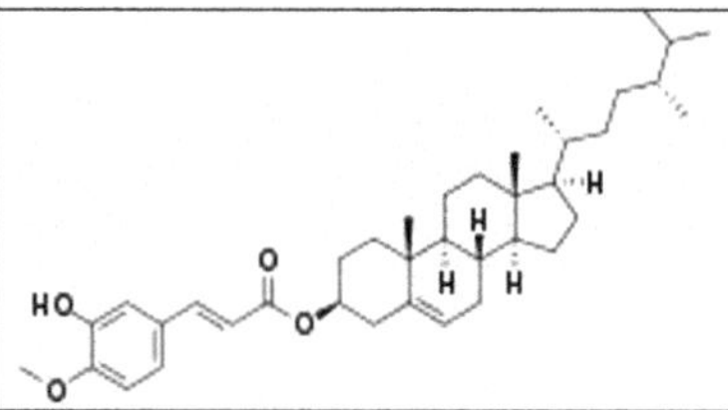	$C_{38}H_{56}O_4$

Fig. 16.1 Structure of oryzanol (Source: https://en.wikipedia.org/wiki/%CE%93-Oryzanol)

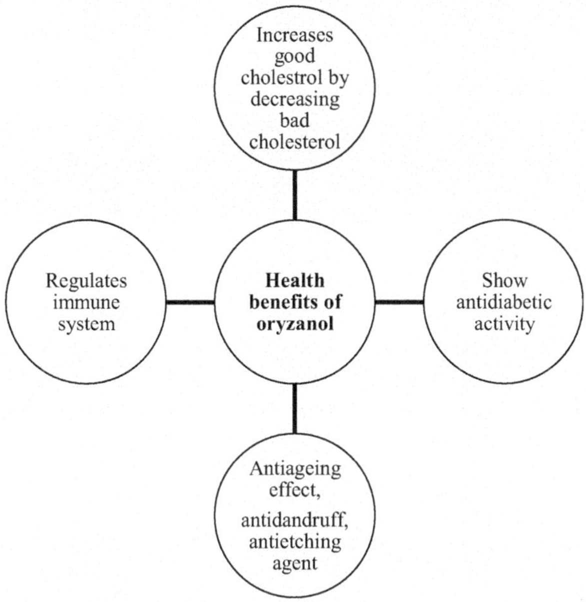

Fig. 16.2 Health benefits of oryzanol

reduce plasma non-HDL-C levels and raise HDL-C (Wilson et al. 2007). Chen and Cheng (2006) stated that consumption of RBO promotes the decomposition of cholesterol by increasing the expression of liver LDL receptors, which in turn is favorable for diminishing LDL-C levels by increasing cholesterol 7-alpha-hydroxylase (CYP7a1) expression. Sierra et al. (2005) studied the effect of diets supplemented with RBO and γ-oryzanol on the immune response of mice and observed that RBO modulates the immune system by boosting B-lymphocyte proliferation.

Studies conducted by various researchers concluded that RBO oryzanol can also prevent the oxidation of linoleic acid and cholesterol even better than vitamin E. Cycloartenyl ferulate and 24-methylenecycloartanyl ferulate present in oryzanol can act as antioxidants in methyl linoleate bulk and multiphase lipid systems and inhibit the oxidation of low-density proteins (Xu et al. 2001; Andreasen et al. 2001; Kikuzaki et al. 2002; Siddiqui et al. 2010). Rice bran oil, like palm oil, also contains substantial amounts (approx. 500 ppm) of tocotrienols. The chemical structure of tocotrienols is similar to the tocopherols and is found in at least four forms. They belong to the vitamin E family and are known to be powerful natural antioxidants, and their ability to prevent chronic disease is comparable to that of γ-oryzanol (Ryan 2011).

16.5 Health Attributes

RBO has lower viscosity and a relatively high smoke point (254 °C) (Rubalya Valantina and Neelameagam 2008), which makes it a healthy cooking oil. RBO is loaded with vitamin E (both tocotrienols and tocopherols) and bioactive phytonutrients, which include phytosterols, γ-oryzanol, squalene, and triterpene alcohols (Ali and Devarajan 2017). All of these compounds exhibit high antioxidant, anti-inflammatory, hypocholesterolemic, antidiabetic, and anticancer activities. The dietary intake of RBO lowers the blood cholesterol level, blood pressure, and blood glucose and can help to reduce inflammation and symptoms of metabolic syndrome. RBO helps to boost the immune system and prevent the process of premature aging and age-related neurodegenerative diseases. Because of its cardiac-friendly phytochemicals and antioxidant potentials, RBO has been categorized as healthy edible oil for human consumption and has attained the status of "heart-healthy oil" (Ali and Devarajan 2017).

16.5.1 Reduction of Dyslipidemia and Cardiovascular Diseases (CVD)

Cholesterol is insoluble in blood and is transported by carriers called lipoproteins in the blood plasma of all the primates. Sizes of lipoprotein vary from high-density lipoprotein (HDL) to low-density lipoprotein (LDL) and very low-density lipoprotein (VLDL). LDL is called bad cholesterol as it gets deposited on the wall of the arteries and form plaque, the major cause of heart attack. A person is at high risk of CVD and arteriosclerosis if LDL cholesterol level is above the normal values. While HDL transports cholesterol from the arteries to the liver to get it metabolized, excreted, or reutilized.

The factors related to the development of CVD are gender, obesity, age, high blood pressure, smoking, and mainly raised cholesterol level in blood. It influences the circulatory system thereby increasing the risk of CVDs like cardiomyopathy, heart attack, and coronary heart disease (Hansson 2005). As per World Health Organization (WHO) report, due to cardiovascular disease over 17 million public died in the year 2008 and it is estimated to increase up to 25 million till 2030 (Celermajer et al. 2012).

An inflammatory reaction occurs in arteriosclerosis influencing large and medium-sized blood vessels that become permeable in the presence of high LDL level and free radicals. The reactions may cause injury in the blood vessel and further block the blood circulation. Apart from the change in lifestyle, increasing antioxidant intake is the right therapeutic approach for preventing the occurrence of CVD (Saklamaz et al. 2005).

Several studies conducted on rabbits, rodents, non-human mammals, and humans demonstrated that the bioactive compound viz., γ-oryzanol, phytosterols, tools, squalene, and octacosanol present in RBO has the ability to reduce total blood plasma cholesterol and triglycerides while raising HDL cholesterol which in turn helps to improve the plasma lipid profile (Cicero and Gaddi 2001; Wang et al. 2003; Wilson et al. 2007). In a study, various unsaturated vegetable oils were evaluated for their effect on serum lipoprotein levels in monkeys. The remarkable decrease of 25% and 30%, respectively, in total serum cholesterol and LDL cholesterol was found in the animals fed with a diet containing RBO (Wilson et al. 2000). In another study, the hypocholesterolemic effect of refined RBO along with coconut oil and canola oil was evaluated in hamsters. The animals were fed on a chow-based diet in addition to 0.03% cholesterol along with 5% oils (w/w) for 8 weeks. A significant reduction ($P < 0.05$) in cholesterol absorption (15–17% reduction), plasma TC and LDL-C (42% and 48%, respectively) was observed in RBO-treated animal groups (Ausman et al. 2005).

The hypolipidemic effect was evaluated for γ-oryzanol fraction of RBO blended with refined groundnut oil gum (3–10%). The result showed a 15% decrease in the cholesterol level in the liver and 7–16% decrease in blood serum cholesterol compared to a diet containing pure groundnut oil (Chandrashekar et al. 2012). To evaluate the antihyperlipidemic effect, the phytochemical γ-oryzanol at different concentrations ranging from 50 and 100 mg/kg was fed to the hyperlipidemic rats for 21 days/3 weeks. Regardless of the dose administered, significant decrease ($P < 0.05$) in the levels of serum cholesterol, triacylglycerides, LDL, VLDL and a significant increase in the level of serum HDL and hepatic antioxidant enzymes were found in the groups treated with the γ-oryzanol compared to control group that was fed on atorvastatin (a pharmaceutical agent administered to lower blood cholesterol) (Ghatak and Panchal 2012). Wilson et al. (2007) evaluated the cholesterol-lowering properties of γ-Oryzanol, trans-ferulic acid, and RBO in hamsters for 10 weeks. In comparison to the control group (hamsters fed on chow-based hypercholesterolemic diet (HCD) containing 10% coconut oil and 0.1% cholesterol), the total plasma cholesterol, LDL and VLDL were significantly ($P < 0.05$) lower in animals fed with γ-oryzanol (0.5%) followed by RBO (10% RBO in place of coconut oil) and trans-ferulic acid (0.5%). Similarly, lipid profile had also been improved using other bioactive components extracted from rice bran oil. Lipid levels were evaluated in rats using regular diet, high cholesterol diet, and diet containing RBO concentrated mixture (composition of the concentrated mixture was 277.7 γ-oryzanol, 89.99 phytosterols, 11.66 tools, 3.36 squalene, and 0.73 octacosanols). Rats fed with a diet containing RBO concentrated mix (15gm/kg diet) revealed significantly ($P < 0.05$) upsurge in HDL level compared with high cholesterol diet. HDL lowers the risk of arteriosclerosis by minimizing the cholesterol deposition in the endothelium and carrying it to the liver for excretion (Ha et al. 2005).

Vitamin E, tocopherols and tocotrienols are synthesized in the plants and are proven antioxidants and exert hypolipidemic effects (Xu et al. 2001). However, prolonged intake of vitamin E and dose higher than RDA could have a pro-oxidant effect (Pearson et al. 2006). In an experimental study, RBO extracted tocotrienol-rich fraction (TRF) was investigated for 1 week in hyperlipidemic rats at a concentration of 0–50 mg TRF/kg/day. TRF-treated rats showed a dose-dependent 38–48%

decreased triglycerides and around 48% decreased total cholesterol content compared to hyperlipidemic rats. Hypocholesterolemic impact of TRF may be due to the repressive role of HMG-CoA reductase enzyme, a rate-limiting enzyme of the mevalonate pathway, the metabolic pathway that produces cholesterol (Frank et al. 2012). Further, administration of 8 mg of TRF/kg/day showed highest antioxidant effect. It may be due to the capacity of TRF to give electrons to free radicals generated during the peroxidation of lipids. And thus, it was concluded that 8 mg TRF/kg/day is appropriate to attain the desired antioxidant and hypocholesterolemic effects (Minhajuddin et al. 2005).

Numerous studies conducted in humans with dyslipidemia have exhibited that γ-oryzanol fraction of RBO in the diet improves the lipid profile, inflammatory and oxidative stress levels (Nicolosi et al. 1991; Sharma and Rukmini 1986; Sasaki et al. 1990; Rong et al. 1997; Rajnarayana et al. 2001; Accinni et al. 2006). Nicolosi et al. (1991) conducted a study on persistent schizophrenic patients having dyslipidemia; they were provided with a diet containing γ-oryzanol at a concentration of 100 mg, three times a day for 16 weeks. The authors reported significantly ($P < 0.05$) decreased LDL-cholesterol (LDL-C) and total cholesterol (up to a 40% reduction), whereas, no change in HDL cholesterol (HDL-C) was observed. Similarly, Sasaki et al. (1990) also reported significantly ($P < 0.05$) decreased triglycerides, lipid peroxides, VLDL-C, and LDL-C after consumption of RBO cooked food for 50 days. Furthermore, as reported by Accinni et al. (2006) dietary supplementation of γ-oryzanol (40.2 mg) and niacin (18 mg), and vitamin E (4 mg) and PUFA (660 mg EPA + 440 mg DHA), could significantly ($P < 0.01$) improve the lipid profile, reactive oxygen species, and total antioxidant capacity of dyslipidemic patients. Studies have reported oryzanol has a hypolipidemic effect possibly either by increasing fecal excretion of cholesterol and its metabolites or by suppressing the HMG CoA reductase activity (Wilson et al. 2007; Seetharamiah and Chandrasekhara 1990).

Further, non-saponifiable lipids of RBO, oryzanol, contributes to the cholesterol-lowering action of RBO associated with significant ($P < 0.01$) reductions in aortic fatty streak formation (67%) (Rong et al. 1997).

Another study was performed in 14 human subjects who consumed RBO diet (contributing one-third of total dietary fat) for 10 weeks substituting oil blends. (The diet contained vegetable oil blends of canola, corn, peanut, olive, and palm oil and butter having the similar fatty acid composition to that of RBO. Both diets were designed to provide 37% of total energy as fat.) Significant decrease in total cholesterol and decrease in LDL cholesterol by 7% ($P < 0.0004$) were found and no changes appeared in the HDL levels during the evaluation on subjects taking RBO diet. Further, an LDL constituent (apolipoprotein B) level was also found lower (0.97 g/L) in the subjects taking RBO as a diet fat (Most et al. 2005). Kuriyan et al. (2005) studied RBO and sunflower oil as cooking oils in 14 subjects of age group 40–60 years with known hyperlipidemia, and their lipid profiles were evaluated for two periods each of 3 months including a washout period of 3 weeks. A steady decline ($P < 0.05$) in serum triglycerides and total cholesterol (30% and 8% respective reduction) were found during both the periods when RBO was used for cooking. The authors suggested that RBO as cooking oil should be included in daily routine because of several advantages and health benefits. The above-mentioned studies suggested that the RBO unsaponifiable fraction contains γ-oryzanol accredited to these health benefits in humans. To ascertain the role of γ-oryzanol concentration in creating any variation in cholesterol levels, another study was conducted wherein 30 hypercholesterolemic subjects were chosen and divided into two groups. γ-oryzanol at a level of 0.05 g per day (low level) or 0.8 g per day (high level) concentrations were given for 4 weeks. Regardless of the concentration used, the LDL and HDL cholesterol, total plasma cholesterol, and triacylglycerol were lowered in both the groups compared to its initial level (assessed at the start of the study). Low and high gamma-oryzanol containing RBO feeding for 4 weeks lowered total plasma cholesterol (6.3%), LDL-C (10.5%), and the LDL-C/HDL-C ratio (18.9%). Methylated sterols (phytosterols) in gamma-oryzanol is ineffective at inhibiting dietary cholesterol absorption, but 4-desmethylsterols could enhance its cholesterol-lowering ability. If all ferulated sterols (esterification of plant sterols) become de-ferulated (de-esterified) in the gut, γ-oryzanol at low and high concentration (level of 0.05 g per day or 0.8 g per day in the present study) in RBOs provided intestinal loads of 453 and 740 mg/day free 4-desmethylsterols, respectively. This intestinal load of 453–740 mg/day of efficacious free plant sterol equivalents had similar effects on lipoproteins. The study further confirmed that γ-oryzanol might improve lipid metabolism (Berger et al. 2005).

The mechanism by which RBO works on the metabolism of lipid has not been completely elucidated. Rice bran oil contains unsaponifiable and saponifiable lipid fractions. The unsaponifiable fraction such as phytosterols, polyphenols, and other RBO extracted bioactive components are supposed to contribute toward the antihyperlipidemic effect, while the saponifiable lipid fractions did not seem to contribute toward lipid metabolism. γ-oryzanol is a natural antioxidant and is composed of four main components (cycloartenyl ferulate, 24-methylenecycloartanyl ferulate, campesteryl ferulate, and sitosteryl ferulate). γ-oryzanol composition and its amount in rice depend upon the rice species and its cultivation environment (Krishna et al. 2001; Bergman and Xu 2003). In vitro scavenging activity of each form of γ-oryzanol had been investigated separately and similar scavenging ability had been found in all of its constituents viz., cycloartenyl ferulate, 24-methylenecycroartanyl ferulate, campesteryl ferulate, and beta-sitosteryl ferulate (Akiyama et al. 2001). γ-oryzanol stops the free radical chain reaction at higher concentrations compared to tocopherol fraction that captures organic radicals at very fewer concentrations. Therefore, tocopherol fraction of RBO had come out to be a more effective antioxidant (Juliano et al. 2005). In general, all the bioactive constituents viz., γ-oryzanol, phytosterols, tocols, squalene, octacosanol, etc., of RBO contribute toward the antioxidant and hypolipidemic effect. In this perspective, the notion that a variety of RBO constituents collectively are liable for the overall effect on lipid balance is the most valid proposition concerning the valuable effects of these bioactive constituents (Cicero and Gaddi 2001).

16.5.2 Prevention of Diabetes Mellitus

Diabetes mellitus (DM), a metabolic disorder, leads to hyperglycemic conditions, resulting in defects in insulin action, insulin secretion or both along with severely disturbing the metabolism of carbohydrate, protein, and fat (American Diabetes Association). The frequency of DM had risen considerably in the last few years and the reason could be lifestyle, inappropriate and unhealthy dietary intake, and an increase in obesity. The multiple complications due to DM include neuropathy, retinopathy, vascular damages, and various heterogeneous diseases. DM is of two types, type 1 called insulin-dependent DM (IDDM) and type 2 referred to as non-insulin-dependent DM (NIDDM). IDDM is found in people having reduced or no endogenous secretion of insulin thereby exogenous insulin therapy is required (Wu et al. 2014).

Whereas, in NIDDM, insulin resistance and insulin secretion had been impaired in individuals due to obesity and lack of physical exercise (Bastak 2005).

The hyperglycemic condition also leads to the fatality of renal cells by stimulating DNA disintegration and thereby the development of renal pathology (nephropathy) (Keim et al. 2001; Allen et al. 2003). Supplementation of Vitamin E (544 mg/kg) in the diet was linked to the reduction of DM problem (Halim and Mukhopadhyay 2006). The RBO-extracted TRF was evaluated for its capacity to improve renal functions in hyperglycemic rats. Compared to the untreated group, the group treated with TRF-RBO at a concentration of 200 mg/kg/day over a period of 8 weeks revealed significant ($P < 0.001$) refinement in renal functions having diabetic nephropathy by virtue of its hypoglycemic status, reduced diabetic proteinuria, blood urea nitrogen (BUN), total nitric oxide levels in both serum and urine, and oxidative stress parameters of rats with type 1 DM (Siddiqui et al. 2010). Jung et al. (2007) studied the type 2 diabetic mice for hypoglycemia by oral administration of ferulic acid (FA) (0.05 g/kg/day). The study was conducted for 17 days and after its completion, there was significantly decreased blood glucose levels (416.8 mg/dL) and increased plasma insulin levels ($p < 0.05$) than the control group. Further, FA-treated groups had significantly ($p < 0.05$) elevated hepatic glycogen content (22.88 mg/g) and glucokinase activity compared with the control group (the difference was not significant).

Humans' adipose tissue synthesizes and produces cytokines; a group of proteins that act as chemical messenger. Among cytokines, adiponectin, a hormonal protein secreted by fat cells, circulates in the blood and directs various metabolic signaling like regulation of blood glucose and breakdown of fatty acid. Too much dietary fat, particularly from meat, results in a condition called insulin resistance in the adipocyte cells which can hinder the production of adiponectin, resulting in type 2 DM (Vettor et al. 2005). Nagasaka et al. (2011) studied the effect of γ-oryzanol in mice for lowering serum cholesterol by measuring adiponectin serum levels. The animals were fed orally on a diet that included palmitate acid (0.17 mg/mL in corn oil, $n = 5$), beef tallow group (0.5 mL), and corn oil group (0.5 mL, control group, $n = 5$) to induce hypo-adiponectinemia. In addition, 0.025 mmol of γ-oryzanol was given along with beef tallow or corn oil to both animal groups. All groups were analyzed after 120 h. Significantly ($P < 0.05$) increased adiponectin levels in mouse plasma were observed in both the groups fed on γ-oryzanol along with beef tallow or corn oil in comparison to mice fed on corn oil. It was concluded that the secretion of insulin could be regulated and the functioning of a peptide hormone (adiponectin) could be normalized by the γ-oryzanol intake thereby the risk of hyperglycemia associated with high-fat diet could be minimized. It was discussed further that adiponectin is one of the most important adipocytokines and is the improvement factor of obesity and insulin resistance. In this study, it was demonstrated that high-fat diet (beef tallow, in particular, palmitate in the present study) could reduce serum adiponectin level in mice. Stimulation of adiponectin production (via γ-oryzanol) could control this condition; therefore, these results could be simulated for a possible drug discovery and the obese-related metabolic diseases therapy. Somsuvra et al. (2012) evaluated the hypoglycemic potential of γ-oryzanol in streptozotocin (STZ)-induced diabetic rats for 11 days. The results exhibited a decline (ns) (initial glucose level of 340–400 mg/dL) in the serum glucose level (13.09% and 12.38% respectively) in these rats 4 h after administration of γ-oryzanol at 100 mg/kg and 50 mg/kg body weight respectively. At the end of the study, γ-oryzanol at higher concentration (100 mg/kg) demonstrated a highly significant ($P < 0.05$) reduction in serum glucose level compared to the STZ-treated control group.

16.5.3 Tumor and Cancer Reducing Effects

Reactive oxygen species (ROS) and free radicals are produced continuously as a by-product of various oxidative metabolic processes. These highly reactive compounds if present in excess concentrations could damage the cellular structure, cell wall and genetic material within the cell, and further trigger the occurrence of several diseases including cancer (Valko et al. 2004). The enzymatic and non-enzymatic antioxidants deactivate the free radicals at the molecular and cellular level. Certain enzymes are naturally produced inside the human body such as catalase, superoxide dismutase, and glutathione peroxidase that controls the free radical production and cell damage to some extent via enzymatic mechanism (Allen and Tresini 2000). Natural non-enzymatic antioxidants such as tocopherols, tocotrienols, vitamin E, vitamin A, and certain phytochemical compounds have known antioxidant activity that further helps to control the occurrence of free radicals (Liu 2003).

Accretion of genetic and molecular defects in epithelial cells of the colon such as variation in normal epithelial cells followed by massive proliferation leads to colon carcinogenesis. Shih et al. (2011) conducted a study in which rats were used as a colon pro-carcinogen model (azoxymethane (AOM) or 1,2-dimethylhydrazine). The rats fed with a high-fat AIN-93G diet containing RBO (5, 10, and 15%) demonstrated significantly lower tumor formation in the colon and preneoplastic bruise. Further, the level of enzymes in the liver such as superoxide dismutase, catalase, glutathione, and thiobarbituric acid reactive substances was also improved

($P < 0.05$) especially in a diet containing 10 and 15% RBO. Similarly, Panala et al. (2009) reported that bioactive constituents present in RBO (7 or 14%) inhibit the growth of the cancer cells; therefore, the tumor formation was decreased in RBO diet fed to azoxymethane-treated rats compared to control diet (formulations of diets based on AIN-93G) after 45 weeks. The outcome was correlated with the presence of tocopherols, tocotrienols, squalene, γ-oryzanol, inositol hexaphosphate (IP6), phytosterols, etc., present in RBO. Phytic acid, a natural antioxidant, present in rice bran is known to inhibit the formation of hydroxyl radicals (HO). Transformation of a normal cell into cancer cells involves a metal-mediated reaction, triggered by the presence of iron. Potentially reactive iron may present in large amount in the intestine (specifically in a population consuming red meat). It probably suppresses the iron-mediated step during the formation of cancer (Graf and Eaton 1993). In an experimental model, rats were treated with azoxymethane (a potent carcinogen) and fed on phytic acid (0.2% w/v) extracted from RBO and it resulted in a reduction in colon tumor formation and thus, showed the potency of RBO in reducing colon cancer (Norazalina et al. 2010). Iqbal et al. (2003) revealed the chemopreventive activity of diet loaded with TRF of RBO against breast cancer carcinogen 7,12-dimethylbenzanthracene (DMBA, a chemical carcinogen known to induce mammary carcinogenesis). Feeding of TRF (10 mg/kg body weight/day) for 6 months to DMBA-administered rats, reduced the severity of neoplastic transformations and was indicative of lower carcinogen-stimulated tissue damage.

Kong et al. (2009) studied the topical application of cycloartenyl ferulate (CF), a phenolic compound extracted from RBO to inhibit cancer (cell lines SW480 and SW620) in mice in vitro growth inhibition on human colorectal adenocarcinoma SW480. The mechanism linked with CF-induced chemoprevention was established by in vitro SW480 CRC cell line. The study exhibited the capacity of CF to elicit apoptosis in SW480 cells via activation of caspase-protein (enhanced activation of caspase-8 and -3) in both extrinsic and intrinsic pathways. Further, the growth and survival of normal colon CCD-18-Co cells were supported by CF.

Physical exercise is commonly predicted to have a positive impact on the physiological functions of the body and it promotes overall well-being. Health benefits of exercises are huge and already known but exhaustive exercise might create oxidative stress. Excessive exercise creates oxidative stress in the body due to the imbalance between antioxidants concentration and free radicals (Urso and Clarkson 2003). Ismail et al. (2010) designed an exercise program to induce oxidative stress in rats. The rats of the same sex, weight, and strain had performed swimming exercise 60 min/day, for 5 days in a week and continued for a period of 10 weeks. The gamma-oryzanol rich fraction (ORF) was extracted and fractionated from rice bran using supercritical fluid extraction (SFE) and given to rats to study the effect on antioxidant and oxidative stress related genes in liver. The results revealed that animals treated with ORF emulsion had strong antioxidant activity (upregulated the antioxidant genes) and had the potential to downregulate gene expression linked with oxidative stress when compared with commercially available gamma-oryzanol preparations.

16.5.4 Other Health Attributes

Nutritional status and intake of quality food have a direct impact on the functioning of the immune system, response to various infections and autoimmunity. The immunological cell membrane is susceptible to damage by free radicals; therefore, a fair supplementary ingestion can assist in safeguarding these cells (Hodin et al. 2012).

RBO is rich in gamma-oryzanol, phytosterols, and sterolins, compounds with proven antioxidant properties that could modulate the immune system and thus has proven immune-stimulatory effects (Sierra et al. 2005). Ghatak and Panchal (2012) evaluated the antioxidant potential of γ-oryzanol at different concentrations viz., 25, 50, and 100 mg/kg for immune responses in experimental rats for 21 days. The results revealed that regardless of the dose administered, γ-oryzanol might induce both cellular and humoral immunity, making γ-oryzanol a potential immunomodulator. Islam et al. (2008) studied the anti-inflammatory response from RBO-extracted components viz. γ-oryzanol (γ-ORZ), which contains a number of phytosteryl ferulates, such as cycloartenyl ferulate (CAF), campesteryl ferulate, b-sitosteryl ferulate, and campestanyl ferulate one of the components of γ-ORZ, and ferulic acid (FA), a possible metabolite of γ-ORZ in vivo. The responses of γ-ORZ component viz. γ-oryzanol and cycloartenyl ferulate at a dosage of 50 mg/kg/day was evaluated for extreme colitis condition stimulated in mice. Significant inhibition ($P < 0.01$) in the inflammatory response was observed in both the components extracted from RBO. It was further concluded that the anti-inflammatory effect of phytosteryl ferulates could be mediated by inhibition of NF-kB activity. The inhibition effect could be partially due to FA moiety in the structure of phytosteryl. Therefore, it could be used as effective therapeutic or defensive agents for gastrointestinal inflammation. Further, Akihisa et al. (2000) isolated six novel feruloyl esters of triterpene alcohols and sterols and six known ferulates from the methanol extract of edible rice bran. The anti-inflammatory activity of all the compounds was evaluated for mouse ear inflammation stimulated by 12-O-tetradecanoyl phorbol-13-acetate (TPA). Marked inhibitory activity had been shown by all of the ferulates at an inhibitory dose (ID50) of 0.1–0.8 mg per ear.

The effect of RBO-derived cycloartenyl ferulate (cycloartenol ferulic acid ester; CAF) on skin allergy was studied in the dorsal skin of rats. The extract was injected intradermally with anti-DNP IgE. The results showed that the non-polar structure of cycloartenyl ferulate proved to be capable of sequestering the immunoglobulin E which usually binds to and activates mast cells and thereby inhibits the allergic reaction mediated by mast cell degranulation. CF attenuates mast cell degranulation and prevents the binding of immunoglobulin E (IgE) with allergy-related signaling receptor FceR1 (Oka et al. 2010).

16.6 Adverse Effects and Individual Concerns

As RBO is used as a cooking oil in various food preparations; it is very important to assess its chemical and nutrient composition, nutrient quality, and toxicological safety. Various studies have been conducted which indicate that consumption of RBO in the diet does not have a severe effect on absorption, utilization, and growth promotion in rats (Sharma and Rukmini 1986; Seetharamiah and Chandrasekhara 1989). The bioactive components (β-sitosterol, campesterol, stigmasterol, squalene, γ-oryzanol, tocopherols, and tocotrienols) of RBO did not reveal any toxicity and carcinogenicity in in-vivo assays in mice and rats (Tamagawa et al. 1992a, b). Toxicological effects of RBO have been evaluated in rats by conducting multigeneration studies in which they were fed RBO at 10% level and reproductive performance was judged by the percentages of consumption, birth weight, litter size, weaning weight, and pre-weaning mortality (Rukmini et al. 1980). Also, RBO did not show any mutagenicity when tested by a bacterial reverse mutation in Ames mutagenicity assays (Polasa and Rukmini 1987; Kusum et al. 2011).

Studies have also revealed that RBO is not a sensitizer and does not cause allergic reactions. Ocular toxicity assays showed negative results for RBO, rice germ oil (RGO), and rice wax. Dermal exposure assay confirmed that RBO and RGO were not phototoxic and had a UV absorption maximum at 315 nm (Amended Final report on Safety Assessment of RBO 2006). No side effects were observed in adults and children even at high doses (1.5–3.3 g/day) of phytosterols from the RBO as they are poorly absorbed and can effectively be excreted via biliary route (Becker et al. 1993; Weststrate and Meijer 1998). Araghi et al. (2016) evaluated the toxicity and safety aspects of RBO in the chicken embryo model. They demonstrated that RBO showed no toxicity in the chicken embryo model, and therefore, it might be regarded as safe for human consumption. Oluremi et al. (2013) observed that crude RBO may contain some heavy metals (Pb, As, Cr, Se, Cd, Ni), and therefore it should be refined to reduce the Fe and Cu overloads (60 mg/kg. 0.7 mg kg, respectively), as they may appear in higher quantities than recommended by the CODEX range (1.5 and 0.1 mg/kg respectively).

16.7 Food Applications

The bran obtained by industrial milling of rice is a powder-like material, containing pericarp, aleurone, pulverized germ, and little endosperm. The rice bran and germ are rich in nutrients like protein, lipids, vitamins, and trace minerals and can help alleviate nutritional deficiencies. The utilization of bran is an economic and sustainable solution for consumers, processors as well as environment. Owing to its high nutritional value and functional properties, a wide number of food applications of rice bran have been identified (Table 16.1). Nowadays, consumers are very enthusiastic about their fitness, well-being, and diet and prefer to consume food with

Table 16.1 Food applications of rice bran, rice bran oil, and rice bran proteins

Samples	Product	Usage	Function/Results	Reference
Rice bran protein isolate	Rice noodles	10%	Improved nutritional quality, dough properties, cooking quality	Kim et al. (2014)
Rice bran protein concentrate	Biscuits	5, 10, 15%	The increased protein content of the product	Yadav et al. (2011)
Rice bran protein concentrate	Bread	1–5%	Less weight loss in the sample (30.42%) than control sample (43.95%)	Jiamyangyuen et al. (2005b)
Rice bran oil	Trans-free margarine	40%	Better physical properties, crystallization, and melting behaviors than control	Ornla-ied et al. (2016)
Rice bran oil	Bakery shortening	50%	Improved the baking and organoleptic properties in bread and cookies	Kaur et al. (2012); Sekhon et al. (1997)
Stabilized rice bran	Bread and cookies	5, 10, 15%	Lysine and dietary fiber contents were increased	Ameh et al. (2013)
Defatted rice bran	Cookies	10–20%	Improved texture and nutritive value	Sharif et al. (2009)
Rice bran oil	Low heat whole milk powder	0.1–0.2%	Improved oxidative stability	Nanua et al. (2000)
Rice bran fiber	Pork meat proteins	0.1, 0.5, 1, 2%	Improved moisture retention, protein solubility, and water-holding capacity in samples having 1% rice bran fiber	Choi et al. (2011)

therapeutic benefits, which has subsequently led to increased functionality of rice bran to be used as a food additive to develop functional foods (Luh and Liu 1980).

Rice bran in the form of full-fat rice bran, defatted rice bran, extracted rice bran oil, and protein concentrates can be used in food (Prakash 1996). RBO is not only known to be a superior culinary oil but can also be utilized in the production of salad dressings and mayonnaise (Swern 1972). Sharma (2002) reported that RBO is more stable at a higher temperature, have a longer shelf life, and offers enhanced taste and flavor to food items compared to other refined oils, and thus, can be used in the preparation of snack foods (Sarkar 1992). Also, frying in RBO takes lesser time which saves energy and absorption of oil is 15% less, making it more economical. Nutritional quality of RBO makes it suitable for nutraceutical products and the antioxidant properties of RBO make it a potential food additive to be used to enhance the storage stability (Nanua et al. 2000; Kim and Godber 2001).

16.7.1 In Bakery Products

Rice bran oil can maintain its nutritive quality even at high temperatures (254 °C smoke point) and thus can be used to make margarine, spreads, and shortenings for use in the baking industry (Eady et al. 2011). Numerous researchers have reported that rice bran is suitable for the development of various baked products like doughnuts, pancakes, waffle mixes, multigrain bread, specialty bread, muffins, pastries, pasta, crackers, and cookies. It not only enhances the nutritive value but also the sensory and functional properties of the product like color, flavor, emulsifying and foaming capacity, water, and fat absorption (Sharif 2009; Hu et al. 2009; Islam et al. 2012; James et al. 1989; Al-Okbi et al. 2016; Esa et al. 2013; Phongthai et al. 2017).

Sekhon et al. (1997) and Singh et al. (1995) reported that defatted rice bran can be used for making cookies by substituting rice bran 10–20% of the wheat flour, without adversely affecting the quality. In another study, Phimolsiripol et al. (2012) developed gluten-free bread with rice bran (10%) having enhanced final bread quality, like darker crust color, higher volume, and a softer crumb. Kaur et al. (2012) reported that the linoleic acid content of bread was enhanced (34.98%) when bakery shortenings made with RBO were used. As RBO is rich in antioxidants, Sharif et al. (2005) incorporated it at various levels in cookies (0, 25, 50, 75, and 100%) by gradually replacing shortening to improve the shelf life. The results indicated that thiobarbituric acid (TBA) number decreased (0.03 in 100% RBO compared to 0.08 in control sample) and the onset of rancidity was hindered with the increase in the percentage of RBO. Moreover, the study suggested that 50% RBO replacement as bakery shortening can produce superior quality cookies in terms of organoleptic properties and shelf life.

Trans-free inter-esterified hard fat (IEHF) can be used as a substitute to partially hydrogenated fat for application as shortening and margarine and are formed by lipase-catalyzed inter-esterification from RBO, fully hydrogenated soybean oil (FHSBO), and coconut oil (CO) (Adhikari et al. 2012). Similar shortenings were made from interesterified RBO with palm stearin-based vegetable shortening, which results in the production of low trans fatty acids (Reshma et al. 2008).

In a study by Shaik et al. (2017), rice bran oil semisolid fraction (RBOSF) spreads were prepared and standardized by edible gelators/hydrocolloids (5% for hard variant and 2% for soft variant), which was then used in baked products such as cakes with desired quality and sensory parameters such as sponginess, moistness, flavor, texture and overall acceptability and very low amount of trans fats. Mayonnaise-type spreads containing 37–43% RBO and 4–7% soy protein concentrate and 52–57% water were successfully prepared (Gracia et al. 2009).

Hydrolytic rancidity is common in rice bran owing to the presence of inherently occurring enzyme activity (lipases), therefore there is a need to stabilize it for effective utilization in the food industry (Sharma and Chauhan 2002). Carroll (1990) incorporated stabilized rice bran up to 20%, in the production of yeast bread. Improvement in moisture retention was observed in bread due to the hygroscopicity of the rice bran and more air incorporation in leavening processes was seen by its ability to foam. Younas et al. (2011) concluded that supplementation of heat stabilized rice bran (HSRB) at 10–15% is suitable for the production of rice bran supplemented cookies with better nutritional quality and sensory acceptability. In another study, Mishra (2017) prepared cookies from wheat flour with supplementation of microwave stabilized defatted rice bran at 5, 10, 15, and 20%. It was found that the average width and spread factor of cookies increased proportionally with the increase in the level of rice bran. The moisture, ash, protein, fat, and fiber content of cookies increases with the supplementation of rice bran from 3.14% to 7.28%, 0.69% to 2.55%,11.23% to 13.84%, 14.27% to 17.03%, 0.16% to 9.5% respectively, whereas carbohydrate content and energy value decreases from 70.51% to 49.16% and 455.39 Kcal to 397.89 Kcal. The results revealed that from the overall acceptability rating, that cookie with 10% RB obtained the highest score at $p \geq 0.05$.

16.7.2 In Dairy Industry

RBO and rice bran have been widely used as animal feed to improve the quality and quantity of the milk produced by cattle. There is the potential for RBO as a natural antioxidant in whole milk powder (WMP). Addition of RBO to milk can also increase the vitamin E content of the milk (3.7 μg/g and 7.5 μg/g, for 0.1% and 0.2% RBO supplementation, respectively).

Nanua et al. (2000) investigated the effect of addition of RBO (0.1 and 0.2%) on the oxidative stability of low-heat and high-heat whole milk powder (WMP). Milk having 3.6% fat was fortified with RBO and was concentrated and dried. The results indicated that RBO, when added at 0.1% of the original milk, was effective in reducing the oxidation of low-heat WMP without imparting a detectable flavor change.

Abbas et al. (2017) successfully developed nutritious yogurt fortified with probiotic bacteria and RBO. RBO was incorporated at 1, 2, and 3% level in skim milk and control was cow's milk with 3% fat content. Starter culture (*L. delbrueckii* subsp., *bulgaricus,* and *S. salivarius* subsp., *thermophilus*) and *B. bifidum* (1:1:1) were added at the rate of 3% (v/v) for yogurt preparation. Results elucidated that the viscosity of resultant yogurt was significantly ($p < 0.05$) increased as rice bran oil percent increased either fresh or at all the storage period up to the seventh day of storage. The sensorial characteristics of yogurt with 3% RBO were almost very close to the control one.

16.7.3 In Meat and Meat Products

Defatted rice bran is extensively used in foods like protein supplements and as a binding constituent for meat and sausage products. Hammond (1994) concluded that chicken coated with stabilized rice bran fiber (RBF) absorbs a low amount of fat during frying. In many studies, RBF was added to meat batters and emulsions to provide better water retention, superior emulsion stability and rheological properties (Choi et al. 2007a, b, 2008¸ 2009; Turhan et al. 2005) and to significantly decrease cooking loss (Paneras and Bloukas 1994; Crehan et al. 2000; Hughes et al. 1997; Yang et al. 2007). Huang et al. (2005) also concluded in their study that pork meatballs with 10% rice bran showed good textural properties. In a study conducted by Kim et al. (2000), crude rice bran oil 0, 1%, and 2% (w/w), was added to restructured beef roasts that were stored at 4 °C and analyzed at 0, 7, and 14 days to determine nutritional properties and oxidative stability. It was observed that TBARs numbers were 50% lower ($p < 0.05$) in roasts with rice bran oil after 7 days of storage compared to control. The addition of 2% rice bran oil (w/w) was effective in improving both oxidative stability and vitamin E levels.

16.7.4 In Confectionary

The industrial application of rice bran in confectionary comprises supplementation of rice bran in a granola bar, and fruit and fiber bars, rice bran cereal, etc. Rice wax is used as coatings for chocolates, cakes, and tablets, a mold-release agent, brightener, a wax coating of vegetables and fruits, and plasticizing material of chewing gum (Nagendra Prasad et al. 2011; Friedman, 2013). Rice bran protein can be used as an outstanding base material for food with high sugar like cake batters, whipped toppings, frozen desserts and confections (Cao et al. 2009; Phongthai et al. 2017).

16.7.5 In Protein Concentrates

By-products from cereals like rice, maize, barley, wheat, and rye can be used to prepare protein concentrates (Lasztity et al. 1995). Beside good amount of oil present in rice bran, it also has 10–15% protein content, out of which 37% are water-soluble, 31% are salt-soluble, 2% are alcohol-soluble, and 27% are alkali-soluble storage proteins (Fabian and Ju 2011). Alkaline extraction and acid or heat precipitation are used for the extraction of proteins from full-fat or defatted rice bran. The amount of protein in concentrates ranges from 19.4 to 76.1% from full-fat rice bran and 17.5 to 85.0% from the latter (Prakash 1996). Rice bran proteins are of superior quality and have various applications in food and pharmaceutical industries due to their functional properties (emulsion properties, solubility, balanced amino acid profile, etc.) (Jiamyangyuen et al. 2005a). Rice bran protein isolates have promising emulsification properties and solubility; consequently, they are used in food products as a natural emulsifier (Lee et al. 2004). Emulsion properties and solubility of rice bran protein hydrolysates make them suitable for use in products like toppings, beverages, coffee whiteners, confectionary, meat, and bakery products (Hamada 2000). Rice bran protein extracts have been added in liquid foods like milk and other drinks and baked goods like bread to increase the protein content of products without compromising their quality (Watchararuji et al. 2008; Jiamyangyuen et al. 2005b). Rice bran proteins can be used in infant foods as it has balanced amino acid profile and hypoallergenic nature (Wang et al. 1999).

Edible films having good tensile strength can be made from rice protein concentrates in amalgamation with polysaccharide pullulan (Shih 1996). Rice bran protein (RBP) was used to make edible protein film in a study conducted by Adebiyi et al. (2008). It was observed that the puncture strength (PS) of RBP films increased up to pH 8.0 and then decreased. PS of protein films depends on the degree of protein purity, quality, and composition. Shin et al. (2011) concluded from their study that composite films, with 4% RBP and 4% gelatin, were the most desirable with regard to the physical property of films, having the highest TS of 28.42 MPa. These results suggest that the RBP/gelatin blend film can be applied to food packaging.

16.8 Alternative Applications

16.8.1 RBO Emulsion

Cold-pressed oils without additional synthetic antioxidants have adequate shelf stability and improved safety as they retain higher levels of natural antioxidants (Singer et al. 2008). Cold-pressed rice bran oil (CPRBO) comprises endogenous antioxidants (oryzanols, tocotrienols, tocopherols), having the potential for use in the prevention of oxidative damages and promotion of health. Thanonkaew et al. (2015) designed an emulsion-based delivery system to protect CPRBO against lipid oxidation, though maintaining good physical stability and oral bioavailability, using glyceryl monostearate (GMS) as an emulsifier. An oil-in-water emulsion was successfully formed with relatively small droplets ($d < 200$ nm) from CPRBO. Bernardi et al. (2011) developed a nanoemulsion which composed of 10% rice bran oil, 10% surfactants sorbitan oleate/PEG-30 castor oil, 0.05% antioxidants, and 0.50% preservatives formulated in distilled water. The nanoemulsion was stable, non-irritating, increased the relative hydration of the skin, the skin oiliness, and maintained normal skin pH values. Such emulsions find its applications in the cosmetic industry in sunscreen formulations (Coppini et al. 2001), anti-aging products (Patel and Naik 2004), topical formulations (Lilitchan et al. 2008), and in the pharmaceutical industry as an alternative treatment for skin diseases like atopic dermatitis and psoriasis (Bernardi et al. 2011; Lerma-García et al. 2009).

16.8.2 Rice Bran Wax and Organogels

The major steps involved in RBO refining are dewaxing, degumming, deacidification, bleaching, deodorization, and winterization (Martini and An 2003; Vali et al. 2005; Ghosh and Bandyopadhyay 2005). Rice bran wax (RBX) is removed in dewaxing step and may have potential application in various industries like cosmetic, pharmaceutical, food, polymer, and leather (Vali et al. 2005). Organogels are viscoelastic materials made up of organic gelators and liquid oils (Abdallah et al. 2000). The liquid oil phase is immobilized by a three-dimensional arrangement of self-assembled, inter-twined gelator fibers (Vintiloiu and Leroux 2008). Organogels are used in cosmetics, deodorants, and hair-care materials (Pernetti et al. 2007). Organogels have desirable properties like easy handling, fast melting, enhanced softness, spreadability, easy film formation, and water-barrier properties (Bot and Agterof 2006; Bot et al. 2008). Dassanayake et al. (2011) studied the properties of organogels made from rice RBX, Candelilla Wax (CLX) and Carnauba Wax (CRX) and observed that the rate of organogel formation, hardness, and the thermal stability of wax crystals were highest for RBX gels compared to CLX and CRX.

16.8.3 RBO as Antifoaming Agent

Foam formation during chemical processes is undesirable and needs to be collapsed, and is easily achieved by chemical, mechanical, or thermal means (Perry and Chilton 1973). Dewaxed RBO can be used as an anti-foaming agent in the chemical and food industry (El-Zanafi and Zaher 1990). They observed that the power of dewaxed RBO to break and control the foam formation in an aqueous solution of sodium dodecyl benzene sulfonate was greater than that of commercial oleic acid. The foam height could be reduced from 17 cm to 14 cm using 0.5 cm^3 oleic acid and to 10 cm using 0.5 cm^3 dewaxed rice bran oil.

16.8.4 In Bio Butanol Production

Increase in global warming is witnessed by researchers due to the escalation of carbon dioxide emissions in the surroundings by incineration of fossil fuels; therefore, search for alternative fuel for transportation is the need of the hour. Defatted rice bran (DRB) is a cheap and renewable agricultural product accessible all over the world. DRB is an appealing feedstock for conversion into ethanol as it contains many carbohydrates (39% cellulose and 31% hemicellulose) and cellulosic polysaccharides and less lignin (4%) (Tanaka et al. 2006; Chandel et al. 2009). Al-Shorgani et al. (2012) found poor butanol fermentation performance in rice bran hydrolysates compared to defatted rice bran hydrolysates which were likely due to the presence of oil in the rice bran. Therefore, the extraction of oil from rice bran could increase the butanol production by 88%.

16.8.5 In Packaging

In a study done by Nihul et al. (2014), a natural-based plasticizer, epoxidized rice bran oil (ERBO), was produced by epoxidation of RBO with peroxy acid generated in situ and then was added to polyvinyl chloride (PVC). It was found that 60% of conventional plasticizer di-(octyl) phthalate (DOP) can be replaced by synthesized ERBO. ERBO presented fairly good incorporation and plasticizing performance, as revealed by the results of mechanical properties, exudation, migration tests, and thermal stability, thus can be effectively used as a plasticizer in the plastic industry. From the results obtained in a study by Kuriakose (1995), it was evident that raw rice bran oil can be advantageously used in the sulfur vulcanization of styrene butadiene rubber (SBR). This oil can very well replace the conventional processing oils, fatty acid, an antioxidant in an SBR compound. Apart from the non-toxic nature of the rice-bran oil, it is comparatively cheaper than the aromatic/naphthenic oils used in rubber compounding.

16.9 Conclusion

Rice bran is a waste product of rice processing industries. This chapter explains how rice bran and its products like rice bran oil, fiber, protein hydrolysates, etc., could become the next simple, yet significant, opportunity in global health, nutrition, and food security. Rice bran is an alternative and economic source of plant-based hypoallergenic and high-quality protein that can be used in weaning and infant foods. It can be used as food ingredients in many food products such as meatball, noodles, biscuit, bread, and gluten-free products. Rice bran oil is also called wonder oil, rich in vitamin E and gamma oryzanol, which contributes functional/health properties making it suitable for nutraceutical products. The antioxidant properties of RBO make it a potential food additive to be used to improve the storage stability. Though challenges are apparent to increase dietary rice bran in the global food market, upcoming technologies and continuous research are needed to utilize it efficiently. Additional research and global funding opportunities are necessary to tackle issues of hunger and nutrition-related disease through dietary rice bran interventions.

References

Abbas HM, Shahein NM, Rabou NSA, Fouad MT, Zaky WM (2017) Probiotic fermented milk supplemented with rice bran oil. Int J Dairy Sci 12:204–210

Abdallah DJ, Sirchio SA, Weiss RG (2000) Hexatriacontane organogels. The first determination of the conformation and molecular packing of a low-molecular-mass organogelator in its gelled state. Langmuir 16:7558–7561

Accinni R, Rosina M, Bamonti F, Della Noce C, Tonini A, Bernacchi F, Campolo J, Caruso R, Novembrino C, Ghersi L, Lonati S, Grossi S, Ippolito S, Lorenzano E, Ciani A, Gorini M (2006) Effects of combined dietary supplementation on oxidative and inflammatory status in dyslipidemic subjects. Nutr Metab Cardiovasc Dis 16:121–127

Adebiyi AP, Adebiyi AO, Jin DH, Ogawa T, Muramoto K (2008) Rice bran protein-based edible film. 43(3):476–483. https://doi.org/10.1111/j.1365-2621.2006.01475.x

Adhikari P, Shin JA, Lee JH, Kim HR, Kim IH, Hong ST, Lee KT (2012) Crystallization, physicochemical properties, and oxidative stability of the Interesterified hard fat from Rice bran oil, fully hydrogenated soybean oil, and coconut oil through lipase-catalyzed reaction. Food Bioproc Tech 5:2474–2487. https://doi.org/10.1007/s11947-011-0544-4

Afinisha D, Arumughan C (2012) Effect of saponification on composition of unsaponifiable matter in rice bran oil. J Oleo Sci 61(7):241–247

Akihisa T, Yasukawa K, Yamaura M, Ukiya M, Kimura Y, Shimizu N, Arai K (2000) Triterpene alcohol and sterol ferulates from rice bran and their anti-inflammatory effects. J Agric Food Chem 48(6):2313–2319

Akiyama Y, Hori K, Hata K, Kawane M, Kawamura Y, Yoshiki Y, Okubo K (2001) Screening of chemiluminescence constituents of cereals and DPPH radical scavenging activity of gamma-oryzanol. Luminescence 16:237–241

Ali A, Devarajan S (2017) Nutritional and health benefits of Rice bran oil. In: Manickavasagan A, Santhakumar C, Venkatachalapathy N (eds) Brown rice. Springer, Cham. https://doi.org/10.1007/978-3-319-59011-0_9

Allen RG, Tresini M (2000) Oxidative stress and gene regulation. Free Radic Biol Med 28:463–499

Allen DA, Harwood S, Varagunam M, Raftery MJ, Yaqoob MM (2003) High glucose-induced oxidative stress causes apoptosis in proximal tubular epithelial cells and is mediated by multiple caspases. FASEB J 17:908–910

Al-Okbi SY, Hamed IM, Mohamed DA, Hussein AMS, Helal AM (2016) Physiochemical, sensory and biological properties of wheat-Rice bran composite biscuits, crackers and pasta. Res J Pharm Biol Chem Sci 7(3):1201–1214

Al-Shorgani NKN, Kalil MS, Yusoff WMW (2012) Biobutanol production from rice bran and de-oiled rice bran by clostridium saccharoperbutylacetonicum N1-4. Bioprocess Biosyst Eng 35:817–826. https://doi.org/10.1007/s00449-011-0664-2

Ameh MO, Gernah DI, Igbabul BD (2013) Physico-chemical and sensory evaluation of wheat bread supplemented with stabilized Undefatted Rice bran. Food Nutr Sci 4:43–48. https://doi.org/10.4236/fns.2013.49A2007

Amended Final Report on the Safety Assessment of Oryza sativa (rice) bran oil, Oryza sativa (rice) germ oil, rice bran acid, Oryza sativa (rice) bran wax, hydrogenated rice bran wax, Oryza sativa (rice) bran extract, Oryza sativa (rice) extract, Oryza sativa (rice) germ powder, Oryza sativa (rice) starch, oryza sativa (rice) bran, hydrolyzed rice bran extract hydrolyzed rice bran protein, hydrolyzed rice extract, and hydrolyzed rice protein (2006) Int J Toxicol 25(Suppl. 2):91–120. https://doi.org/10.1080/10915810600964626

Andreasen MF, Landbo AK, Christensen LP, Hansen A, Meyer AS (2001) Antioxidant effects of phenolic rye (Secale cereal L.) extracts, monomeric hydro-xycinamates and ferulic acid ehydrodimers on human low-density lipoproteins. J Agric Food Chem 49:4090–4096

Araghi A, Sei S, Sayra R, Sadighara P (2016) Safety assessment of rice bran oil in a chicken embryo model. Avicenna J Phytomed 6(3):351–356

Ausman LM, Rong N, Nicolosi RJ (2005) Hypocholesterolemic effect of physically refined rice bran oil: studies of cholesterol metabolism and early atherosclerosis in hypercholesterolemic hamsters. J Nutr Biochem 16(9):521–529

Bastak S (2005) Diabetes mellitus and its treatment. Int J Diabetes Metab 13:111–134

Becker M, Staab D, von Bergmann K (1993) Treatment of severe familial hypercholesterolemia in childhood with sitosterol and sitostanol. J Pediatr 122:292–296

Berger A, Rein D, Schäfer A, Monnard I, Gremaud G, Lambelet P, Bertoli C (2005) Similar cholesterol lowering properties of rice bran oil, with varied γ-oryzanol, in mildly hyperchlestrolemic men. Eur J Nutr 44:163–173. https://doi.org/10.1007/s00394-004-0508-9

Bergman CJ, Xu Z (2003) Genotype and environment effects on Tocopherol, Tocotrienol, and γ-Oryzanol contents of southern US Rice. Cereal Chem 80:446–449

Bernardi DS, Pereira TS, Maciell NR, Bortoloto J, Viera GS, Oliveira GC, Rocha-Filho PA (2011) Formation and stability of oil-in-water nanoemulsions containing rice bran oil: in vitro and in vivo assessments. J Nanobiotechnol 9:44

Bot A, Agterof WGM (2006) Structuring of edible oils by mixtures of γ-oryzanol with β-sitosterol or related phytosterols. J Am Oil Chem Soc 83:513–521. https://doi.org/10.1007/s11746-006-1234-7

Bot A, Adel DR, Roigers E (2008) Fibrils of γ-oryzanol + β-sitosterol in edible oil organogels. J Am Oil Chem Soc 85:1127–1134. https://doi.org/10.1007/s11746-008-1298-7

CAC (Codex Alimentarius Commission) (2003) The need for inclusion of rice bran oil in the standards for named vegetable oils. In: Joint FAO/WHO food standards programme. Committee on fats and oils, Codex

Cao X, Wen H, Li C, Gu Z (2009) Differences in functional properties and biochemical characteristics of congenetic rice proteins. J Cereal Sci 50:184–189

Carroll LE (1990) Functional properties and applications of stabilized rice bran in bakery products. Food Technol 44:74–76

Celermajer DS, Chow CK, Marijon E, Anstey NM, Woo KS (2012) Cardiovascular disease in the developing world Prevalences, patterns, and the potential of early disease detection. J Am Coll Cardiol 60:1207–1216

Chandel AK, Narasu ML, Rudravaram R, Pogaku R, Rao LV (2009) Bioconversion of de-oiled rice bran (DORB) hemicellu- losic hydrolysate into ethanol by Pichia stipitis NCM3499 under optimized conditions. Int J Food Eng 5:1–13

Chandrashekar P, Kumar PKP, Ramesh HP, Lokesh BR, Krishna AGG (2012) Hypolipidemic effect of oryzanol concentrate and low temperature extracted crude rice bran oil in experimental male wistar rats. J food Sci Technol 1–8

Chen CW, Cheng HH (2006) Arice bran oil diet increases LDL receptor and HMG-Coareductase mRNa expressions and insulin sensitivity in rats with streptozotocin/nicotinamide-induced type 2 diabetes. J Nutr 136(6):1472–1476

Choi YS, Jeong JY, Choi JH, Han DJ, Kim HY, Lee MA et al (2007a) Quality characteristics of meat batters containing dietary fiber extracted from rice bran. Korean J Food Sci Anim Resour 27(3):228–234

Choi YS, Lee MA, Jeong JY, Choi JH, Han DJ, Kim HY, Lee ES, Kim CJ (2007b) Effects of wheat fiber on the quality of meat batter. Korean J Food Sci An 27:22–28. https://doi.org/10.5851/kosfa.2007.27.1.22

Choi YS, Jeong JY, Choi JH, Han DJ, Kim HY, Lee MA, Kim HW, Paik HD, Kim CJ (2008) Effects of dietary fiber from rice bran on the quality characteristics of emulsion-type sausages. Korean J Food Sci An 28:14–20. https://doi.org/10.5851/kosfa.2008.28.1.14

Choi YS, Choi JH, Han DJ, Kim HY, Lee MA, Kim HW, Jeong JY, Kim CJ (2009) Characteristics of low-fat meat emulsion systems with pork fat replaced by vegetable oils and rice bran fiber. Meat Sci 82:266–271

Choi YS, Choi JH, Han DJ, Kim HY, Lee MA, Kim HW, Jeong JY, Kim CJ (2011) Effects of rice bran fiber on heat-induced gel prepared with pork salt-soluble meat proteins in model system. Meat Sci 88:59–66

Cicero AF, Gaddi A (2001) Rice bran oil and Gama-oryzanol in the treatment of hyperlipoproteinaemias and other conditions. Phytother Res 15:277–289

Coppini D, Paganizzi P, Santi P, Ghirardini A (2001) Capacitá protettiva Nei confronti delle radiazioni Solari di derivati di origine vegetable. Cosmetic News 136:15–20

Crehan CM, Hughes E, Troy DJ, Buckley DJ (2000) Effects of fat level and maltodextrin on the functional properties of frankfurters formulated with 5, 12 and 30% fat. Meat Sci 55:463–469. https://doi.org/10.1016/S0309-1740(00)00006-1

Dassanayake LSK, Kodali DR, Ueno S, Sato K (2011) Physical properties of Organogels made of Rice bran wax and vegetable oils. In: Edible oleogels: structure and health implications. Elsevier, pp 149–172. https://doi.org/10.1016/B978-0-9830791-1-8.50010-3

Eady S, Wallace A, Willis J, Scott R, Frampton C (2011) Consumption of a plant sterol-based spread derived from rice bran oil is effective at reducing plasma lipid levels in mildly hypercholesterolaemic individuals. Br J Nutr 15:1–12

El-Zanafi EM, Zaher FA (1990) Utilization of Rice bran oil to produce an antifoamer. JAOCS 67 (1):61–63

Esa NM, Ling TB, Peng LS (2013) By-products of Rice processing: An overview of health benefits and applications. J Rice Res 1:107. https://doi.org/10.4172/jrr.1000107

Fabian C, Ju YH (2011) A review on rice bran protein: its properties and extraction methods. Crit Rev Food Sci Nutr 51(9):816–827. https://doi.org/10.1080/10408398.2010.482678

Frank J, Chin W, Schrader C, Eckert GP, Rimbach G (2012) Do tocotrienols have potential as neuroprotective dietary factors? Ageing Res Rev 11(1):163–180

Friedman M (2013) Rice bran, rice bran oils, and rice hulls: composition, food and industrial uses, and bioactivities in humans, animals, and cells. J Agric Food Chem 61(45):10626–10641. https://doi.org/10.1021/jf403635v

Ghatak SB, Panchal SJ (2012) Anti-hyperlipidemic activity of oryzanol, isolated from crude rice bran oil, on triton WR-1339-induced acute hyperlipidemia in rats. Rev Bras Farm 22:642–648

Ghosh M, Bandyopadhyay S (2005) Studies on the crystal growth of rice bran wax in a hexane medium. J Am Oil Chem Soc 82:229–231

Gingras L (2002) N-3 fatty acid dietary recommendations and food sources to achieve essentiality and cardiovascular benefits. Am J Clin Nutr 83:1526–1535

Gopala K, Hemakumar KH, Khatoon S (2006) Study on the composition of rice bran oil and its higher free fatty acids value. J Am Oil Chem Soc 83(2):117–120

Gosh M (2007) Review on recent trends in rice bran oil processing. J Am Oil Chem Soc 84:315–324

Gracia K, Sriwattana S, No HK, JAH C, Prinyawiwatkul W (2009) Sensory optimization of a mayonnaise-type spread made with Rice bran oil and soy protein. J Food Sci 74(6):S248–S254. https://doi.org/10.1111/j.1750-3841.2009.01203.x

Graf E, Eaton JW (1993) Suppression of colonic cancer by dietary phytic acid. Nutr Cancer 19:11–19

Gul K, Yousuf B, Singh AK, Singh P, Wani AA (2015) Rice bran: nutritional values and its emerging potential for development of functional food – a review. Bioact Carbohydr Diet Fibre 6:24–30

Ha TY, Han S, Kim SR, Kim IH, Lee HY, Kim HK (2005) Bioactive components in rice bran oil improve lipid profiles in rats fed a high-cholesterol diet. Nutr Res 25:597–606

Halim E, Mukhopadhyay AK (2006) Effect of Ocimum sanctum (Tulsi) and vitamin E on biochemical parameters and retinopathy in streptozotocin induced diabetic rats. Indian J Clin Biochem 21:181–188

Hamada JS (2000) Characterization and functional properties of rice bran proteins modified by commercial exoproteases and endoproteases. J Food Sci 65(2):305–310

Hammond N (1994) Functional and nutritional characteristics of rice bran extracts. Cereal Foods World 39:752–754

Hansson GK (2005) Inflammation, atherosclerosis, and coronary artery disease. New Engl J Med 352:1685–1695

Hodin CM, Visschers RGJ, Rensen SS, Boonen B, Damink SWMO, Lenaerts K (2012) Total parenteral nutrition induces a shift in the Firmicutes to Bacteroidetes ratio in association with Paneth cell activation in rats. J Nutr 142:2141–2147
Hu G, Huang S, Cao S, Ma Z (2009) Effect of enrichment with hemicellulose from rice bran on chemical and functional properties of bread. Food Chem 115:839–842. https://doi.org/10.1016/j.foodchem.2008.12.092
Huang SC, Shiau CY, Liu TE, Chu CL, Hwang DF (2005) Effects of rice bran on sensory and physico-chemical properties of emulsified pork meatballs. Meat Sci 70(4):613–619. https://doi.org/10.1016/j.meatsci.2005.02.009
Hughes E, Cofrades S, Troy DJ (1997) Effects of fat level, oat fiber and carrageenan on frankfurters formulated with 5, 12 and 30% fat. Meat Sci 45(3):273–281
Inform article (2014). www.oilsandfatsinternational.com
Iqbal J, Minhajuddin M, Beg ZH (2003) Suppression of 7,12-dimethylbenz[alpha] anthracene-induced carcinogenesis and hypercholesterolaemia in rats by tocotrienol-rich fraction isolated from rice bran oil. Eur J Cancer Prev 12:447–453
Islam MS, Murata T, Fujisawa M, Nagasaka R, Ushio H, Bari AM, Hori M, Ozaki H (2008) Anti-inflammatory effects of phytosteryl ferulates in colitis induced by dextran sulphate sodium in mice. Br J Pharmacol 154:812–824
Islam MZ, Taneya MLJ, Shams-Ud-Din SM, Hoque MM (2012) Physicochemical and functional properties of brown rice (Oryza sativa) and wheat (Triticum aestivum) flour and quality of composite biscuit made thereof. Agriculturists 10(2):20–28
Ismail M, Al-Naqeeb G, Mamat W, Ahmad Z (2010) Gamma-oryzanol rich fraction regulates the expression of antioxidant and oxidative stress related genes in stressed rat's liver. Nutr Metab 7:23
James C, Courtney DLD, Lorenz K (1989) Rice bran-soy blends as protein supplements in biscuits. Int J Food Sci Technol 24:495–502
Jiamyangyuen S, Harper WJ, Srijesdaruk V, Kumthonglang K (2005a) Study of extraction and functional properties of rice bran protein concentrate. Milchwissenschaft 60(2):192–195
Jiamyangyuen S, Srijesdaruk V, Harper WJ (2005b) Extraction of rice bran protein concentrate and its application in bread. Songklanakarin J Sci Technol 27(1):55–64
Juliano C, Cossu M, Alamanni MC, Piu L (2005) Antioxidant activity of gamma-oryzanol: mechanism of action and its effect on oxidative stability of pharmaceutical oils. Int J Pharm 299:146–154
Jung EH, Kim SR, Hwang IK, Ha TY (2007) Hypoglycemic effects of a phenolic acid fraction of rice bran and ferulic acid in C57BL/KsJdb/db mice. J Agric Food Chem 55:9800–9804
Kaneto H, Katakami N, Matsuhisa M, Matsuoka T (2010) Role of reactive oxygen species in the progression of type 2 diabetes and atherosclerosis. Mediators Inflamm:1–11
Kaur A, Jassal V, Thind SS, Aggarwal P (2012) Rice bran oil an alternate bakery shortening. J Food Sci Technol 49(1):110–114. https://doi.org/10.1007/s13197-011-0259-6
Keim AL, Chi MMY, Moley KH (2001) Hyperglycemia-induced apoptotic cell death in the mouse blastocyst is dependent on expression of p53. Mol Reprod Dev 60:214–224
Kikuzaki H, Hisamoto M, Hirose K (2002) Antioxidant properties of ferulic acid and its related compounds. J Agric Food Chem 50:2161–2168
Kim JS, Godber JS (2001) Oxidative stability and vitamin E levels increased in restructured beef roast with added rice bran oil. J Food Qual 24:17–26
Kim JS, Godber JS, Prinaywiwatkul W (2000) Restructured beef roasts containing rice bran oil and fiber influences cholesterol oxidation and nutritional profile. J Muscle Foods 11(2):111–127
Kim Y, Kee JI, Lee S, Yoo SH (2014) Quality improvement of rice noodle restructured with rice protein isolate and transglutaminase. Food Chem 145:409–416
Kong CK, Lam WS, Chiu LC, Ooi VE, Sun SS, Wong YS (2009) A rice bran polyphenol, cycloartenyl ferulate, elicits apoptosis in human colorectal adenocarcinoma SW480 and sensitizes metastatic SW620 cells to TRAIL-induced apoptosis. Biochem Pharmacol 77:1487–1496

Krishna AGG, Khatoon S, Shiela PM, Sarmandal CV, Indira TN, Mishra A (2001) Effect of refining of crude rice bran oil on the retention of oryzanol in the refined oil. J Am Oil Chem Soc 78:127–131

Kuriakose AP, Rajendran G (1995) Use of rice-bran oil in the compounding of styrene butadiene rubber. J Mater Sci 30:2257–2262

Kuriyan R, Gopinath N, Vaz M, Kurpad AV (2005) Use of rice bran oil in patients with hyperlipidaemia. Natl Med J India 18:292–296

Kusum R, Bommayya H, Fayaz Pasha P, Ramachandran HD (2011) Palm oil and rice bran oil: Current status and future prospects. Int J Plant Physiol Biochem 3(8):125–132

Lasztity R, Tomoskozi S, Nagy J, Bajkai T, Sarkadi L (1995) Investigation of some functional and nutritive properties of cereal germ proteins. Per Pol Chem Eng 39(1):63–76

Lee SK, Jang IS, Kim MK, Park SK, Lee WY, Youn KS, Bae DH (2004) Changes in functional properties of rice bran and sesame meal proteins through chemical modifications. Food Sci Biotechnol 13:555–560

Lerma-García MJ, Herrero-Martínez JM, Simó-Alfonso EF, Mendonça CRB, Ramis-Ramos G (2009) Composition industrial processing and applications of rice bran gamma-oryzanol. Food Chem 115(389–404):389–404. https://doi.org/10.1016/j.foodchem.2009.01.063

Liang Y, Gao Y, Lin Q, Luo F, Wu W, Lu Q, Liu Y (2014) A review of the research progress on the bioactive ingredients and physiological activities of rice bran oil. Eur Food Res Technol 238:169–176

Lilitchan S, Tangprawat C, Aryusuk K, Krisnangkura S, Chokmoh S, Krisnangkura K (2008) Partial extraction method for the rapid analysis of total lipids and gamma-oryzanol contents in rice bran. Food Chem 106:752–759. https://doi.org/10.1016/j.foodchem.2007.06.052

Liu RH (2003) Health benefits of fruit and vegetables are from additive and synergistic combinations of phytochemicals. Am J Clin Nutr 78:517S–520S

Lu BS, Barber S, Benedito DBC (1991) Rice bran: chemistry and technology. In: Luh BS (ed) Rice production and utilization, vol II. Van Nostrand Reinhold, New York

Luh BS, Liu YK (1980) Rice ours in baking. In: Juliano BO (ed) Rice: chemistry and technology. St Paul, American Association of Cereal Chemists, pp 470–485

Martini S, An MC (2003) Crystallization of sunflower oil waxes. J Am Oil Chem Soc 80:525–532

Mingyai S, Kettawan A, Srikaeo K, Singanusong R (2017) Physicochemical and antioxidant properties of rice bran oils produced from colored rice using different extraction methods. J Oleo Sci 66:565–572

Minhajuddin M, Beg ZH, Iqbal J (2005) Hypolipidemic and antioxidant properties of tocotrienol rich fraction isolated from rice bran oil in experimentally induced hyperlipidemic rats. Food Chem Toxicol 43:747–753

Mishra N (2017) Utilization of waste defatted rice bran in formulation of functional cookies and its effect on physiochemical characteristic of cookies. Int J Adv Sci Res 2(5):64–68

Most MM, Tulley R, Morales S, Lefevre M (2005) Rice bran oil, not fiber, lowers cholesterol in humans. Am J Clin Nutr 81:64–68

Nagasaka R, Yamsaki T, Uchida A, Ohara K, Ushio H (2011) Gamma-Oryzanol recovers mouse hypoadiponectinemia induced by animal fat ingestion. Phytomedicine 18:669–671

Nagendra Prasad MN, Sanjay KR, Shravya Khatokar M, Vismaya MN, Nanjunda Swamy S (2011) Health benefits of Rice bran - a review. J Nutr Food Sci 1:108. https://doi.org/10.4172/2155-9600.1000108

Nanua JN, McGregor JU, Godber JS (2000) Influence of high-Oryzanol Rice bran oil on the oxidative stability of whole milk powder. J Dairy Sci 83:2426–2431

Nayik GA, Majid I, Gull A, Muzaffar A (2015) Rice bran oil, the future edible oil of India: a mini review. J Rice Res. https://doi.org/10.4172/2375-4338.1000151

Nicolosi RJ, Ausman LM, Hegsted DM (1991) Rice bran oil lowers serum total and lowdensity lipoprotein cholesterol and apo B levels in nonhuman primates. Atherosclerosis 88:133–142

Nihul PG, Mhaske ST, Shertukde VV (2014) Epoxidized rice bran oil (ERBO) as a plasticizer for poly-vinyl chloride (PVC). Iran Polym J. https://doi.org/10.1007/s13726-014-0254-7

Norazalina S, Norhaizan ME, Hairuszah I, Norashareena MS (2010) Anticarcinogenic efficacy of phytic acid extracted from rice bran on azoxymethane-induced colon carcinogenesis in rats. Exp Toxicol Pathol 62:259–268

Oka T, Fujimoto M, Nagasaka R, Ushio H, Hori M, Ozaki H (2010) Cycloartenyl ferulate, a component of rice bran oil-derived γ-oryzanol, attenuates mast cell degranulation. Phytomedicine 17(2):152–156

Oliveira MS, Feddern V, Kupsk L, Cipolatti EP, Furlong EB, Soares LAS (2011) Changes in lipid, fatty acids and phospholipids composition of whole rice bran after solid-state fungal fermentation. Bioresour Technol 102:8335–8338

Oluremi OI, Solomon AO, Saheed AA (2013) Fatty acids, metal composition and physico-chemical parameters of Igbemo Ekiti rice bran oil. J Environ Chem Ecotoxicol 5(3):39–46

Ornla-ied P, Sonwai S, Lertthirasuntorn S (2016) Trans-free margarine fat produced using enzymatic interesterification of rice bran oil and hard palm stearin. Food Sci Biotechnol 25 (3):673–680

Orthoefer FT (2005) Rice bran oil. In: Shahidi, F Bailey's industrial oil and fat products 6 edn. Wiley, p 465

Pal YP, Pratap AP (2017) Rice bran oil: a versatile source for edible and industrial applications. J Oleo Sci 66:551–556

Pali V (2013) Rice bran oil-unique gift of nature: a review. Agri Rev 34:288–294

Panala V, Verghese M, Boateng J, Field R, Shackelford L, Walker LT (2009) A comparison of rice bran, corn oil and soybean oil against azoxymethane induced colon cancer in a fisher 344 rat model. Int J Cancer Res 5:25–35

Paneras ED, Bloukas JG (1994) Vegetable oils replace pork backfat for low-fat frankfurters. J Food Sci 59:725–733

Patel M, Naik SN (2004) Gamma-oryzanol from rice bran oil—a review. J Sci Ind Res 63:569–578

Pearson P, Lewis SA, Britton J, Young IS, Fogarty A (2006) The prooxidant activity of high-dose vitamin E supplements in vivo. BioDrugs 20:271–273

Pernetti M, van Malssen K, Kalnin D, Flo ter E (2007) Structuring of edible oil with lecithin and sorbitan tri-stearate. Food Hydrocoll 21:855–861

Perry RH, Chilton CH (1973) Chemical engineers handbook, 5th edn. McGraw-Hill, Kogakusha

Phimolsiripol Y, Mukprasirt A, Schoenlechner R (2012) Quality improvement of rice-based gluten-free bread using different dietary fibre fractions of rice bran. J Cereal Sci 56:389–395. https://doi.org/10.1016/j.jcs.2012.06.001

Phongthai S, Homthawornchoo W, Rawdkuen S (2017) Preparation, properties and application of rice bran protein: a review. Int Food Res J 24(1):25–34

Polasa K, Rukmini C (1987) Mutagenicity tests of cashewnut shell liquid, rice-bran oil and other vegetable oils using the salmonella typhimurium/microsome system. Food Chem Toxicol 25:763–766

Prakash J (1996) Rice bran proteins: properties and food uses. Crit Rev Food Sci Nutr 36 (6):537–552. https://doi.org/10.1080/10408399609527738

Rajnarayana K, Prabhakar MC, Krishna DR (2001) Influence of rice bran oil on serum lipid peroxides and lipids in human subjects. Indian J Physiol Pharmacol 45:442–444

Reshma MV, Saritha SS, Balachandran C, Arumughan C (2008) Lipase catalyzed interesterification of palm stearin and rice bran oil blends for preparation of zero trans shortening with bioactive phytochemicals. Bioresour Technol 99:5011–5019

Rong N, Ausman LM, Nicolosi RJ (1997) Oryzanol decreases cholesterol absorption and aortic fatty streaks in hamsters. Lipids 32:303–309

Rubalya Valantina S, Neelameagam P (2008) A study of rheological behavior and oxidative stability of rice bran and corn oil using FTIR spectra. Int J Pure Appl Phys 4(1):77–86

Rukmini C, Prasad JS, Rao K (1980) Effects of feeding T-2 toxin to rats and monkeys. Food Cosmet Toxicol 18:267–269

Ryan EP (2011) Bioactive food components and health properties of rice bran. J Am Vet Med Assoc 238:593–600

Saklamaz A, Comlekci A, Temiz A, Caliskan S, Ceylan C, Alacacioglu A, Yasil S (2005) The beneficial effects of lipid-lowering drugs beyond lipid-lowering effects: a comparative study with pravastatin, atorvastatin, and fenofibrate in patients with type IIa and type IIb hyperlipidemia. Metabolism 54:677–681

Sanghi DK, Tiwle R (2015) A review of comparative study of rice bran oil and rice bran wax. Int J Pharm Rev Res 5:403–410

Sarkar S (1992) Nutrition of rice bran oil in relation to its Puri cation. J Am Oil Chem Soc 68:956–962

Sasaki J, Takada Y, Handa K, Kusuda M, Tanabe Y, Matsunaga A, Arakawa K (1990) Effects of Gama-oryzanol on serum lipids and apolipoproteins in dyslipidemic schizophrenics receiving major tranquilizers. Clin Ther 12:263–268

Seetharamiah GS, Chandrasekhara N (1989) Studies on hypocholesterolemic activity of rice bran oil. Atherosclerosis 78:219–224

Seetharamiah GS, Chandrasekhara N (1990) Effect of oryzanol on cholesterol absorption and biliary and fecal bile acid in rats. Indian J Med Res 92:471–475

Sekhon KS, Dhillon SS, Singh N, Singh B (1997) Functional suitability of commercially milled rice bran in India for use in different food products. Plant Foods Hum Nutr 50:127–140

Shaik R, Kuna A, Azam M, Tilathoo R, Kanuri M, Samala G (2017) Effect of rice bran oil spread on the physical, sensory and fatty acid profile of cake. J Food Sci Technol 54(7):2126–2134. https://doi.org/10.1007/s13197-017-2652-2

Sharif MK (2009) Rice industrial by-products management for oil extraction and value-added products. Degree of Doctor of Philosophy of Food Technology, National Institute of Food Science and Technology, University of Agriculture, Faisalabad, Pakistan, Doctor of Philosophy thesis, 206 p

Sharif K, Butt MS, Anjum FM, Nasir (2005) Improved quality of baked products by rice bran oil. Internet J Food Safety 5:1–8

Sharif MK, Butt MS, Anjum FM, Nawaz H (2009) Preparation of fiber and mineral enriched defatted rice bran supplemented cookies. Pak J Nutr 8:571–577

Sharif MK, Butt MS, Anjum FM, Khan SH (2014) Rice bran: a novel functional ingredient. Crit Rev Food Sci Nutr 54:807–816

Sharma AR (2002) Edible rice bran oil - consumer awareness programme. Rice bran oil promotion committee. Solvent Extractors Association of India, Mumbai

Sharma HR, Chauhan GS (2002) Effect of stabilized rice bran-fenugreek blends on the quality of breads and cookies. J Food Sci Nutr 39:225–233

Sharma RD, Rukmini C (1986) Rice bran oil and hypocholesterolemia in rats. Lipids 21:715–717

Shih FF (1996) Edible films from rice protein concentrate and pullulan. Cereal Chem 73 (3):406–409

Shih CK, Ho CJ, Li SC, Yang SH, Hou WC, Cheng HH (2011) Preventive effects of rice bran oil on 1,2-dimethylhydrazine/dextran sodium sulphate-induced colon carcinogenesis in rats. Food Chem 126:562–567

Shin YJ, Jang SA, Song KB (2011) Preparation and mechanical properties of rice bran protein composite films containing gelatin or red algae. Food Sci Biotechnol 20:703. https://doi.org/10.1007/s10068-011-0099-1

Siddiqui S, Rashid KM, Siddiqui WA (2010) Comparative hypoglycemic and nephroprotective effects of tocotrienol rich fraction (TRF) from palm oil and rice bran oil against hyperglycemia induced nephropathy in type 1 diabetic rats. Chem Biol Interact 188:651–658

Sierra S, Lara-Villoslada F, Olivares M, Jiménez J, Boza J, Xaus J (2005) Increased immune response in mice consuming rice bran oil. Eur J Nutr 44:509–516

Singer A, Nogala-Kalucka M, Lampart-Szczap E (2008) The content and antioxidant activity of phenolic compounds in cold-pressed plant oils. J Food Lipids 15:137–149

Singh B, Sekhon KS, Singh N (1995) Suitability of full fat and defatted rice bran obtained from Indian rice for use in food products. Plant Foods Hum Nutr 47:191–200

Somsuvra B, Ghatak S, Panchal S (2012) Anti-diabetic activity of oryzanol and its relationship with the antioxidant property. Int J Diab Dev Ctries 32:185–192

Sookwong P, Mahatheeranont S (2017) Supercritical CO_2 extraction of Rice bran oil – the technology, manufacture, and applications. J Oleo Sci 66:557–564

Sookwong P, Nakagawa K, Murata K, Kojima Y, Miyazawa T (2007) Quantitation of tocotrienol and tocopherol in various rice brans. J Agric Food Chem 55:461–466

Statistica (2016). https://www.statista.com/topics/1443/rice/. Accessed Dec 2016. https://www.worldriceproduction.com/

Swern O (1972) Bailey's industrial oil and fat products. 3rd ed. inter. Sc. Pub. pp 196 & 225

Tamagawa M, Otaki Y, Takahashi T, Otaka T, Kimura S, Miwa T (1992a) Carcinogenicity study of gamma-oryzanol in B6C3F1 mice. Food Chem Toxicol 30:49–56

Tamagawa M, Shimizu Y, Takahashi T, Otaka T, Kimura S, Kadowaki H, Uda F, Miwa T (1992b) Carcinogenicity study of gamma-oryzanol in F344 rats. Food Chem Toxicol 30:41–48

Tanaka T, Hoshina M, Tanabe S, Sakai K, Ohtsubo S, Taniguchi M (2006) Production of D-lactic acid from defatted rice bran by simultaneous saccharification and fermentation. Bioresour Technol 97(2):211–217

Thanonkaew A, Wongyai S, Decker EA, McClements DJ (2015) Formation, antioxidant property and oxidative stability of cold pressed rice bran oil emulsion. J Food Sci Technol 52(10):6520–6528. https://doi.org/10.1007/s13197-015-1743-1

Turhan S, Sagir I, Ustun NS (2005) Utilization of hazelnut pellicle in low-fat beef burgers. Meat Sci 71:312–316

Urso ML, Clarkson PM (2003) Oxidative stress, exercise, and antioxidant supplementation. Toxicology 189:41–54

Vali SR, Ju YH, Kaimal TNB, Chern YT (2005) A process for the preparation of food-grade rice bran wax and the determination of its composition. J Am Oil Chem Soc 82:57–64

Valko M, Izakovic M, Mazur M, Rhodes CJ, Telser J (2004) Role of oxygen radicals in DNA damage and cancer incidence. Mol Cell Biochem 266:37–56

Vettor R, Milan G, Rossato M, Federspil G (2005) Review article: adipocytokines and insulin resistance. Aliment Pharm Ther 22:3–10

Vintiloiu A, Leroux JC (2008) Organogels and their use in drug delivery—a review. J Control Release 125:179–192

Wang M, Hettiarachchy NS, Qi M, Burks W, Siebenmorgen T (1999) Preparation and functional properties of rice bran protein isolate. J Agric Food Chem 47(2):411–416

Wang YW, Jones PJ, Pischel I, Fairow C (2003) Effects of policosanols and phytosterols on lipid levels and cholesterol biosynthesis in hamsters. Lipids 38:165–170

Watchararuji K, Goto M, Sasaki M, Shotipruk A (2008) Value-added subcritical water hydrolysate from rice bran and soybean meal. Bioresour Technol 99:6207–6213

Weststrate JA, Meijer GW (1998) Plant sterol-enriched margarines and reduction of plasma total- and LDL-cholesterol concentrations in normocholesterolaemic and mildly hypercholesterolaemic subjects. Eur J Clin Nutr 52(5):334–343

White PJ (2008) Fatty acids in oilseeds (vegetable oils). In: Chow CK (ed) Fatty acids in foods and their health implications, 3rd edn. CRC, New York, pp 229–230

Wilson TA, Ausman LM, Lawton CW, Hegsted DM, Nicolosi RJ (2000) Comparative cholesterol lowering properties of vegetable oils: beyond fatty acids. J Am Coll Nutr 19:601–607

Wilson TA, Nicolosi RJ, Woolfrey B, Kritchevsky D (2007) Rice bran oil and oryzanol reduce plasma lipid and lipoprotein cholesterol concentrations and aortic cholesterol ester accumulation to a greater extent than ferulic acid in hypercholesterolemic hamsters. J Nutr Biochem 18(2):105–112

Wu Y, Ding Y, Tanaka Y, Zhang W (2014) Risk factors contributing to type 2 diabetes and recent advances in the treatment and prevention. Int J Med Sci 11(11):1185–1200. https://doi.org/10.7150/ijms.10001

Xu Z, Hua N, Godber JS (2001) Antioxidant activity of tocopherols, tocotrienols, and gamma-oryzanol components from rice bran against cholesterol oxidation accelerated by 2,2′-azobis (2-methylpropionamidine) dihydrochloride. J Agric Food Chem 49:2077–2081

Yadav RB, Yadav BS, Chaudhary D (2011) Extraction, characterization and utilization of rice bran protein concentrate for biscuit making. Br Food J 113:1173–1182

Yang HS, Choi SG, Jeon JT, Park GB, Joo ST (2007) Textural and sensory properties of low fat pork sausages with added hydrated oatmeal and tofu as texture-modifying agents. Meat Sci 75:283–289

Younas A, Bhatti MS, Ahmed A, Randhawa MA (2011) Effect of rice bran supplementation on cookie baking quality. Pak J Agri Sci 48(2):129–134

Zhang MW, Zhang RF, Zhang FX, Liu RH (2010) Phenolic profiles and antioxidant activity of black rice bran of different commercially available varieties. J Agric Food Chem 58:7580–7587

Chapter 17
Safflower (*Carthamus tinctorius*) Seed

Misbah Nazir, Sania Arif, Iftikhar Ahmed, and Nauman Khalid

Abstract Safflower is the oldest cultivated crop, typically grown at a small-scale level. Each part of this crop has multiple uses ranging from oil, foods, flavors, coloring agents, dyes, and medicinal benefits. Safflower crop can tolerate salinity and drought effectively. Various techniques are used to harvest the oil ranging from mechanical to supercritical fluid extraction; the extracted oil is rich in oleic and linoleic acids. The functional attributes of safflower include treatment of atherosclerosis, menopause, skin infections, and bone-related disorders. Recently, safflower oil is used to encapsulate sensitive bioactive compounds for nutrients delivery systems. This chapter highlights safflower chemical characteristics in combination with health benefits.

Keywords Safflower oil · Nutrients · Fatty acids · Health benefits · Extraction methods · Stability profile

---

M. Nazir
School of Mechanical and Manufacturing Engineering, National University of Sciences and Technology, Islamabad, Pakistan

S. Arif
Institute of Microbiology and Genetics, Georg-August-Universität, Göttingen, Germany

I. Ahmed
National Culture Collection of Pakistan (NCCP), Bioresources Conservation Institute (BCI), National Agricultural Research Centre, Islamabad, Pakistan

N. Khalid (✉)
School of Food and Agricultural Sciences, University of Management and Technology, Lahore, Pakistan
e-mail: nauman.khalid@umt.edu.pk

B. Tanwar, A. Goyal (eds.), *Oilseeds: Health Attributes and Food Applications*, https://doi.org/10.1007/978-981-15-4194-0_17

17.1 History and Origin

Safflower, also known as false saffron, is a thistle-like, diploid, annual herbaceous crop that flourishes in hot and dry climates. Safflower is believed to have been domesticated over 4000 years ago (Shirwaikar et al. 2010). The Scientific name of safflower is *Carthamus tincturious*. It belongs to family Asteraceae in the order Asterales (Asgarpanah and Kazemivash 2013). After domestication, cultivation of safflower spreads throughout the Middle East, Far East, India, and Northern Africa. During the late 1890s, safflower was introduced to North America and commercial production was started in 1950s (Pearl and Burke 2014).

Vavilov reported six centres of origins of safflower in 1951 that are as follows. Centre I—Chinese: It includes China, Japan, and Korea, referred to the Far East in later studies. Centre II—India: It includes India and both east and west Pakistan. Centres III and IV—Central Asiatic and Near Eastern: It includes areas from Afghanistan to Turkey and from southern Union of Soviet Socialist Republics to the Indian Ocean, referred to the Middle East in later studies. Centre V—Mediterranean: It includes the areas that border the Nile north of Aswan, referred to Egypt in later studies. The Southern reach of Centre V: It includes areas that border the Nile in northern Sudan and southern Egypt, referred to Sudan in later studies. The Western portion of Vavilov's Centre V: It includes Algeria, France, Italy, Morocco, Portugal, Romania, and Spain, referred to Europe. Centre VI—Ethiopian: referred to Ethiopia (Vavilov 1951; Singh and Nimbkar 2006). For cultivated safflower, three centres of origin were proposed by Vavilov (1951) including India (based upon diversity and ancient culture of production), Afghanistan (based on diversity and proximity to wild species), and Ethiopia (based upon the presence of wild species of safflower) (Vavilov 1951; Singh and Nimbkar 2006).

17.2 Production

Safflower is a bushy, herbaceous plant which possesses several branches, that are categorized as primary, secondary, and tertiary. The taproot system of *C. tincturious* spreads to two to three meters in soil with sufficient depth. This deep root system allows safflower to extract nutrient from deep down the soil making it an ideal crop for rainfed systems (Asgarpanah and Kazemivash 2013).

Safflower is cultivated mainly for its seeds which are employed to extract oil and the resulting oil is utilized as edible oil due to its rich unsaturated fatty acids content and nutritional value. *C. tincturious* is also grown for its flowers that can be used as medicines in several cultures and are also used as coloring and flavoring agents in foods (Kumar and Kumari 2011; Pearl and Burke 2014). Interest in safflower production has increased in last few years mainly due to the huge shortage of oilseed production due to scarce rainfall, healthy eating choice of consumers for oils having

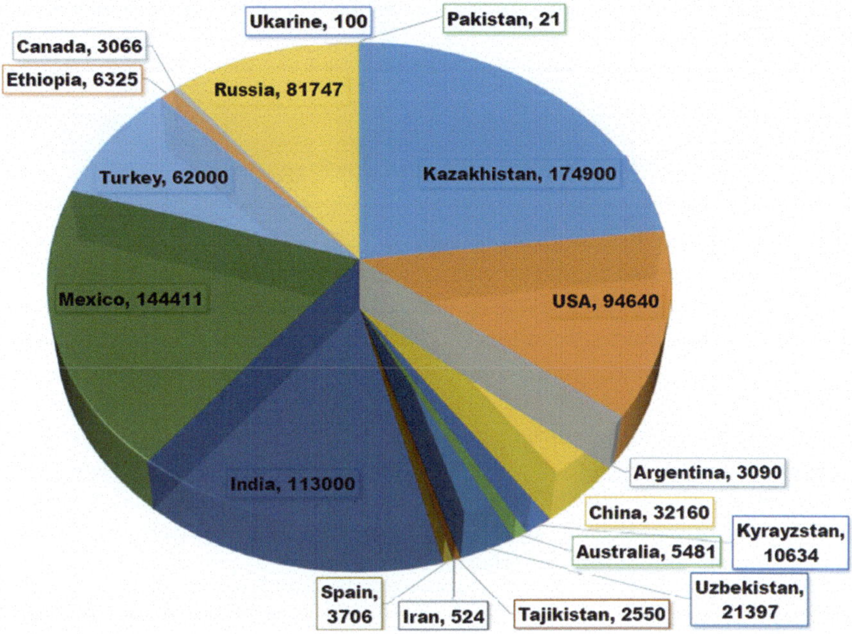

Fig. 17.1 Safflower production (in tons) in different countries of the world. The data presented from FAOSTAT during 2014 season production profile

the larger amount of unsaturated fatty acids and medicinal properties of safflower (Pearl and Burke 2014; Khalid et al. 2017).

Land cultivating safflower has grown at a rate of 4.9% per annum. With an annual growth rate of 0.97%, the average global yield is 805–872 kg per hectare. Safflower production reached 867,659 tons in 2014 and is increasing at a rate of 5.6% per annum. More than 60 countries are growing safflower, but the countries (Fig. 17.1) leading in the production of safflower include India, Kazakhstan, Mexico, United States of America, and Argentina. With a share of 93% of cumulative production, America and Asia are the major production areas for safflower. Moreover, areas of Africa and Oceania exhibit strong potential for the successful growth and production of this crop (Khalid et al. 2017).

17.3 Composition

Oilseeds serve as a principal source of raw materials including carbohydrates, fats, and proteins and find important application in food and nutraceuticals. They are potentially low-cost renewable resources of important compounds including tocopherols and phytochemicals (Bozan and Temelli 2008). To date, more than 200 compounds have been found and isolated from safflower including polysaccharides, glycosides, steroids, fatty acids, flavonoids, coumarins, and alkaloids (Asgarpanah and Kazemivash 2013).

17.3.1 Carbohydrates, Proteins, and Fats

Main components of the seed include carbohydrates, fats, and proteins. Percentage composition of the seed varies in different cultivars. Safflower seeds consist of 33–45% of the hull while 55–67% of the seeds consist of kernel (Hall Iii 2016). Total carbohydrate and fiber content is 52% of the seed (Bozan and Temelli 2008). Percentage of fiber in the seeds varies in different cultivars and is found higher in thickly walled cultivars (approximately 34%) while the lower amount is found in thin-walled varieties (approximately 11%) (Hall Iii 2016). Sugar content ranges from 3.2 to 9.2%, while protein content is between 14.9 and 17%. Extractable lipids account for 25–40% of the seed (Hamrouni-Sellami et al. 2007). The oil content of the seed also varies depending upon the thick- or thin-hulled cultivars and is found between 27 and 60% of the seed (Hall Iii 2016).

17.3.2 Tocopherols, Phytosterols, and Carotenoids

Tocopherols, phytosterols, and carotenoids make up the nonsaponifiable lipid component. Safflower seeds are reported to contain tocopherol content of 50–100 μg/g of the seed. Whereas, in the case of traditional and high oleic acid safflower oil, the amount is even higher, 684 μg/g and 941 μg/g respectively. Tocopherol content varies in different cultivars and mean concentrations in seven different regions have been reported to be 676–827 μg/g of the oil (Hall Iii 2016).

In general, α-tocopherol is found in the highest amount and varies between 93 and 98% of the total tocopherol content. β-, γ-, and δ-tocopherols make up the remaining content of tocopherols. Carotenoid amount observed in the oil is 2.5 μg/g (Hall Iii 2016). In the case of sterols, significant variation exists in the literature regarding total, free, and esterified sterols. Safflower seeds consist of approximately 2000–4500 μg/g of phytosterols. Variation in phytosterol data may exist due to seed maturity stage depending when the analysis was performed. Phytosterols content in safflower oil has been reported to be 1520–5745 μg/g. Approximately 48% of the phytosterols are free sterols while the remaining 39% are esterified sterols (Hall Iii 2016).

Irrespective of the sterol form, β-sitosterol accounts for 50–70% of phytosterols whereas stigmasterol, Δ5-avenasterol, Δ7-stigmasterol and campesterol make up 6–12% each of the phytosterols. Maturity and refining of seeds have been reported to reduce the total sterols content while relatively maintain the percentage composition of the type of phytosterols (Hall Iii 2016). However, the percentage of each phytosterol remains relatively consistent, for example, β-sitosterol accounting for the greatest percentage (Hall Iii 2016).

17.3.3 Phenolic Compounds

Phenolic acids, derivatives of phenolic acids flavonoids, are prominent phenolic compounds found in oil-rich seeds (Bozan and Temelli 2008). Most of the phenolic components are located in flowers and seed meal of safflower (Hall Iii 2016). Seed meal or cake of safflower is the by-product left after the extraction of safflower seed oil. Seed meal has been reported to contain 3800–5700 μg/g of phenolic glycosides (Hall Iii 2016). These compounds appear bitter in taste to humans but are less bothersome to animals and are thus used as the protein source in animal feeds. Total phenolic content of safflower seeds has been reported to be 126.0 mg gallic acid equivalence (GAE) per gram of seed, while only 26 μg GAE per gram has been observed in the oil (Hall Iii 2016).

Phenolic compounds impart nutritional and sensory qualities to the seeds. They contribute in protecting seeds against oxidative deterioration at even lower concentration, while at higher concentrations they impart dark color, bitter taste, and off-flavor to some oilseeds and nuts (Tanwar et al. 2018, 2019; Shahidi 2000). Antioxidant capability of phenolic compounds depends upon the number of hydroxyl groups in the molecule. Literature reports the enhancement in the antioxidant capacity of phenolic acids esterified with other groups, i.e., sugar molecules (Liyana-Pathirana and Shahidi 2006; Amarowicz et al. 2003).

Seed oils also contain phenolic compounds that contribute to their stability (Tovar et al. 2001). According to a study, safflower seeds contain 5590 μg/g of free phenolic compounds while 9670 μg/g of esterified phenolic compounds (Bozan and Temelli 2008).

Other phenolic compounds including 2 amino-3,4-dimethylbenzoic acids, 4-hydroxy-benzhydrazide derivative, and phenolic acids; trans-ferulic, chlorogenic, *p*-coumaric, and syringic acids have been found in variable amounts in water extracts of safflower (Hall Iii 2016). Water extracts of safflower have also been reported to contain various flavonoid compounds including quercetin, gallocatechin, epigallocatechin, quercetin dehydrate, rutin hydrate, naringin, luteolin, and kaempferol in variable amounts (Hall Iii 2016). Different concentrations of phenolics have been observed in different cultivars of safflower but the general trends observed for individual phenolics have been relatively similar regardless of different cultivars and locations of safflower seed source (Hall Iii 2016). In addition to phenols, it also contains various other bioactive components including lignin, flavonoids, and serotonin derivatives (Zhang et al. 1997).

17.3.3.1 Serotonin Derivatives

Many serotonin derivatives have been isolated from safflower seeds and seed meals and several pure serotonin derivatives have been found to exhibit strong antioxidant activities. Serotonin derivatives in seeds include *N*-feruloylserotonin, *N*-feruloylserotonin-*O*-β-D-glucopyranoside, *N*-(*p*-coumaroyl)serotonin, *N*-(*p*-coumaroyl)tryptamine, *N*-(*p*-coumaroyl)serotonin-*O*-β-D-glucopyranoside, *N*-feruloyltryptamine, Serotobenine, 4,4″-bis(*N*-*p*-coumaroyl)serotonin, *N*-feruloylserotonin-*O*-β-D-glucopyranoside, *N*-(*p*-coumaroyl)serotonin-*O*-β-D-glucopyranoside, 4,4″-bis(*N*-*p*-coumaroyl)serotonin, 4-[*N*-(*p*-coumaroyl)serotonin-4″-yl]-*N*-feruloyl serotonin, and 4,4″-bis(*N*-*p*-feruloy)serotonin (Zhou et al. 2014). In addition to seeds, serotonin derivatives have also been isolated from safflower oil (Zhang et al. 1997).

Serotonin-derived phenylpropanoid amides including safflomide (trans-*N*-caffeoyltryptamine) and serotomide (trans-*N*-caffeoylserotonin) have also been found in safflower (Park 2008). In safflower seeds, most of the serotonin derivatives are found as a family of molecules containing seven to ten members containing a serotonin moiety bound to a phenylpropanoid moiety via an amide bond. Apart from the antioxidative effect and radical-scavenging activities, many health-promoting properties have been attributed to them including anti-tyrosinase activity, anti-tumor activities, fibroblast growth-promoting properties, immunomodulatory and analgesic properties (Jin et al. 2008). Due to strong antioxidant and therapeutic properties, safflower has great potential to be used as prophylactic and treatment option as well as in functional foods for its health benefitting properties (Kang et al. 2009).

17.3.4 Oil

The oil content of the safflower seeds varies from 23 to 36% of the seed depending upon the different cultivars (Matthaus et al. 2015). Safflower oil contains a high amount of linoleic acid and a considerable amount of oleic acid (Asgarpanah and Kazemivash 2013).

The major fatty acid component is the linolenic acid having a mean value of 70.6% (Matthaus et al. 2015). The safflower seed oil also contains considerable amounts of saturated fatty acids, particularly palmitic (5.7–6.81%) and stearic acids (1.88–2.57%) (Matthaus et al. 2015). According to another report by Gambacorta and Leone (1997), safflower oil contained 7.4% palmitic, 2.9% stearic, 14.3% oleic and 74.3% linoleic acids (Gambacorta and Leone 1997). Kim and colleagues reported the safflower seed oil composition as 63–72% linoleic acid, 16–25% oleic acid, and 1–6% α-linolenic acid (Kim et al. 2000).

Fatty acid composition is influenced greatly by environmental conditions as well as genotypic variations (Robertson et al. 1978). Apart from essential fatty acids, safflower seed oils are also good sources of vitamin E (Robertson et al. 1978).

Amount of α-, β-, and γ tocopherols ranges from 46.05 to 70.93 mg/100 g, 0.85 to 2.16 mg/100 g and trace amount to 0.45 mg/100 g oils, respectively. Total tocopherols content ranged between 47.29 and 73.09 mg per 100 g (Matthaus et al. 2015). Moreover, fatty acid and tocopherols content vary in different cultivars.

17.4 Medicinal Properties

Safflower extract is important as a therapeutic agent and has been used as medicine since ancient times. They have been employed in traditional medicines for their effectivity as purgatives, antipyretic and anti-dotes, analgesics, antioxidants, anti-inflammatory, antidiabetics, bone-strengthening agents and immune modulators.

17.4.1 Skin Emollient and Anti-Melanogenic Activities

The high linolenic acid content of safflower can be employed in improving the quality and appearance of the skin (Roh et al. 2004). Vitamin E content of safflower oil also promotes the health of skin and hair and light texture of the oil aids in the efficient absorption of the oil into skin and scalp (Maru 2014). Safflower oil has been employed in the formulation intended for skin conditioning purposes and for acne vulgaris treatment (Anderson 2005; Toombs 2005).

Numerous patents have been filed describing the use of safflower oil separately or in combination with other oils in many herbal and cosmetic products (Maru 2014). Moreover, *N*-feruloylserotonin and *N*-(*p*-coumaroyl) found in safflower seeds have been found to strongly inhibit the melanin production, unveiling the potential of safflower to be developed as potent melanogenesis inhibiting formulations (Zhou et al. 2014). In another study, methanolic extracts of safflower seeds have been shown to exhibit strong anti-melanogenic activities and promote skin whitening in B16 melanoma cell lines and *Streptomyces bikiniensis*. Active compounds isolated from these methanolic extracts consisted of *N*-(*p*-coumaroyl) serotonin, *N*-feruloylserotonin, and acacetin (Roh et al. 2004).

17.4.2 Analgesic and Neuromodulatory Activity

The safflower seed oil has been used for many years as an analgesic remedy in Korean medicinal preparations for herbal acupuncture (Popov et al. 2009). Sterile oil from safflower seeds has been produced in Korea and tested for its anesthetic effect on sciatic nerve of the mouse. Oil hence produced has been shown to exhibit efficient local anesthetic properties along with moderate analgesic properties.

Serotonergically controlled reduction in pain reactions was observed (Popov et al. 2009).

More than 70% of the safflower seed oil content is linolenic acid. Linolenic acid accelerates the proteolytic degradation of the tyrosinase enzymes involved in the synthesis of monoaminergic neurotransmitters that are involved in transmitting pain signals (Popov et al. 2009). Safflower seeds contain low amounts of 5HT and serotonin conjugates. Representative serotonin derivatives include 4-coumaroylserotonin and feruloylserotonin. They possess strong antioxidant and other therapeutic properties and have great potential to be used as a prophylactic and treatment option for various ailments (Popov et al. 2009).

Analgesic properties of safflower oil are also attributed to the presence of serotonin (5HT). Administration of the oil to the region of sciatic nerve leads to an increase in serotonin quantity in the synaptic clefts of the sciatic nerve and also partially blocks the functional activity of this nerve. This effect is also beneficial in treating other neurological problems, i.e., depression. Also, it is a safer option, when compared to tricyclic antidepressant (TCA) amitriptyline, safflower seed oil exhibited no neurotoxicity (Popov et al. 2009).

Serotonin conjugates can have a direct analgesic effect. *N*-feruloylserotonin has been shown to reduce anxiety in the high pain-threshold rats. Scientists have observed selective stress-reducing effects of *N*-feruloylserotonin in stress-sensitive animals (Yamamotova et al. 2007). The analgesic effect of safflower seed oil can also be attributed to strong antioxidant activity and inhibitory effect on tyrosinase by serotonin conjugates. Serotonin derivatives have been reported to exert a neuroprotective effect on high glucose-induced nerve cell death. It has been shown to inhibit the overproduction of the mitochondrial superoxide produced by hyperglycemia by acting as a scavenger of the superoxide radical (Popov et al. 2009).

In pharmacopuncture, safflower seed oil can be used as a local anesthetic for treatment of pain related to various etiologies. Pharmacopuncture is widely employed in eastern medicine and it directly administers the drugs into acupuncture point which increases the therapeutic efficacy and requires a very small amount of drug with few or no side effects of the therapy (Popov et al. 2009).

17.4.3 Anti-Ulcerogenic Activity

Gastric ulcers pose a significant problem in the medical field with significant morbidity and development of complications in gastric ulcers exhibit significant mortality rates globally (Toma et al. 2014). The essential oil of safflower seeds has been shown to exhibit efficient antiulcerogenic activity. After administration of hydrochloric acid-ethanol in mice to induce ulcers, a safflower seed oil dose of 750 mg/kg was shown to decrease the ulcerogenic lesions (Toma et al. 2014). Also, significant antiulcerogenic properties were observed against non-steroidal anti-inflammatory drug (NSAID)-induced gastric ulcers at a dose of 187.5 mg/kg. The same dose was seen to increase the gastric volume, gastric mucus production, and gastric pH levels along with decreasing the total gastric acid secretion (Toma et al. 2014). Further, the acute toxicity testing showed no effect of safflower seed oil at a dose of 5 g/kg on mortality, feeding, water consumption, bodyweight, or any other physiological factors when compared to the control group and safflower seed oil was found to be non-toxic (Toma et al. 2014). The anti-ulcerogenic activity of safflower seed oil might be due to the cytoprotective activity in gastric mucosa. The cytoprotective effect might be imparted by a lot of linolenic acids present in the oil. New scientific studies related to this pathology need to be carried out (Toma et al. 2014).

17.4.4 Estrogen Modulating Activity

Safflower seeds contain various polyphenolic compounds including flavonoids, glycosides, and lignins. Some of these compounds have been known to exhibit weak estrogenic or anti-estrogenic activities in mammals and are thus termed as phytoestrogens (Dixon 2004).

Tracheloside, a lignan glycoside isolated from safflower seeds, has been reported to significantly decrease the activity of an estrogen-inducible marker enzyme, alkaline phosphatase. Tracheloside exhibited efficient anti-estrogenic activity comparable to the commercially available drug tamoxifen, that is used to treat breast cancer by inhibiting the estrogenic effect on breast cancer cells (Yoo et al. 2006). Lignans, flavones, and serotonin extracts in safflower work as phytoestrogens and also regulate the metabolism of bone protection and bone formation (Kang et al. 1999).

Safflower seed flavonoids and lignans representing the physiological effects similar to that of estrogen (phytoestrogens) have also exhibited anti-carcinogenic properties apart from antioxidant, serum cholesterol regulation, and bone health improvement properties (Draper et al. 1997; Anderson and Garner 1997).

Phytoestrogens have been reported to reduce bone resorption and bone loss in oophorectomized rat models. Administration of phytoestrogens led to decreased excretion of calcium via urine and greatly reduced the bone resorption (Draper et al. 1997). Phytoestrogens extracted from defatted safflower seed meal have been shown to improve blood lipid status and cholesterol excretion in estrogen-deficient ovariectomized animal models without affecting the uterotropic action. Phytoestrogens imparted the estrogenic effects to the estrogen-deficient rat models and ameliorated the negative effects related to the deficiency of estrogen (Cho et al. 2004). Phytoestrogens along with estrogenic effects also possess anti-androgenic activity. Reduction in prostate weight and in testosterone levels has been observed in castrated rats following the administration of safflower seed extracts at a dose of 300 mg/kg body weight (Rashed et al. 2014).

17.4.5 Antidiabetic Activity

Safflower seeds have been shown to exhibit antidiabetic effects and prevent complications related to diabetes. Various compounds in safflower seeds have been shown to produce therapeutic effects in diabetes. Serotonin derivatives, *N-p*-coumaroyl serotonin, and *N*-feruloyl serotonin isolated from Safflower seed have been evaluated for their inhibitory activity against α-glucosidase. α-glucosidase is involved in the digestion of carbohydrates and α-glucosidase inhibitors prevent the digestion of carbohydrates by α-glucosidase and help in controlling glucose concentration. The potent inhibitory activity of serotonin derivatives against α-glucosidase was observed with 50% inhibitory concentration (IC_{50}) values of 47.2 μmol/L and 99.8 μmol/L for *N-p*-coumaroyl serotonin and *N*-feruloyl serotonin, respectively. IC_{50} values of serotonin derivatives were even lower than reference α-glucosidase inhibitor drugs. IC_{50} values of reference drugs were 907.5 μmol/L and 278.0 μmol/L for acarbose and 1-deoxynojirimycin, respectively. Results suggest the use of safflower seed as a treatment remedy for diabetes. Apart from medicinal preparations, safflower seeds can also be utilized for developing functional foods and supplements intended for the diabetic population (Takahashi and Miyazawa 2012).

Metabolic impairments in maternal diabetes create a pro-inflammatory environment in the uterus and lead toward alterations in placental development and functions. Matrix metalloproteinases (MMPs), when produced in excess, are one of the factors producing pro-inflammatory environment. Martinez and colleagues reported that dietary supplementation with 6% of olive oil or 6% of safflower oil to the diabetic rats prevented the MMPs' over-activities in the placenta, and beneficial effects were reflected in rat sera. Placentas from diabetic rats exhibited increased protein concentrations and activities of MMP and were decreased when diabetic rats received the olive and safflower dietary treatments (Martinez et al. 2012).

Safflower seed oil is a rich source of polyunsaturated fatty acids, when supplemented to Sprague-Dawley rat's model resulted in increased concentration of arachidonic acid, which in turn resulted in the reduced incidence of diabetic embryopathy and prevented the malformation of off-springs in diabetic rats (Reece et al. 1996). Higa and colleagues also reported the reduced malformation rates during maternal diabetes in rats when supplemented with safflower. Reduced malformation was attributed to the regulation of nitric oxide homeostasis and arachidonic acid in embryos that subsequently prevented the developmental damage during organogenesis (Higa et al. 2010).

Supplementation with safflower oil for 30 days at a dose of 200 mg/kg of the body to hyperglycemic, type-1 diabetic rats produced hypoglycemic effects along with hypolipidemic effects as well (Rahimi et al. 2014).

Asp et al. (2011) reported the effect of safflower seed oil supplementation on post-menopausal, obese women with type 2 diabetes. Supplementation with 8 g of safflower oil in diet improved the glycemia, inflammation, and blood lipids over a period of 16 weeks. Authors concluded that small change in the quality of dietary can augment diabetes treatment as well as improve other risk factors for complications related to diabetes (Asp et al. 2011).

17.4.6 Anti-Oxidant Activity

The antioxidant can be defined as any substance that can significantly delay or prevent the oxidation of an oxidize-able substrate instead of being present at a low concentration compared to that of the substrate. Antioxidants are of great importance as they help in protecting the human body against damage by reactive free radicals generated in Alzheimer's disease, atherosclerosis, the ageing process, cancer, ischemic heart disease, and Parkinson's disease. Apart from synthetic compounds, natural products and their derivatives possess effective antioxidative properties hence imparting anticancer, anti-aging, anti-inflammatory, and hypolipidemic properties to these compounds (Koyama et al. 2006)

A sufficient number of antioxidants have been reported in safflower seeds. Matairesinol, 8 hydroxyarctigenin (lignans), acacetin and acacetin 7-*O*-glucoside (flavones), feruloylserotonin, and *N*-*p*-coumaroyl (serotonin derivatives) have been isolated from safflower seeds showing strong antioxidant activity in vitro (Kang et al. 1999; Kim et al. 2007). Ethanol-ethyl acetate extracts of safflower seeds have been reported to inhibit low-density lipoprotein (LDL) oxidation induced in vitro by copper ions. Major phenolic and active constituents of the extract were found to be serotonin derivatives (Fig. 17.2), *N*-feruloylserotonin and *N*-(*p*-coumaroyl) serotonin and their glucoside derivatives. These serotonin derivatives were reported to be absorbed into the blood circulation and prevent the atherosclerotic lesion development by inhibiting the formation of oxidized LDL because of their strong antioxidative activity (Koyama et al. 2006).

Furthermore, serotonin derivatives have been shown to exhibit stronger radical scavenging and lipid peroxidation activities as compared to α-tocopherol, acacetin, and matairesinol (Kang et al. 1999). Seven antioxidant serotonin derivatives have been isolated from safflower seed meal or cake, and have exhibited comparatively strong antioxidant activities when tested by ferric thiocyanate and α,-α-diphenyl-β-picrylhydrazyl (DPPH) assays (Zhang et al. 1997). Similarly, the serotonin derivative, *N*-*p*-coumaroyl serotonin has been shown to possess antioxidant and anti-inflammatory action for fibroblasts and also helps in cell proliferation in fibroblast (Takii et al. 1999).

17.4.7 Anti-Inflammatory Activity

Inflammation is a type of immune response and serves as a vital mechanism in host defenses against external damage, infections, or harmful physical or chemical

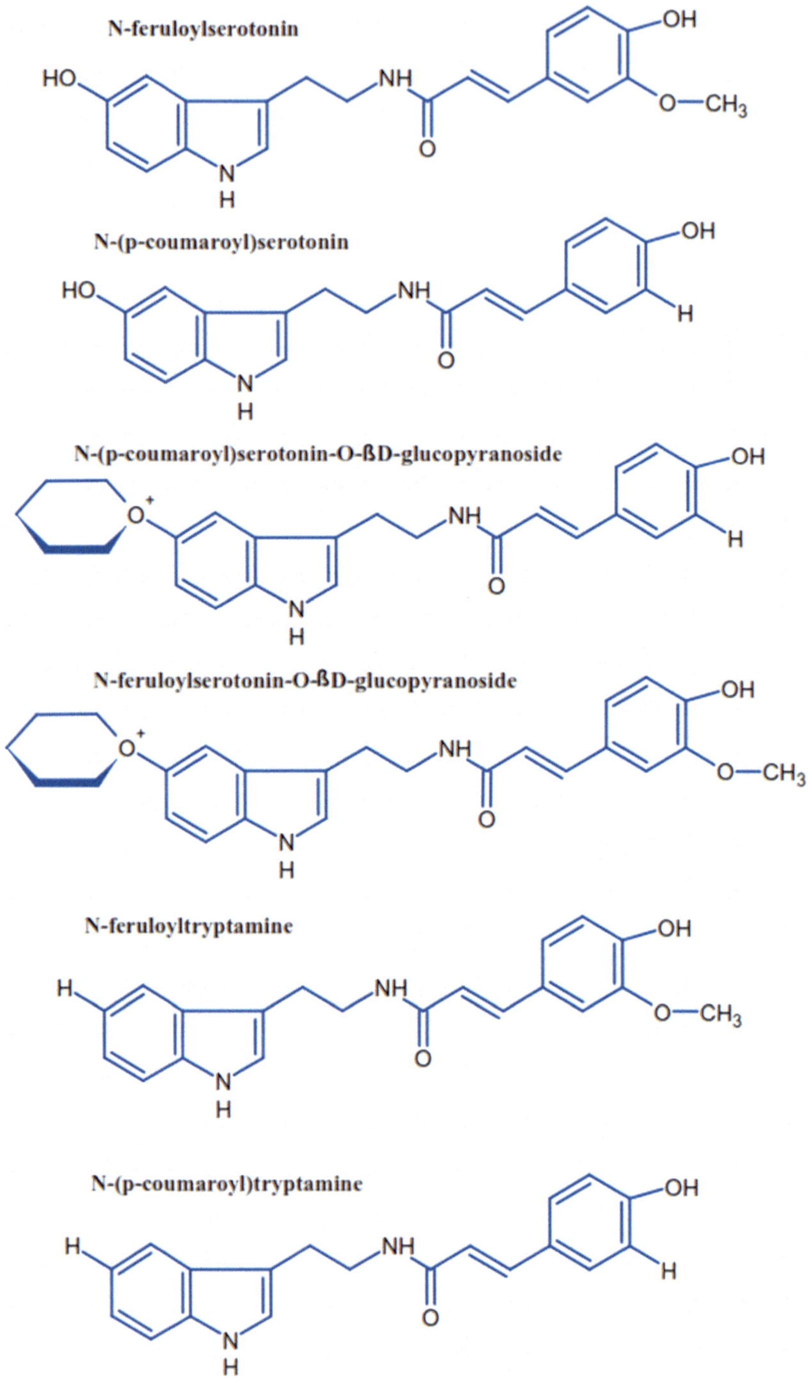

Fig. 17.2 Various alkaloids present in safflower oil. These alkaloids represent the serotonin derivatives and contributes to functional properties of safflower oil

Serotobenine

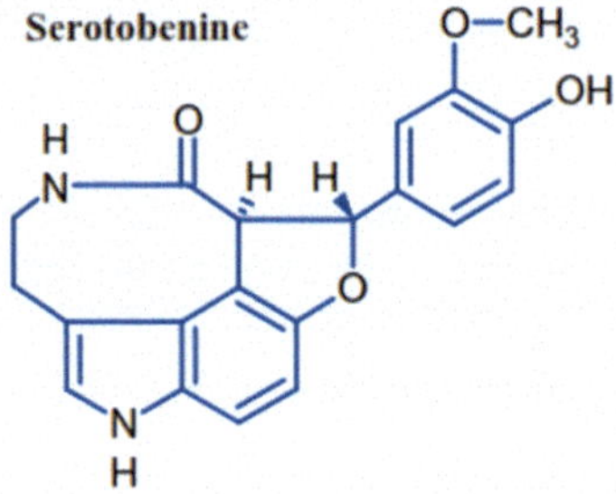

4,4*-bis(N-p-coumaroyl)serotonin

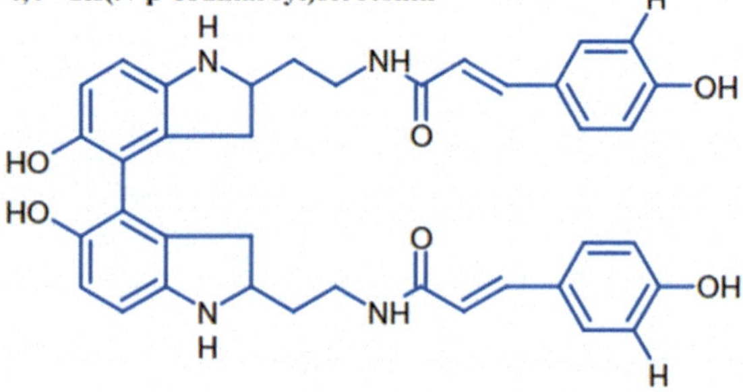

4-[N-(p-coumaroyl)serotonin-4*-yl]-N-feruloylserotonin

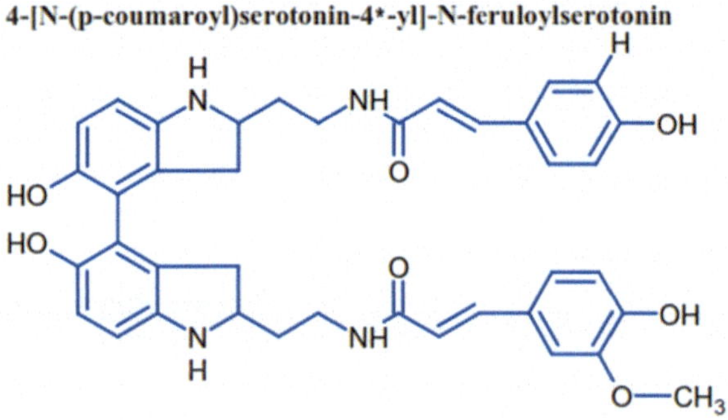

4,4*-bis(N-p-feruloy)serotonin

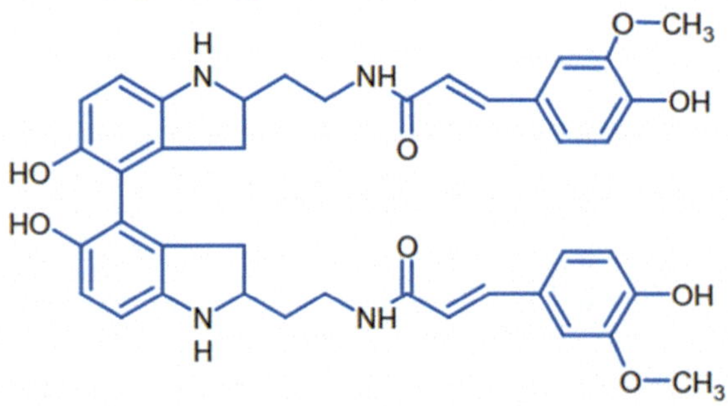

Fig. 17.2 (continued)

stimuli (Cho et al. 2009). Many cells and mediators are instrumental in the proceeding of inflammation (Asgarpanah and Kazemivash 2013).

Macrophages are crucial in the immune responses and participate in both innate and acquired immunity (Asgarpanah and Kazemivash 2013; Woods et al. 2009). When activated by stimulants, macrophages secrete pro-inflammatory products including nitric oxide, prostaglandin E2, interleukin (IL)-6, and IL-1β, tumor necrosis factor (TNF)-α, granulocyte/macrophage colony-stimulating factor (GM-CSF) and other inflammatory mediators in response to cell signaling (Tetley 2005; Asgarpanah and Kazemivash 2013).

An excessive immune response is undesirable and can be involved in the pathogenesis of various diseases including arthritis, atherosclerosis, inflammatory bowel diseases, etc. (Moilanen and Vapaatalo 1995; Guzik et al. 2003). Therefore, apposite countermeasures are required against the overproduction of inflammatory mediators. Despite the presence of many steroidal or non-steroidal anti-inflammatory drugs, researchers are looking for natural products to develop them into anti-inflammatory agents because of the side-effects produced by chemical drugs.

Serotonin derivatives from safflower seeds have been reported to exhibit anti-inflammatory effects. Serotonin derivatives for ethyl acetate extract of safflower seeds have been isolated and among them, *N*-feruloylserotonin has been found to be most potent in inhibiting lipopolysaccharide (LPS)-induced production of nitric oxide and prostaglandin E2, (IL)-6, and IL-1β in RAW 264.7 macrophages (Jo et al. 2017).

70% ethanol extracts of safflower seeds inhibited the production of nitric oxide and prostaglandin E2 in LPS-stimulated RAW 246.7 macrophages. While the expression of TNF-α, IL-1β, and IL-6 was significantly reduced. Due to the significant effects on inflammatory factors, the study suggested safflower as a potential anti-inflammatory therapeutic agent (Kim et al. 2013).

Ethyl acetate extracts of safflower, mainly employing cosmosiin, acacetin, *N*-(*p*-coumaroyl) serotonin, and *N*-feruloyl serotonin, were found to exhibit effective anti-inflammatory effects against LPS-stimulated RAW 264.7 macrophages. All of the four compounds were found to significantly inhibit the production of LPS-stimulated nitric oxide and pro-inflammatory cytokines (Kim et al. 2015).

Apart from above-mentioned mediators, other mediators can also be involved in inflammation and can be a target of anti-inflammatory compounds. Jun and colleagues have reported the anti-inflammatory effects of methanol extracts of safflower. These extracts were reported to produce anti-inflammatory effects by inducing heme oxygenase-1 expression and inhibiting nuclear factor kappa B (NF-κB) activity (Jun et al. 2011). Serotonin derivatives, *N*-(*p*-coumaroyl) serotonin and *N*-(*p*-coumaroyl) tryptamine, from safflower seed extracts have been shown to inhibit the production of pro-inflammatory cytokines including IL-1α, IL-1β, IL-6, IL-8, and TNF-α, produced by lipopolysaccharide-stimulated human monocytes (Takii et al. 2003). Seeds of safflower also exhibit neuroprotective effects through inhibition of the secretion of pro-inflammatory cytokines including nitric oxide and prostaglandin E2 (Asgarpanah and Kazemivash 2013).

17.4.8 Antitumor and Anti-Cancerous Activity

Seed extracts of safflower together with dendritic cell (DC)-based vaccine have been shown to surge the levels of TNF-α and IL-1β in mouse CD117+ (c-kit)-derived DCs and exhibit strong antitumor activity (Chang et al. 2011). The safflower seed oil has been reported to contain the compound alkane-6,8-diols, that has been shown to inhibit 12-*O*-tetradecanoylphorbol-13-acetate-induced tumor promotion in mouse skin (Yasukawa et al. 1996).

The phenolic compounds from methanol extracts of safflower seeds have been shown to exhibit strong cytotoxic effects in a dose-dependent manner against three cancer cell lines including HepG2, MCF-7, and HeLa (Bae et al. 2002b). Acacetin and luteolin found in safflower seeds have shown strong cytotoxic effects against in MCF-7, HepG2, and HeLa cell lines (Bae et al. 2002b).

Herbal formulation, called Zhu-Xiang, consists of herbal extracts from ginseng and *C. tinctorius*. Zhu-Xiang has been reported to treat the MDA-MB-231 breast cancer cell and normal human mammary gland cell lines. Inhibitory effect of Zhu-Xiang on cell proliferation has been found significantly greater as compared to standard cytotoxic drugs cyclophosphamide, epirubicin, and 5-fluorouracil. Also, different concentrations have been reported to inhibit proliferation activity in solid tumors (Loo et al. 2004). Furthermore, *N*-feruloylserotonin and *N*-(*p*-coumaroyl) serotonin from safflower seeds have been reported to strongly inhibit melanin production in *Streptomyces bikiniensis* and B16 melanoma cell lines (Roh et al. 2004).

17.4.9 Bone Protecting Activity

The loss in the bone mass can be a consequence of conditions including immobilization, calcium deficiency, endocrinological and nutritional changes, age and postmenopause. In postmenopausal osteoporosis, increased bone loss is proposed to be the result of estrogen deficiency (Bae et al. 2002a). The safflower seed oil contains high levels of the linoleic acid level, which exhibits anti-inflammatory activity, correcting bone loss due to ovariectomy or estrogen deficiency and increased intestinal absorption of calcium (Bae et al. 2002a).

In Korea and Japan, safflower seeds have long been used for preventing the onset of osteoporosis in postmenopausal women (Cho et al. 2011). Seeds and extract of safflower have been reported to stimulate the differentiation of bone-forming cells, osteoblasts in bone fractures and found instrumental in speedy recovery and healing in bone fracture incidents (Chung et al. 1999; Kim et al. 1998; Seo et al. 2000).

Seed powder of safflower contains many minerals especially calcium, potassium, and magnesium and has been shown to effectively protect against the osteoporotic process induced by ovariectomy in the rat model. Safflower seed powder exhibited effective inhibition of bone loss induced by estrogen deficiency (Bae et al. 2002a).

Safflower seeds contain phytoestrogens and have demonstrated bone-protecting effects without any significant consequence on the uterus in estrogen deficiency ovariectomized rats. Ovariectomy leads to estrogen deficiency which consequently leads to bone loss. The beneficial effect of safflower seeds might have been mediated, at least in part, by the stimulating effect of polyphenolic compounds on the proliferation of osteoblasts (Kim et al. 2002). Aqueous extract of safflower seeds has been reported to significantly accelerate the osteoblast differentiation rate in the experimental group as compared to the control group in the murine osteoblastic MC3T3-E1 cell line (Jang et al. 2007).

While mixed polyphenolic compounds from ethanolic extracts of safflower seed meals have been found to exert stimulatory effects on proliferation of ROS 17/2.8 osteoblast-like cells in the ovariectomized rat (Cho et al. 2007). Trachelogenin found in germinating seeds of safflower has been reported to exert differentiation and proliferation effects on calvarial bone cells in mice models (Kim et al. 2009).

Safflower extracts have huge potential to be developed into a drug for bone and periodontal defects regeneration. They exert a protective effect on bones caused by estrogen deficiency, without having any detrimental intrauterine effects (Kim et al. 2008). Apart from seeds, safflower oil has also demonstrated bone-protecting effects. The oral dose of 1 mL/kg to ovariectomized rats for 30 days showed positive effects as compared to ovariectomized control rats treated with vehicle only (Alam et al. 2006). Also, safflower bud has been reported to inhibit the differentiation of bone-dissolving cells osteoclasts, and prevent the bone loss in ovariectomized mice (Choi et al. 2017).

17.4.10 Cardioprotective and Anti-Lipidogenic Activities

Low-density lipoproteins (LDL) or bad cholesterols have been shown to be a strong, consistent, and independent risk factor for coronary heart disease. Fat is a vital component of the diet and it regulates serum and hepatic levels of lipids. Serum total cholesterol levels are greatly influenced by the amount and type of dietary fats. Saturated fatty acid-rich diets increase the LDL cholesterol levels in contrast to diets rich in the polyunsaturated fatty acids (Ihara-Watanabe et al. 2000).

The high level of polyunsaturated fatty acids in safflower oil has been shown to suppress the level of LDL or bad cholesterols and subsequently prevent against their harmful effects on the vasculature. Being full of unsaturated fatty acids, safflower oil has been reported to play important roles in lipid metabolism and decreasing total cholesterol levels (Nunes and Samanta 2014).

Ethanol-ethyl acetate extract of safflower seeds has been reported to exert strong antioxidant effects and prevent the development of atherosclerotic lesion by inhibiting the oxidation of low-density lipoprotein (LDL). Serotonin derivatives *N*-feruloylserotonin, *N*-(*p*-coumaroyl) serotonin, and their glucoside derivatives have been identified as major phenolic and active constituents of the extract (Koyama et al. 2006). Moreover, ethanolic extracts of safflower seed meal powder or defatted seed powder has been reported to increase the plasma levels of high-density lipoproteins (HDL) or good cholesterol in ovariectomized rats. Six polyphenolic compounds mainly categorized as lignans, flavones, and serotonin derivatives were found as the active ingredients that inhibited the cholesterol synthesis in liver Hep G2 cells and reduced cellular contents of total cholesterol (Cho et al. 2000, 2004).

Water and ethanol extracts of safflower seeds containing phenolic compounds have been reported to positively affect triglyceride levels in male rats and resulted in increased HDL cholesterol ratio in total cholesterol. Plasma and hepatic lipids were positively affected by supplementation of the diet with safflower extracts (Moon et al. 2001). Cholesterol modifying and lowering effects of safflower have been mainly attributed to serotonin derivatives, flavones, and lignans that are present in safflower seeds. These polyphenolics have also been reported to improve the overall status of lipids altered or weakened due to estrogen deficiency. The serotonin derivatives in safflower seed extract have been reported to prevent atherosclerosis and LDL oxidation in apolipoprotein-E deficient mice both ex vivo and in vitro, respectively (Koyama et al. 2006).

Supplementation with safflower seed extracts in the diet has also been reported to decrease inflammation and/or arterial stiffness and oxidative stress in healthy individuals (Koyama et al. 2008). Safflower seed polyphenols *N*-(*p*-coumaroyl)serotonin and the *N*-feruloylserotonin have also been reported to improve the elasticity of the aorta wall in the Kurosawa and Kusanagi-hypercholesterolemic (KHC) rabbits and ameliorate atherosclerosis (Katsuda et al. 2009). Polyphenols are also reported to inhibit the progress of atherosclerosis and improve wall distensibility of arteries, reduction in arterial stiffness and reduced cardiovascular risks were reported as the therapeutic benefits of these extracts (Takimoto et al. 2011).

17.4.11 Hepatoprotective Activity

Safflower seeds are reported to contain or limit the damage or toxicity to the liver by reducing total hepatic and plasma cholesterols, plasma triglycerides, and atherogenic index in high cholesterol-fed rats. Moreover, the supplementation significantly increased the activity of hepatic 3-hydroxy-3-methylglutaryl-coenzyme A (HMG-CoA) reductase while significantly decreased the activity of hepatic acyl-coenzyme A cholesterol acyltransferase (ACAT) (Moon et al. 2001).

Methanolic extracts of safflower seeds exhibited significant liver protection effects in rats with carbon tetrachloride (CCl_4)-induced hepatotoxicity. The antioxidant compound, dehydroabietylamine isolated from safflower seeds and leaves displayed strong hepatoprotective effects in rats with CCl_4-induced hepatotoxicity (Paramesha et al. 2011). Dichloromethane extracts of safflower seeds have been reported to decrease body weight and total cholesterol in hypercholesteremic rats fed with 2% cholesterol diet (Arpornsuwan et al. 2010)

Another study by Rahimi et al. (2014) reported the decrease in levels of blood glucose, triglycerides, LDL, total cholesterol, alanine aminotransferase (ALT), alkaline phosphatase (ALP), and aspartate aminotransferase (AST) in alloxan-induced diabetic rats when treated by safflower oil as diet supplementation at a dose of 200 mg/kg (Rahimi et al. 2014).

17.4.12 Anti-Obesity/Anti-Adipogenic Activity

Safflower due to its composition, the effect on lipid profile and hypoglycemic effects carries the potential to reduce obesity and fat accumulation. Studies have reported the antiadipogenic effects of the safflower supplementation. A study reported the reduction in body fat accumulation in meal-fed Sprague–Dawley rats following a diet rich in high oleic acid-rich safflower oil (Takeuchi et al. 1995). In another study, Zhang and colleagues reported the altered expression of adiposity-related genes in adipocytes following a diet supplemented with safflower oil. Altered gene expression in adipocytes further improved the effectivity of safflower in ameliorating diet-induced obesity (Zhang et al. 2010).

Consumption of safflower diet has been shown to increase the activity of lipoprotein lipase in both heart and skeletal muscles, resulting in the increased rate of fat oxidation and decreased serum triglycerol levels (Shimomura et al. 1990). Moreover, safflower oil protects against fat-induced insulin resistance by increasing the concentration of peroxisomal acyl CoA oxidase and 3-ketoacyl-CoA thiolase enzymes in the liver (Khalid et al. 2017).

Another study reports the efficient potential of safflower oil in treating cellulitis and localized fat. Reduction in body mass index and tight circumference along with improvement in the degree of cellulitis was observed after 90 days of daily supplementation with 1.6 g of oleic acid and 9 g of linoleic acids (Nunes and Samanta 2014). In addition to oil extracts, safflower seed and seed meal have also been reported to exhibit anti audiogenic properties.

Defatted safflower seeds or safflower meal are rich in polyphenols that can be effective against adipogenesis and other problems related to obesity. The anti-adipogenic effects of ethanolic extracts of defatted safflower seeds were evaluated by Hwang and colleagues in 3T3-L1 cultured preadipocytes and in C57BL/6J ob/ob mice that were fed a high-fat diet. In 3T3-L1 preadipocytes, safflower extracts were shown to inhibit their differentiation into adipocytes and decrease the expression of adipogenic transcription factors and target genes of these factors. While in C57BL/6J ob/ob mice, body fat, plasma low-density lipoprotein cholesterol level, and plasma and hepatic triglyceride levels reduced significantly suggesting the strong potential of safflower in the treatment of obesity and obesity-associated metabolic disorders, improving plasma and hepatic lipid profiles (Hwang et al. 2016).

Another study evaluated the antioxidant and antiadipogenic activity of hot water extracts of safflower seeds. Anti-adipogenic effects were evaluated through the accumulation of lipids in 3T3-L1 adipocytes. Phenolic components of safflower seed extract were reported to exhibit antiadipogenic activity and suppress the accumulation of lipids in 3T3-L1 adipocytes (Yu et al. 2013).

17.5 Adverse Effects and Reactions

Safflower has been reported to be non-toxic and non-allergic in nature (Goodman 1964). No adverse effects or reactions related to the biologically active substances in safflower seeds have been registered. Oil of safflower seeds has been used for many years in Korean herbal acupuncture, but till date, no serious side effects have been reported. Safflower oil administration at acupuncture site along the sciatic nerve in mice neither produced adverse effects on the general state of the animal nor produced any irritation at the injection site (Popov et al. 2009).

Similarly, no toxic effects have been reported of safflower seed extracts in cytotoxic assays (Zhao et al. 2009). Thus, owing to its non-toxic and non-allergic properties safflower oil has been extensively used in cosmetics and for human consumption as edible oil (Khalid et al. 2017).

17.6 Applications in Pharmaceutical and Food Industry

Widespread easy cultivation and effective biological activities of safflower have made it an attractive target for food and medicine applications in many parts of the world (Asgarpanah and Kazemivash 2013).

Safflower is mainly cultivated for its oil. Efficient therapeutic activities, easy cultivation and harvesting and widespread of safflower have made it an effective food and medical product around the world (Asgarpanah and Kazemivash 2013). Safflower seed oil possesses various beneficial properties regarding therapeutic and food applications. Unsaturated fatty acids component, high stability at elevated temperature, and very little aroma and taste make safflower oil as an ideal commodity in food products (Khalid et al. 2017).

Higher oleic safflower oil contents have been found very stable and do not produce smoke or smell at elevated temperatures (Gyulai 1996). Safflower oil is found stable at lower temperatures as well, which makes it suitable for developing products intended to be stored at lower temperatures. Safflower oil in salad dressings has been found to maintain its stability even at −12 °C (Weiss 1971). Safflower oil is also employed in infant food formulations and food coatings. It is also used for the production of margarine and as the prophylactic treatment for coronary heart disease (Emongor 2010).

Also, the cost-effectiveness along with nutritional and pharmaceutical efficacy comparable to that of olive oil and other costly nutritional oils make safflower oil an ideal product in the edible oil market. Safflower ~~is~~ oil is incorporated into a diet to reduce hyperlipidemia and prevent against atherosclerosis and other coronary heart diseases (Abidi 2001).

Safflower oil is also effective in reducing fat-induced insulin reduction and hence can be incorporated into a diet to prevent cardiovascular complications related to obesity and diabetes (Neschen et al. 2002). Safflower oil also possesses the ability to serve as a natural vehicle for storing and stabilizing other biological lipophilic compounds and can be instrumental in developing functional foods or can be used as a vehicle for drug delivery. Oxidative stability of polyunsaturated fatty acids in the cod liver oil can be enhanced by encapsulation in safflower oil (Fischer et al. 2014).

Safflower oil emulsions containing linoleic fatty acid can be employed as a good source of dietary supplements intended for increasing metabolism in healthy weight program. These supplements have the ability to increase metabolism rate and support the healthy weight program containing linoleic fatty acid diets (Maru 2014). Apart from oral formulations safflower oil can also be used in paternal emulsions and injectable medications owning to its non-allergic nature and find clinical applications for nutritional and medical purposes (Floyd 1999; Yeh et al. 2000).

Owing to the diverse benefits and oxidative stability of safflower oil it can be developed into different products, for example, organo-gels that are described as semisolid systems with self-assembled gelation capabilities. Organo-gels containing safflower are reported to be non-cytotoxic and capable of holding lipophilic compounds for a longer period of time along with efficient targeted delivery of active components to the desired location (Khalid et al. 2017).

In dairy and cattle industry, safflower seeds are employed for increasing body mass in milk production along with improving fatty acids profile in the milk of various animals including sheep, goats, and cows (Alizadeh et al. 2012; Dschaak et al. 2010). The whole seed of safflower has been used as a food source for poultry (Goodman 1964). While safflower seed meal or cake is actively used as animal feed. Safflower seed meal or cake has the potential to be used as human food if the bitter parts are removed (Nagaraj 1995). Safflower seed meal in combination with all-purpose flour in 1:3 proportion has been reported to be highly suitable for manufacturing of protein-enriched biscuits consisting of 22% of the protein (Singh and Abidi 2005).

Extracts from safflower seed meal have also been reported to act as natural antimicrobial agents in fresh fruits and vegetables (Son et al. 2017). Extracts inhibited the spread and growth of the bacterial pathogen *Listeria monocytogenes* on lettuce. Bacteria have been reported to contaminate the fruits and vegetables contributing to the rotting of fruits and vegetables and on consumption producing disease in humans.

Another interesting and important application of safflower is its utilization for the production of recombinant products. Oil bodies obtained from oilseeds have been exploited for a variety of biotechnology applications in past, mostly for producing recombinant products especially proteins as it reduces purification steps and costs associated with the production of heterogonous proteins. Recombinant proteins are targeted to be attached to the oil bodies and such attachment permits the target protein to be purified along with the oil body fraction, which upon centrifugation floats to the surface of ground seeds/water slurry (Huang et al. 2017; Mündel and Centre 2004).

Fibroblast growth factor 10 (FGF10) has been successfully expressed in safflower seeds. FGF10 exhibits diverse biological functions and is extensively investigated in fundamental research and clinical applications involving hair growth, tissue repair, and burn wounds. The plant expression vector was created and introduced into safflower plants using *Agrobacterium tumefaciens*. Recombinant FGF10 was successfully expressed in safflower seeds and MTT assays demonstrated dose-dependent effects of recombinant FGF10 on cellular proliferation (Huang et al. 2017).

Another product produced from safflower seeds was recombinant Apolipoprotein A1 Milano. Apolipoprotein A1 is naturally found in HDL and functions in protecting against cardiovascular disease by transporting the cholesterol from tissues to the liver. Apolipoprotein A1 Milano has been produced in bacterial vectors and shown to reduce atherosclerosis in humans and murine models. But production in bacterial vectors and purification is difficult and costly. The study reported the successful production of recombinant apoA1Milano in safflower seeds and successful extraction subsequently. Intravenous administration of the recombinant product obtained was shown to effectively mobilize tissue cholesterol and reduce atherosclerosis in murine models (Chyu et al. 2010).

Recombinant insulin has been produced in safflower seeds by the SemBioSys_a Canadian company based in Calgary. Safflower tissues are transformed genetically by inserting human insulin genes and insulin product accumulates in the seeds of the mature transgenic safflower plant. The type of insulin produced by safflower is proinsulin that is further converted by enzymes into the insulin called SBS-1000. SBS-1000 has been found identical to human insulin. Transgenic safflower is hoped to provide a cheaper source of insulin for people with diabetes (Mündel and Centre 2004).

17.7 Future Considerations

Safflower has gained considerable importance in the edible oil market. The trend toward developing new and better varieties of safflower is increasing and is a focus of research for improving safflower. Varieties with higher polyunsaturated and monounsaturated fatty acids are of great interest and are being produced. More varieties need to be developed as well using hybrid production and molecular techniques as well.

Oil composition and nutritional profile are reported to change upon the change in environmental conditions and type of soil along with genotype of the safflower. More research is required in evaluating the effects of stressed environments such as high salinity and low moisture conditions on safflower. New varieties need to be developed that can thrive better in dry climates and are less affected by environmental changes specifically improving the yield under stress conditions.

A broad spectrum of functional and bioactive properties of safflower have been reported in the literature. However, the active components are not fully characterized. Many studies have been conducted although but there is still a need to assess the bioactivity of safflower. Also, most of the studies are conducted in vitro or on animal models. Studies on humans evaluating the functionality and bioavailability of safflower seed oil, mechanism of action, toxicological effects, and recommended dosage of intake need to be carried out to set the recommendations for regulatory bodies and approval of safflower products as bioactive and functional compounds. Moreover, further studies on toxicological effects also need to be carried out.

Last but not the least, safflower seed oil meal or cake contains various nutritional and bioactive compounds and is used for animal feed but is not suitable for human consumption due to its strong and bitter taste. Methods to eliminate or reduce the bitterness of the seed meal need to be deduced. Some of the bitter compounds, however, possess biological activities and thus should be evaluated for their alternative use or by masking their taste.

17.8 Conclusion

Safflower is a well-known crop cultivated for its edible oil. Safflower possesses nutritional and therapeutic properties. It contains various nutrients and bioactive compounds. Various pharmacological functions including anti-oxidation, anti-inflammation, anti-diabetic, antihyperlipidemic, analgesic, and osteogenic are exhibited by safflower. Safflower seed oil, seed meal, and extracts can be employed in developing different therapeutic and functional food products.

References

Abidi S (2001) Chromatographic analysis of plant sterols in foods and vegetable oils. J Chromatogr A 935(1):173–201

Alam MR, Kim SM, Lee JI, Chon SK, Choi SJ, Choi IH, Kim NS (2006) Effects of safflower seed oil in osteoporosis induced-ovariectomized rats. Am J Chin Med 34(04):601–612

Alizadeh A, Alikhani M, Ghorbani G, Rahmani H, Rashidi L, Loor J (2012) Effects of feeding roasted safflower seeds (variety IL-111) and fish oil on dry matter intake, performance and milk fatty acid profiles in dairy cattle. J Anim Physiol Anim Nutr 96(3):466–473

Amarowicz R, Raab B, Shahidi F (2003) Antioxidant activity of phenolic fractions of rapeseed. J Food Lipids 10(1):51–62

Anderson RR (2005) Topical aminolevulinic acid-photodynamic therapy for the treatment of acne vulgaris. Google Patents,

Arpornsuwan T, Changsri K, Roytrakul S, Punjanon T (2010) The effects of the extracts from Carthamus tinctorius L. on gene expression related to cholesterol metabolism in rats. Songklanakarin J Sci Technol 32(2)

Asgarpanah J, Kazemivash N (2013) Phytochemistry, pharmacology and medicinal properties of Carthamus tinctorius L. Chin J Integr Med 19(2):153–159

Asp ML, Collene AL, Norris LE, Cole RM, Stout MB, Tang S-Y, Hsu JC, Belury MA (2011) Time-dependent effects of safflower oil to improve glycemia, inflammation and blood lipids in obese, post-menopausal women with type 2 diabetes: a randomized, double-masked, crossover study. Clin Nutr 30(4):443–449

Bae C-S, Park C-H, Cho H-J, Han H-J, Kang S-S, Choi S-H, Uhm C-S (2002a) Therapeutic effects of safflower (Carthamus tinctorius L.) seed powder on osteoporosis. Appl Microscopy 32 (3):285–290

Bae S-J, Sun-Mi Shim Y-JP, Lee J-Y, Chang E-J, Choi S-W (2002b) Cytotoxicity of phenolic compounds isolated from seeds of safflower (Carthamus tinctorius L.) on cancer cell lines. Food Sci. Biotechnol 11:140–146

Bozan B, Temelli F (2008) Chemical composition and oxidative stability of flax, safflower and poppy seed and seed oils. Bioresour Technol 99(14):6354–6359

Chang J-M, Hung L-M, Chyan Y-J, Cheng C-M (2011) Wu R-Y (2011) Carthamus tinctorius enhances the antitumor activity of dendritic cell vaccines via polarization toward Th1 cytokines and increase of cytotoxic T lymphocytes. Evid Based Complement Alternat Med 2011:274858. https://doi.org/10.1093/ecam/nen068

Cho S-H, Choi S-W, Choi Y-S, Kim H-J, Park Y-H, Bae Y-C, Lee W-J (2007) Effect of ethanol extract of safflower seed on bone loss in ovariectomized rat. Food Sci Biotechnol 16(3):392–397

Cho S-H, Choi S-W, Choi Y, Lee W-J (2000) Effects of defatted safflower and perilla seed powders on lipid metabolism in ovariectomized female rats fed high cholesterol diets. J Kor Soc Food Sci Nutr 30(1):112–118

Cho S-H, Jang J-H, Yoon JY, Han C-D, Choi Y, Choi S-W (2011) Effects of a safflower tea supplement on antioxidative status and bone markers in postmenopausal women. Nutr Res Pract 5(1):20–27

Cho S-H, Lee H-R, T-B KIM, Choi S-W, Lee W-J, Choi Y (2004) Effects of defatted safflower seed extract and phenolic compounds in diet on plasma and liver lipid in ovariectomized rats fed high-cholesterol diets. J Nutr Sci Vitaminol 50(1):32–37

Cho W, Nam J-W, Kang H-J, Windono T, Seo E-K, Lee K-T (2009) Zedoarondiol isolated from the rhizoma of Curcuma heyneana is involved in the inhibition of iNOS, COX-2 and pro-inflammatory cytokines via the downregulation of NF-κB pathway in LPS-stimulated murine macrophages. Int Immunopharmacol 9(9):1049–1057

Choi J-H, Lim S-K, Kim D-I, Park M-J, Kim Y-K, Lee A-C, Kim Y-M, Yang S-J, Park J-H (2017) Safflower bud inhibits RANKL-induced osteoclast differentiation and prevents bone loss in ovariectomized mice. Phytomedicine 34:6–13

Chung S, Choi H, Chung M, Ahn M, Yoo T, Rheu H, Yang J (1999) Effects of safflower seed on the fracture healing in rat tibia. Yakhak Hoeji 43(4):526–534

Chyu K-Y, Yano J, Li X, Zhou J, Lio WM, Chan C, Dimayuga PC, Cercek B, Nykiforuk CL, Shen Y (2010) Safflower seed-derived des-1, 2-apoA1Milano reduces atherosclerosis in hypercholesterolemic apolipoprotein e (−/−) mice. J Am Coll Cardiol 55(10):A50.E478

Dixon RA (2004) Phytoestrogens. Annu Rev Plant Biol 55:225–261

Draper CR, Edel MJ, Dick IM, Randall AG, Martin GB, Prince RL (1997) Phytoestrogens reduce bone loss and bone resorption in oophorectomized rats. J Nutr 127(9):1795–1799

Dschaak C, Eun J-S, Young A, Bergman J (2010) Nutritive merits of whole Nutrasaff safflower seed when fed to Holstein dairy cows during midlactation. Anim Feed Sci Technol 156 (1):26–36

Emongor V (2010) Safflower (Carthamus tinctorius L.) the underutilized and neglected crop: a review. Asian J Plant Sci 9(6):299–306

Fischer JJ, Nykiforuk CL, Chen X, Shen Y, Zaplachinksi S, Murray EW, Kuhlman P, Chen R, Harry I, Boothe JG (2014) Delayed oxidation of polyunsaturated fatty acids encapsulated in safflower (Carthamus tinctorius) oil bodies. Int J Eng Sci Innov Technol 3:512–522

Floyd AG (1999) Top ten considerations in the development of parenteral emulsions. Pharm Sci Technol Today 2(4):134–143

Gambacorta G, Leone A, Cazzato E Seed composition of some safflower cultivars tested in South Italy. In: proceedings of the IVth international safflower conference. Bari, Italy, 1997. pp 353–356

Goodman DH (1964) Safflower: utilization and significance in nutrition and allergy. J Allergy 35 (1):38–42

Guzik T, Korbut R, Adamek-Guzik T (2003) Nitric oxide and superoxide in inflammation. J Physiol Pharmacol 54(4):469–487

Gyulai J. Market outlook for safflower. In: proceedings of north American safflower conference, Great Falls, Montana, January, 1996. pp 17–18

Hall Iii C (2016) Overview of the oilseed safflower (Carthamus tinctorius L.). Ref Module Food Sci. https://doi.org/10.1016/B978-0-08-100596-5.00030-5

Hamrouni-Sellami I, Salah HB, Kchouk ME, Marzouk B (2007) Variations in phytosterol composition during the ripening of Tunisian safflower (Carthamus tinctorius L.) seeds. Pak J Biol Sci 10:3829–3834

Higa R, White V, Martinez N, Kurtz M, Capobianco E, Jawerbaum A (2010) Safflower and olive oil dietary treatments rescue aberrant embryonic arachidonic acid and nitric oxide metabolism and prevent diabetic embryopathy in rats. Mol Hum Reprod 16(4):286–295

Huang J, Yang J, Guan L, Yi S, Du L, Tian H, Guo Y, Zhai F, Lu Z, Li H (2017) Expression of bioactive recombinant human fibroblast growth factor 10 in Carthamus tinctorius L. seeds. Protein Expr Purif 138:7–12

Hwang E-Y, Yu M-H, Jung Y-S, Lee S-P, Shon JH, Lee S-O (2016) Defatted safflower seed extract inhibits adipogenesis in 3T3-L1 preadipocytes and improves lipid profiles in C57BL/6J Ob/Ob mice fed a high-fat diet. Nutr Res 36(9):995–1003

Ihara-Watanabe M, Umekawa H, Takahashi T, Furuichi Y (2000) Comparative effects of safflower oil and perilla oil on serum and hepatic lipid levels, fatty acid compositions of serum and hepatic phospholipids, and hepatic mRNA expressions of 3-hydroxy-3-methylglutaryl CoA reductase, LDL receptor, and cholesterol 7alpha-hydroxylase in young and adult rats. Food Res Int 33 (10):893–900

Jang H-O, Park Y-S, Lee J-H, Seo J-B, Koo K-I, Jeong S-C, Jin S-D, Lee Y-H, Eom H-S, Yun I (2007) Effect of extracts from safflower seeds on osteoblast differentiation and intracellular calcium ion concentration in MC3T3-E1 cells. Nat Prod Res 21(9):787–797

Anderson JJB, Garner SC (1997) The effects of phytoestrogens on bone. Nutr Res 17 (10):1617–1632

Jin Q, Shan L, Yue J, Wang X (2008) Spectrophotometric determination of total serotonin derivatives in the safflower seeds with Ehrlich's reagent and the underlying color reaction mechanism. Food Chem 108(2):779–783

Jo A-R, Han H-S, Seo S, Shin J-S, Lee JY, Kim HJ, Lee K-T (2017) Inhibitory effect of moschamine isolated from Carthamus tinctorius on LPS-induced inflammatory mediators via AP-1 and STAT1/3 inactivation in RAW 264.7 macrophages. Bioorg Med Chem Lett 27 (23):5245–5251. https://doi.org/10.1016/j.bmcl.2017.10.035

Jun MS, Ha YM, Kim HS, Jang HJ, Kim YM, Lee YS, Kim HJ, Seo HG, Lee JH, Lee SH (2011) Anti-inflammatory action of methanol extract of Carthamus tinctorius involves in heme oxygenase-1 induction. J Ethnopharmacol 133(2):524–530

Kang G-H, Chang E-J, Park S-W (1999) Antioxidative activity of phenolic compounds in roasted safflower (Carthamus tinctorius L.) seeds. Prev Nutr Food Sci 4(4):221–225

Kang K, Park S, Kim YS, Lee S, Back K (2009) Biosynthesis and biotechnological production of serotonin derivatives. Appl Microbiol Biotechnol 83(1):27–34

Katsuda S-I, Suzuki K, Koyama N, Takahashi M, Miyake M, Hazama A, Takazawa K (2009) Safflower seed polyphenols (N-(p-coumaroyl) serotonin and N-feruloylserotonin) ameliorate atherosclerosis and distensibility of the aortic wall in Kurosawa and Kusanagi-hypercholesterolemic (KHC) rabbits. Hypertens Res 32(11):944–949

Khalid N, Khan RS, Hussain MI, Farooq M, Ahmad A, Ahmed I (2017) A comprehensive characterisation of safflower oil for its potential applications as a bioactive food ingredient-a review. Trends Food Sci Technol 66:176–186

Kim D-H, Hwang E-Y, Son J-H (2013) Anti-inflammatory activity of Carthamus tinctorious seed extracts in raw 264.7 cells. J Life Sci 23(1):55–62

Kim D-H, Moon Y-S, Park T-S, Son J-H (2015) Serotonins of safflower seeds play a key role in anti-inflammatory effect in lipopolysaccharide-stimulated RAW 264.7 macrophages. J Plant Biotechnol 42(4):364–369

Kim E-O, Oh J-H, Lee S-K, Lee J-Y, Choi S-W (2007) Antioxidant properties and quangification of phenolic compounds from safflower (Carthamus tinctorius L.) seeds. Food Sci Biotechnol 16 (1):71–77

Kim EO, Kim KS, Lee WJ, Choi SW (2009) Proliferative and differentiative effects of trachelogenin isolated from germinated safflower (Carthamus tinctorius) seeds on calvarial bone cells. Food Sci Biotechnol 18(3):689–693

Kim H, Bae Y, Park R, Choi S, Cho S, Choi Y, Lee W (2002) Bone-protecting effect of safflower seeds in ovariectomized rats. Calcif Tissue Int 71(1):88–94

Kim J-H, Jeon S-M, An M-Y, Ku S-K, Lee J-H, Choi M-S, Moon K-D (1998) Effects of diet of Korean safflower (Carthamus tinctorious L.) seed powder on bone tissue in rats during the recovery of rib fracture. J Kor Soc Food Sci Nutr 27:698–704

Kim K-W, Suh S-J, Lee T-K, Ha K-T, Kim J-K, Kim K-H, Kim D-I, Jeon JH, Moon T-C, Kim C-H (2008) Effect of safflower seeds supplementation on stimulation of the proliferation, differentiation and mineralization of osteoblastic MC3T3-E1 cells. J Ethnopharmacol 115(1):42–49

Kim S, Cha J, Jeong S, Chung C, Choi Y, Cho Y (2000) Properties of the chemical composition of safflower (Carthamus tinctorius L.) sprout. Korean J Life Sci 10:68–73

Koyama N, Kuribayashi K, Seki T, Kobayashi K, Furuhata Y, Suzuki K, Arisaka H, Nakano T, Amino Y, Ishii K (2006) Serotonin derivatives, major safflower (Carthamus tinctorius L.) seed antioxidants, inhibit low-density lipoprotein (LDL) oxidation and atherosclerosis in apolipoprotein E-deficient mice. J Agric Food Chem 54(14):4970–4976

Koyama N, Suzuki K, Furukawa Y, Arisaka H, Seki T, Kuribayashi K, Ishii K, Sukegawa E, Takahashi M (2008) Effects of safflower seed extract supplementation on oxidation and cardiovascular risk markers in healthy human volunteers. Br J Nutr 101(4):568–575

Kumar SP, Kumari BR (2011) Factors affecting on somatic embryogenesis of safflower (Carthamus tinctorius L) at morphological and biochemical levels. World J Agric Sci 7:197–205

Liyana-Pathirana CM, Shahidi F (2006) Importance of insoluble-bound phenolics to antioxidant properties of wheat. J Agric Food Chem 54(4):1256–1264

Loo WT, Cheung MN, Chow LW (2004) The inhibitory effect of a herbal formula comprising ginseng and carthamus tinctorius on breast cancer. Life Sci 76(2):191–200

Martinez N, Sosa M, Higa R, Fornes D, Capobianco E, Jawerbaum A (2012) Dietary treatments enriched in olive and safflower oils regulate seric and placental matrix metalloproteinases in maternal diabetes. Placenta 33(1):8–16

Maru R (2014) Safflower oil emulsion as dietary supplement and preparation thereof. Google Patents,

Matthaus B, Özcan M, Al Juhaimi F (2015) Fatty acid composition and tocopherol profiles of safflower (Carthamus tinctorius L.) seed oils. Nat Prod Res 29(2):193–196

Moilanen E, Vapaatalo H (1995) Nitric oxide in inflammation and immune response. Ann Med 27 (3):359–367

Moon K-D, Back S-S, Kim J-H, Jeon S-M, Lee M-K, Choi M-S (2001) Safflower seed extract lowers plasma and hepatic lipids in rats fed high-cholesterol diet. Nutr Res 21(6):895–904

Mündel H-H, Centre LR (2004) Safflower production on the Canadian prairies: revisited in 2004. Lethbridge Research Station, Agriculture and Agri-Food Canada, Lethbridge, Alta

Nagaraj G (1995) Quality and utility of oilseeds. J Oilseeds Res 7:47–55

Neschen S, Moore I, Regittnig W, Yu CL, Wang Y, Pypaert M, Petersen KF, Shulman GI (2002) Contrasting effects of fish oil and safflower oil on hepatic peroxisomal and tissue lipid content. Am J Physiol Endocrinol Metab 282(2):E395–E401

Nunes S, Samanta MD (2014) Evaluation of safflower oil on cellulitis and localized fat. J Am Acad Dermatol 70(5):AB17

Paramesha M, Ramesh CK, Krishna V, Kumar YSR, Parvathi KM (2011) Hepatoprotective and in vitro antioxidant effect of Carthamus tinctorious L, var Annigeri-2-, an oil-yielding crop, against CCl4-induced liver injury in rats. Pharmacogn Mag 7(28):289
Park JB (2008) Serotomide and safflomide modulate forskolin-stimulated cAMP formation via 5-HT1 receptor. Phytomedicine 15(12):1093–1098
Pearl SA, Burke JM (2014) Genetic diversity in Carthamus tinctorius (Asteraceae; safflower), an underutilized oilseed crop. Am J Bot 101(10):1640–1650
Popov A, Li I, Kang D-I (2009) Analgesic properties of "CF" extracted from safflower (Carthamus tinctorius L.) seeds and potential for its use in medicine. Pharm Chem J 43(1):41–44
Rahimi P, Asgary S, Kabiri N (2014) Hepatoprotective and Hypolipidemic effects of Carthamus tinctorius oil in Alloxan-induced type 1 diabetic rats. J HerbMed Pharmacol 3(2)
Rashed MM, Shallan M, Mohamed DA, Fouda K, Hanna LM (2014) Biological evaluation of anti-androgenic effect of some plant foods. J Food Nutr Res 2(9):645–651
Reece EA, Wu Y-K, Wiznitzer A, Homko C, Yao J, Borenstein M, Sloskey G (1996) Dietary polyunsaturated fatty acid prevents malformations in offspring of diabetic rats. Am J Obstet Gynecol 175(4):818–823
Robertson J, Chapman G, Wilson R (1978) Relation of days after flowering to chemical composition and physiological maturity of sunflower seed. J Am Oil Chem Soc 55(2):266–269
Roh JS, Han JY, Kim JH, Hwang JK (2004) Inhibitory effects of active compounds isolated from safflower (Carthamus tinctorius L.) seeds for melanogenesis. Biol Pharm Bull 27 (12):1976–1978
Seo H, Kim J, Kwak D, Jeon S, Ku S, Lee J, Moon K, Choi M (2000) The effects of safflower seed powder and its fraction on bone tissue in rib-fractured rats during the recovery. Kor J Nutr 33 (4):411–420
Shahidi F (2000) Antioxidants in food and food antioxidants. Mol Nutr Food Res 44(3):158–163
Shimomura Y, Tamura T, Suzuki M (1990) Less body fat accumulation in rats fed a safflower oil diet than in rats fed a beef tallow diet. J Nutr 120:1291–1296
Shirwaikar A, Khan S, Kamariya YH, Patel BD, Gajera FP (2010) Medicinal plants for the management of post menopausal osteoporosis: a review. Open Bone J 2:1–13
Singh R, Abidi A (2005) Protein enriched biscuits from safflower (Carthamus tinctorius L.) cake. Bev Food World 32:46. https://doi.org/10.1201/9781420005363.ch6
Singh V, Nimbkar N (2006) Safflower (Carthamus tinctorius L.). Oil Seed Crops 6:167–194
Son H-J, Kang J-H, Song KB (2017) Antimicrobial activity of safflower seed meal extract and its application as an antimicrobial agent for the inactivation of Listeria monocytogenes inoculated on fresh lettuce. LWT Food Sci Technol 85:52–57
Takahashi T, Miyazawa M (2012) Potent α-Glucosidase inhibitors from safflower (Carthamus tinctorius L.) seed. Phytother Res 26(5):722–726
Takeuchi H, Matsuo T, Tokuyama K, Shimomura Y, Suzuki M (1995) Diet-induced thermogenesis is lower in rats fed a lard diet than in those fed a high oleic acid safflower oil diet, a safflower oil diet or a linseed oil diet. J Nutr 125(4):920
Takii T, Hayashi M, Hiroma H, Chiba T, Kawashima S, Zhang HL, Nagatsu A, Sakakibara J, Onozaki K (1999) Serotonin derivative, N-(p-coumaroyl) serotonin, isolated from safflower (Carthamus tinctorius L.) oil cake augments the proliferation of normal human and mouse fibroblasts in synergy with basic fibroblast growth factor (bFGF) or epidermal growth factor (EGF). J Biochem 125(5):910–915
Takii T, Kawashima S, Chiba T, Hayashi H, Hayashi M, Hiroma H, Kimura H, Inukai Y, Shibata Y, Nagatsu A (2003) Multiple mechanisms involved in the inhibition of proinflammatory cytokine production from human monocytes by N-(p-coumaroyl) serotonin and its derivatives. Int Immunopharmacol 3(2):273–277
Takimoto T, Suzuki K, Arisaka H, Murata T, Ozaki H, Koyama N (2011) Effect of N-(p-coumaroyl) serotonin and N-feruloylserotonin, major anti-atherogenic polyphenols in safflower seed, on vasodilation, proliferation and migration of vascular smooth muscle cells. Mol Nutr Food Res 55(10):1561–1571

Tanwar B, Modgil R, Goyal A (2018) Antinutritional factors and hypocholesterolemic effect of wild apricot kernel (Prunus armeniaca L.) as affected by detoxification. Food Funct 9(4):2121–2135

Tanwar B, Modgil R, Goyal A (2019) Effect of detoxification on biological quality of wild apricot (Prunus armeniaca L.) kernel. J Sci Food Agric 99(2):517–528

Tetley TD (2005) Inflammatory cells and chronic obstructive pulmonary disease. Curr Drug Targets Inflam Allergy 4(6):607–618

Toma W, Guimarães LL, Brito AR, Santos AR, Cortez FS, Pusceddu FH, Cesar A, S Júnior L, Pacheco MT, Pereira CD (2014) Safflower oil: an integrated assessment of phytochemistry, antiulcerogenic activity, and rodent and environmental toxicity. Revista Brasileira de Farmacognosia 24 (5):538–544

Toombs EL (2005) Cosmetics in the treatment of acne vulgaris. Dermatol Clin 23(3):575–581

Tovar MJ, Motilva MJ, Romero MP (2001) Changes in the phenolic composition of virgin olive oil from young trees (Olea europaea L. cv. Arbequina) grown under linear irrigation strategies. J Agric Food Chem 49(11):5502–5508

Vavilov NI (1951) The origin, variation, immunity and breeding of cultivated plants. Soil Sci 72 (6):482

Weiss E (1971) Castor, sesame and safflower. Leonard Hill Books, London

Woods JA, Vieira VJ, Keylock KT (2009) Exercise, inflammation, and innate immunity. Immunol Allergy Clin North Am 29(2):381–393

Yamamotova A, Pometlova M, Harmatha J, Rašková H, Rokyta R (2007) The selective effect of N-feruloylserotonins isolated from Leuzea carthamoides on nociception and anxiety in rats. J Ethnopharmacol 112(2):368–374

Yasukawa K, Akihisa T, Kasahara Y, Kaminaga T, Kanno H, Kumaki K, Tamura T, Takido M (1996) Inhibitory effect of alkane-6, 8-diols, the components of safflower, on tumor promotion by 12-O-tetradecanoylphorbol-13-acetate in two-stage carcinogenesis in mouse skin. Oncology 53(2):133–136

Yeh S-L, Chao C-H, Lin M-T, Chen W-J (2000) Effects of parenteral infusion with medium-chain triglycerides and safflower oil emulsions on hepatic lipids, plasma amino acids and inflammatory mediators in septic rats. Clin Nutr 19(2):115–120

Yoo HH, Park JH, Kwon SW (2006) An anti-estrogenic lignan glycoside, tracheloside, from seeds of Carthamus tinctorius. Biosci Biotechnol Biochem 70(11):2783–2785

Yu S-Y, Lee Y-J, Kim J-D, Kang S-N, Lee S-K, Jang J-Y, Lee H-K, Lim J-H, Lee O-H (2013) Phenolic composition, antioxidant activity and anti-adipogenic effect of hot water extract from safflower (Carthamus tinctorius L.) seed. Nutrients 5(12):4894–4907

Zhang HL, Nagatsu A, Watanabe T, Okuyama H (1997) Antioxidative compounds isolated from safflower (Carthamus tinctorius L.) oil cake. Chem Pharm Bull 45(12):1910–1914

Zhang Z, Li Q, Liu F, Sun Y, Zhang J (2010) Prevention of diet-induced obesity by safflower oil: insights at the levels of PPARα, orexin, and ghrelin gene expression of adipocytes in mice. Acta Biochim Biophys Sin 42(3):202–208

Zhao G, Gai Y, Chu W-J, Qin G-W, Guo L-H (2009) A novel compound N 1, N 5-(Z)-N 10-(E)-tri-p-coumaroylspermidine isolated from Carthamus tinctorius L. and acting by serotonin transporter inhibition. Eur Neuropsychopharmacol 19(10):749–758

Zhou X, Tang L, Xu Y, Zhou G, Wang Z (2014) Towards a better understanding of medicinal uses of Carthamus tinctorius L. in traditional Chinese medicine: a phytochemical and pharmacological review. J Ethnopharmacol 151(1):27–43

Chapter 18
Camelina (*Camelina sativa*) Seed

Sushil K. Singh, Bipin Rajpurohit, and Poonam Singha

Abstract Camelina [*Camelina sativa* (L.) Crantz., Brassicaceae] is also known by the names of false flax and gold of pleasure. Historically, camelina oil was used in cooking and as fuel. Additionally, the seed meal was used as livestock feed. Currently, the crop is produced commercially in the United States and utilized as biodiesel. The phytochemical potential of camelina has renewed the interest in this crop for health and food applications. Mechanical extraction is the oldest method for oil extraction. Newer and greener methods like the supercritical-CO_2 extraction method have also been studied for oil extraction from camelina. The oil content in camelina seed ranges from 30 to 50%, which is rich in omega-3 fatty acids (25–50% of total fatty acids) and antioxidants. The health-promoting attributes of camelina oil arise due to the presence of alpha-linolenic acid and antioxidants. The major antinutritional compounds in camelina are glucosinolates, tannins, and erucic acid. The oil can be used for skin ailments, cardiovascular diseases, cancer, and chronic diseases. Additionally, camelina oil has the potential in food applications such as edible oil and as a functional ingredient, feed applications, and industrial applications.

Keywords *Camelina sativa* · Edible oil · Fatty acids · Oilseed crop · Phytochemicals

S. K. Singh · P. Singha
Department of Food Process Engineering, National Institute of Technology, Rourkela, Odisha, India

B. Rajpurohit (✉)
Department of Dairy & Food Science, South Dakota State University, Brookings, SD, USA
e-mail: bipin.rajpurohit@jacks.sdstate.edu

B. Tanwar, A. Goyal (eds.), *Oilseeds: Health Attributes and Food Applications*,
https://doi.org/10.1007/978-981-15-4194-0_18

18.1 Origin and History

Camelina [*Camelina sativa* (L.) Crantz., Brassicaceae] was originated in Central Asia and the Mediterranean. Its cultivation is believed to have begun in the Neolithic times and was used in the preparation of porridge and bread (Hatt 1937). For centuries, it has been referred to as the "gold-of-pleasure" (USA) and "false flax" (UK). In Europe, the production of *Camelina sativa* almost doubled during the Iron age and it was mainly grown as an oil-producing crop (Knorzer 1978). The press cake after oil extraction was used as an animal feed. Evidence of seeds of camelina was found in the archaeological excavations of the Bronze Age in the regions surrounding the North Sea (Knorzer 1978). Applications of camelina oil were known to the Romans and they utilized camelina oil as lamp fuel, cooking oil, and massage oil; the meal was used as livestock feed (Pilgeram et al. 2007). It was exclusively used as an oilseed crop in the Eastern and Northern Europe until the early 1940s before the widespread use of rapeseed. Camelina is still popular as a weed species in Europe and North America (Fig 18.1).

18.2 Production

In recent years, the demand for food and biofuel oils has led to an increase in the production of camelina. In North America and Europe, a number of pilot-scale productions have been developed. Camelina can be grown at very low input cost. It can grow under varying climatic conditions and can resist drought to some extent. However, it does not grow in organic soil and heavy clay. It can be grown both as a summer crop (annually) and as a winter crop (biannually). Camelina requires less water, pesticide, and fertilizer for its cultivation compared to other oilseed crops such as soybean, sunflower, and canola. Camelina is cultivated in the northern regions of Europe, Asia, and North America (Moser 2010). It is commonly cultivated with other crops (especially with wheat followed by soy) in the mid and southern United States (Marra and Carlson 1986; Crabtree et al. 1990).

The per hectare yield of camelina seeds ranges from 336 to 2240 kg. In a study at the University of Minnesota, camelina yield was reported to be in the range of 600–1700 kg/ha at Rosemount, Minnesota (45°N latitude), with an average in the range of 1100–1200 kg/ha over several years of trials. Typically, the oil content in camelina seeds is 35%. In Montana, where dry land conditions existed, camelina yield was reported 1800–2000 pounds and 900–1700 pounds per acre in regions with 16–18 in. and 13–15 in. precipitation, respectively. In Idaho, yields of 1700–2200 pounds per acre of camelina have been reported.

Fig. 18.1 Camelina (*Camelina sativa*) plant, flowers, fruit, and seeds. Gerhard Nitter, Reutlingen (https://commons.wikimedia.org/wiki/File:Camelina_sativa_0_(G._Nitter).jpg, "Camelina sativa 0 (G. Nitter)", https://creativecommons.org/licenses/by-sa/3.0/legalcode). Bliesgauoele (https://commons.wikimedia.org/wiki/File:Leindotter_im_Saarland2.jpg, https://creativecommons.org/licenses/by-sa/4.0/legalcode). Marie Portas (https://commons.wikimedia.org/wiki/File:Camelina_sativa_fruit_seed_(2).jpg, "*Camelina sativa* fruit seed (2)", https://creativecommons.org/licenses/by-sa/2.0/fr/deed.en)

18.3 Chemical Composition

18.3.1 Camelina Oil Extraction

Broadly, there are three techniques of extracting oil from the camelina seeds, namely mechanical, solvent, and enzymatic extraction. Mechanical extraction is the oldest method for oil extraction. Unlike solvent extraction, it does not utilize organic solvents. Additionally, the oil extracted retains bioactive compounds like phenols, flavonoids, fatty acids, and minerals. The oil extracted utilizing this method needs further processing like filtration and degumming. The method has low extraction efficiency. However, a better yield is possible with engine or manual mechanical presses. The organic solvent extraction method utilizes organic solvents like hexane. High pressure and temperature with ethanol can also be utilized as greener alternatives. Solvent extraction results in a powdered meal that has less than 1% residual oil. Mechanical extraction may leave as much as 15% residual oil in seed meal (Berhow et al. 2014). The method has a higher extraction efficiency as compared to cold mechanical pressing and thus gained popularity for industrial-scale oil extraction. The supercritical-CO_2 extraction method has also been studied for oil extraction (Moslavac et al. 2014). CO_2 is an economical solvent that is readily available and can be easily removed from the products. However, a disadvantage with this method is the high equipment cost. The enzymatic extraction method is time-consuming, but it does not produce harmful solvents or chemicals (Popa et al., 2017). The cold mechanical pressing of camelina seed results in oil and a by-product called seed meal. Chemical composition of camelina seed and oil has been discussed in the following sections.

18.3.2 Camelina Oil Composition

The oil content in camelina seed ranges from 300 to 490 g kg^{-1} (Berti et al. 2016). The camelina oil has two major fractions: the unsaponifiable includes tocopherols and sterols, and the saponifiable includes fatty acids. The major fatty acids are oleic (18:1, 14–16%), linoleic (LA) (18:2, 15–23%), alpha-linolenic (ALA) (18:3, 31–40%), and eicosenoic (20:1, 12–15%) acid. The minor fatty acids are palmitic (16:0), stearic (18:0), and erucic (22:1) acids (Berti et al. 2016). The fatty acid composition of camelina oil obtained from several literature sources (Moser and Vaughn 2010; Budin et al. 1995; Leonard 1998; Zubr and Matthaus 2002; Fröhlich and Rice 2005) is shown in Table 18.1.

The factors that influence the composition of oil are environment, location, cultivar, and oil extraction method (Berti et al. 2016). Camelina oil is rich in omega-3 fatty acids (approximately 35–45%) which is essential for human and animal nutrition and has positive health benefits. The oil is also rich in potential

Table 18.1 Chemical composition of camelina seed (Berhow et al. 2014)

Component	Percentage
Camelina oil	
Fat	100
Seed meal	
Residual oil	10
Crude protein	45
Soluble sugars	10
Crude fiber	13
Minerals	5
Phytochemicals	10

Table 18.2 Composition of fatty acids in camelina oil (% of total fatty acids) from several literature sources

Fatty acid	Carbon number	Budin et al. (1995)	Leonard (1998)	Moser and Vaughn (2010)	Zubr and Matthaus 2002	Fröhlich and Rice 2005
Palmitic	16:0	6.4	5.3	6.8	5.4	5.4
Stearic	18:0	2.8	2.5	2.7	2.5	2.6
Oleic	18:1	15.9	12.6	18.6	14.9	14.3
Linoleic	18:2	20.9	15.6	19.6	15.2	14.3
Linolenic	18:3	30.7	37.5	32.6	36.8	38.4
11-Eicosenoic	20:1	13.6	15.5	12.4	15.5	16.8
Erucic	22:1	3.0	2.9	2.3	2.8	2.9
Others	—	6.6	8.1	5	6.9	5.3

antioxidants (such as gamma tocopherol or vitamin E) which inherently increases the stability and shelf life of camelina oil (Dubois et al. 2007; Tripathi and Mishra 2007).

18.3.3 Camelina Meal Composition

Camelina meal that is obtained after the typical mechanical extraction contains approximately 10% residual oil, 45% protein, 15% insoluble fiber, 10% soluble carbohydrates, 5% minerals, 0.2% nucleic acids, and 10% of a mixture of phytochemicals comprising of terpenoids, phenolics, glycosylated glucosinolates, and flavonoids (Berhow et al. 2014). Detailed explanation of the composition of camelina meal is given below:

18.3.3.1 Carbohydrates

The carbohydrates present are mono-, di-, tri-, tetra-, and oligosaccharides (Berhow et al. 2014). The mono- and di-saccharide levels are very low (sucrose content is about 5.5%). Oligosaccharides namely raffinose and stachyose are below 1% (Ibrahim and El Habbasha 2015). The polysaccharides present are pectin, starch, and mucilage. The mucilage is produced after the addition of water, which forms a gel and the content is about 6.7% (Berhow et al. 2014; Ibrahim and El Habbasha 2015). The starch and pectin content is less than 1% each (Ibrahim and El Habbasha 2015).

18.3.3.2 Dietary Fiber

Cellulose is the most abundant form of dietary fiber. The content of lignin, a polyphenolic compound associated with dietary fiber, is 6.7%. However, the levels of β-glucan are significantly low (Berhow et al. 2014; Ibrahim and El Habbasha 2015).

18.3.3.3 Proteins

Proteins are least characterized in camelina seed meal (Berhow et al. 2014). The most abundant proportion of seed meal protein is in the form of seed storage proteins and they are about 60% of total seed meal protein. These constitute about 20% of total proteins in mature whole seeds (Ibrahim and El Habbasha 2015).

Table 18.3 Composition of vitamins and minerals in camelina seed (Berhow et al. 2014)

Nutrient	Content (μg/g)
Vitamin	
Thiamin (B1)	18.8
Riboflavin (B2)	4.4
Niacin (B3)	194
Pantothenic acid (B5)	11.3
Pyridoxine (B6)	1.9
Biotin (B7)	1.0
Folate (B9)	3.2
Mineral	
Potassium	16,000
Phosphorus	14,000
Calcium	10,000
Magnesium	5100
Sulfur	2400
Sodium	600
Chlorine	400
Iron	329
Copper	99
Zinc	69
Manganese	40
Nickel	1.9

18.3.3.4 Vitamins

The detectable vitamins present in camelina meal are Vitamins B1 (thiamin), B3 (niacin), B5 (panthothenic acid), B2 (riboflavin), B6 (pyridoxine), B7 (biotin), and B9 (folate) (Table 18.3). The meal is considered a good source of vitamin B1, B3, and B5 (Berhow et al. 2014).

18.3.3.5 Minerals

The minerals present in camelina meal include calcium, magnesium, sodium, potassium, chlorine, phosphorus, sulfur, iron, copper, manganese, nickel, and zinc (Berhow et al. 2014). Content of calcium, potassium, and phosphorus is between 1.0 and 1.6% (Table 18.3). The iron, manganese, and zinc contents are 329 μg/g, 40 μg/g and 69 μg/g, respectively (Ibrahim and El Habbasha 2015).

18.4 Antinutritional Factors

Camelina contains glucosinolates, tannins, and erucic acid (C22:1) (Abramovič et al. 2007; Matthaus and Zubr 2000; Zubr 1997). These affect the nutritional content of the seeds. Glucosinolates and erucic acids are the major concerns when feeding camelina because of their effect on the thyroid and the cardiovascular system (Tripathi and Mishra 2007). However, the glucosinolate levels in camelina meal are lower as compared to other brassicaceous species. Camelina seeds contain three main glucosinolates: glucoarabin (9-methyl-sulfinylnonyl glucosinolate; GS9), glucocamelinin (10-methylsulfinyl-decyl glucosinolate; GS10), and 11-methyl-sulfinyl-undecyl glucosinolate (GS11) (Schuster and Friedt 1998). Not much is known about the biological role of these glucosinolates and the toxicity of their hydrolysis products. It is the degraded products from glucosinolates that contributes to the toxicity of glucosinolates. The degraded products, specifically thiocyanates, oxazolidinethiones, and nitriles, are bitter in taste (Mithen et al. 2000), and are shown to decrease thyroid function in rainbow trout (Burel et al. 2001). Phytic acid reduces the availability of minerals, such as phosphorus, zinc, and magnesium (Denstadli et al. 2006). Tannins inhibit amylase, lipase, and protease activities (Tanwar et al. 2018, 2019; Mandal and Ghosh 2010). The presence of sinapine in aquafeed reduces the feed palatability due to its bitter taste. The presence of erucic acid in camelina oil may make its consumption in higher quantities unsafe. High erucic acid has been linked to heart disease (Kok et al. 2018). Camelina lines devoid of erucic acid were identified and cultivars with very low erucic acid are being developed.

18.5 Phytonutrients/Phenolics

The phenolic compounds are produced in plants mainly for protection against insects, bacteria, fungi and offer resistance against UV light. Camelina oil fractions contain the non-volatile terpenes namely sterols, tocopherols, low content of uncyclized terpenes squalene and phytol. In addition, some other degradation products are also present, the amount of which depends on the extraction process utilized (Berhow et al. 2014).

During the oil extraction process, a part of the total phenolics is transferred into the oil. The nonpolar phenolics particularly tocopherols present in camelina oil extracted from various cultivars grown in Scandinavia, Northern Europe, and Central Europe have been analyzed (Dubois et al. 2007; Matthaus and Zubr 2000). The polar phenolics content of fat-free camelina seeds was found to be 1.8 mg/g seed (Gallic acid equivalent) (Matthäus 2002). The total phenolic content of camelina oil utilizing the Folin–Ciocalteu method was determined to be 128 mg chlorogenic acid equivalent/kg. Extraction procedures at higher pressures and temperatures assist the extraction of the higher amount of phenolics from seeds (Abramovič et al. 2007). Therefore, the total phenol content of cold-pressed oils is generally lower than oils extracted by alternative extraction methods. Other factors affecting the phenolic production in plants are variety, location, ripeness at harvest, growing climate, storage conditions, etc. (Günç Ergönül and Aksoylu 2018). Camelina oil has a high content of tocopherols and the total tocopherol content is estimated between 700 and 800 mg/kg of seed and γ-tocopherol being 90% of the total (Murphy 2016). The major tocopherols present in the oil are α-tocopherol, δ-tocopherol, and γ-tocopherol. Additionally, low levels of β-tocopherol have also been detected (Berhow et al. 2014).

Antioxidant compounds identified in *Camelina sativa* seed include sinapine, protocatechuic acid, quercetin glucoside, quercetin, p-hydroxybenzoic acid, salicylic acid, sinapic acid, rutin, catechin, and ellagic acid. The cake possesses a similar phenolic profile. Studies have shown that camelina oil-based spreads are more stable than fish oil and flax oil but less stable than sunflower oil, sesame, corn, and olive oil (Eidhin et al. 2003b). Camelina cake had a high reducing power and free radical scavenging ability (Terpinc et al. 2012).

18.6 Health Attributes

Several studies and clinical trials have suggested a positive correlation between the consumption of camelina oil and positive health effects (Rokka et al. 2002; Waraich et al. 2013; Zubr 1997). Camelina oil is known for its medicinal value and considered as a curative measure for many health problems. It has many therapeutic and healing properties such as anti-inflammatory, antioxidant, anti-arthritic, and immunomodulatory. For medical purposes, the oil is used orally against colic attacks, stomach and intestinal ulcers, digestive problems, and gastritis.

As an external remedy, camelina oil is useful in curing skin diseases (acne), skin scratches, has healing effect on bruises, sprains, and inflammations (Rode 2002). The antimicrobial efficacy of camelina seeds also makes it suitable as natural preservatives (Kumar et al. 2017). Camelina oil is also useful for slenderness recovery, skin elasticity, and regeneration of skin (Vollman et al. 2007). Because of their fatty acid profile, camelina oil is generally included in many of the cosmetic products such as lipsticks, lotions, bar soaps, balms, and skin creams. Cold-pressed (when seed material is pressed mechanically in a controlled environment) camelina oil has a high content of alpha-linolenic acid or ALA (25–40%). The health-promoting attributes of camelina oil arise due to high contents of ALA, tocopherols, and antioxidants.

The omega-3 fatty acids are essential fatty acids. They are important for the retina and nerve functions. ALA cannot be synthesized by the human body and their deficiency may cause suboptimal growth and neurological abnormalities. Additionally, ALA can improve the omega-6/omega-3 fatty acid ratio in food (Waraich et al. 2013). The health effects of ALA have been comprehensively described by Goyal et al. (2014). Consumption of camelina oil for 12 weeks increased ALA in erythrocyte membranes (EM), plasma phospholipids (PL), cholesteryl esters (CE), and triglycerides (TG) in human subjects with impaired glucose mechanism (Manninen et al. 2019a).

The effect of ALA of camelina oil on fatty acid profile and serum lipids has been studied (Karvonen et al. 2002). Diet rich in camelina oil has shown to improve the lipid profile by decreasing LDL and total cholesterol (Karvonen et al. 2002). In a study that compared the metabolic effect of camelina oil, fatty fish and lean fish on subjects with impaired glucose metabolism, it was found that camelina oil was most effective in improving the serum lipid profile. However, there was no significant difference in glucose metabolism and anti-inflammatory response (Schwab et al. 2018). Researchers conducted studies to determine the effect of camelina oil and fish oil on atherogenic and anti-atherogenic functions of LDL and HDL, respectively. Camelina oil decreased the binding of lipoproteins to aortic proteoglycans by decreasing serum LDL cholesterol concentration (Manninen et al. 2019b). The anti-inflammatory effects of ALA-enriched diets have also been studied. Currently, diseases like obesity and atherosclerosis are regarded as low-grade chronic inflammatory diseases. Therefore, ALA-rich diet may be useful in decreasing inflammatory markers associated with obesity (Barceló-Coblijn and Murphy 2009). In a recent study, researchers concluded that *Camelina sativa* oil may be beneficial in reducing inflammation. A 12 week diet rich in *Camelina sativa* oil resulted in a decrease of peripheral blood mononuclear cells (PBMC) interferon-gamma (*IFNG*) mRNA expression in 17 human subjects with impaired glucose mechanism (de Mello et al. 2019). IFN-γ is an inflammatory marker (Shu et al. 2016).

Docosahexaenoic acids or DHA are vital for the development of the brain. An increase in the levels of DHA in the brain was observed in rats fed on ALA-rich diet. This suggested the potential of ALA to limit or reverse neurogenerative disease in the animal models studied (Barceló-Coblijn and Murphy 2009). Owing to the health benefits of ALA mentioned above, it is believed that food sources with omega-3 fatty acids have benefits for health. Among all, the omega-3 long-chain polyunsaturated fatty acids (PUFA) had the most conclusive results obtained for rheumatoid arthritis (Miles and Calder 2012). Additionally, epidemiological studies indicate that omega-3 fatty acids may be beneficial (i.e., lower the associated risk) in breast, prostate, and colorectal cancers (Gerber 2012). Phytonutrients are associated with reducing the risk of inflammation, thrombotic tendency, carcinogenesis, and cardiovascular disease (Pietta 2000).

18.7 Food Applications

In January 2010, Health Canada approved camelina oil as a novel food. Since then, camelina oil is sold in the Canadian food market as an edible oil (http://www.hc-sc.gc.ca/fn-an/gmf-agm/appro/index-eng.php). However, the pressed camelina meal after oil extraction has not been approved safe for human consumption. It is being said that camelina oil is the new go-to cooking oil. Camelina oil can be used in food as a replacement for other omega-3 oils. In the GRAS exemption claim of camelina oil submitted to the FDA, it was proposed that the oil can be used in several food products. Camelina oil can be added during the production of cookies (excluding low fat and dietetic), meal replacement beverages, sports drinks, energy drinks, cereal and granola bars (excluding low fat types), yogurt (excluding fat-free types), fruit-flavored drinks, fruit juices, vegetable juices, chips, and other savory snacks, salad dressings (excluding mayonnaise, fat-free and low-fat types), soft candies, nut-based chocolate confectioners, etc. The high omega-3 content in camelina oil makes it suitable to be used as specialty oils such as omega-3 enriched butter, margarine, salad dressings, and spreads.

18.7.1 As an Edible Oil

The health benefits associated with camelina oil makes it a potential choice for food application.

Camelina oil has been consumed in Europe for a long time. The smoke point of the oil is high (246 °C). The smoke point is the temperature at which an oil starts to burn, producing smoke and harmful chemical by-products. The smoke point of coconut oil and extra-virgin olive oil is 176 °C approximately. Therefore, camelina oil is a better choice for cooking at high temperatures than these oils; however, deep frying in camelina oil develops an undesirable paint like flavor (Crowley and Frohlich 1998). Therefore, camelina oil can be used for sautéing, baking, and moderate heat cooking applications (Fan and Eskin 2013). Other than cooking, camelina oil can be used in cream spreads, salad dressings, margarines, and specialty oil. The important parameter considering oil would be its oxidative stability. As mentioned earlier, camelina contains a considerable high amount of tocopherols and phenols making it oxidatively stable oil. Camelina oil has almond-like taste and aroma. This makes it versatile as a cooking oil in comparison to flax oil which is generally used for omega 3 supplementation (Fan and Eskin 2013). The oil is commercially available; however, it does not have FDA GRAS (Generally Recognized as Safe) status. Camelina oil was utilized as a source of PUFA to produce human milk fat substitutes for babies (Popa et al. 2017).

Salo and Kuusisto (2016) investigated the effects of plant stanol ester and camelina oil in hypercholesterolemic human subjects, when administered in a yogurt mini-drink together with a meal. Their aim was to study the cholesterol-lowering effect of camelina oil although low quantity (2 g) of camelina oil neither enhanced the cholesterol-lowering effects of plant stanol ester nor had an effect on the serum triglyceride levels. Similar studies are needed to be conducted to find out the health benefit of camelina oil in dairy products. In another study, camelina oil was added at a concentration equivalent to 10% of the recommended dietary intake of 2 g/day of ALA as per EC regulation no 432/2012 to increase the *n*-3 polyunsaturated fatty acids (PUFA) in yogurt. The yogurt was found to have acceptable sensory characteristics after 21 days of storage at 4 °C (Dal Bello et al. 2015).

18.7.2 Animal Feed Applications

Compared to other oil crops, camelina is rich in Omega 3 fatty acids. In every oilseed crop extraction process, a press cake remains which contains almost three-fourth of seed nutrients. Utilizing this press cake, the profitability of the crop can be improved. Camelina meal used for animal feed is currently available in the form of pellets or cylindrical shaped cakes. The press cakes have economic value and can be utilized as a nutritive supplement, due to their high crude protein content (43%) (Tripathi and Mishra 2007). Several researchers concluded that camelina had beneficial effects on animals when incorporated in the feed. However, the presence of glucosinolates, their degradation products, and the presence of non-starch polysaccharides (NSPs) is a major disadvantage for brasicca crops for inclusion in diets of the animals.

18.7.2.1 As an Aquaculture Feed

The press cake can be an effective replacement for fish feed. Camelina meal consists of a considerable amount of crude protein level (38–43%), chiefly of methionine and phenylalanine (Hixson and Parrish 2014). The fish feed mainly includes a huge percentage of fish meal and fish oil. Most of the fish meal or oil comes from fish industrial waste. The options of availability of fish meal are very rare and also expensive (Singh and Muthukumarappan 2016). In fish industries, they cannot increase the production of fish oils, as there are limits on established fishing quotas and Total Allowable Catches. This affects the production of fish oil to a great extent, as the supply rates cannot meet the demand which ultimately results in the decline of fish meal production. This infers that in future the fish farms have to find an alternative for feeding fish. Thus, replacing the fish feed with some plant sources of the same nutrient source like camelina meal can avoid the scarcity in raw materials as well as a good supplement of nutrients for fish. Oil containing *n*-3 LC-PUFA was produced from transgenic *Camelina sativa* to replace fish oil in salmon feeds (Betancor et al. 2017). This did not result in any negative effect on fish performance, metabolic responses, and nutritional attributes of the farmed fish. However, the camelina meal alone cannot replace the exact nutrients of the fish meal but a combination of various other oilseed crops can provide the same effect as fish meal.

18.7.2.2 As Feed in Dairy Industry

Camelina meal holds FDA GRAS status for being utilized as an ingredient in ruminant (beef, cattle, and goats) and monogastric (chicken and swine) feed rations. However, the approval for inclusion in dairy rations is pending. Feeding camelina seeds and meal to lactating dairy cattle tended to decrease dry matter intake but did not significantly affect milk production. Camelina meal was found to decrease milk fat yield and content, with changes in fatty acid composition that resulted in the modification of butter spreadability (Hurtaud and Peyraud 2007).

When camelina cake was supplemented in goats diet, an increase in polyunsaturated fatty acids, including conjugated linoleic acid and omega-3 fatty acids, was observed in the raw milk (Pikul et al. 2014). The increased bioactive compounds were also found in Kefir after production and in storage. Additionally, the Kefir was acceptable in terms of quality parameters like texture, flavor, aroma, color, and consistency (Cais-Sokolińska et al. 2015).

18.7.2.3 As Poultry Feed

When camelina meal was added in poultry feed, the omega-3 fatty acid content increased in eggs (Tripathi and Mishra 2007), which aided in upgrading the quality of broiler chickens and laying hens. The eggs obtained had a pleasant odor which is not often seen in the case of flax seeds. In another study conducted by Frame et al. (2007) on including camelina in starter diets of turkeys (*Meleagris gallopavo*), more than 5% inclusion in the diets resulted in reduced weight and poor feed conversion.

18.7.2.4 As Feed in Beef and Pork Industry

It was reported that the camelina is suitable for pigs (*Sus domesticus*) and ruminants for their rich protein and energy content (Matthaus and Zubr 2000). In a research conducted by Eidhin et al. (2003a), camelina oil reduced the plasma omega-6 fatty acid content and serum triglyceride levels, and increased the plasma omega-3 fatty acid content in a controlled pig feeding trial. Moreover, some other researchers reported that omega-3 fatty acid content in beef and animal meat can be improved by the addition of flaxseed and other enriched sources of *n*-3 fatty acids like camelina cake in animal feed (Barceló-Coblijn and Murphy 2009; Woods and Fearon 2009).

18.8 Alternative Applications

18.8.1 Camelina as a Biofuel

Camelina can withstand drought conditions and grow in soil with low fertility (Gehringer et al. 2006; Zanetti et al. 2009). Therefore, it is suitable for biofuel production in regions that do not support oil crop production, such as the Great Plains of the United States. Breeding programs such as the Agrobacterium-based floral infiltration method make camelina highly potential in producing novel, high-value industrial oils. The oil stability index of biodiesel obtained from camelina is lower as compared to the biofuel derived from other feedstocks. Studies have demonstrated that the cost of production of camelina oil is lower than the production cost of rapeseed oil (Bansal and Durrett 2016). For more than 1500 commercial flights, jet fuel blends containing up to 50% biofuel derived from camelina, waste cooking oil, and algae have been used.

18.8.2 Chemical Derivatives of Camelina Oil

Chemical modification of camelina oil was limited to the production of biodiesel. Some of the newer derivatives include pressure sensitive adhesives, monomers, nanocomposites, ultraviolet polymers, and alkyd resins. Another host of derivatives can be produced from epoxidized camelina oil (Berti et al. 2016).

18.8.3 Non-food Applications of Camelina Meal

Although the major application of camelina meal is animal feed, some alternative applications are soil fungicides, bio-oils, bio-herbicides, and adhesives (Berti et al. 2016).

18.9 Future Challenges

The recent interest and research in camelina are impressive; however, further research is needed in several areas. Some of the major challenges with world-wide acceptability of camelina as food and biofuel feedstock are: poor yield, low price, lack of awareness in farmers, and presence of antinutritional factors. Availability and consistent supply of refined oil is a major hindrance for developing new products including food products. The oil fraction of camelina has been well characterized and its applications have been studied to a great extent. However, the potential of camelina meal has not been studied well. The meal may have potential as livestock feed, but more research is needed to determine optimum levels of inclusion. Since camelina meal has a significant amount of protein, it might be worth characterizing the protein and finding ways to include camelina meal in human food. Solutions for logistics and storage of seeds and oil need to be devised. The specifications of the oil for use in jet fuel and biodiesel, specific to the cracking process need to be developed.

Utilization of modern biotechnology tools to identify genes associated with yield, quality, and other agronomic traits may be considered. It will be beneficial to understand the mechanism of drought and freeze resistance, and camelina's ability to resist abiotic stress. The knowledge may be applied to other important oilseed crops. Additionally, it will be beneficial to study the impact of camelina on soil quality. Competition by weed is a major challenge in camelina production. The development of herbicides will be beneficial for the commercial production of camelina.

18.10 Conclusions

Camelina is a highly adaptable crop with relatively low inputs and multiple benefits. Camelina meal and oil have great potential to be used as feed and food. Camelina meal has a high content of protein and fiber. Therefore, it has a huge potential to be developed as a sustainable raw material for the production of biodegradable plastics, sheets, films, and fibers. Currently, the major focus has been utilizing camelina oil for biofuel. Camelina oil has potential in multiple health and food applications due to the unique fatty acid profile and antioxidant potential. Improvement in plant breeding will play a major role in providing novel and sustainable sources of camelina and open a new market for novel product development. Breeding programs to improve grain yield and oil content will help *Camelina sativa* to become a more competitive oilseed crop.

References

Abramovič H, Butinar B, Nikolič V (2007) Changes occurring in phenolic content, tocopherol composition and oxidative stability of Camelina sativa oil during storage. Food Chem 104:903–909

Bansal S, Durrett TP (2016) Camelina sativa: an ideal platform for the metabolic engineering and field production of industrial lipids. Biochimie 120:9–16

Barceló-Coblijn G, Murphy EJ (2009) Alpha-linolenic acid and its conversion to longer chain n-3 fatty acids: benefits for human health and a role in maintaining tissue n-3 fatty acid levels. Prog Lipid Res 48:355–374

Berhow MA, Vaughn SF, Moser BR et al (2014) Evaluating the phytochemical potential of Camelina: an emerging new crop of old world origin. In: Phytochemicals—biosynthesis, function and application, Recent advances in phytochemistry, vol 44. Springer, Cham, pp 129–148

Berti M, Gesch R, Eynck C et al (2016) Camelina uses, genetics, genomics, production, and management. Ind Crop Prod 94:690–710

Betancor MB, Li K, Sprague M, Bardal T, Sayanova O, Usher S, Han L, Måsøval K, Torrissen O, Napier JA, Tocher DR (2017) An oil containing EPA and DHA from transgenic Camelina sativa to replace marine fish oil in feeds for Atlantic salmon (Salmo salar L.): effects on intestinal transcriptome, histology, tissue fatty acid profiles and plasma biochemistry. PLoS One 12(4): e0175415

Budin JT, Breene WM, Putnam DH (1995) Some compositional properties of camelina (Camelina sativa L. Crantz) seeds and oils. J Am Oil Chem Soc 72:309–315

Burel C, Boujard T, Kaushik SJ et al (2001) Effects of rapeseed meal-Glucosinolates on thyroid metabolism and feed utilization in rainbow trout. Gen Comp Endocrinol 124:343–358

Cais-Sokolińska D, Pikul J, Wójtowski J et al (2015) Evaluation of quality of kefir from milk obtained from goats supplemented with a diet rich in bioactive compounds. J Sci Food Agric 95:1343–1349

Crabtree RJ, Prater JD, Mbolda P (1990) Long-term wheat, soybean, and grain sor- ghum double-cropping under rainfed conditions. Agron J 82:683–686

Crowley J, Frohlich A (1998) Factors affecting the composition and use of Camelina, Dublin, Teagasc

Dal Bello B, Torri L, Piochi M et al (2015) Healthy yogurt fortified with n-3 fatty acids from vegetable sources. J Dairy Sci 98:8375–8385

Denstadli V, Skrede A, Krogdahl Å et al (2006) Feed intake, growth, feed conversion, digestibility, enzyme activities and intestinal structure in Atlantic salmon (Salmo salar L.) fed graded levels of phytic acid. Aquaculture 256:365–376

Dubois V, Breton S, Linder M et al (2007) Fatty acid profiles of 80 vegetable oils with regard to their nutritional potential. Eur J Lipid Sci Technol 109:710–732

Eidhin DN, Burke J, Lynch B et al (2003a) Effects of dietary supplementation with Camelina oil on porcine blood lipids. J Food Sci 68:671–679

Eidhin DN, Burke J, O'Beirne D (2003b) Oxidative stability of ω3-rich Camelina oil and Camelina oil-based spread compared with plant and fish oils and sunflower spread. J Food Sci 68:345–353

Fan L, Eskin M (2013) Camelina oil: chemistry, properties and utilization. Rec Res Dev Lipids 9:125–137

Frame D, Palmer M, Peterson B (2007) Use of Camelina sativa in the diets of young turkeys. J Appl Poultry Res 16:381–386

Fröhlich A, Rice B (2005) Evaluation of Camelina sativa oil as a feedstock for biodiesel production. Ind Crop Prod 21:25–31

Gehringer A, Scoles GJ, Friedt W et al (2006) Genetic mapping of agronomic traits in false flax (Camelina sativa subsp. sativa). Genome 49:1555–1563

Gerber M (2012) Omega-3 fatty acids and cancers: a systematic update review of epidemiological studies. Br J Nutr 107:S228–S239

Goyal A, Sharma V, Upadhyay N, Gill S, Sihag M (2014) Flax and flaxseed oil: an ancient medicine & modern functional food. J Food Sci Technol 51(9):1633–1653

GRAS. (n.d.) GRAS Notice (GRN) No. 642. http://www.fda.gov/Food/IngredientsPackagingLabeling/GRAS/NoticeInventory/default.htm

Günç Ergönül P, Aksoylu ÖZ (2018) Identification of bioactive compounds and total phenol contents of cold pressed oils from safflower and camelina seeds. J Food Measure Characteriz 12:2313–2323

Hatt G. (1937) Landbrug i Danmarks Oldtid, Udvalget for folkeoplysnings fremme, Koebenhavn (in Danish).

Hixson SM, Parrish CC (2014) Substitution of fish oil with camelina oil and inclusion of camelina meal in diets fed to Atlantic cod (Gadus morhua) and their effects on growth, tissue lipid classes, and fatty acids1. J Anim Sci 92:1055–1067

Hurtaud C, Peyraud JL (2007) Effects of feeding Camelina (seeds or meal) on milk fatty acid composition and butter spreadability. J Dairy Sci 90:5134–5145

Ibrahim F, El Habbasha ES (2015) Chemical composition, medicinal impacts and cultivation of camelina (Camelina sativa): Review. Int J PharmTech Res 8:114–122

Karvonen HM, Aro A, Tapola NS et al (2002) Effect of [alpha]-linolenic acid[ndash]rich Camelina sativa oil on serum fatty acid composition and serum lipids in hypercholesterolemic subjects. Metabolism 51:1253–1260

Knorzer KH (1978) Evolution and spread of gold of pleasure (Camelina sativa S.L.). Bererichte der Deutschen Botanischen Gesellschaft 91:187–195

Kok W-M, Mainal A, Chuah C-H et al (2018) Content of erucic acid in edible oils and mustard by quantitative ^{13}C NMR. Eur J Lipid Sci Technol. https://doi.org/10.1002/ejlt.201700230

Kumar K, Gupta SM, Arya MC et al (2017) In vitro antimicrobial and antioxidant activity of Camelina seed extracts as potential source of bioactive compounds. Proc Natl Acad Sci India Sect B Biol Sci 87:521–526

Leonard EC (1998) Camelina oil: α-linolenic source. Inform 9:830–838

Mandal S, Ghosh K (2010) Inhibitory effect of Pistia tannin on digestive enzymes of Indian major carps: an in vitro study. Fish Physiol Biochem 36:1171–1180

Manninen S, Lankinen M, de Mello V et al (2019a) The effect of Camelina sativa oil and fish intakes on fatty acid compositions of blood lipid fractions. Nutr Metab Cardiovasc Dis 29:51–61

Manninen S, Lankinen M, Erkkila A et al (2019b) The effect of intakes of fish and Camelina sativa oil on atherogenic and anti-atherogenic functions of LDL and HDL particles: a randomized controlled trial. Atherosclerosis 281:56–61

Marra MC and Carlson GA. (1986) Double-cropping wheat and soybeans in the southeast: input use and patterns of adoption. Natural resource economics division, eco- nomic research service, U.S. Department of Agriculture, Washington, D.C., USA.

Matthaus B, Zubr J (2000) Variability of specific components in Camelina sativa oilseed cakes. Ind Crop Prod 12:9–18

Matthäus B (2002) Antioxidant activity of extracts obtained from residues of different oilseeds. J Agric Food Chem 50(12):3444–3452

de Mello VD, Dahlman I, Lankinen M et al (2019) The effect of different sources of fish and Camelina sativa oil on immune cell and adipose tissue mRNA expression in subjects with abnormal fasting glucose metabolism: a randomized controlled trial. Nutr Diabetes 9:1

Miles EA, Calder PC (2012) Influence of marine n-3 polyunsaturated fatty acids on immune function and a systematic review of their effects on clinical outcomes in rheumatoid arthritis. Br J Nutr 107:S171–S184

Mithen RF, Dekker M, Verkerk R et al (2000) The nutritional significance, biosynthesis and bioavailability of glucosinolates in human foods. J Sci Food Agric 80:967–984

Moser BR (2010) Camelina (Camelina sativa L.) oil as a biofuels feedstock: Golden opportunity or false hope? Lipid Technol 22:270–273

Moser BR, Vaughn SF (2010) Evaluation of alkyl esters from Camelina sativa oil as biodiesel and as blend components in ultra low-sulfur diesel fuel☆. Bioresour Technol 101:646–653

Moslavac T, Jokić S, Šubarić D et al (2014) Pressing and supercritical CO_2 extraction of Camelina sativa oil. Ind Crop Prod 54:122–129
Murphy EJ (2016) Chapter 8—Camelina (Camelina sativa). In: TA MK, Hayes DG, Hildebrand DF et al (eds) Industrial oil crops. AOCS, Champaign, pp 207–230
Pietta PG (2000) Flavonoids as antioxidants. J Nat Prod 63:1035–1042
Pikul J, Wójtowski J, Danków R, Teichert J, Czyżak-Runowska G, Cais-Sokolińska D, Cieślak A, Szumacher-Strabel M, Bagnicka E (2014) The effect of false flax (Camelina sativa) cake dietary supplementation in dairy goats on fatty acid profile of kefir. Small Rumin Res 122(1–3):44–49
Pilgeram AL, Sands DC, Boss D et al (2007) Camelina sativa, a Montana omega-3 and fuel crop. In: Janick J, Whipkey A (eds) New Crops and New Uses. ASHS, Alexandria, pp 129–131
Popa AL, Jurcoane S, Dumitriu B (2017) Camelina sativa oil—a review. Sci Bull Ser F Biotechnol 21:233–238
Rode J (2002) Study of autochthon Camelina sativa (L.) Crantz in Slovenia. Int J Geogr Inf Syst 9:313–318
Rokka T, Alén K, Valaja J et al (2002) The effect of a Camelina sativa enriched diet on the composition and sensory quality of hen eggs. Food Res Int 35:253–256
Salo P, Kuusisto P (2016) Cholesterol lowering effect of plant stanol ester yoghurt drinks with added Camelina oil. Cholesterol 2016:1–12
Schuster A, Friedt W (1998) Glucosinolate content and composition as parameters of quality of Camelina seed. Ind Crop Prod 7:297–302
Schwab US, Lankinen MA, de Mello VD et al (2018) Camelina sativa oil, but not fatty fish or lean fish, improves serum lipid profile in subjects with impaired glucose metabolism-a randomized controlled trial. Mol Nutr Food Res 62(4). https://doi.org/10.1002/mnfr.201700503
Shu C-C, Hsu C-L, Lee C-Y et al (2016) Inflammatory markers and clinical characteristics for predicting persistent positivity of interferon gamma release assay in dialysis population. Sci Rep 6:34577
Singh SK, Muthukumarappan K (2016) Effect of feed moisture, extrusion temperature and screw speed on properties of soy white flakes based aquafeed: a response surface analysis. J Sci Food Agric 96:2220–2229
Tanwar B, Modgil R, Goyal A (2018) Antinutritional factors and hypocholesterolemic effect of wild apricot kernel (Prunus armeniaca L.) as affected by detoxification. Food Funct 9(4):2121–2135
Tanwar B, Modgil R, Goyal A (2019) Effect of detoxification on biological quality of wild apricot (Prunus armeniaca L.) kernel. J Sci Food Agric 99(2):517–528
Terpinc P, Polak T, Makuc D et al (2012) The occurrence and characterisation of phenolic compounds in Camelina sativa seed, cake and oil. Food Chem 131:580–589
Tripathi MK, Mishra AS (2007) Glucosinolates in animal nutrition: a review. Anim Feed Sci Technol 132:1–27
Vollman J, Moritz T, Kargl C et al (2007) Agronomic evaluation of camelina genotypes selected for seed quality characteristics. Ind Crop Prod 26:270–277
Waraich E, Ahmed Z, Ahmad R et al (2013) Camelina sativa, a climate proof crop, has high nutrtive value and multiple-uses: a review. Aust J Crop Sci 7:1551–1559
Woods VB, Fearon AM (2009) Dietary sources of unsaturated fatty acids for animals and their transfer into meat, milk and eggs: a review. Livest Sci 126:1–20
Zanetti F, Vamerali T, Mosca G (2009) Yield and oil variability in modern varieties of high-erucic winter oilseed rape (Brassica napus L. var. oleifera) and Ethiopian mustard (Brassica carinata a. Braun) under reduced agricultural inputs. Ind Crop Prod 30:265–270
Zubr J (1997) Oil-seed crop: Camelina sativa. Ind Crop Prod 6:113–119
Zubr J, Matthaus B (2002) Effects of growth conditions on fatty acids and tocopherols in Camelina sativa oil. Ind Crop Prod 1:155–162

Chapter 19
Pumpkin (*Cucurbita pepo*) Seed

Shraddha Adsul and Vaishali Madkaikar

Abstract A pumpkin seed, also known as a "*pepita*", means "little seed of squash". *Cucurbita pepo* is the pumpkin species of the genus: Cucurbita, family: Cucurbits, sub-family: Cucurbitaceae, genera: Cucurbita L. *Cucurbita pepo* L. is the species among *Cucurbitaceae* family having the greatest monetary value of the genus. The seeds are typically rather flat, asymmetrically oval, light green in color, and may have a white outer hull. The *Cucurbita pepo* seeds have potential application and can be used as an alternative oil and protein source in novel food formulations such as cooking oils, as an ingredient in margarine blends, flours for instant soups, cookies, etc.

Keywords Pumpkin seed · *Cucurbita pepo* · Food applications · Cardiovascular diseases · Phytonutrients · Pharmacological applications · Health benefits

19.1 Origin and History

The word Pumpkin is derived from a Greek word "Pepon" meaning large melon. French changed the word "Pepon" to "Pompon" and English modified the word to "Pompion". Later, the American migrants replaced the "ion" with "kin" giving rise to "Pumpkin" (Anon 2007). Pumpkin is found to be an inhabitant to steppe and temperate regions of North America. Pumpkin (*Cucurbita pepo*) was domesticated at least two times, initially in Mexico more than 10,000 years ago and later in the United States more than 4000 years ago. The oldest evidence for the existence of Cucurbita in the World is in 1492 (Paris 2008, 2015). Remains of stems, seeds, and fruits of *Cucurbita pepo* and *Cucurbita moschata* have been identified and found in

S. Adsul (✉)
Department of Food Science and Nutrition, SNDT College of Home Science, Pune, Maharashtra, India

V. Madkaikar
Deenanath Mangeshkar Hospital and Research Center, Pune, Maharashtra, India

B. Tanwar, A. Goyal (eds.), *Oilseeds: Health Attributes and Food Applications*,
https://doi.org/10.1007/978-981-15-4194-0_19

the cliff of the rocks local to the southwestern United States confirming its origin in America (Orzolek et al. 2017).

Columbus noted them on his first expedition, while Jacques Cartier accounted that the crop was grown in Canada in the 1530s. Cabeza de Vaca and Hernando de Soto mentioned about the cultivation of the crop in Florida during 1540s–1550s. Also, numerous historical evidences emphasize cucurbit production for fruits and seeds in 1697 in Styria though it may have an earlier origin. However, the production of seed for oil extraction became popular during the 1700s and 1800s. Comprehensive statistical data pertaining to harvest has been available from the year 1874 onwards. In the late 1800s hot water was used to soak the pumpkin seeds before peeling and pressing whereas, in the North-east area of Styria and some nearby areas seeds were pounded with their seed coat. The earliest appearance of thin-coated seed was evidenced during 1870–1880. Naked or hull-less cultivar of pumpkin suitable for oil production was grown locally in Styria since 1915 (Teppner 2004) and natural chance mutations which produced hull-less pumpkin seeds were apparent during the 1880s. But scientific documentation of the same happened around 1925. These mutations have not affected the quality of oil in the seeds and the growth of vining form. Systematic breeding activities and genetic studies were initiated in the 1930s based on local plant material. After this, pumpkins are cultivated extensively for making seed oil in Europe. While breeding to enhance the oil concentration of naked seeds of the Styrian vining was reported to be initiated in 1940 at the breeding station located at the Lamberg in east Styria. Latest breeding programs focus on disease resistance and enhancement of seed yield, decrease in the length of vines, increase of the seed oil content, and more recently to increase its tocopherols content (Winkler 2000).

19.2 Production

As there is limited data on exclusive pumpkin plantation for seed production it is difficult to state the exact worldwide pumpkin seed production. Seeds are sold in the market as a by-product of pumpkin pulp production and hence, data gets lost in the major vegetable category. The pumpkin seeds are generally categorized as "seed general" or "oilseed" in import/export data which gets lost further in trade. Moreover, when broader statistics are applied this data becomes insignificant (Baxter et al. 2012).

China is the leading country in the production of pumpkin with 7,241,409 tons (2014), accounting for 40% of total world production followed by India (4,987,123 tons, 27%, Fig. 19.1). The statistical database was endorsed by the Food and Agriculture Organization Corporate (FAOSTAT 2017).

With sufficient crop and pollinator management procedures 3–4 quintal seed production can be achieved from 1-ha area (Choudhary 2017). In Central Europe area where "oil pumpkin" is grown, the estimated production is around 500–800 kg dry seeds/ha area which may increase up to a maximum of 1200 kg/ha area (Bavec

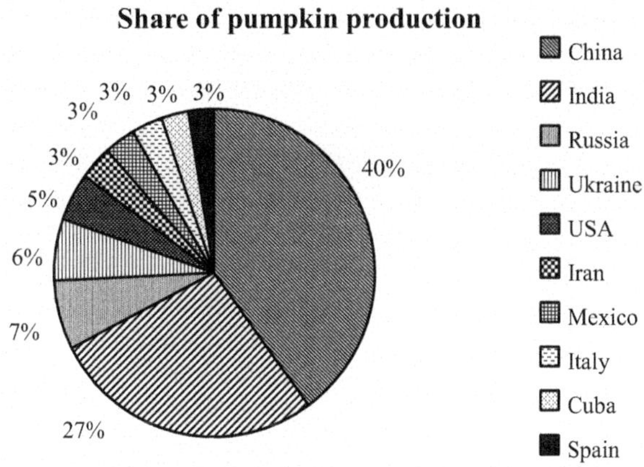

Fig. 19.1 Share of production of pumpkin of different countries in 2014 (FAOSTAT 2017)

et al. 2007). Around 1 L of oil can be recovered from 2.5 kg of dried seeds (GAIN 2004).

19.3 Processing of Pumpkin Seeds

19.3.1 Seed

The types of seeds desired from pumpkins can be broken down into two types: confectionery seeds for snacks application and oilseeds for oil extraction. Seeds are sorted based on size and color, it can further be roasted, salted, sugar coated, or can be used as spices or flavorings to enhance the acceptability of the supplemented food products in the market (Baxter et al. 2012).

19.3.2 Oil Pressing

In the traditional method of hot press process for extracting oil, pumpkin seeds are milled through a stone mill, with the addition of water and salt, followed by roasting in pans at a temperature of 100–130 °C. The hot seed mixture is then passed through the press to extract oil. The amount of salt and water are added with particular time-temperature combinations in order to extract pumpkin seed oil. In Central Europe, pumpkin seeds oil is extracted using a hot pressing method. However, in cold pressing, dry seeds are mechanically pressed without thermal processing (Bavec et al. 2007).

Oil yield and quality are vital to determine the viability of commercial production. Traditionally solvent extraction method is used with n-hexane as the most widely used solvent owing to its efficacy and availability (Prámparo et al. 2003).

Application of heat before or during extraction results in the rupturing of cell emulsion and decreases oil viscosity, which facilitates oil fluidity, movement, and lowers oil surface tension. At the same time, it can have a negative effect on chemical composition and susceptibility to oxidation (Rodríguez-Miranda et al. 2013). A special type of pumpkin (*Cucurbita pepo* var.) seed oil is manufactured in southeastern Austria and adjacent regions using press method after roasting the hulled seeds having an oil concentration of around 38%. The characteristic nutty and roasted aroma results from heat application before pressing where the temperature goes as high as 130 °C. In general, roasting of seed meal at the temperature of 110 °C for 50–60 min imparts dark green color to the oil (Teppner 2004).

The naked seed oil has dark color with green to red tinge due to the presence of protochlorophyll which is present in the innermost layer of seed-chlorenchyma. Clear or straw color oil is extracted from pumpkin seeds having seed coat (Lelley et al. 2009). Protochlorophyll is light sensitive and initiates lipid oxidation; hence this oil should be stored in dark. Dark green color and the foam formation in the oil limit its use in cooking (Murkovic 2009).

19.3.3 Seed Meal

A valuable byproduct of pumpkin seed oil production is seed meal (Makni et al. 2011), also referred to as defatted pumpkin flour (Lazos 1992). Due to its high protein content, it can be used for substituting breadcrumbs or flour in the bakery industry. In Central Europe, it is mainly used as a high protein stock feed (Baxter et al. 2012).

19.4 Chemical Composition

Proximate composition of seeds varies with characteristics of seeds, genus, species, conditions of harvest, fruit maturity level, and cultivation zone (Rodríguez-Miranda et al. 2013).

19.4.1 Carbohydrates

Pumpkin seeds contain approximately 25.19–28.03% carbohydrates (Elinge et al. 2012; Gohari Ardabili et al. 2011). However, Kim et al. (2012) observed 12.22 ± 7.47 g/100 g carbohydrates (on dry weight basis) in pumpkin seeds. Total carbohydrates in defatted pumpkin seed meal extracted with hexane were found to be 9.13 g/100 g, and the amount of sucrose, raffinose, and stachyose was 1.79 g/100 g, 0.41 g/100 g, and 0.81 g/100 g, respectively (Mansour et al. 1993). The crude fiber content of the seeds varies greatly and has been reported to be 3.80% (Sharma and Lakhawat 2017), 1.0% (Elinge et al. 2012), 5.34% (Gohari Ardabili et al. 2011), and 14.82 ± 0.55 g/100 g (Kim et al. 2012)

19.4.2 Proteins

Protein content in pumpkin seeds ranged from 24.50 to 32.00 g/100 g (Kim et al. 2012; Elinge et al. 2012; Aliu et al. 2016; Nyam et al. 2009) and approximately 55.40 g/100 g in defatted seed flour (Lazos 1986). Sharma and Lakhawat (2017) have reported 29.65% protein content in roasted pumpkin seed flour. Pumpkin seed's protein contains all 9 essential amino acids (Table 19.1) (El-Adawy and Taha 2001; Patel 2013). *Cucurbita pepo* seeds have the highest content of arginine (63.99 ± 0.88 mg/kg) as compared to other species such as *Cucurbita maxima* and *Cucurbita moschata* (Kim et al. 2012).

Higher concentrations of essential amino acids were reported in protein isolates (40.92–42.59 g per 16 g N) than concentrate (36.13 g per 16 g N) and meal (35.03 g per 16 g N). During isolate preparation, a few protein fractions are lost which may result in loss of non-essential amino acids and increase levels of essential amino acids (Mansour et al. 1993). Table 19.2 provides a summary of the estimated biological quality of pumpkin (*Cucurbita pepo*) seed flours/protein by different researchers.

After oil extraction, the protein content of oil cake can increase up to 65%. Various fractions of this protein are cucurbitin and albumin. The major fraction is represented by cucurbitins, 12S globulins composed of six similar subunits (hexamer) with molecular weights of 54 kDa each and are the major storage proteins in pumpkin seeds (Bučko et al. 2016). Zdunczyk et al. (1999) conducted a comparative study on the chemical composition as well as the nutritive value of pumpkin seed cake (PSC), soybean meal, and casein. The authors reported higher content of protein (59.8 g/100 g) in PSC. The amino acid profile of PSC showed substantial amounts of tryptophan (1.54 g/16 g N) and low levels of lysine (3.21) and isoleucine (3.83 g/16 g N). True digestibility coefficient (TD) of PSC protein was found to be 83.1% which was not significantly different than soybean meal. Though protein efficiency ratio (PER) of pumpkin seed was lower (1.01) than soybean meal, it can be at par after supplementing with lysine. PER of the diet containing PSC and soybean meal in equal proportion of protein demonstrated the higher value of PER (1.98) in comparison with only soybean meal (PER 1.50).

19.4.3 Lipids

Stevenson et al. (2007) studied the oil content of 12 different cultivars of pumpkin seeds, which was found to be ranging from 22 to 50% depending on variety and

Table 19.1 Amino acid composition of pumpkin seeds, seed flour, and seed meal

Amino acid	Raw seeds (Kim et al. 2012) mg/kg*	Pumpkin seed flour (El-Adawy and Taha 2001) g/16 g of *N*	Seed meal (Mansour et al. 1993) g/16 g of *N*	Defatted seed flour (Atuonwu and Akobundu 2010) g/100 g protein
Isoleucine	13.96 ± 0.74	3.21	2.66	2.98
Leucine	24.14 ± 0.96	6.49	6.13	5.71
Lysine	13.14 ± 0.48	4.17	5.20	4.30
Cysteine	NR	1.17	1.52	0.79
Methionine	4.20 ± 0.37	1.88	1.25	1.43
Total sulfur amino acid	NR	3.05	2.77	2.22
Tyrosine	8.18 ± 0.01	3.17	2.94	3.06
Phenylalanine	15.52 ± 0.53	4.47	4.00	3.47
Total aromatic amino acids	NR	7.64	6.94	6.53
Threonine	7.56 ± 0.07	3.30	2.75	2.11
Tryptophan	NR	0.86	1.56	0.99
Valine	17.43 ± 0.69	4.71	3.40	5.40
Essential to total amino acid (%)	NR	33.43	35.03	43.27
Histidine	18.37 ± 0.08	3.26	3.62	2.33
Arginine	63.99 ± 0.88	19.03	16.70	4.85
Aspartic acid	29.95 ± 0.25	9.61	10.19	8.9
Glutamic acid	60.26 ± 0.04	17.33	18.13	9.50
Serine	14.99 ± 0.21	5.41	5.46	2.19
Proline	11.98 ± 0.37	3.37	4.34	2.12
Glycine	18.70 ± 0.36	4.32	5.86	3.80
Alanine	17.76 ± 0.03	4.24	4.29	4.76

*Values are mean ± SD
NR not reported

species of pumpkin seed. The gas chromatography (GC) and gas chromatography/mass spectrophotometry (GC/MS) analysis of oil from six varieties of pumpkin seeds (*Cucurbita pepo* L.) grown in Serbia and extracted using cold press method indicated higher amount of monounsaturated fatty acids (37.1 ± 0.70 to 43.6 ± 0.69 g/100 g) in oil (Rabrenović et al. 2014). A variety of *Cucurbita pepo* that do not fully develop seed coats, found in the southeastern region of Austria and known as Styrin pumpkin, was showed to have approximately 26.6 and 54.2% oleic acid and linoleic acid, respectively. The concentration of triacylglycerol components namely dioleopalmitin, dipalmitolinolein, triolein, palmitoleolinolein, dioleolinolein, dilinoleopalmitin, and dilinoleolein was reported to be ranging

Table 19.2 Protein quality of pumpkin (*Cucurbita pepo*) seed flour documented by various researchers

Method of evaluation	Pumpkin seed flour (El-Adawy and Taha 2001)	Seed meal (Mansour et al. 1993)	Defatted seed flour (Atuonwu and Akobundu 2010)
In vitro protein digestibility	90.01%	87.78%	77.91%
Chemical score	75.85%	77%	62.0%
1st limiting AA	Lysine	Isoleucine, Valine	Threonine
2nd limiting AA	Isoleucine	Threonine	Lysine
Essential AA index (EAAI)	85.41	NR	57.31
Protein efficiency ratio (PER)	2.15	NR	1.80

NR not reported

from 5.8 to 18.8, 8.1–8.8, 6.3–20.5, 15.0–16.1, 16.7 to 23.0, 4.6–15.4, and 6.7–19.4%, respectively (Yoshida et al. 2004) (Table 19.3).

19.4.4 Vitamins

A very little information is available on the vitamin content of pumpkin seeds. Among B-complex vitamins, the concentration of niacin was found to be highest (61.43 mg/kg) and least for riboflavin (2.47 mg/kg) in pumpkin seed meal (Mansour et al. 1993). Total tocopherol in pumpkin seed oil was found to be ranging from 38.03 ± 0.25 to 64.11 ± 0.07 mg/100 g (Rabrenović et al. 2014). A great variation in tocopherol content was observed, for example, γ-tocopherol has been reported predominant (41–620 mg/kg) with 5–10 times more than α-tocopherol (140 mg/kg) dry material. The concentrations of β-tocopherol and δ-tocopherol were comparatively low (Murkovic et al. 1996). Gohari Ardabili et al. (2011) reported total tocopherols 882.65 ± 18.32 mg/kg in styrin pumpkin seed oil. Carotenoid fraction concentrations in pumpkin seed have been reported to be 17.46 ± 18.29 mg of β-carotene and 0.16 ± 0.16 mg of β-cryptoxanthin per kg raw weight of seeds (Kim et al. 2012). A few minor pigment fractions also have been isolated such as luteoxanthin, chrysanthemaxanthin, auroxanthin epimers, flavoxanthin, α-cryptoxanthin, β-cryptoxanthin, violaxanthin, lutein epoxide, and α-carotene (Matus et al. 1993).

Table 19.3 Fatty acid compositions of pumpkin (*C. pepo*) seeds and seed oil by various authors

Fatty acid	Raw seeds (Kim et al. 2012)	Seed oil (Bardaa et al. 2016)	Styrin pumpkin oil (Gohari Ardabili et al. 2011)
Myristic acid (14:00)	0.23 ± 0.06	0.233 ± 0.023	NR
Palmitic acid (16:00)	12.97 ± 0.72	14.828 ± 0.145	10.68 ± 0.42
Heptadecanoic acid (17:00)	ND	0.084 ± 0.004	NR
Stearic acid (18:00)	4.67 ± 0.15	6.676 ± 0.024	8.67 ± 0.27
Oleic acid (18:1)	32.40 ± 0.56	25.87 ± 0.27	38.42 ± 0.37
Linoleic acid (18:2)	36.40 ± 0.82	50.88 ± 0.106	39.84 ± 0.08
Arachidic acid (20:00)	0.39 ± 0.06	0.433 ± 0.053	NR
Eicosenoic acid (20:1n-9)	ND	0.0858 ± 0.017	NR
α-Linolenic acid (18:3n-3)	ND	0.183 ± 0.004	NR
Behenic acid (22:00)	0.37 ± 0.06	0.058 ± 0.057	NR
Palmitoleic acid (16:7)	NR	0.151 ± 0.003	0.58 ± 0.14
Vaccenic acid (18:7)11-Octadecenoic acid	NR	0.501 ± 0.11	NR
Erucic acid (22:9)	NR	0.055 ± 0.09	NR
Gadoleic acid (20:1)	NR	NR	1.14 ± 0.00
SFA	18.62 ± 0.64	NR	19.35 ± 0.16
MUFA	32.40 ± 1.66	NR	80.65 ± 0.16
PUFA	36.40 ± 0.82	NR	–

NR not reported, *ND* not detected

19.4.5 Minerals

According to Aliu et al. (2016), potassium (K) is found to be the most abundant element in pumpkin seed. In radiochemical analysis of thin husked pumpkin (*Cucurbita pepo* L.) seed oil and oil cake from Slovenia showed the selenium content of 0.023–0.037 mg/kg in seeds and 0.034–0.047 mg/kg in oil cake. The iodine content was reported as 0.005–0.013 mg/kg, 0.002–0.003, and 0.007–0.032 mg/kg in seeds, oil, and oil cake, respectively (Kreft et al. 2002). Pumpkin seeds contain good amounts of zinc. However, the presence of phytate in the seed may form complex with zinc in small intestine reducing zinc bioavailability (Ovca et al. 2011). The mineral content of pumpkin seeds as reported by various authors has been summarized in Table 19.4.

Table 19.4 Mineral content of pumpkin seeds (*Cucurbita pepo*)

Mineral	mg/100 g (Elinge et al. 2012)	mg/100 g (Aliu et al. 2016)
Potassium	273.00	4646.05
Sodium	170.00	NR
Calcium	9.78	106.74
Magnesium	67.41	NR
Phosphorus	47.68	NR
Iron	3.75	5.39
Cobalt	2.17	ND
Manganese	0.06	NR
Zinc	14.14	7.91

ND not detected, *NR* not reported

Table 19.5 Antinutritional factors in *Cucurbita pepo* seeds (g/100 g of dry matter)

Antinutritional factors	Seed flour (El-Adawy and Taha 2001)	Defatted seed meal (Atuonwu and Akobundu 2010)
Raffinose	0.29 ± 0.13	0.80
Stachyose	0.52 ± 0.19	3.00
Verbascose	0.23 ± 0.10	NR
Phytate	2.37 ± 0.29	0.44
Tannins	0.17 ± 0.09	0.69
Trypsin inhibitor (TIU /g of protein)	1.39 ± 0.19	2.07
Saponin	NR	0.56
Hydrogen cyanide (mg/100 g)	NR	4.08

NR not reported

19.5 Antinutritional Factors

The antinutritional factors (ANFs) may be defined as "metabolic compounds" synthesized in plants to protect against insects. The different mechanisms are involved in the generation of these secondary metabolites viz. inactivation of some nutrients, metabolic utilization of feed, or diminution of the digestive process, which can result in undesirable physiological effects (Emire et al. 2013; Khokhar and Owusu Apenten 2003). ANFs have been reported to show negative effects such as reduced digestion of nutrients, decreased bioavailability of minerals, and inhibition of the enzymatic activity of several digestive enzymes (Tanwar et al. 2018, 2019a). However, recent numerous studies have supported the beneficial effects of ANFs such as antimicrobial, anticancer, antioxidant activities, if consumed at lower concentrations, and thus, are termed as phytonutrients (Tanwar et al. 2019b). Also, the processing treatments like soaking, fermentation, blanching, roasting, steaming, and extrusion may affect the level of antinutrients in these foods (Khokhar and Owusu Apenten 2003; Sihag et al. 2015). The concentration of several antinutritional factors in pumpkin seed flour is shown in Table 19.5.

Elinge et al. (2012) reported the presence of oxalates, phytate, hydrocyanic acid, and nitrate in pumpkin (*Cucurbita pepo* seeds). The values stated for these ANFs were oxalates: 0.023 ± 0.02 mg, phytate: 35.06 ± 1.10 mg, hydrocyanic acid: 0.22 ± 0.02 mg, and nitrate: 2.27 ± 0.03 mg per 100 g of sample. The concentration of these ANFs was below the safe level for human consumption.

19.6 Phytonutrients/Phenolics

Phytonutrients are the bioactive components with specific biological actions that help to maintain human health. The common examples of phytonutrients are phytosterols, polyphenols, glucosinolates, phytoestrogens, flavonoids, carotenoids, and terpenoids. Several pharmacological studies have revealed their antioxidant, antimicrobial, anticancer, antiaging, antidiabetic, hypolipidemic, hypotensive, and anti-inflammatory effects (Gupta and Prakash 2014). Mandl et al. (1999) isolated Δ^{7}-sterols as major sterols component from pumpkin seed oil which comprised of $\Delta^{7,22.25}$-stigmasatrienol (326 mg/kg), $\Delta^{7,22}$-stigmastadien3ß-ol (spinasterol) (300 mg/kg), $\Delta^{7},^{25}$-stigmastadienol plus Δ^{7}-stigmastenol (310 mg/kg), Δ^{7}-avenasterol (164 mg/kg), and β-sitosterol (58 mg/kg). Generally, β-sitosterol is the major phytosterol contributing 75.7–87.3% followed by campesterol (22.60 ± 3.90 mg/100 g) and stigmasterol (18.02 ± 1.18 mg/100 g). Ergosterol and squalene content in pumpkin seed oil were observed to be 22.40 ± 1.46 and 590.69 ± 26.58 mg/100 g, respectively (Nyam et al. 2009). Rabrenović et al. (2014) reported the squalene concentration in pumpkin seed oil ranging from 583.2 ± 23.6 to 747 ± 16 mg/100 g. The phenolic content of pumpkin seed oil is shown in Table 19.6.

Total phenolic and phytosterol content in five different varieties of pumpkin seeds was found to be 0.96–2.52 and 2.02–2.65 mg/g, respectively. The squalene concentration was reported as 0.92 to 1.29 mg/g, whereas α-tocopherol was ranging from 83.09 to 98.57 μg/g. The results of the study also indicated the presence of a significant amount of γ-amino butyric acid (GABA) which was noted as 3.71–15.53 mg/100 g (Qi et al. 2012). Phytosterol composition of pumpkin seeds is provided in Table 19.7.

Table 19.6 Phenolic content in pumpkin seed (Nyam et al. 2009)

Phenolic acids	Amount (mg/100 g)
Gallic acid	0.26 ± 0.02
Protocatechuic acid	0.08 ± 0.02
p-hydroxybenzoic acid	0.20 ± 0.02
Vanillic acid	0.60 ± 0.04
Caffeic acid	0.41 ± 0.01
Syringic acid	Trace
p-coumaric acid	0.17 ± 0.03
Ferulic acid	0.15 ± 0.01

Table 19.7 Phytosterol composition of pumpkin seeds (mg/100 g)

Product	References	
	Bardaa et al. (2016)	Phillips et al. (2005)
β-Sitosterol	44.405 ± 0.148	13.1
Campesterol	2.8 ± 0.014	3.4
Stigma-sterol	2.92 ± 0.028	ND
Δ^5-avena-sterol	1.85 ± 0.014	3.5
Sitostanol	7.75 ± 0.085	3.5
Campestanol	0.195 ± 0.007	0.8
Other sterols	NR	241.0
Total phytosterols	NR	265 (252–278)
Cholesterol	0.15 ± 0.014	NR
24-methylene-cholesterol	0.265 ± 0.007	NR
Δ^7-Campesterol	1.275 ± 0.021	NR
Δ^{5-23}-Stigmastadienol	0.045 ± 0.007	NR
Clerosterol	2.475 ± 0.035	NR
Δ^{5-24}-Stigmastadienol	17.325 ± 0.063	NR
Δ^7- Stigmastenol	7.75 ± 0.085	NR
Δ^7- Avenasterol	14.51 ± 0.035	NR
Total sterols (ppm)	2086.5 ± 19.092	NR
β-Sitosterol (ppm)	959 ± 21.213	NR

ND not detected, NR not reported

The squalene content of pumpkin seed oil is affected by the method of extraction and was found to be high in oils extracted from husk seeds than in de-hulled/naked seeds (Nakić et al. 2006). Nawirska-Olszańska et al. (2013) reported that pumpkin seeds contain cardiac glycosides, terpenoids, and resins. Pumpkin seeds also contain glycosides viz. cucurbitacin L 2-*O*-β-D glucopyranoside and cucurbitacin K 2-*O*-β-D-glucopyranoside (Wang et al. 2007). Besides these, 16-hydroxy 22,23,24,25,26,27-hexanorcucurbit-5-en-11, 20-dione3-*O*-α-L-rhamnopyranosyl-(1 → 2)-β-D-glucopyranoside, and 2,16-dihydroxy22,23,24,25,26,27 hexanorcucurbit-5-en-11,20-dione 2-*O*-β-D-glucopyranoside have also been reported by Rahman (2013). Lariciresinol and secoisolariciresinol are the types of lignin which have been extracted from pumpkin seeds using GC-MS technique (Sicilia et al. 2003).

19.7 Effect of Processing on Characteristics and Chemical Composition

The pumpkin seed oil extracted after roasting was analyzed using high-performance liquid chromatography (HPLC) with fluorescence and GC-MS technique to identify polycyclic aromatic hydrocarbons (PAHs) and volatile compounds, respectively. Roasting at higher temperatures (150 °C) showed the presence of light PAHs precisely phenanthrene; and volatile compounds composition was also affected, whereas at low temperatures aldehydes and alcohols were observed. Pyrazines formed during roasting are thought to be responsible for the aroma which differs from fresh to roasted depending on the degree of roasting (Potočnik and Košir 2016). Procida et al. (2012) conducted a study to analyze the fatty acid, carotenoid namely zeaxanthine, lutein, α- and γ-tocopherol concentrations in commercially available pumpkin seed oil samples and to determine the effect of roasting prior to oil pressing on volatile fractions. The study showed that the aroma of the commercial oils was in proportion to the thermal treatment. Also, the volatile compounds formed during chemical reactions from lipid peroxidation, Strecker degradation and Maillard reactions and their formation are affected by temperature used for roasting. Roasting at higher temperatures resulted in the strong aroma. On the other hand, lower temperatures enhanced the therapeutic quality (α- and γ-tocopherol and carotenoids profile) of oils. Siegmund and Murkovic (2004) noted that in order to derive typical nutty roasted aroma of pumpkin seed oil, prior roasting at a temperature of at least 90 °C is essential. For commercial production of oil, seeds are roasted up to 130 °C. Murkovic et al. (2004) demonstrated an increase in total sterol content in pumpkin seed after roasting from 1710 to 1930 mg/kg. Van Hoed et al. (2017) suggested that roasting of seeds prior to oil extraction not only enhanced the oil appearance and flavor but also resulted in higher tocopherol and phenolic compounds without significant formation of polycyclic aromatic hydrocarbons (PAH).

The germination of pumpkin seeds resulted in elevated moisture content, crude ether extract, crude protein, and total ash contents by 0.05, 0.37, 0.78, and 0.18 folds, respectively. It also resulted in increased crude fiber content, mineral content specifically iron, zinc, and manganese. Among vitamins, germination resulted in a marked increase in β-carotene and riboflavin (vitamin B_2) contents (Moustafa 2013). The effects of variation in particle size, meal to solvent ratio, and contact time on oil extraction method using n-hexane were studied using response surface methodology. The oil yield negatively correlated with particle size. The optimum yield of 422 g/kg was obtained at 0 h contact time, with particle size of 0.59 mm and pumpkin seed meal to solvent ratio of 1:20 (w/v). A short heat treatment at 60 °C for 10 min reduced oil extraction time and saved energy as well as operation cost without affecting oil quality. Fatty acid composition of extracted pumpkin seed oil was found to be 755 g/kg of unsaturated fatty acids with 431 g/kg of linoleic acid and 324 g/kg of oleic acid. With its high linoleic-oleic acids oil, pumpkin seed oil can be a good alternative to other edible oils like corn, sesame, sunflower, soybean, or cotton (Rodríguez-Miranda et al. 2013).

In a study conducted by Vujasinovic et al. (2012) on chemical composition and oxidative stability of the extracted oil, de-hulled pumpkin seeds post roasting with variation in time (30 and 60 min) and temperatures (90–130 °C) before oil extraction resulted in enhanced concentrations of phospholipids (from 0.005 to 0.463%), total phenolic compounds (from 4.63 to 19.60 mg/kg), and total tocopherols (from 265.79 to 350.98 mg/kg) in oil. Although the presence of these functional compounds improved the oxidative stability of the oil, it lowered the quality of oil owing to elevated levels of primary and secondary oxidation products exemplified by high Totox values.

19.8 Health Attributes

Pumpkin seed oil has been used in many countries namely China, India, Brazil, Mexico, America, Argentina, and Yugoslavia as a part of traditional medicine therapy. The seeds are widely used in the treatment of prostate gland and urinary bladder disorders. The health benefits of pumpkin seeds are mainly attributed to its nutrient and phytochemical composition. Studies have indicated higher concentrations of proteins, polyunsaturated fatty acids, carotenoids, tocopherol, minerals, and bioactive compounds like phytosterols, antioxidative, phenolic compounds, lignans, and triterpenes in pumpkin seed. The extracts of pumpkin have also been found to be effective against tumor, bacteria, mutations and are reported to have hypocholestremic, hypolipidemic, and antioxidant properties (Caili et al. 2006).

19.8.1 In Hyperglycemia

Hyperglycemia is the condition of high blood glucose (blood sugar). High blood sugar happens when the body has too little insulin or when the body cannot use insulin properly. Bharti et al. (2013) demonstrated the antidiabetic effect of *Cucurbita pepo* extracts on Wistar rats with loss of β-cell sensitivity and impaired glucose response. *Cucurbita pepo* seed extract (CPSE) when administered for 3 weeks at a dose of 5 g CPSE/kg body weight caused marked decrease in blood glucose levels accompanied by a reduction in hyperinsulinemia and improved sensitivity to insulin, which was assessed using Homeostasis Model Assessment (HOMA) index for insulin resistance (IR). Administration of CPSE resulted in significant increase in glucagon like polypeptide-1(GLP-1) concentration in the cecum of diabetic rats, enlargement of cecal tissue, notable increase in islets number, and improvements in the insulitis grade. Though there was fasting hyperglycemia, CPSE treatment showed a significant increase in fasting plasma insulin levels which could be due to increase in insulin-positive cell mass. The researchers concluded that hypoglycemic effect of pumpkin seed extract could be attributed to the presence of tocopherol, which is also responsible for reducing oxidative markers and improvements in cecal and pancreatic characteristics.

The supplementation of flax and pumpkin seed mixture exhibited improvement in glycemic control and decrease in elevated plasma liver enzyme levels (aspartate transaminase and alanin transaminase) in alloxan-induced diabetic Wistar rats (Makni et al. 2011). In a similar setup, another animal study showed a marked decrease in plasma and kidney malonaldialdehyde levels (MDA). The observation of histological sections of the kidney showed glomerular hypertrophy and tubular dilatation to a lesser extent in comparison with the diabetic group without supplementation (Makni et al. 2010).

Li et al. (2001) showed oil extracted from ungerminated pumpkin seeds possessed blood glucose-lowering activity and also resulted in improvements in blood glucose tolerance. The proteins from germinated pumpkin seed resulted in increased serum insulin and better blood glucose. However, the ungerminated seed protein did not exhibit hypoglycemic activity.

19.8.2 In Cardiovascular Diseases

Cardiovascular diseases (CVDs) are a group of disorders of the heart and blood vessels, which include mainly the narrowing of the blood vessel due to plaque deposition (atherosclerosis). Pumpkin seed oil (PSO) was administered to spontaneously hypertensive rats to study its hypotensive potential. In this study, PSO was fed to rats (40 mg/kg body weight) alone or in combination with antihypertensive drugs like calcium antagonist Felodipine (0.45 mg/kg body weight) or angiotensin-converting enzyme (ACE) inhibitor, and captopril (9 mg/kg body weight) for 4 weeks. Pretreatment with PSO for 4 weeks prior to intravenous (IV) administration of antihypertensive drugs resulted in the betterment of blood pressure and improvement in antioxidant activity in heart and kidney of spontaneously hypertensive rats and thus, concluded that PSO-based formulations can be administered as an adjuvant in the treatment of hypertension (Zuhair et al. 2000). In another study, the administration of flax and pumpkin seed mixture to rats fed with 1% cholesterol diet showed a significant reduction in malondialdehyde concentration and improvement in antioxidant defense mechanism with a considerable increase in the poly and mono-unsaturated fatty acids. This signifies the anti-atherogenic activity of the flax and pumpkin seed mixture (Makni et al. 2008).

Barakat and Mahmoud (2011) examined the effect of administration of diet supplemented with either flax/pumpkin or purslane/pumpkin seed mixture (at a ratio of 5:1) on increased blood lipid levels, kidney function, and as immune-modulators in rats fed with cholesterol-rich (2%) diets. Administration of flax/pumpkin or purslane/pumpkin seed mixture prevented excess weight gain in hypercholesterolemic rats. Hypolipidemic and antiatherogenic effects, and weight loss results were more prominent for purslane/pumpkin seed mixture. The diets with added seed mixture implicated an increase in HDL cholesterol and a decrease in LDL cholesterol. Atherogenic index was significantly reduced owing to a decrease in LDL/HDL ratio. In a study conducted to demonstrate hypolipedimic and hypotensive activity of pumpkin seed oil (PSO) in non-ovariectomized and ovariectomized Sprague-Dawley rats, the rats fed with PSO had low levels of systolic and diastolic blood pressure in both the groups. The lipid profile also improved with supplementation of PSO which was deranged possibly due to inadequate estrogen availability (Gossell-Williams et al. 2008).

Ristic-Medic et al. (2014) conducted a cross-sectional and follow-up dietary intervention study to observe the effects of milled sesame/pumpkin/flaxseed mixture dietary treatments on nutrition status, plasma lipids, phospholipids fatty acid profile, inflammatory markers, and symptoms of pruritus in patients on hemodialysis. Patients were asked to consume 30 g of milled seed mixture containing sesame, pumpkin, and flax seeds in the ratio of 1:1:3 in a single dose at the evening for 12 weeks. Dietary addition of seed mixture not only resulted in a significant decrease in blood pressure (BP) of around 11 mm systolic and 6 mm Hg diastolic but also showed improvement in serum lipid levels, inflammatory markers, and serum fatty acid composition.

19.8.3 Effects on Menopausal Symptoms

In a randomized, double-blinded and placebo-controlled study, supplementation of 2 g of pumpkin seed oil for 12 weeks to 35 women (natural menopause or benign surgical history resulting in menopause) resulted in a drop in diastolic blood pressure, increased plasma high-density lipoprotein cholesterol (HDL) levels, and reduced menopausal symptoms score with a marked reduction in hot flushes severity, lower episodes of headaches, and decreased pain in joints (Gossell-Williams et al. 2011). In another clinical study on 82 perimenopausal women diagnosed with urinary incontinence, the effectiveness of dietary supplement formula containing soybean extract and pumpkin seed extract was evaluated. The subjects received two tablets of the supplement per day for 4 weeks, followed by one tablet per day for an additional 4 weeks. The combined phytoestrogen pyrogallol found in pumpkin in addition to genistein in soybean extract resulted in safe and effective estrogenic and androgenic activity, consequently resulting in lowering the risk of urinary incontinence and improvement in the quality of life in perimenopausal women (Marañón 2017). Faloon (2008) has also reported that aqueous pumpkin seed extracts are effective in inhibiting the aromatase enzyme and strengthens the pelvic muscles through increase in testosterone levels and by binding to the androgen receptor present on pelvic muscle cells. Hence, helps in maintaining the structural integrity of pelvic floor and less incidences of urinary tract disorders in females. These studies thus suggest the possible protective role of pumpkin seed extract in treating urinary incontinence arising due to deranged levels of hormones after menopause.

19.8.4 Anticarcinogenic Activity

Tomar et al. (2014) studied the anticancer activity of pumpkin 2S albumins using human cancer-derived cell lines (used as a fundamental model to study the cancer biology and therapeutic effects of anticarcinogen in a laboratory set up). The protein showed evidence of a strong anticancer activity toward ovarian teratocarcinoma

(PA-1), breast cancer (MCF-7), hepatocellular carcinoma (HepG2), and prostate cancer (PC-3 and DU-145) cell lines in culture. Studies using acridine orange staining and DNA fragmentation methods revealed the induction of apoptosis and cytotoxic effect of pumpkin 2S albumin. DNase activity of pumpkin protein was exhibited against linear as well as super-coiled double-stranded DNA. Circular dichroism (CD) spectroscopy used to analyze secondary structures showed pumpkin 2S albumin to be highly stable even at 90 °C and its alpha helical structure was retained. In an in vitro study hydroalcoholic extracts of pumpkin seed inhibited the growth of both cancer as well as to lesser extent hyperplastic cells (Medjakovic et al. 2016). Bardaa et al. (2016) evaluated the cytotoxicity of cold-pressed pumpkin seeds (*Cucurbita pepo* L.) oil to HeLa cells. Results indicated that pumpkin seed oil at a concentration of 100 mg/mL destroyed almost more than half of the cancerous HeLa cells.

19.8.5 Protection Against Cellular Damage

In an animal study on Wistar rats, protective effects of pumpkin seed extract (PSE) were evaluated on the adverse effects of tramadol on the testes of adult and adolescent rats. The animals were divided into four groups. Control group received 1 mL of 0.9% saline. Second group received 20, 40, or 80 mg/kg body weight of tramadol in 1 mL of distilled water during the first, second, and third week, respectively. Third group received oral dose of PSE (40 mg/kg body weight in distilled water (1 mL) and fourth group received daily oral dose of tramadol (20, 40, and 80 mg/kg body weight during the first, second, and third week) plus PSE (40 mg/kg body weight). Owing to the antioxidant activity and free radical scavenging ability, PSE expressed prophylactic effect on tramadol-induced testicular damage which was comparatively less in the testes of the adolescent group than adult group (Minisy et al. 2017).

19.8.6 In Kidney Disease and Liver Injury

Traditionally pumpkin seed has been used to treat the overactive bladder (OAB) symptoms. A randomized, double-blind, placebo-controlled study was conducted to evaluate the effectiveness and safety of Cucuflavone (supplement containing extracts of pumpkin seed and soy germ extracts). Total of 120 subjects with OAB received the supplement for 12 weeks. A drastic decrease in urinary frequency, urgency, urinary incontinence, nocturea, and OAB was observed in the test group when compared with control, suggesting the potential of the combination of pumpkin seed and soy germ extract to be used as an alternative treatment to alleviate OAB symptoms and also to improve quality of life (Shim et al. 2014).

In an in-vivo study on supplementation of pumpkin seed protein isolate revealed its intense antioxidant activity in male Sprague-Dawley rats with acetaminophen-induced acute liver injury. In an in vitro model the pumpkin isolate exhibited radical scavenging activity (about 80%), chelating activity of Fe^{2+} ions (around 64%), and an inhibition of xanthine oxidase (around 10%). The supplementation of isolate and low-protein fed diet showed alleviated plasma activity levels of liver enzymes (AST and ALT). The supplementation also resulted in reducing the harmful effects of protein malnutrition and toxicity of acetaminophen (Nkosi et al. 2006a). Nkosi et al. (2006b) demonstrated antiperoxidative properties of pumpkin seed protein isolates with an animal study model. The Sprague-Dawley rats fed on low protein diet for 5 days were divided into three subgroups. Carbon tetrachloride (CCl_4) injections were administered to two subgroups, while one group received an equal amount of olive oil. After a period of 2 h one of the subgroups who received CCl_4 injection was fed with diets containing 20% pumpkin protein isolate. The rest two subgroups were continued (24, 48, and 72 h) on low protein diets until the investigations were carried out. Supplementation with pumpkin seed protein isolate showed promising results in reducing the damage caused due to protein malnutrition and CCl_4 intoxication. The study also showed a significant reduction in lipid peroxidation level.

Pretreatment of rats with pumpkin seed oil resulted in alleviating the effects of alcohol-induced liver injury. The protective role of pumpkin seed oil could be attributed to antioxidant compounds in pumpkin seeds. The male albino rats (30 numbers) were categorized into three groups. Group 1 which served as a control received distilled water, Group 2 received 10% ethyl alcohol in drinking water for 4 weeks named as positive control. Third group was administered with pumpkin oil (50 mg/kg) for a period of 4 weeks after pretreatment with 10% ethyl alcohol in drinking water for 15 days. The pumpkin seed oil treatment resulted in significant improvements in augmenting the elevated levels of serum alanine aminotransferase (ALT) and alkaline phosphatase (ALP) though it did not affect aminotransferase (AST) and gamma glutamyl transferase (γGT) activities. It also showed a significant reduction in total serum proteins as compared to alcohol-treated group. The levels of antioxidant enzymes (glutathione level, glutathione-S-transferase, and catalase) in pumpkin seed oil-treated group increased ($p < 0.01$; LSD) significantly as compared to alcohol-treated group. The same group also demonstrated a significant decrease in liver lipid peroxidation as compared to the positive control group (Abou Seif 2014).

Makni et al. (2008) demonstrated anti-atherogenic and protective effects of flax pumpkin seed mixture on the liver. The study was performed on 30 male Wistar rats divided into three groups: a control group (CD), CD-chol group fed diet with 1% cholesterol, and MS-chol group fed diet enriched with flax and pumpkin seed mixture. The results revealed that plasma and liver total cholesterol and triacylglycerol levels were reduced by 25%; 15% and 39%; 18%, respectively in MS-chol group compared to those of CD-chol group. Marked improvement in mono unsaturated fatty acids such as oleic acid and eicosaenoic acid (+1.2 and + 1.3 fold in plasma and + 1.6 and + 1.5 fold in liver) and a decrease in saturated fatty acids such as palmitic and stearic acid content (−0.6 and −0.7 fold in plasma and −0.4 and −0.6 fold in liver) were also observed in MS-chol group as compared to those of CD-chol group. Improvement in glutathione levels in plasma (54%) and liver (54%) along with significant decreases of catalase activity (58%, 69%), superoxide dismutase (28%, 34%) and glutathione peroxidase (36%, 37%) were observed as compared to those of control group. The histological studies revealed that the control and MS-chol groups showed normal cell architecture viz. less cells and lipid vacuolization, whereas in CD-chol rats significant morphological changes were observed.

19.8.7 In Prostate Health

Proposed health benefits of pumpkin seeds on prostate health have been documented to increase the strength of prostate gland, support hormone function, and improve muscle contraction in men (Lim 2017). A multicentric clinical trial was conducted on 2245 patients with stage I and II benign prostate to evaluate the efficacy and safety of pumpkin seed extract (1–2 capsules per day for 12 weeks) suffering from benign prostate hyperplasia (BPH). The effectiveness of treatment was measured with the International Prostate Symptom Score given by American Urological Association (I-PSS), and quality of life questionnaire (LQ Index). The results of the study showed a drop in I-PSS by 41.4% and improvement in the life quality by 46.1% during therapy. In more than 96% of patients the extract was helpful in alleviating BPH symptoms without any adverse side effects (Friederich et al. 2000).

A study by Gossell-Williams et al. (2006) has demonstrated that testosterone-induced prostate hyperplasia (0.3 mg/100 g body weight) can be subdued by the inclusion of 2.0 or 4.0 mg *Cucurbita pepo* oil in rats. The study was conducted in three groups of Sprague-Dawley male rats, wherein Group 1 received testosterone and pumpkin seed oil (2.0 or 4.0 mg/100 g of body weight) simultaneously for 20 days; Group 2 received testosterone and corn oil for 20 days; and Group 3 received only corn oil for 20 days (weekdays, i.e., Monday to Friday for 4 weeks). The results showed increased protective effects against testosterone-induced prostate hyperplasia and it was also noticed that the protective effects of pumpkin seed oil were more prominent at a higher dosage of 4.0 mg/kg ($P < 0.02$).

19.8.8 In Traditional Medicine

Numerous ethno-pharmacological study data reveals the traditional use of pumpkin seeds in many countries for treating various diseases. Table 19.8 summarizes the traditional use of pumpkin seeds in different parts of the world.

Table 19.8 Usage of pumpkin seeds (*Cucurbita pepo*) in traditional medicines

Place/Country	Ethno medical uses	Form
Mexico	Treat irritable bladder	Seeds
Egypt	Prostatic complains	Seeds
Europe	Treatment of urinary and prostate disease	Seeds
China	Deworming medicine, anti-parasitic, spleen and lung health	Seeds
Central and North America	Taenicide, diuretic, gastritis, burns, enteritis and febrile diseases	Seed oil
Centro America	Headaches and neuralgia	Compression
Northern Mexico	Diuretic, bronchitis and fever, anti-parasitic	Infusion/ decoction

Source: Perez Gutierrez (2016)

19.8.9 For Better Sleep and Antidepressant Activity

In *Ayurveda*, consumption of 1–2 spoons pumpkin seeds—"*Kushmanda Beeja*" is advised for sound sleep. It is considered as "Sleep Inducing Substance" ("*Nidra Janaka Dravya*") due to the presence of high amounts of magnesium and zinc (Jambli et al. 2017). Halson (2013, 2014) suggested the inclusion of ~200 g of pumpkin seeds equivalent to 1 g of tryptophan to enhance sleep latency and subjective sleep quality.

Traditionally pumpkin seeds are used to treat depression (Lim 2017). Antidepressant qualities are attributed to mood-enhancing amino acid tryptophan which is present in considerable amounts in pumpkin seeds (Yadav et al. 2010). Umadevi et al. (2011) studied the effect of aqueous and alcoholic *Cucurbita pepo* seeds extract on methyl isobutyl ketone-induced depression in male albino Wistar rats. Forced Swim Test (FST) was conducted to examine the antidepressant activity. Rats were randomly categorized into 5 groups as Group I – control and Group II – depressive control, Group III and IV treated with aqueous and alcoholic seed extracts (100 mg/kg body weight) for 30 days, and Group V treated with the standard drug (imipramine) at a dose of 30 mg/kg body weight for 15 days. The forced swim test (FST) results showed a marked decrease in immobility time in comparison with depressed rats after receiving seed extracts. The activity of antioxidant enzymes namely superoxide dismutase, glutathione peroxidase, and catalase was found to increase in the brain and serum of rats treated with extracts, in addition to improvement in the concentration of reduced glutathione, vitamin C and decreased lipid peroxidation in the brain and serum of test rats in comparison with depressed rats. The extracts were found to be as effective as antidepressant drug – imipramine (used as control).

19.8.10 Anti-Inflammatory Activity

Fahim (1995) carried out an investigation to study the efficacy of pumpkin seed oil (PSO) against inflammation. Freund's complete adjuvant was used to induce arthritis in rats and indomethacin, a classic anti-inflammatory drug, was used as control. The study was conducted in two phases: acute and chronic, wherein PSO was used to treat acute and PSO along with indomethacin in chronic phase treatment. The supplementation of PSO had promising effects in reducing the arthritic symptoms, especially during the chronic phase with significant inhibition of paw edema. The treatment with control showed similar results; however, it resulted in increased lipid peroxide levels.

Wound healing performance of the topical application of pumpkin seed oil was experimented in uniform wound-induced 18 adult Wistar male rats divided into three groups namely the control group topically treated with saline solution, treated with reference drug (0.13 mg/mm^2), and target group treated with 0.52 μL/mm^2 of pumpkin seed oil which was applied every 2 days till the control group showed complete healing. The results demonstrated the enhanced cutaneous wound healing in the target group than in the control group. The topical application of cold-pressed pumpkin seeds oil provided a connective tissue matrix which assists migration of the fibroblasts during re-epithelialization process and healthy structure of collagen fibers and absence of inflammatory cells. This therapeutic potential of pumpkin seed oil could be related to the presence of substantial amounts of inflammatory mediators such as polyunsaturated fatty acids, an antioxidant vitamin E, and phytonutrient β-sitosterol which has strong angiogenic activity, and also promotes fibroblasts multiplication (Bardaa et al. 2016).

19.8.11 Antimicrobial Activity

Antimicrobial assay of pumpkin seed oil was carried out against the pathogenic bacteria *Staphylococcus aureus* by the disk diffusion technique. The authors observed strong antimicrobial activity of pumpkin seed extract against pathogenic gram-positive *Staphylococcus aureus* with an average inhibition zone of 15.33 $\pm$ 1.52 mm (Rabia et al. 2014; Chonoko and Rufai 2011). Bardaa et al. (2016) reported antimicrobial activity against *Bacillus subtilis* and also highlighted the efficacy of oil against gram-positive than gram-negative species of bacteria. Pumpkin seeds can be used to treat parasitic infections like acute schistosomiasis and tapeworm infection and this antiparasitic activity could be attributed to cucurbitacins in pumpkin seeds (Younis et al. 2000; Patel 2013). Sood et al. (2012) studied antifungal activities of *Cucurbita pepo* and noticed that it has high potency against *Fusarium oxysporium* and *Trichoderma reesei*.

19.8.12 Antioxidant Activity

Popović et al. (2013) investigated the antioxidant activity and functional properties of cucurbitin extracted from pumpkin oil cake and enzymatically hydrolyzed by pepsin, alcalase, and flavourzyme enzymes. The results of the study revealed that cucurbitin hydrolysate obtained by alcalase and pepsin showed antioxidant potential when compared with flavourzyme hydrolysates. The alcalase hydrolysate (at 25.6% degree of hydrolysis) showed reducing power of 0.25 ± 0.01 A700 nm and ABTS (2,2′-azino-bis (3-ethylbenzothiazoline-6-sulfonic acid)) scavenging activity was found to be 3.34 ± 0.02 mmol/L Trolox equivalent antioxidant coefficients (TEAC).

In a study conducted to demonstrate the antioxidant and lipoxygenase inhibitory activities of pumpkin seeds, extracts were prepared using solvents of decreased polarity (water, methanol, acetone, and ethylacetate) from four varieties of pumpkin seeds. The lipid fractions were extracted using the Folch method, polar and neutral lipids were separated using a counter-current distribution extraction method. The total extractable phenolic content was highest (85–92%) in water extracts with 54–64 μmol gallic acid equivalents followed by 7–15% of total extractable phenolics and 5–11 μmol gallic acid in methanol extract. The highest 2, 2-diphenyl-1-picrylhydrazyl (DPPH) radical scavenging activity was recorded in water (4.63–6.71 mg/mL expressed as E50 values) and methanol extracts (4.40–5.42 mg/mL expressed as E50 values) whereas ethylacetate extracts demonstrated lowest scavenging activity. The polar lipid fractions were found to have more potent antioxidant activity compared to neutral lipid fractions. The inhibitions of soybean lipoxygenase activity were stated as 34–60% and 50% for water and methanol extracts, respectively. The lipid peroxidation inhibition was found to be greater in acetone extracts (34.71–100%) (Xanthopoulou et al. 2009).

Saavedra et al. (2013) studied the effect of different drying methods and extraction solvents on antioxidant and total phenolic content of the seeds and shell of squash pumpkin. The results revealed that among the drying methods, the oven-dried samples were found to have higher antioxidant and phenolic content, whereas among the shell and seeds, the shell samples reportedly had higher antioxidant (1.47–70.96% inhibition of DPPH radicals) and phenol content values (2.00–10.69 mg GAE/g DW).

Protein isolate from pumpkin oil cake (POCPI) obtained using food-grade enzymes namely trypsin and alcalase was studied for antioxidative activities by Nourmohammadi et al. (2017). The authors observed that alcalase-hydrolyzed $POCPI_1$ exhibited better antioxidant activities in terms of total antioxidative potential, ferrous ion chelating activities, and DPPH radical scavenging activity. Also, the amount of hydrophobic amino acids (33.49%) demonstrated a direct relationship between alcalase-hydrolyzed $POCPH_1$ and the antioxidative activity.

In another study by Bardaa et al. (2016), β-carotene/linoleic acid bleaching test was conducted to ascertain the inhibition of lipid peroxidation of pumpkin seed oil. The results revealed that pumpkin seed oil showed higher lipid peroxidation inhibitory activity than BHT which could be attributed to its tocopherol content.

In a study design 40 male albino rats were divided into four groups; control group rats instilled with normal saline (2 mL/kg), 2 mL/kg of 0.1 N HCL at pH 1.25, oral pumpkin seed oil at a dose of ~1375 mg/kg/day for a period of 7 days and 2 mL/kg of 0.1 N HCL at pH 1.25 in addition to the same amount of pumpkin seed oil. The results revealed that lung tissue samples of the rats treated with pumpkin oil showed decreased interalveolar septa thickness, inflammatory cell infiltration, less collagenous fiber area, and immune expression of inducible nitric oxide synthase (iNOS) as compared to control group. Clinical symptoms of disturbed alveolocapillary membrane, degeneration of type II pneumocytes, and plentiful alveolar macrophages were also reduced significantly with the treatment of pumpkin seed oil (Omar and Sarhan 2017).

19.8.13 As Immune-Suppressant

In an in vitro study, the efficacy of pumpkin seeds extracts as immune-suppressant was assessed in stimulated and unstimulated peripheral blood mononuclear cells (PBMC) of a healthy donor. The tryptophan degradation and neopterin production increased in stimulated PBMC, which got suppressed with extracts of pumpkin seeds in a dose-dependent manner. The pumpkin seed extracts showed potent immune-suppressant activity due to cytokine IFN-γ-induced biochemical changes. In mitogen-stimulated PBMCs, the formation and release of cytokine IFN-γ was suppressed due to the seed extract. Data obtained from this study revealed that immunosuppressive effect of pumpkin seeds extracts could be attributed to the antioxidant compounds, which are responsible for detoxification of reactive oxygen species (ROS) and so also, they may lower the formation of ROS initiated by IFN-γ (Winkler et al. 2005).

19.9 Adverse Effects and Individual Concerns

Though pumpkin seeds are used in a variety of food preparations, allergic reactions to these seeds are rare and limited data is available on allergens in the seeds. A research trial was conducted in an Out Patient Department Setup in Spain. Among the three patients who were vulnerable for allergic reactions, pumpkin seed products were administered to observe the allergenic response. Two out of three patients reported anaphylactic reactions, whereas the third patient suffered from an episode of oral allergy syndrome (OAS) and abdominal pain. It was also noted that three of them had the tolerance to fruit as well as to other cucurbits but were allergic to some other foods (nuts like cashews, pistachio, pine seeds, lipid transfer protein (LTP) from nuts, peach allergen – Pru p 3). The diagnosis of allergic reactions was confirmed with a skin prick test (SPT), prick by prick test (PP), and laboratory testing of serum allergen-specific IgE antibodies using ImmunoCAP method. All the diagnostic tests were positive for the pumpkin seeds and negative for pumpkin fruit and profilin in all the three patients. IgE immunoblotting from patients having anaphylaxis demonstrated a protein band of 6 kDa similar to that of pine seeds along with other bands of 34 kDa and 48 kDa, whereas patients with OAS revealed 26 kDa and 30 kDa in pumpkin seed extract which could be possible allergens (Valverde-Monge et al. 2017).

19.10 Food Applications

Pumpkin seeds (raw or roasted) are incorporated in cakes, bread, cereals, and are also added to salads whereas; seed oil is used as cooking oil. Both seed and seed oil are rich in natural phytosterols and demonstrate nutraceutical potential also (Phillips et al. 2005). Pumpkin seed oil can also be added to culinary preparations like pasta, salad, and soups as dressing along with honey, olive oil, and vinegar (Bavec et al. 2007).

19.10.1 As Cooking Oil

In Africa and the Middle East countries pumpkin seed oil is used as cooking oil whereas, in the south part of Austria and in nearby areas of Slovenia and Hungary it is used as salad oil (Wenzl et al. 2002). Lower concentration of linoleic acids and other highly unsaturated fatty acids gives oxidative stability to pumpkin seed oil and makes it suitable for use in human diet, culinary, and industry (Aliu et al. 2016; Stevenson et al. 2007). However, in spite of all these characteristics the industrial exploitation of pumpkin seed oil is limited. Prevc et al. (2015) have reported two orders of higher magnitude conductivity for roasted pumpkin oil (RPO) than refined sunflower oil and extra virgin olive oil with the application of dielectric spectroscopy at low frequencies. The conductivity of pumpkin seed oils is influenced by phospholipid and metal contents and hence it can serve as a cheap method to determine adulteration of RPO with other vegetable oils (Table 19.9).

19.10.2 In Bakery Industry

The pumpkin seed flour has great potential to be fortified with wheat flour and thus can improve its protein and mineral quality. Additionally, as reported by El-Adawy and Taha (2001) pumpkin seed kernel flour exhibits excellent water and oil absorption capacities, emulsification properties, as well as foaming capacity and thus can be used in various bakery and meat preparations. Moustafa (2013) developed biscuits and cake fortified with mixed raw and/or germinated flax and pumpkin seeds. The

Table 19.9 Physicochemical properties of pumpkin (*Cucurbita pepo*) seed oil

Product	El-Adawy and Taha (2001)	Gohari Ardabili et al. (2011)
Refractive index	1.4706 ± 0.001 (at 25 °C)	0.9151 ± 0.0002 (at 30 °C)
Specific gravity (25 °C)	0.917 ± 0.006 (at 25 °C)	1.4662 ± 0.0002 (at 30 °C)
Acid value (mg of KOH/g of oil)	2.88 ± 0.18	0.78 ± 0.02
Saponification value (mg of KOH/g of oil)	206 ± 2.56	190.69 ± 1.40
Ester value (mg of KOH/g of oil)	203.12 ± 3.28	NR
Iodine value (g of I/100 g of oil)	109 ± 2.29	104.36 ± 0.04
Peroxide value (mEq/kg)	3.60 ± 0.41	10.85 ± 0.62
Unsaponifiable matters content (% of oil)	NR	5.73 ± 0.82

NR Not reported

results revealed that the sensory properties (color, texture, taste, odor, and overall acceptability) of developed biscuits were not significantly different than control whereas, sensory properties of developed cakes were significantly different. The study indicated that germination of seeds further improves its nutritional quality, by enhancing its micronutrient content, fatty acid, and macronutrients composition.

Atuonwu and Akobundu (2010) observed that wheat flour can be replaced up to 10% with defatted pumpkin seed flour to yield cookies with enhanced nutritional and acceptable sensory qualities of cookies. Białek et al. (2015) developed muffins by partial substitution of wheat flour with pumpkin seed flour (33%). Incorporation of pumpkin seed flour not only resulted in a general decrease in short and long chain saturated fatty acids (especially myristic, palmitic and stearic) but also marked reduction in the concentration of odd and branched chain fatty acids independent of storage time was observed. El-Soukkary (2001) evaluated bread quality prepared with the fortification of wheat flour with various pumpkin seed products like raw, roasted, germinated, fermented protein concentrates and isolates. He reported that the addition of pumpkin seed products to wheat flour resulted in an enhancement in the protein and mineral concentration of bread compared with that of control without any detrimental effects on dough or loaf quality.

Revathy and Sabitha (2013) developed and analyzed various bakery products viz. bread, bun (plain, coconut), biscuits (sweet, salt, and coconut), and cake (plain, plum) using pumpkin seed flour with partial replacement of wheat flour at 10, 20, and 30% level. The findings of the study revealed that incorporation of pumpkin seed flour resulted in an enhancement in the nutritive value and their acceptability was at par with that of standard products.

19.10.3 In Meat and Meat Products

Pumpkin seed oil-in-water emulsions (PSO/W) were used to replace beef fat in raw and cooked model system chicken meat emulsion (MSME) treatments which resulted in a significant change in chemical composition and pH values. The addition of PSO/W to the MSME showed a reduction in total expressible fluid values along with enhancement in cooking yield and oxidative stability on storage. Thus, revealing the potential of pumpkin seed oil in developing meat products with a healthy lipid profile (Serdaroğlu et al. 2017).

19.10.4 As Protein Sources

After oil extraction the remaining defatted pumpkin cake is mainly used as animal feed though it has about 60% of protein and a nutritional significance (Peričin et al. 2009). Naked pumpkin (*Cucurbita pepo* L.) oil press-cake and cold-pressed hemp oil was used to develop a stable oil-base spread. In order to develop optimum spread stabilizer −1.25% (w/w) and 80% of hemp oil (w/w, of the total added oil) was used. The results of the study revealed that developed spread has 0.97 g/serve of ω-3 fatty acids; 68.4 mm of penetration depth; 9.2% of oil separation after 3 months of storage; with 17.5 as a sensory score. The appearance, texture, and spreadability of developed spread were similar to commercial peanut butter; owing to the presence of ω-3 fatty acids it has functional nutritional value. The shelf life study indicated that it can be successfully stored for 1 month without visible oil separation (Radočaj et al. 2012).

19.10.5 As a Fortifying Agent

Sharma and Lakhawat (2017) developed gravy mixes using pumpkin seed flour. The acceptability of gravies prepared with incorporation of 5% pumpkin seed flour was found to be most acceptable though it was acceptable even at 8 and 10%.

19.11 Alternative Applications

19.11.1 In Developing Packaging Films for Food Industry

Fahmy et al. (2008) suggested that esterase found in pumpkin seed extract can be useful in food processing industry as they exhibit property like microbial esterase. Popović et al. (2011) illustrated the possibility of biodegradable film formation using pumpkin oil cake (PuOC). The study indicated highest tensile strength (TS) of

68.08 MPa and elongation to break (EB) of 36.62% in developed films at pH 12 and 90 °C. Film which was formulated at pH 10 and 60 °C exhibited prominent free radical scavenging activity using ABTS test. In another study, Popović et al. (2012) developed biodegradable films using pumpkin oil cake protein isolate (PuOC PI) at various pH (2–12) levels and plasticizer concentration, i.e., 0.3–0.6 g glycerol/g of PuOC PI. The study revealed the possibility of PuOC PI film formation over a broad range of pH with the only exception at pH 4–8. Gas permeability analysis of films formulated at 0.4 g glycerol/g PuOC PI concentration demonstrated superior barrier activity for gases like oxygen, nitrogen, carbon-di-oxide, and air. The authors concluded that due to enhanced break elongation and superior gas barrier activity pumpkin oil cake can be used to develop stretchable coating having gas barrier activity.

19.11.2 In Pharmaceutical Industry

In a randomized, placebo-controlled, double-blind study positive effect on hair growth was observed after supplementation of pumpkin seed oil for 24 weeks in male patients having androgenetic alopecia of mild to moderate level (Cho et al. 2014). Cold press pumpkin seed oil found to have tocopherols, sterols, and polyunsaturated fatty acids which also demonstrated strong antioxidant and antimicrobial activity. It may offer valuable protection against skin diseases like dermatological wounds. These properties of pumpkin seed oil can be exploited in pharmaceutical and cosmetic industry (Bardaa et al. 2016).

Hataminia and Farhadian (2017) have offered safe and controllable process for adsorption of fatty acids (FA) of pumpkin seed on iron oxide nanoparticles to facilitate its application in pharmacotherapy. The study showed that particles had superparamagnetic property and net charge of 52 mv, making it appropriate for its application in the pharmaceutical industry. Under optimum conditions TEM analysis with spherical morphology demonstrated about 20 nm estimated particle size. The predictable thickness of fatty acids on nanoparticles surface was around 5 nm. Fatty acid layers were formed around nanoparticles which was obvious in Fourier transform infrared spectroscopy (FTIR). Oil can be used in the treatments of prostate and skin cancer. The loading of substrate having therapeutic potential provides newer drug delivery mechanism.

19.11.3 In Cattle-Feed Industry

Klir et al. (2017) evaluated the quality of goat milk by replacing the soybean meal in goat's diet with that of pumpkin seed cake (PSC) or extruded linseed (ELS). The pumpkin seed cake (PSC) diet reduced total ω-6 fatty acids (2.96 vs. 3.54 g/100 g fatty acids, $P < 0.05$) and thus, PSC can be used to replace soyabean meal without affecting milk yield.

19.11.4 To Enhance Nutritive Value of Poultry Feed

After oil extraction the remaining defatted pumpkin cake is mainly used as animal feed though it has about 60% of protein and thus, holds great nutritional significance (Peričin et al. 2009). Achilonu et al. (2017) observed that the inclusion of pumpkin meals in poultry feeds have a promising impact on the general poultry performance, enhanced feed conversion ratio, rate of growth, and egg production both in terms of quality and quantity. Furthermore, it also helps to maintain balanced gut microflora which positively affects the health of poultry. Hajati et al. (2011) studied the effect of supplementation of corn-soybean meal-wheat based diet with pumpkin seed oil on blood lipid profile and performance of broiler chicken. They concluded that supplementation resulted in the reduction in mortality rates of broiler chicken along with a decrease in plasma cholesterol and triglyceride concentration and without any significant effect on the carcass composition of the abdominal fat pad at 49 days of age. Machebe et al. (2013) studied the impact of feeding seeds and root extracts of various plants to poultry birds. In this study, breeder turkey hens supplemented with pumpkin seed supplement (PSS) showed a significant increase in the number of eggs laid and fertile egg percentage. The study also demonstrated that PSS extracts due to the phytochemical agents may protect turkey egg embryo to some extent during the incubation period which was not seen in eggs of hens in the control group. Herkeľ et al. (2014) observed that the supplementation of feed mixture with pumpkin oil (3%) resulted in improvements in average weight of eggs and also showed better laying tendency compared to the control group.

19.12 Future Challenges

There are ample opportunities and priorities for improving the quality of crops at the farm level; including formulation of disease-resistant species, increasing seed production and also to yield more oil, and enhancing nutritional quality of seeds specifically the tocopherol content. Also there is a scope for growing the pumpkins in an organic way which may have higher nutritional properties for health benefits. Various studies have revealed that the method of oil extraction and processing influences the therapeutic and sensory attributes of the pumpkin seed oil, so there is scope for finding the best method of oil extraction and processing treatments required for improving the therapeutic characteristics of the oil besides having better sensory acceptance.

Numerous data is available on the health benefits of pumpkin seeds in the prevention and treatment of different disorders. Still the data is limited in view of specific component responsible for its medicinal/remedial potential, long-term use, safety trials, and side effects on humans. The studies can be undertaken to investigate the bioactive substances found in pumpkin seeds which work in synergistic ways. Furthermore, recent market studies show that there is a constant increase in demand for pumpkin seed and its products. Although pumpkin seed hydrolysate has been shown to have a very good potential use in the food industry, the evidence about its application is limited. Based on the available literature there are numerous opportunities to develop pumpkin seed-enriched health food products and thus, branding and promotion of pumpkin seed and its various products seems to be a promising field.

19.13 Conclusion

Pumpkin seeds exhibit numerous characteristics potentially beneficial in terms of its application in food, nutraceutical, cosmetics, and fodder industry. The high content of bioactive compounds has rendered "superfood" title to pumpkin seeds. Snacks manufactured from these seeds have gained considerable demand due to its excellent palatability. Moreover, indigenous therapeutic usage (both preventive and curative aspects) has made it more popular among scientists and therapist. Further research and scientific studies as well as documentation are required to authenticate its vast and relevant applications.

References

Abou Seif H (2014) Ameliorative effect of pumpkin oil (*Cucurbita pepo* L.) against alcohol-induced hepatotoxicity and oxidative stress in albino rats. Beni-Suef Univ J Basic Appl Sci 3:178–185. https://doi.org/10.1016/j.bjbas.2014.08.001

Achilonu M, Nwafor I, Umesiobi D et al (2017) Biochemical proximates of pumpkin (Cucurbitaeae spp.) and their beneficial effects on the general Well-being of poultry species. J Anim Physiol Anim Nutr 102:5–16. https://doi.org/10.1111/jpn.12654

Aliu S, Rusinovci I, Salihu S et al (2016) Nutritive values and seed chemical variability of pumpkin (*Cucurbita pepo* L.) populations. J Food Agric Environ 14(2):28–33. https://doi.org/10.1234/4.2016.3756

Anon (2007) Pumpkin patch—everything you ever wanted to know about pumpkins! Pumpkin-patch.com, In. http://www.pumpkin-patch.com/. Accessed 17 Nov 2017

Atuonwu A, Akobundu E (2010) Nutritional and sensory quality of cookies supplemented with defatted pumpkin (*Cucurbita pepo*) seed flour. Pak J Nutr 9:672–677. https://doi.org/10.3923/pjn.2010.672.677

Barakat L, Mahmoud R (2011) The antiatherogenic, renal protective and immunomodulatory effects of purslane, pumpkin and flax seeds on hypercholesterolemic rats. N Am J Med Sci 411–417. doi: https://doi.org/10.4297/najms.2011.3411

Bardaa S, Ben Halima N, Aloui F et al (2016) Oil from pumpkin (*Cucurbita pepo* L.) seeds: evaluation of its functional properties on wound healing in rats. Lipids Health Dis. https://doi.org/10.1186/s12944-016-0237-0

Bavec F, Mlakar G, Rozman Č et al (2007) Oil pumpkins: niche for organic producers. In: Janick J, Whipkey A (eds) Issues in new crops and new uses. ASHS Press, Alexandria, VA
Baxter GG, Murphy K and Paech A (2012). The potential to produce pumpkin seed for processing in north East Victoria. RIRDC publication no. 11/145 RIRDC project no. PRJ-005518
Bharti S, Kumar A, Sharma N et al (2013) Tocopherol from seeds of *Cucurbita pepo* against diabetes: validation by in vivo experiments supported by computational docking. J Formos Med Assoc 112:676–690. https://doi.org/10.1016/j.jfma.2013.08.003
Białek M, Rutkowska J, Adamska A et al (2015) Partial replacement of wheat flour with pumpkin seed flour in muffins offered to children. CyTA J Food 14(3):391–398. https://doi.org/10.1080/19476337.2015.1114529
Bučko S, Katona J, Popovic L et al (2016) Functional properties of pumpkin (*Cucurbita pepo*) seed protein isolate and hydrolysate. J Serb Chem Soc 81:35–46. https://doi.org/10.2298/jsc150615081b
Caili F, Huan S, Quanhong L (2006) A review on pharmacological activities and utilization technologies of pumpkin. Plant Foods Hum Nutr 61:70–77. https://doi.org/10.1007/s11130-006-0016-6
Cho Y, Lee S, Jeong D et al (2014) Effect of pumpkin seed oil on hair growth in men with androgenetic alopecia: a randomized, double-blind, placebo-controlled trial. Evid Based Complement Alternat Med 2014:1–7. https://doi.org/10.1155/2014/549721
Chonoko UG, Rufai AB (2011) Antibacterial activity of *Cucurbita pepo* (pumpkin) against *Staphylococcus aureus* and *Salmonella typhi*. Bayero J Pure Appl Sci 4:145–147
Choudhary B (2017) How to produce quality seed of pumpkin—Krishisewa. Krishisewa.com, In. http://www.krishisewa.com/articles/seed-production/247-pumpkin-quality-seed.html. Accessed 18 Nov 2017
El-Adawy T, Taha K (2001) Characteristics and composition of watermelon, pumpkin, and paprika seed oils and flours. J Agric Food Chem 49:1253–1259. https://doi.org/10.1021/jf001117+
Elinge CM, Muhammad A, Atiku FA, Itodo AU, Peni IJ, Sanni OM, Mbongo AN (2012) Proximate, mineral and anti-nutrient composition of pumpkin (*Cucurbita pepo* L.) seeds extract. Int J Plant Res 2:146–150. https://doi.org/10.5923/j.plant.20120205.02
El-Soukkary F (2001) Evaluation of pumpkin seed products for bread fortification. Plant Foods Hum Nutr 56:365–384. https://doi.org/10.1023/a:1011802014770
Emire SA, Jha YK, Mekam M (2013). Role of antinutritional factors in food industry. Bev Food World 23–28
Fahim A (1995) Effect of pumpkin-seed oil on the level of free radical scavengers induced during adjuvant-arthritis in rats. Pharmacol Res 31:73–79. https://doi.org/10.1016/1043-6618(95)80051-4
Fahmy A, Abo-Zeid A, Mohamed T et al (2008) Characterization of esterases from *Cucurbita pepo* cv. "Eskandrani". Bioresour Technol 99:437–443. https://doi.org/10.1016/j.biortech.2006.11.062
Faloon, W. (2008). A breakthrough in the relief of overactive bladder and urinary incontinence. LE Magazine
Food and Agriculture Organization Corporate Statistical Database (FAOSTAT) (2017) Food and agriculture organization. FAO, Data base. Food and Agriculture Organization of the United Nations. Retrieved from: http://www.fao.org/faostat/en/#data/QC. Accessed 1 Dec 2017
Friederich M, Theurer C, Schiebel-Schlosser G (2000) Prosta fink forte capsules in the treatment of benign prostatic hyperplasia. Multicentric surveillance study in 2245 patients. Forsch Komplementarmed Klass Naturheilkd 7(4):200–204. https://doi.org/10.1159/000021344s
GAIN (2004). Pumpkin Seed Oil—The "Green Gold of Styria", an Austrian specialty. In: USDA Foreign Agricultural Service Global Agriculture Information Network Report Number AU4011, 2 pp. 204–210. [online], http://www.australianoilseeds.com/data/assets/pdf_file/0014/7025/bavec,franc00220.pdf. Accessed 17 Nov 2017

Gohari Ardabili A, Farhoosh R, Haddad Khodaparast M (2011) Chemical composition and physicochemical properties of pumpkin seeds (*Cucurbita pepo* Subsp. *pepo* Var. *Styriaka*) grown in Iran. J Agric Sci Technol 13(Suppl Issue):1053–1063

Gossell-Williams M, Davis A, O'Connor N (2006) Inhibition of testosterone-induced hyperplasia of the prostate of Sprague-Dawley rats by pumpkin seed oil. J Med Food 9:284–286. https://doi.org/10.1089/jmf.2006.9.284

Gossell-Williams M, Lyttle K, Clarke T et al (2008) Supplementation with pumpkin seed oil improves plasma lipid profile and cardiovascular outcomes of female non-ovariectomized and ovariectomized Sprague-Dawley rats. Phytother Res 22(7):873–877. https://doi.org/10.1002/ptr.2381

Gossell-Williams M, Hyde C, Hunter T et al (2011) Improvement in HDL cholesterol in postmenopausal women supplemented with pumpkin seed oil: pilot study. Climacteric 14(5):558–564. https://doi.org/10.3109/13697137.2011.563882

Gupta C, Prakash D (2014) Phytonutrients as therapeutic agents. J Compl Integr Med 11 (3):151–169. https://doi.org/10.1515/jcim-2013-0021

Hajati H, Hasanabadi A, Waldroup P (2011) Effects of dietary supplementation with pumpkin oil (*Cucurbita pepo*) on performance and blood fat of broiler chickens during finisher period. Am J Anim Vet Sci 6:40–44. https://doi.org/10.3844/ajavsp.2011.40.44

Halson S (2013) Nutritional interventions to enhance sleep. Sports Sci Exchange 26(116):1–5

Halson S (2014) Sleep in elite athletes and nutritional interventions to enhance sleep. Sports Med 44:13–23. https://doi.org/10.1007/s40279-014-0147-0

Herkeľ R, Gálik B, Bíro D et al (2014) The effect of pumpkin and flaxseed oils on selected parameters of laying hens performance. Acta Fytotechnica et Zootechnica 17:96–99. https://doi.org/10.15414/afz.2014.17.03.96-99

Jambli S, Chavan S, Prashanth S (2017) The management of primary insomnia through pancha sugandha sadhita takra dhara—a pilot study. J Ayurveda Integr Med Sci. https://doi.org/10.21760/jaims.v2i1.7488

Khokhar S, Owusu Apenten RK (2003) Antinutritional factors in food legumes and effects of processing. In: The role of food, agriculture, forestry and fisheries in human nutrition, vol IV. Encyclopedia Of Life Support Systems (EOLSS), Oxford, pp 82–116

Kim M, Kim E, Kim Y et al (2012) Comparison of the chemical compositions and nutritive values of various pumpkin (Cucurbitaceae) species and parts. Nutr Res Pract 6:21. https://doi.org/10.4162/nrp.2011.5.6.21

Klir Z, Castro-Montoya J, Novoselec J et al (2017) Influence of pumpkin seed cake and extruded linseed on milk production and milk fatty acid profile in alpine goats. Animal 11:1772–1778. https://doi.org/10.1017/s175173111700060x

Kreft I, Stibilj V, Trkov Z (2002) Iodine and selenium contents in pumpkin (*Cucurbita pepo* L.) oil and oil-cake. Eur Food Res Technol 215:279–281. https://doi.org/10.1007/s00217-002-0563-5

Lazos E (1986) Nutritional, fatty acid, and oil characteristics of pumpkin and melon seeds. J Food Sci 51:1382–1383. https://doi.org/10.1111/j.1365-2621.1986.tb13133.x

Lazos E (1992) Certain functional properties of defatted pumpkin seed flour. Plant Foods Hum Nutr 42:257–273. https://doi.org/10.1007/bf02193934

Lelley T, Loy B, Murkovic M (2009) Hull-less oil seed pumpkin. In: Vollmann I, Rajcan I (eds) Oil crops, handbook of plant breeding, vol 4. Springer, New York, pp 469–492. https://doi.org/10.1007/978-0-387-77594-4

Li QH, Tian Z, Cai TY (2001) Study on the hypoglycemic action of pumpkin extract in diabetic rat. Ying Yang Xue Ba 25:34–36

Lim P (2017) Asian herbals and aphrodisiacs used for managing ED. Transl Androl Urol 6:167–175. https://doi.org/10.21037/tau.2017.04.04

Machebe N, Ugwu S, Atu C, Mbunwen N (2013) Intake of some biological seeds and root extracts of plants improves fertility and hatchability of Turkey eggs. J Basic Appl Sci. https://doi.org/10.6000/1927-5129.2013.09.69

Makni M, Fetoui H, Gargouri N et al (2008) Hypolipidemic and hepatoprotective effects of flax and pumpkin seed mixture rich in ω-3 and ω-6 fatty acids in hypercholesterolemic rats. Food Chem Toxicol 46:3714–3720. https://doi.org/10.1016/j.fct.2008.09.057

Makni M, Sefi M, Fetoui H et al (2010) Flax and pumpkin seeds mixture ameliorates diabetic nephropathy in rats. Food Chem Toxicol 48:2407–2412. https://doi.org/10.1016/j.fct.2010.05.079

Makni M, Fetoui H, Gargouri N et al (2011) Antidiabetic effect of flax and pumpkin seed mixture powder: effect on hyperlipidemia and antioxidant status in alloxan diabetic rats. J Diabetes Complications 25:339–345. https://doi.org/10.1016/j.jdiacomp.2010.09.001

Mandl A, Reich G, Lindner W (1999) Detection of adulteration of pumpkin seed oil by analysis of content and composition of specific Δ7-phytosterols. Eur Food Res Technol 209:400–406. https://doi.org/10.1007/s002170050516

Mansour E, Dworschák E, Lugasi A et al (1993) Nutritive value of pumpkin (*Cucurbita pepo* Kakai 35) seed products. J Sci Food Agric 61:73–78. https://doi.org/10.1002/jsfa.2740610112

Marañón J (2017) Clinical study: effect of supplementation with high genistein soybean isoflavones and pumpkin standardized extract on urinary incontinence in western perimenopausal women. J Gynecol Womens Health. https://doi.org/10.19080/jgwh.2017.04.555627

Matus Z, Molnar P, Szabo LG (1993) Main carotenoids in pressed seeds (*Cucurbitae semen*) of oil pumpkin (*Cucurbita pepo convar. pepo var. styriaca*). Acta Pharm Hung 63(2):247–256

Medjakovic S, Hobiger S, Ardjomand-Woelkart K et al (2016) Pumpkin seed extract: cell growth inhibition of hyperplastic and cancer cells, independent of steroid hormone receptors. Fitoterapia 110:150–156. https://doi.org/10.1016/j.fitote.2016.03.010

Minisy F, Massoud A, Omara E et al (2017) Protective effect of pumpkin seed extract against testicular toxicity induced by tramadol in adolescent and adult male albino rats: a light and electron microscopic study. Egypt Pharm J 16:43. https://doi.org/10.4103/1687-4315.205828

Moustafa R (2013) Chemical, technological and biological evaluation of raw and germinated flax and pumpkin seed mixtures. J High Inst Publ Health 43:98–111. https://doi.org/10.21608/jhiph.2013.19993

Murkovic M (2009) Pumpkin seed oil. Gourmet and health-promoting specialty oils 345-358. https://doi.org/10.1016/b978-1-893997-97-4.50018-8

Murkovic M, Hillebrand A, Winkler J, Pfannhauser W (1996) Variability of vitamin E content in pumpkin seeds (*Cucurbita pepo* L). Z Lebensm Unters Forsch 202:275–278. https://doi.org/10.1007/bf01206096

Murkovic M, Mullender U, Neunteufl H (2004) Carotenoid content in different varieties of pumpkins. J Food Compos Anal 15:633–638

Nakić S, Rade D, Škevin D et al (2006) Chemical characteristics of oils from naked and husk seeds of *Cucurbita pepo* L. Eur J Lipid Sci Technol 108:936–943. https://doi.org/10.1002/ejlt.200600161

Nawirska-Olszańska A, Kita A, Biesiada A et al (2013) Characteristics of antioxidant activity and composition of pumpkin seed oils in 12 cultivars. Food Chem 139:155–161. https://doi.org/10.1016/j.foodchem.2013.02.009

Nkosi C, Opoku A, Terblanche S (2006a) In vitro antioxidative activity of pumpkin seed (*Cucurbita pepo*) protein isolate and its in vivo effect on alanine transaminase and aspartate transaminase in acetaminophen-induced liver injury in low protein fed rats. Phytother Res 20:780–783. https://doi.org/10.1002/ptr.1958

Nkosi C, Opoku A, Terblanche S (2006b) Antioxidative effects of pumpkin seed (*Cucurbita pepo*) protein isolate in CCl4-induced liver injury in low-protein fed rats. Phytother Res 20:935–940. https://doi.org/10.1002/ptr.1977

Nourmohammadi E, SadeghiMahoonak A, Alami M, Ghorbani M (2017) Amino acid composition and antioxidative properties of hydrolysed pumpkin (*Cucurbita pepo* L.) oil cake protein. Int J Food Prop 20:3244–3255. https://doi.org/10.1080/10942912.2017.1283516

Nyam K, Tan C, Lai O et al (2009) Physicochemical properties and bioactive compounds of selected seed oils. LWT—Food Sci Technol 42:1396–1403. https://doi.org/10.1016/j.lwt.2009.03.006

Omar N, Sarhan N (2017) The possible protective role of pumpkin seed oil in an animal model of acid aspiration pneumonia: light and electron microscopic study. Acta Histochem 119:161–171. https://doi.org/10.1016/j.acthis.2017.01.002

Orzolek MD, Elkner TE, Lamont WJ, Kime LF and Harper JK (2017) Pumpkin production. Article for small scale and part-time farming project at Penn state with support from USDA—extension service. doi: https://extension.psu.edu/pumpkin-production. Accessed 18 Nov 2017

Ovca A, van Elteren J, Falnoga I, Šelih V (2011) Speciation of zinc in pumpkin seeds (*Cucurbita pepo*) and degradation of its species in the human digestive tract. Food Chem 128:839–846. https://doi.org/10.1016/j.foodchem.2011.03.102

Paris HS (2008) Summer squash. In: Prohens J, Nuez F (eds) Handbook of plant breeding, Vegetables, vol I. Springer, New York, pp 351–379

Paris H (2015) Germplasm enhancement of *Cucurbita pepo* (pumpkin, squash, gourd: Cucurbitaceae): progress and challenges. Euphytica 208:415–438. https://doi.org/10.1007/s10681-015-1605-y

Patel S (2013) Pumpkin (Cucurbita sp.) seeds as nutraceutic: a review on status quo and scopes. Mediterr J Nutr Metab 6(3):183–189. https://doi.org/10.3233/s12349-013-0131-5

Perez Gutierrez R (2016) Review of *Cucurbita pepo* (pumpkin) its phytochemistry and pharmacology. Med Chem. https://doi.org/10.4172/2161-0444.1000316

Peričin D, Radulović-Popović L, Vaštag Ž et al (2009) Enzymatic hydrolysis of protein isolate from hull-less pumpkin oil cake: application of response surface methodology. Food Chem 115:753–757. https://doi.org/10.1016/j.foodchem.2008.12.040

Phillips K, Ruggio D, Ashraf-Khorassani M (2005) Phytosterol composition of nuts and seeds commonly consumed in the United States. J Agric Food Chem 53:9436–9445. https://doi.org/10.1021/jf051505h

Popović S, Peričin D, Vaštag Ž et al (2011) Evaluation of edible film-forming ability of pumpkin oil cake; effect of pH and temperature. Food Hydrocoll 25:470–476. https://doi.org/10.1016/j.foodhyd.2010.07.022

Popović S, Peričin D, Vaštag Ž et al (2012) Pumpkin oil cake protein isolate films as potential gas barrier coating. J Food Eng 110:374–379. https://doi.org/10.1016/j.jfoodeng.2011.12.035

Popović L, Peričin D, Vaštag Ž et al (2013) Antioxidative and functional properties of pumpkin oil cake globulin hydrolysates. J Am Oil Chem Soc 90:1157–1165. https://doi.org/10.1007/s11746-013-2257-5

Potočnik T, Košir I (2016) Influence of roasting temperature of pumpkin seed on PAH and aroma formation. Eur J Lipid Sci Technol 119:1500593. https://doi.org/10.1002/ejlt.201500593

Prámparo M, Mattea M, Gregory S (2003) Influencia del tipo de contacto sólido líquido en la extracción de aceites vegetales. Grasas Aceites 53(4):592–597

Prevc T, Šegatin N, Kralj P et al (2015) Influence of metal ions and phospholipids on electrical properties: a case study on pumpkin seed oil. Food Control 54:287–293. https://doi.org/10.1016/j.foodcont.2015.01.040

Procida G, Stancher B, Cateni F, Zacchigna M (2012) Chemical composition and functional characterisation of commercial pumpkin seed oil. J Sci Food Agric 93:1035–1041. https://doi.org/10.1002/jsfa.5843

Qi H, Yang Y, Wang L et al (2012) Determination of bioactive ingredients in five pumpkin seeds cultivars. Adv Mat Res 554-556:985–989. https://doi.org/10.4028/www.scientific.net/amr.554-556.985

Rabia A, Asma S, Tariq M (2014) Characterization and antibacterial study of pumpkin seed oil (*Cucurbita pepo*). Life Sci Leaflets 49:053–064

Rabrenović B, Dimić E, Novaković M et al (2014) The most important bioactive components of cold pressed oil from different pumpkin (*Cucurbita pepo* L.) seeds. LWT—Food Sci Technol 55:521–527. https://doi.org/10.1016/j.lwt.2013.10.019

Radočaj O, Dimić E, Vujasinović V (2012) Development of a hull-less pumpkin (*Cucurbita pepo* L.) seed oil press-cake spread. J Food Sci 77:C1011–C1017. https://doi.org/10.1111/j.1750-3841.2012.02874.x

Rahman AHMM (2013) Ethno-medico-botanical studies on cucurbits of Rajshahi division. Bangl J Med Plants Stud 1(3):118–125

Revathy MN, Sabitha N (2013) Development, quality evaluation and popularization of pumpkin seed flour incorporated bakery products. Int J Food Nutr Sci 2(2):40–45

Ristic-Medic D, Perunicic-Pekovic G, Rasic-Milutinovic Z et al (2014) Effects of dietary milled seed mixture on fatty acid status and inflammatory markers in patients on hemodialysis. Sci World J 2014:1–9. https://doi.org/10.1155/2014/563576

Rodríguez-Miranda J, Hernández-Santos B, Herman-Lara E et al (2013) Effect of some variables on oil extraction yield from Mexican pumpkin seeds. CyTA J Food 12:9–15. https://doi.org/10.1080/19476337.2013.777123

Saavedra M, Aires A, Dias C et al (2013) Evaluation of the potential of squash pumpkin by-products (seeds and shell) as sources of antioxidant and bioactive compounds. J Food Sci Technol 52:1008–1015. https://doi.org/10.1007/s13197-013-1089-5

Serdaroğlu M, Nacak B, Karabıyıkoğlu M et al (2017) Effects of replacing beef fat with pre-emulsified pumpkin seed oil on some quality characteristics of model system chicken meat emulsions. IOP Conf Ser Earth Environ Sci 85:012045. https://doi.org/10.1088/1755-1315/85/1/012045

Sharma G, Lakhawat S (2017) Development, quality evaluation and acceptability of pumpkin seed flour incorporated in gravy. J Nutr Food Sci. https://doi.org/10.4172/2155-9600.1000613

Shim B, Jeong H, Lee S et al (2014) A randomized double-blind placebo-controlled clinical trial of a product containing pumpkin seed extract and soy germ extract to improve overactive bladder-related voiding dysfunction and quality of life. J Funct Foods 8:111–117. https://doi.org/10.1016/j.jff.2014.03.010

Sicilia T, Niemeyer H, Honig D, Metzler M (2003) Identification and stereochemical characterization of lignans in flaxseed and pumpkin seeds. J Agric Food Chem 51:1181–1188. https://doi.org/10.1021/jf0207979

Siegmund B, Murkovic M (2004) Changes in chemical composition of pumpkin seeds during the roasting process for production of pumpkin seed oil (part 2: volatile compounds). Food Chem 84:367–374. https://doi.org/10.1016/s0308-8146(03)00241-3

Sihag MK, Sharma V, Goyal A, Arora S, Singh AK (2015) Effect of domestic processing treatments on iron, β-carotene, phytic acid and polyphenols of pearl millet. Cogent Food Agric 1 (1):1109171. https://doi.org/10.1080/23311932.2015.1109171

Sood A, Kaur P, Gupta R (2012) Phytochemical screening and antimicrobial assay of various seeds extract of Cucurbitaceae family. Int J Appl Biol Pharm Techol 3(3):401–409

Stevenson D, Eller F, Wang L et al (2007) Oil and tocopherol content and composition of pumpkin seed oil in 12 cultivars. J Agric Food Chem 55:4005–4013. https://doi.org/10.1021/jf0706979

Tanwar B, Modgil R, Goyal A (2018) Antinutritional factors and hypocholesterolemic effect of wild apricot kernel (Prunus armeniaca L.) as affected by detoxification. Food Funct 9 (4):2121–2135

Tanwar B, Modgil R, Goyal A (2019a) Effect of detoxification on biological quality of wild apricot (Prunus armeniaca L.) kernel. J Sci Food Agric 99(2):517–528

Tanwar B, Goyal A, Irshaan S, Kumar V, Sihag MK, Patel A, Kaur I (2019b) Quinoa. In: Johnson J, Wallace T (eds) Whole grains and their bioactives: composition and health, vol 4. Wiley, Hoboken, pp 269–305. https://doi.org/10.1002/9781119129486.ch10

Teppner H (2004) Notes on *Lagenaria* and *Cucurbita* (*Cucurbitaceae*)—review and new contributions. Phyton (Horn, Austria) 44(2):245–308

Tomar P, Nikhil K, Singh A et al (2014) Characterization of anticancer, DNase and antifungal activity of pumpkin 2S albumin. Biochem Biophys Res Commun 448:349–354. https://doi.org/10.1016/j.bbrc.2014.04.158

Umadevi P, Murugan S, Jennifer S, Subakanmani S (2011) Evaluation of antidepressant like activity of *Cucurbita pepo* seed extracts in rats. Int J Curr Pharm Res 3(1):108–113

Valverde-Monge M, Bartolome B, Custa-Herranz J, De las Heras M (2017) Allergy to pumpkin seeds. J Allergy Clin Immunol 139:AB138. https://doi.org/10.1016/j.jaci.2016.12.454

Van Hoed V, Sampaio K, Felkner B et al (2017) Tocopherols and polyphenols in pumpkin seed oil are moderately affected by industrially relevant roasting conditions. Eur J Lipid Sci Technol 119:1700110. https://doi.org/10.1002/ejlt.201700110

Vujasinovic V, Djilas S, Dimic E et al (2012) The effect of roasting on the chemical composition and oxidative stability of pumpkin oil. Eur J Lipid Sci Technol 114:568–574. https://doi.org/10.1002/ejlt.201100158

Wang D, Pan H, Deng X et al (2007) Cucurbitane and hexanorcucurbitane glycosides from the fruits of *Cucurbita pepo* cv dayangua. J Asian Nat Prod Res 9:525–529. https://doi.org/10.1080/10286020600782538

Wenzl T, Prettner E, Schweiger K, Wagner F (2002) An improved method to discover adulteration of Styrian pumpkin seed oil. J Biochem Biophys Methods 53:193–202. https://doi.org/10.1016/s0165-022x(02)00108-2

Winkler J (2000) The origin and breeding of the hull-less seeded styrian oil-pumpkin varieties in Austria. Cucurbit Genet Coop Rep 23:101–104

Winkler C, Schroecksn K, Wirleitner B, Schennach H, Fuchs D (2005) Extracts of pumpkin (*Cucurbita pepo* L.) seeds suppress stimulated peripheral blood mononuclear cells in vitro. Am J Immunol 1(1):6–11

Xanthopoulou M, Nomikos T, Fragopoulou E, Antonopoulou S (2009) Antioxidant and lipoxygenase inhibitory activities of pumpkin seed extracts. Food Res Int 42:641–646. https://doi.org/10.1016/j.foodres.2009.02.003

Yadav M, Jain S, Tomar R et al (2010) Medicinal and biological potential of pumpkin: an updated review. Nutr Res Rev 23:184–190. https://doi.org/10.1017/s0954422410000107

Yoshida H, Shougaki Y, Hirakawa Y et al (2004) Lipid classes, fatty acid composition and triacylglycerol molecular species in the kernels of pumpkin (Cucurbita spp) seeds. J Sci Food Agric 84:158–163. https://doi.org/10.1002/jsfa.1623

Younis Y, Ghirmay S, Al-Shihry S (2000) African *Cucurbita pepo* L.: properties of seed and variability in fatty acid composition of seed oil. Phytochemistry 54:71–75. https://doi.org/10.1016/s0031-9422(99)00610-x

Zdunczyk Z, Minakowski D, Frejnagel S, Flis M (1999) Comparative study of the chemical composition and nutritional value of pumpkin seed cake, soybean meal and casein. Nahrung/Food 43:392–395. https://doi.org/10.1002/(sici)1521-3803(19991201)43:6<392::aid-food392>3.0.co;2-2

Zuhair H, Abd El-Fattah A, El-Sayed M (2000) Pumpkin-seed oil modulates the effect of felodipine and captopril in spontaneously hypertensive rats. Pharmacol Res 41:555–563. https://doi.org/10.1006/phrs.1999.0622

Chapter 20
Cumin (*Cuminum cyminum* L.) Seed

Rudra Pratap Singh Rajput, N. Paramakrishnan, and H. V. Gangadharappa

Abstract Spices are bionutrients that boost the aroma, taste, and flavor of foods and treat several diseases. Cumin (*Cuminum cyminum*) is one of the most widely used spices and is used in cookery for its individual fragrant property. It is an ancient and far used spice from ages; as a result, it was absolutely associated with a degree icon of fidelity and affection. Cumin is offered in numerous appearances like fennel, black cumin, and anise. The proximate composition of the seeds reveals that it contains fatty oil, phenolic acids, essential oils, protein, and other ingredients. The very important elements in cumin are cymene, pinene, thymol, oleoresin, cuminaldehyde, terpinene, etc., which have exhibited their applications according to an ailment. It is a very essential part of iron for lactation, immune systems, skin diseases, and energy.

Keywords Cuminaldehyde · Cumin · Cymene · Thymol · Spice

20.1 Origin and History

20.1.1 History

Cumin (*Cuminum cyminum*) is a small herbaceous annual plant. Initially, cumin seeds are believed to be grown in Iran and Mediterranean province. Cumin is also mentioned within the Bible (Isaiah 28:27) and the New witness (Matthew 23:23). Spanish colonists introduced cumin to the Americans (Srivastava 1989). Nowadays, it is largely grown in Islamic Republic of Iran, Morocco, Tajikistan, Uzbekistan,

R. P. S. Rajput
Department of Pharmaceutics, Columbia Institute of Pharmacy, Raipur, Chhattisgarh, India

N. Paramakrishnan · H. V. Gangadharappa (✉)
Department of Pharmacognosy, JSS Academy of Higher Education and Research, JSS College of Pharmacy, Mysuru, India
e-mail: nparamakrishnan@jssuni.edu.in; hvgangadharappa@jssuni.edu.in

B. Tanwar, A. Goyal (eds.), *Oilseeds: Health Attributes and Food Applications*,
https://doi.org/10.1007/978-981-15-4194-0_20

Turkey, Egypt, India, Chile, Syria, and Mexico. In the middle Ages, cumin was one of the most familiar spices, an era when spices were relatively rare. Cumin was thought to promote fidelity and love. People carried cumin to weddings and walk around with it in their pockets. It was supposed to keep on lovers and chickens from drifting. Thus, married troopers were sending scuffle with a recently baked loaf of bread prepared using cumin.

20.1.2 Etymology

The word "cumin" in English comes as of the French "cumin" that was taken from Arabic language "Kammon" through Spanish "comino" throughout the Arab decree and the Kingdom of Spain within the fifteenth century. This spice is also indigenous to Syrian Arab Republic (an Arabic talking nation) wherever cumin thrives in arid and hot conditions. These forms of words are genuine in many ancient Syro-Arabian languages, including kamūnu in Akkadian. The definitive term could be from Syrian language that would be the Sumerian word gamun. Etymology connects the statement with the Persian town, Kerman. Where the story goes to most of the ancient cumin was produced by Persians. In Asian countries like Pakistan, it is called as jira/jeera or zira, and in Islamic Republic of middle Asia, whereas in Turkey it is popularly recognized as "zira", "kimyon" in north-western China, and "ziran" in Iran. In Arabic, cumin is identified as "al-kamuwn". In Ethiopia, it is known as "kemun". In French, it is Cumin; in German—Romische Kümmel, Kreuzkümmel, in Italian language, it is called Cumino. Komijn in Dutch, Comino in Spanish, Cominho in Portuguese, Kmin in Russian, Arabic: Kammun, Kemouyn, Machin in Chinese, Jeera, Jeraka, Jira, Zeera, Zira in India, Spiskummin in Swedish, and Jintan puteh in Malay.

Cumin belongs to the Apiaceae family and is a common cookery spice that is utilized as a medical aid due to the presence of aromatic constituents within the fruit. Cumin seeds have a small furry, brown color, spermatophyte formed by a boat, a sweet-spicy aroma, and a slightly bitter and pungent flavor (Rebey et al. 2012). The cumin oil produced by steam distillation is added to flavor the desserts, beverages (alcohol), and condiments. It is also used in perfumes, lotions, poultice, and as a sweeter in creams. It is also used as diuretics, carminatives, etc., and to treat toothaches, epilepsy, dyspepsia, diarrhea, jaundice, indigestion, and flatulence.

20.1.3 Useful Parts of Cumin Plant

Leaves: Cumin leaves are a multifid segment with long filaments.

Flowers: They are little, white or pink in color and are exceeded in height by the bracts. The umbels, each partial and general, have five rays with two or three filament-like, one-sided bracts consisting of the involucres.

Fruits: They are ovate with pointed end, lightweight and brown with grayish tint. The fruit of cumin resemble like caraway fruit, but is bigger and about two lines in length, very much longer than the pedicels and little contracted at the sides, fusiform, nearly tapering, crowned by the short teeth of the calyx, the covered densely with short and rough hair. Seeds of the fruits are two in number and are ovoid; plane surfaces together with plano convex. The taste and odor of fruits are comparable to caraway fruits, but is warmer and not so delightful.

Seeds: Seeds are yellowish to brownish gray, elongated in the form of nine protuberances and possess various health benefits. The cumin seeds are used as aromatic stimulant, carminative, stomachic, and astringent. Aromatic and astringent properties of cumin herb are beneficial to the digestive system. It also used in the management of intricate biological disorders, as a eupeptic and carminative, and as astringent in mustang respiratory organ disorders, a cough medication, and as an analgesic (Hanif et al. 2012; De et al. 2003). The volatile oil obtained from the seeds exhibits a massive antibacterial property against *K. pneumonia*.

20.2 Production

Cumin is mainly cultivated in Europe, Asia, the Middle East, and the geographical area of the Asian countries, which are the major cumin exporters. It is currently largely grown in Asian countries like Uzbekistan, Turkey, Tajikistan, Morocco, India, Syria, Egypt, Mexico, and Chile. Cultivation of cumin needs an extended warm summer for 3–4 months with the day-time temperature approximately 30 °C and is usually grown in Mediterranean climates, which is drought tolerant. In general, sandy soil is the best for cultivation. The cumin plants blossom during the month of June and July. Like the other Umbelliferae members, these plants are also harvested once the seeds become brown, threshed, and dried (Singh et al. 2017). The seeds must be planted in tiny pots, packed with loose soil, and pushed into an exceptionally modest hotbed to bring up the seedlings. Weeding could be done and therefore the plants can flower fine.

20.2.1 Description

Cumin belongs to the Apiaceae family. The Apiaceae is also called as Umbelliferae (both of these Umbelliferae and Apiaceae are acceptable by the International System of Botanical Nomenclature). These families include usually aromatic plants having hollow stems, like parsley, parsnip, carrot, fennel, caraway, dill, etc.

Some extremely poisonous plants like hemlock belong to these families. Apiaceae is a massive family consisting of around 300 genera and over 3000 numbers of species. The former name Apiaceae is derived from the term "inflorescence" being within the style of composite "umbel" flowers, which is symmetric with 5 tiny sepals, 5 stamens, and 5 petals. Cumin or *Cuminum cyminum* are the only living species in their genus.

It is a little perennial herb having a thin branched stem and 20-30 cm height. The leaves of the cumin are 5–10 cm in length, bipinnate/pinnate and with leaflets like threads. The flowers of cumin are tiny, pink or white, and bear in umbels. The fruit (normally referred to as seed) could be a lateral cigar-shaped or ovoid, 4–5 mm in length, consisting of one seed in it. These seeds almost look like fennel seeds, however are dark in color and small in length (Singh et al. 2017).

20.2.2 Cultivation Areas

The key producers of cumin are the Asian nation and China, which produces around 70% of the global supply and consumes 90%, whereas India consumes around 63% of the world's cumin. The other major producer is Mexico. In general, around 300,000 ton of cumin is annually produced worldwide.

20.3 Chemical Composition

Cumin seed contains approximately 22.27–23.80% and 2.4–5.0% total lipids and essential (volatile) oil, respectively. The proximate composition, major fatty acids profile, minerals, and vitamin content of cumin seed are shown in Tables 20.1 and 20.2. Cumin seed also contains amino acids, glycosides, and flavonoids, with cuminaldehyde as its main constituent (Iacobellis et al. 2005; El-Kani et al. 2007). Cuminaldehyde, α- and β-pinene, limonene, 1, 8-cineole, o- and p-cymene, α- and γ-terpinene, linalool, and safranal are the major components of Cumin fruit (Fig. 20.1).

The organic acids such as aspartic, benzoic acid, citric, malic, propionic, tartaric, ascorbic, maleic, oxalic and fumaric acids, the phenolics such as salicylic, cinnamic, gallic acids, p-hydroxybenzoic acid, hydroquinone, resorcinol and the flavonoids like rutin, quercetin and coumarin are also present in cumin seeds.

The cumin volatile oil comprises of high level of synthetic resin compounds, mostly organic compounds and para-cymene. It consists of β-pinene, γ-terpinene,

Table 20.1 Proximate composition of cumin seed (Amin 2012; Donna and Antony 1993; USDA 2019)

Parameter	Amount (g/100 g)
Moisture	6.00–8.06
Carbohydrates	44.24–44.60
Total lipids	22.27–23.80
Crude protein	17.81–18.00
Crude fiber	10.50
Ash	7.70–9.50

Table 20.2 Fatty acids composition, mineral, and vitamin content of cumin seed (USDA 2019)

Fatty acids	Amount (% of total fats)	Mineral	Amount (mg/100 g)	Vitamin	Amount (mg/100 g)
Capric acid (C10:0)	0.018	Calcium	931.00	Vitamin C	7.70
Lauric acid (C12:0)	0.018	Iron	66.36	Thiamin	0.63
Myristic acid (C14:0)	0.018	Magnesium	366.00	Riboflavin	0.33
Palmitic (C16:0)	1.137	Phosphorus	499.00	Niacin	4.58
Palmitoleic (C16:1)	0.212	Potassium	1788.00	Vitamin B_6	0.44
Stearic (C18:0)	0.344	Sodium	168.00	Folate (μg)	10.00
Oleic (C18:1)	13.618	Zinc	4.80	Choline	24.70
Linoleic (C18:2)	3.103	Copper	0.86	Vitamin B_{12}	0
Total SFAs	1.54	Manganese	3.33	Vitamin A (IU)	1270.00
Total MUFAs	14.04	Selenium (μg/100 g)	5.20	β-Carotene (μg)	762.00
Total PUFAs	3.27				

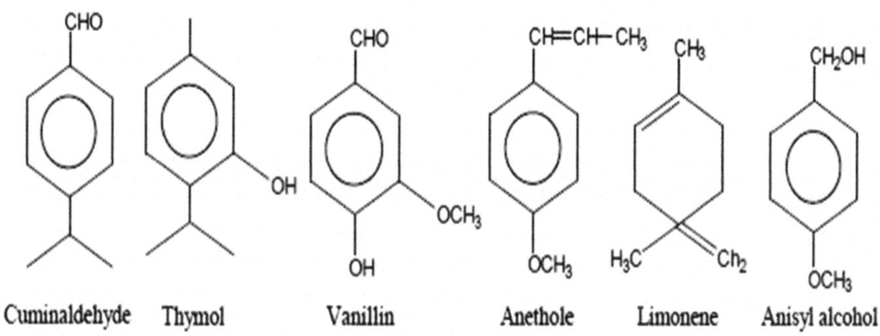

Fig. 20.1 Structure of active compounds of cumin

p-cymene, cuminaldehyde, etc. The cuminaldehyde and para-cymene are two most common active phyto-constituents of cumin used to shield the liver against aerophilic strain and illness (Singh et al. 2017). The variety of phytochemicals of cumin seeds exhibits carminative and anti-flatulent effect. Cumin seeds contain cuminaldehyde (4-isopropylbenzaldehyde), 2-methoxy-3-methylpyrazine, pyrazines, 2-ethoxy 3-iso propyl pyrazine, and 2-methoxy 3-secbutylpyrazine.

20.3.1 Health Attributes

Cumin seeds have been used in food preparation for a long time. This traditional herb is used for many remedies. It is a stimulant and an excellent herb for many digestive disorders, and also used as an antiseptic. The recent studies on cumin seeds revealed that it may also have anti-carcinogenic activities (Srivastava 1989). The anti-tumoral activity of seed on animals showed reduction in the risk of stomach and liver tumors. A few of the medicinal uses of cumin seeds are given below:

- As a whole medicinal plant—used as an aromatic stimulant, anti-spasmodic, sedative and carminative.
- Anti-flatulent in an individual digestion.
- As a medication for colic and dyspeptic headache.
- Oil is used as anti-bacteria.

Cumin is employed as an antiseptic in the treatment of the cuts associated with hemorrhage. Cumin may be a stimulant and furthermore as a noble herbal medicine for biological progression ailments, stimulate the discharge of enzymes from the duct gland which might facilitate absorption of nutrients, increase the liver's capacity to detoxification, anti-carcinogenic property, cut back the chance of abdomen and liver tumor and additionally boost the immune system (Bennett et al. 1982; Johri 2011).

20.3.2 In Anemia and Lactation

Cumin contains a significant amount of iron (approximately 66.00 mg/100 g), which is five times the requirement of iron per day for an adult. Iron is a vital part of Hb, which carries the oxygen from the respiratory organ to the various cells, and is additionally a part of basic accelerator for metabolism and production of energy. Therefore, cumin offers a daily dietary nutritional supplement for anemic individuals.

Developing adolescents, children, lactating women, and pregnant ladies require more quantity of iron. Furthermore, the presence of thymol in cumin is expected to increase the secretion of milk in lactating mothers and camphor that tends to extend secretions from secretory organs, as well as from mammary glands. The fruits have outstanding quantity of metallic elements (above 900 mg/100 g) that accounts around 90% of our day-to-day demand of metallic element.

20.3.3 Skin Disorders

It is used as antioxidants for the protection of skin. It makes the skin younger and glowing. It possesses disinfectant and anti-fungal properties due to the presence of essential oils. The regular use of cumin in food keeps the skin free from rashes, boils, pimples, etc. Components of cumin are efficient detoxicants, which facilitate the removal of toxins from the body.

20.3.4 Anti-oxidant Activity

Because of the presence of flavonoids, monoterpene alcohols, basic flavors, and other poly-phenolic compounds, cumin oils have potent antioxidant properties (Leporatti and Ghedira 2009; De et al. 2009; Najda et al. 2008; Samojlik et al. 2010; Ruberto and Baratta 2000).

20.3.5 Anti-microbial

The cumin seed oil is effective against a wide range of virulent, gram-negative and gram-positive microbial pathogens. It is also used in biofilm-formation against *Streptococcus pyogenes* and *Streptococcus mutans* (Rodov et al. 2010; Derakhshan et al. 2010).

20.3.6 Anti-carcinogenic/Anti-mutagenic Property

The cumin seeds have been reported to inhibit the incidences of carcinoma evoked by a colon-specific substance in the rats and conjointly diminish the activity of the enzymes mucinase and β-glucuronidase. The colon of the rats treated with cumin showed decrease in 3-methylglutaryl COA reductase activity (Skrinjar et al. 2009; Nalini et al. 1998) and benzopyrene-induced tumorigenesis in the abdomen in mice has exhibited inhibitory activities.

20.3.7 Anti-diabetic Activity

Orally administrated cumin has also shown hypoglycemic effect in rabbits (Nalini et al. 2006). The therapeutically active constituents of cuminaldehyde showed inhibitory effect on aldose reductase and α-glucosidase isolated from rat

(Roman-Ramos and Flores-Saenz 1995). Orally administered cumin to alloxan-induced diabetic rats reduced the body weight, tissue, and cholesterol in the plasma, fatty acids, triglycerides, and phospholipids, decreases aspartate transferase, basic enzyme (ALP) and gamma-glutamyl enzyme actions, and diminishes sterol, phospholipids, and triglycerides levels in the tissues of kidney and liver.

20.4 Adverse Effects and Individual Concerns

Cumin is an ancient and favored herb which has numerous benefits and curable effects. However, it has shown some undesirable effects such as dermatitis, metabolic reactions, carcinoma (high dietary concentration), lowering glucose level, and also increasing the risk of hemorrhage (Bennett et al. 1982).

A patient with liver problems and ulcer and lactating or pregnant women must use cumin with attention. Cumin constituents may alter the activity when co-administered with some drugs, which is given in Table 20.3.

20.5 Food Applications

20.5.1 General Uses

Cumin fruits are used as a flavoring agent for cooking foodstuffs and also as a spice. The cumin may be the main part of curries and red chili powders, which is added to a range of food products to improve flavor. It is also used to improve the flavor of beverages (alcoholic), condiments, and desserts.

Table 20.3 Side effect and herbal interactions of cumin (Kaur et al. 2019; Yimer et al. 2019)

Sr no	Active constituents	Interactive drug/herb	Alter/effects
1.	Cumin	*Ginkgo biloba*, garlic, aspirin, warfarin, clopidogrel (Plavix®),	Increase the risk of bleeding
2.	Cumin	Antibiotics, anti-inflammatory agents, anti-neoplastic drugs, anti-fungal, anti-seizure agents, lipid and cholesterol-lowering drugs, estrogens, drugs for osteoporosis, GI agents, phytoestrogens, and opioids	Immune system
3.	Cumin	Glyburide, glimepiride, pioglitazone, rosiglitazone, insulin, chlorpropamide, tolbutamide, glipizide	Decrease the blood sugar level
4.	Cumin	Anti-tuberculosis drugs	Improved rifampin level
5.	Cumin	Cytochrome P-450	Increase the blood glucose level

20.5.2 *Pharmacological Uses*

Active phyto-constituents of cumin are used as a medicament such as antifungal, inhibitor, astringent, coronary artery disease (hardening of the arteries), cancer, bone loss, blood diluents, cataract (eye illness), carpal tunnel syndrome, diabetes, plaque, cavities, and indigestion.

20.6 Conclusion

Spices are very important bionutrients as food ingredients and organic process supplements. Spices are employed as food additives to boost the style and the flavor of the food products from ages. Additionally, spices also have several medicinal properties and are used to cure many ailments in the Indian System of Medicines (Ayurvedic system). Cumin is well-liked as preparation spice and employed in traditional knowledge medical care as a result of the existence of aromatic constituents within the fruits.

References

Amin G (2012) Cumin. In: Peter KV (ed) Handbook of herbs and spices. Woodhead, Cambridge, pp 250–259. https://doi.org/10.1533/9780857095671.250

Bennett MA, Huang TN, Matheson TW, Smith AK, Nickerson W (1982) (η6- Hexamethylbenzene) ruthenium complexes. Inorg Synth 21:74–78

De M, De AK, Mukhopadhvay R, Banerjee AB, Micro M (2003) Antimicrobial activity of Cuminum cyminum L. Ars Pharm 44:257–269

De ML, De FV, Fratianni F, Nazzaro F (2009) Chemistry, antioxidant, antibacterial and antifungal activities of volatile oils and their components. Nat Prod Commun 4:1741–1750

Derakhshan S, Sattari M, Bigdeli M (2010) Effect of cumin (Cuminum cyminum L.) seed essential oil on biofilm formation and plasmid integrity by Klebsiella pneumoniae. Pharmacog Mag 6:57–61

Donna RT, Antony TG (1993) Spices & seasonings: a food technology hand book. VCH, New York, pp 71–74

El-Kani M, Fereshteh G, Mehdi M, Soosan R (2007) Extraction of volatile oil from cumin (Cuminum cyminum L.) with superheated water. J Food Process Eng 30(2):255–266

Hanif C, Ayesha T, Adila S, Saeed M, Tanveer A, Ashfaq M (2012) Physico-chemical investigation and antimicrobial activity of essential oil of Cuminum cyminum L. World Appl Sci J 19 (3):330–333

Iacobellis NS, Cantore PL, Capasso F, Senatore F (2005) Antibacterial activity of Cuminum cyminum L. and Carum carvi L. essential oils. J Agric Food Chem 53:57–61

Johri RK (2011) Cuminum cyminum and Carum carvi: An update. Pharmacogn Rev 5(9):63–72

Kaur G, Upadhyay N, Tharappel LJ, Invally M (2019) Pharmacodynamic interaction of cumin seeds (Cuminum cyminum L.) with glyburide in diabetes. J Compl Integr Med 26:16(4). https://doi.org/10.1515/jcim-2018-0080

Leporatti ML, Ghedira K (2009) Comparative analysis of medicinal plants used in traditional medicine in Italy and Tunisia. J Ethnobiol Ethnomed 5:31–39
Najda A, Dyduch J, Brzozowski N (2008) Flavonoid content and antioxidant activity of caraway roots (Carum carvi L.). Veg Crops Res Bull 68:127–133
Nalini N, Sabitha K, Vishwanathan P, Menon VP (1998) Influence of spices on the bacterial (enzyme) activity in experimental colon cancer. J Ethnopharmacol 62:15–24
Nalini N, Manju V, Menon VP (2006) Effect of spices on lipid metabolism in 1, 2-dimethylhydrazine-induced rat colon carcinogenesis. J Med Food 9:237–245
Rebey IB, Jabri-Karoui I, Hamrouni-Sellami I, Bourgou S, Limam F, Marzouk B (2012) Effect of drought on the biochemical composition and antioxidant activities of cumin (Cuminum cyminum L.) seeds. Ind Crop Prod 36:238–245
Rodov V, Vinokur Y, Gogia N, Chkhikvishvili I (2010) Hydrophilic and lipophilic antioxidant capacities of Georgian spices for meat and their possible health implications. Georgian Med News 179:61–66
Roman-Ramos R, Flores-Saenz JL (1995) Alarcon-Aguilar, F.J., Antihyperglycemic effect of some edible plants. J Ethnopharmacol 48: 25-32
Ruberto G, Baratta MT (2000) Antioxidant activity of selected essential oil components in two lipid model systems. Food Chem 69:167–174
Samojlik I, Lakic N, Mimica-Dukic N, Dakovic-Svajcer K, Bozin B (2010) Antioxidant and hepatoprotective potential of essential oils of coriander (Coriandrum sativum L.) and caraway (Carum carvi L.) (Apiaceae). J Agric Food Chem 58:8848–8853
Singh RP, Gangadharappa HV, Mruthunjaya K (2017) Cuminum cyminum—a popular spice: an updated review. Pharm J 9(3):292–301
Skrinjar MM, Mandi AI, Misan AC, Sakac MB, Saric LC, Zec MM (2009) Effect of mint (Mentha piperita L.) and caraway (Carum carvi L.) on the growth of some toxicogenic Aspergillus species and aflatoxin B1 production. Proc Nat Sci Matica Srpska Novi Sad 116:131–139
Srivastava KC (1989) Extracts from two frequently consumed spices-cumin (Cuminurn cyminum) and turmeric (Curcuma longa)—inhibit platelet aggregation and alter eicosanoid biosynthesis in human blood platelets. Prostaglandins Leukot Essent Fatty Acids 137:57–64
USDA (2019). Spices, cumin seed. Food data central. Available from https://fdc.nal.usda.gov/fdc-app.html#/food-details/170923/nutrients. Last accessed on 12/12/2019
Yimer EM, Tuem KB, Karim A, Ur-Rehman N, Anwar F (2019) Nigella sativa L. (black cumin): a promising natural remedy for wide range of illnesses. Evid Based Complement Alternat Med. 2019:1528635. doi: https://doi.org/10.1155/2019/1528635.

The manufacturer's authorised representative in the EU is Springer Nature Customer Service Centre GmbH, Europaplatz 3, 69115 Heidelberg, Germany. If you have any concerns regarding our products, please contact ProductSafety@springernature.com

Printed and bound by CPI Group (UK) Ltd, Croydon, CR0 4YY
30/09/2026
02225317-0015